NONSEGMENTED NEGATIVE STRAND VIRUSES

Academic Press Rapid Manuscript Reproduction

Proceedings of a Symposium on the
Molecular Biology of Negative Strand Viruses
Sponsored by the University of Alabama in Birmingham
Held in Hilton Head, South Carolina
September 11–18, 1983

NONSEGMENTED NEGATIVE STRAND VIRUSES
Paramyxoviruses and Rhabdoviruses

Edited by

David H. L. Bishop
Richard W. Compans
Department of Microbiology
University of Alabama in Birmingham
Birmingham, Alabama

ACADEMIC PRESS, INC.
(Harcourt Brace Jovanovich, Publishers)
Orlando San Diego San Francisco New York London
Toronto Montreal Sydney Tokyo São Paulo

COPYRIGHT © 1984, BY ACADEMIC PRESS, INC.
ALL RIGHTS RESERVED.
NO PART OF THIS PUBLICATION MAY BE REPRODUCED OR
TRANSMITTED IN ANY FORM OR BY ANY MEANS, ELECTRONIC
OR MECHANICAL, INCLUDING PHOTOCOPY, RECORDING, OR ANY
INFORMATION STORAGE AND RETRIEVAL SYSTEM, WITHOUT
PERMISSION IN WRITING FROM THE PUBLISHER.

ACADEMIC PRESS, INC.
Orlando, Florida 32887

United Kingdom Edition published by
ACADEMIC PRESS, INC. (LONDON) LTD.
24/28 Oval Road, London NW1 7DX

Library of Congress Cataloging in Publication Data
Main entry under title:

Nonsegmented negative strand viruses.

    Proceedings of the 5th International Symposium on
Negative Strand Viruses, held at Hilton Head, S.C.,
Sept. 11-17, 1983.
    Includes index.
    1. Paramyxoviruses--Congresses. 2. Rhabdoviruses--
Congresses. I. Bishop, David H. L. III. Compans,
Richard W. III. International Symposium on Negative
Strand Viruses (5th : 1983 : Hilton Head, S.C.) [DNLM:
1. Paramyxoviridae--Congresses. 2. Rhabdoviridae--Con-
gresses. QW 168.5.P2 N814 1983]
QR404.N66 1984   576'.64   84-6281
ISBN 0-12-102480-6 (alk. paper)

PRINTED IN THE UNITED STATES OF AMERICA

84 85 86 87   9 8 7 6 5 4 3 2 1

Contents

Contributors	*xiii*
Preface	*xxiii*
Contents of Segmented Negative Strand Viruses	*xxv*

Genome Structure

Analysis of the Sendai Virus Genome by Molecular Cloning Benjamin M. Blumberg, Colomba Giorgi, Peter C. Dowling, Lyle A. Dethlefsen, Jean L. Patterson, Laurent Roux, and Daniel Kolakofsky	3
Non-Coding Regulatory Sequences of the Sendai Virus Genome K. C. Gupta, G. G. Re, and D. W. Kingsbury	11
Structure of a Complete Clone of the Sendai Virus *NP* Gene E. M. Morgan, K. C. Gupta, G. G. Re, and D. W. Kingsbury	17
cDNA Cloning, Mapping, and Translation of Ten Respiratory Syncytial Virus mRNAs P. L. Collins, L. E. Dickens, and G. W. Wertz	21
Structural Analysis of Human Respiratory Syncytial Virus Genome Sundararajan Venkatesan, Narayansamy Elango, Masanobu Satake, and Robert M. Chanock	27
The Polymerase Gene of VSV George G. Harmison, Ellen Meier, and Manfred Schubert	35
Cloning and Sequencing of M mRNA of Spring Viremia of Carp Virus Akio Kiuchi and P. Roy	41
Characterization of Measles Virus RNA J. Tucker, A. Ramsingh, G. Lund, D. Scraba, W. C. Leung, and D. L. J. Tyrrell	49

The Cloning of Morbillivirus Specific RNA 55
    J. D. Hull, S. E. H. Russell, E. M. Hoey, B. K. Rima,
    and S. J. Martin

Transcription and Replication

Specificity of Interaction of L, NS, and M Proteins with N–RNA
Complex of Vesicular Stomatitis Virus in a Heterologous *in Vitro*
Reconstitution System 63
    Bishnu P. De, Angeles Sanchez, and Amiya K. Banerjee
Does Modification of the Template N Protein Play a Role
in Regulation of VSV RNA Synthesis? 71
    Jacques Perrault, Patrick W. McClear, Gail M. Clinton,
    and Marcella A. McClure
Binding Studies of NS1 and NS2 of Vesicular Stomatitis Virus 79
    Paul M. Williams and Suzanne Urjil Emerson
Dyad-Symmetry in VSV RNA May Determine the Availability
of the RNA–Polymerase Binding Site 87
    Anneka C. Ehrnst and Alice S. Huang
The Effect of the β–γ Bond of ATP on Transcription of Vesicular
Stomatitis Virus 95
    Todd L. Green and Suzanne Urjil Emerson
Interactions between Cellular La Protein and Leader RNAs 103
    Jack D. Keene, Michael G. Kurilla, Jeffrey Wilusz,
    and Jasemine C. Chambers
Effects of Cell Extracts on Transcription by Virion
and Intracellular Nucleocapsids of Vesicular Stomatitis Virus 109
    Helen Piwnica-Worms and Jack D. Keene
Synthesis of the Various RNA Species in Cells Infected
with the Temperature-Sensitive Mutants of VSV New Jersey 115
    J. F. Szilágyi, R. G. Paterson, and C. Cunningham
Characterization of a Mutant of Vesicular Stomatitis Virus
with an Aberrant *in Vitro* Polyadenylation Activity 123
    D. Margaret Hunt
Role of the Viral Leader RNA in the Inhibition of Transcription
by Vesicular Stomatitis Virus 131
    Brian W. Grinnell and R. R. Wagner
Temperature-Sensitive Mutants of VSV Interfere with the Growth
of Wild-Type Virus at the Level of RNA Synthesis 139
    Debra W. Frielle and Julius S. Youngner
Role of Vesicular Stomatitis Virus Proteins in RNA Replication 147
    John T. Patton, Nancy L. Davis, and Gail W. Wertz

Vesicular Stomatitis Virus Proteins Required for the *in Vitro* Replication of Defective-Interfering Particle Genome RNA   153
    Richard W. Peluso and Sue A. Moyer
Characterization of Polycistronic Transcripts in Newcastle Disease Virus-Infected Cells   161
    A. Wilde and T. Morrison
Molecular Studies on Canine Distemper Virus Replication   167
    T. Barrett, N. T. Gorman, R. C. Patterson, and B. W. J. Mahy
RNA-Dependent RNA Polymerase Associated with Respiratory Syncytial Virus   175
    Gustave N. Mbuy and Olga M. Rochovansky
The Effects of Interferon on Measles Virus RNA Synthesis   183
    J. B. Milstien, A. S. Seifried, M. J. Klutch, and B. Bhatia

Gene Expression, Protein Synthesis, and Protein Modification

Construction and Expression of a Chimeric Gene of Glycoprotein G and Matrix Protein M of Vesicular Stomatitis Virus   193
    J. Capone and H. P. Ghosh
Intracellular Processing of Vesicular Stomatitis Virus and Newcastle Disease Virus Glycoproteins   201
    Trudy G. Morrison and Lori J. Ward
Rescue of VSV in Persistently Infected L Cells by Superinfection with Vaccinia   207
    Patricia Whitaker-Dowling and J. S. Youngner
The Effect of Vesicular Stomatitis Virus on Adenovirus Type 2 Replication   215
    James Remenick and John J. McGowan
Inhibition of Adenovirus and SV40 DNA Synthesis by Vesicular Stomatitis Virus   223
    M. E. Reichmann, Harlan B. Scott II, and Delphine Krantz
Expression of Measles Virus RNA in Brain Tissue   233
    Knut Baczko, Martin Billeter, and Volker ter Meulen
Temporal Changes of Measles Viral Proteins in HeLa Cells Acutely and Persistently Infected with Measles Virus   239
    Karen K. Y. Young and Steven L. Wechsler
Modulation of Measles Virus-Specific Protein Synthesis by Cyclic Nucleotides and an Inducer of Adenylate Cyclase   247
    Steven J. Robbins, Jessica Randle, and Jon Eagle

Viral Proteins: Antigenic and Functional Analyses

Fine Structural Analysis and Phosphorylation Site Determination in VSV NS Protein — 255
    Lorraine L. Marnell and Donald F. Summers

Mapping Phosphate Residues Required for VSV Transcription on the NS Protein Molecule — 265
    C.-H. Hsu and D. W. Kingsbury

Matrix (M) Protein Requirement for the Binding of Vesicular Stomatitis Virus Ribonucleocapsid to Sonicated Phospholipid Vesicles — 271
    John R. Ogden and Robert R. Wagner

Comparative Nucleotide Sequence Analysis of the Glycoprotein Gene of Antigenically Altered Rabies Viruses — 279
    W. H. Wunner, C. L. Smith, M. Lafon, J. Ideler, and T. J. Wiktor

Variation in Glycosylation Pattern of G Proteins among Antigenic Variants of the CVS Strain of Rabies Virus — 285
    B. Dietzschold, W. H. Wunner, M. Lafon, C. L. Smith, and A. Varrichio

Change in Pathogenicity and Amino Acid Substitution in the Glycoprotein of Several Spontaneous and Induced Mutants of the CVS Strain of Rabies Virus — 295
    I. Seif, M. Pepin, J. Blancou, P. Coulon, and A. Flamand

How Many Forms of the Newcastle Disease Virus P Protein Are There? — 301
    L. E. Hightower, G. W. Smith, and P. L. Collins

Four Functional Domains on the HN Glycoprotein of Newcastle Disease Virus — 309
    Ronald M. Iorio and Michael A. Bratt

Mapping Mutant and Wild-Type M Proteins of Newcastle Disease Virus (NDV) by Repeated Partial Proteolysis — 315
    Mark E. Peeples and Michael A. Bratt

Structural Characterization of Human Parainfluenza Virus 3 — 321
    Douglas G. Storey and C. Yong Kang

Characterization of Structural Proteins of Parainfluenza Virus 3 and mRNAs from Infected Cells — 329
    George B. Thornton, Jayasri Roy, and Amiya K. Banerjee

Characterization of Mumps Virus Proteins and RNA — 333
    E. J. B. Simpson, J. A. Curran, S. J. Martin, E. M. Hoey, and B. K. Rima

Analysis of the Antigenic Structure and Function of Sendai Virus

Protein NP ... 339
    K. L. Deshpande and A. Portner

Monoclonal Antibodies as Probes of the Antigenic Structure
and Functions of Sendai Virus Glycoproteins ... 345
    A. Portner

Variations in Antigenic Determinants of Different Strains
of Measles Virus ... 351
    Hooshmand Sheshberadaran, Erling Norrby,
    and Shou-Ni Chen

Positive Identification and Molecular Cloning of the
Phosphoprotein (P) of Measles Virus ... 359
    William J. Bellini, George Englund, Chris D. Richardson,
    R. Nick Hogan, Shmuel Rozenblatt, Chester A. Meyers,
    and Robert A. Lazzarini

Identification of a New Envelope-Associated Protein of Human
Respiratory Syncytial Virus ... 365
    Y. T. Huang, P. L. Collins, and G. W. Wertz

Characterization of the Glycoproteins of Respiratory Syncytial
Virus ... 369
    Dennis M. Lambert and Marcel W. Pons

Biology

Characterization of Rabies Virus Receptor-Rich Regions
at Peripheral and Central Synapses ... 379
    Abigail L. Smith, Thomas G. Burrage,
    and Gregory H. Tignor

Early Interactions of Rabies Virus with Cell Surface Receptors ... 387
    Kevin J. Reagan and William H. Wunner

Microinjection of Monoclonal Antibodies to Vesicular Stomatitis
Virus Nucleocapsid Protein into Host Cells: Effect on Virus
Replication ... 393
    Heinz Arnheiter, Monique Dubois-Dalcq, Manfred Schubert,
    Nancy Davis, John Patton, and Robert Lazzarini

The Coiling of Vesicular Stomatitis Virus Nucleocapsids
at the Inner Surface of Plasma Membranes: Immunolocalization
of the Matrix Protein ... 399
    Ward F. Odenwald, Heinz Arnheiter,
    Monique Dubois-Dalcq, and Robert Lazzarini

Interactions of Viral Proteins with Murine Lymphocytes ... 405
    James J. McSharry, Gail Goodman-Snitkoff,
    and Shinae Kizaka

Host Range Mutants of Piry Virus: A New Type of Mutant
in *Drosophila* ... 413
   G. Brun

Early Appearance and Colocalization of Individual Measles Virus
Proteins using Double-Label Fluorescent Antibody Techniques ... 421
   R. N. Hogan, F. Rickaert, W. J. Bellini, C. Richardson,
   M. Dubois-Dalcq, and D. E. McFarlin

Cross-Reaction of Measles Virus Phosphoprotein with a Human
Intermediate Filament: Molecular Mimicry during Virus Infection ... 427
   Robert S. Fujinami and Michael B. A. Oldstone

Measles Virus Infection of Human Peripheral Blood Lymphocytes:
Importance of the $OKT4^+$ T-Cell Subset ... 435
   Steven Jacobson and Henry F. McFarland

Treatment of Experimental Mumps Meningoencephalitis
using Monoclonal Antibodies ... 443
   Jerry S. Wolinsky, M. Neal Waxham, Alfred C. Server,
   and David C. Merz

Biochemical Aspcts of Chemiluminescence Induced by Sendai
Virus in Mouse Spleen Cells ... 451
   Bernard Semadeni, Maurice J. Weidemann,
   and Ernst Peterhans

Defective Viruses and Virus Persistence

Transcribing VSV_{NJ} DI Particle and Its Biological Activities ... 461
   C. Yong Kang, Ruth Park, John McCulloch,
   and Jeong Sun Seo

Recombination Events during the Generation of DI RNAs of VSV ... 469
   Ellen Meier, George G. Harmison, Jack D. Keene,
   and Manfred Schubert

Structure and Generation of Deletion Mutants of Vesicular
Stomatitis Virus ... 475
   Ronald C. Herman

Sendai Virus DI RNA Species Containing 3'-Terminal Genome
Fragments ... 483
   G. G. Re, E. Morgan, K. C. Gupta, and D. W. Kingsbury

Long-Term Persistence by Vesicular Stomatitis Virus in Hamsters ... 489
   Patricia N. Fultz, John J. Holland, Robert Knobler,
   and M. B. A. Oldstone

Persistent Infections of BHK-21 Cells with Rabies Virus ... 497
   Christine Tuffereau, Florence Lafay, and Anne Flamand

Measles Virus Persistent Infections: Modification of Fatty Acid
Metabolism 505
    T. F. Wild, P. Giraudon, P. Anderton, and G. Zwingelstein
Assembly of Measles Virus Nucleocapsids during Lytic
and Persistent Infections 513
    Linda E. Fisher and Elliott Bedows
Synthesis of Matrix Protein in a Subacute Sclerosing
Panencephalitis Cell Line 521
    Michael J. Carter, Margaret M. Willcocks,
    and Volker ter Meulen
Matrix (M) Protein Alterations Induced by TLCK in Cells Acutely
and Persistently Infected with Measles Virus 529
    Steven L. Wechsler
Mechanisms of RSV DI Particle Interference 537
    Mary W. Treuhaft
The Effect of Virus Persistence on Plasma Membrane-Bound
Functions in CNS-Derived Cell Lines 545
    P. N. Barrett, P. Münzel, C. Winkelkötter, and K. Koschel

Index 553

Contributors

Numbers in parentheses indicate the pages on which the authors' contributions begin.

P. Anderton (505), *Unité de Virologie, INSERM, 69008 Lyon, France*
Heinz Arnheiter (393, 399), *Laboratory of Molecular Genetics, IRP, National Institute of Neurological and Communicative Disorders and Stroke, National Institutes of Health, Bethesda, Maryland 20205*
Knut Baczko (233), *Institute of Virology, University of Würzburg, D-8700 Würzburg, Federal Republic of Germany*
Amiya K. Banerjee (63, 329), *Department of Cell Biology, Roche Institute of Molecular Biology, Roche Research Center, Nutley, New Jersey 07110*
P. N. Barrett (545), *Institute of Virology and Immunobiology, Würzburg University, D-8700 Würzburg, Federal Republic of Germany*
T. Barrett[1] (167), *Division of Virology, Department of Pathology, University of Cambridge, Cambridge CB2 2QQ, United Kingdom*
Elliott Bedows (513), *Departments of Anesthesiology and Epidemiology, University of Michigan, Ann Arbor, Michigan 48109*
William J. Bellini (359, 421), *Laboratory of Molecular Genetics, National Institute of Neurological and Communicative Disorders and Stroke, National Institutes of Health, Bethesda, Maryland 20205*
B. Bhatia (183), *Division of Virology, Office of Biologics, National Center for Drugs and Biologics, Bethesda, Maryland 20205*
Martin Billeter (233), *Institute for Molecular Biology I, University of Zürich, CH-8093 Zürich, Switzerland*
J. Blancou (295), *Centre National d'Etudes sur la Rage, 54220 Malzeville, France*
Benjamin M. Blumberg (3), *Department of Microbiology, University of Geneva, 1205 Geneva, Switzerland*
Michael A. Bratt (309, 315), *Department of Molecular Genetics and Microbiology, University of Massachusetts Medical School, Worcester, Massachusetts 01605*
G. Brun (413), *Laboratoire de Génétique des Virus, C.N.R.S., 91190 Gif sur Yvette, France*

[1]Present address: IMMUNO Research Centre, A-2304 Orth a.d. Donau, Austria.

Thomas G. Burrage (379), *Department of Epidemiology and Public Health, Yale University School of Medicine, New Haven, Connecticut 06510*

J. Capone[2] (193), *Department of Biochemistry, McMaster University, Hamilton, Ontario L8N 3Z5, Canada*

Michael J. Carter[3] (521), *Institute of Virology, University of Würzburg, D-8700 Würzburg, Federal Republic of Germany*

Jasemine C. Chambers (103), *Department of Microbiology and Immunology, Duke University Medical Center, Durham, North Carolina 27710*

Robert M. Chanock (27), *Laboratory of Infectious Diseases, National Institute of Allergy and Infectious Diseases, National Institutes of Health, Bethesda, Maryland 20205*

Shou-Ni Chen[4] (351), *Department of Virology, School of Medicine, Karolinska Institute, Stockholm, Sweden*

Gail M. Clinton (71), *Department of Biochemistry, Medical Center, Louisiana State University, New Orleans, Louisiana 70112*

P. L. Collins (21, 301, 365), *Department of Microbiology and Immunology, School of Medicine, University of North Carolina, Chapel Hill, North Carolina 27514*

P. Coulon (295), *Université Paris-Sud, 91405 Orsay Cedex, France*

C. Cunningham (115), *MRC Virology Unit, Institute of Virology, University of Glasgow, Glasgow G11 5JR, Scotland*

J. A. Curran (333), *Department of Biochemistry, The Queen's University of Belfast, Belfast BT9 7BL, Northern Ireland*

Nancy L. Davis (147, 393), *Department of Microbiology and Immunology, University of North Carolina, Chapel Hill, North Carolina 27514*

Bishnu P. De (63), *Roche Institute of Molecular Biology, Roche Research Center, Nutley, New Jersey 07110*

K. L. Deshpande (339), *Division of Virology and Molecular Biology, St. Jude Children's Research Hospital, Memphis, Tennessee 38101*

Lyle A. Dethlefsen (3), *University of Utah Medical Center, Salt Lake City, Utah 84132*

L. E. Dickens (21), *Department of Microbiology and Immunology, University of North Carolina School of Medicine, Chapel Hill, North Carolina 27514*

B. Dietzschold (285), *The Wistar Institute, Philadelphia, Pennsylvania 19104*

Peter C. Dowling[5] (3), *New Jersey Medical School, Newark, New Jersey 07103, and VA Medical Center, East Orange, New Jersey 07019*

Monique Dubois-Dalcq (393, 399, 421), *Laboratory of Molecular Genetics, National Institute of Neurological and Communicative Disorders and Stroke, National Institutes of Health, Bethesda, Maryland 20205*

---

[2]Present address: Department of Biology, Massachusetts Institute of Technology, Cambridge, Massachusetts 02139.

[3]Present address: Department of Virology, Royal Victoria Infirmary, Newcastle upon Tyne NE1 4LP, United Kingdom.

[4]Present address: National Institute of Biological Products, Chengdu, Sichuan, People's Republic of China.

[5]Present address: Department of Microbiology, University of Geneva, 1205 Geneva, Switzerland.

Jon Eagle (247), *Virology and Immunology Section, Queensland Institute of Medical Research, Herston, Brisbane, Queensland 4006, Australia*

Anneka C. Ehrnst[6] (87), *Department of Microbiology and Molecular Genetics, Harvard Medical School, and Division of Infectious Disease, The Children's Hospital, Boston, Massachusetts 02115*

Narayansamy Elango (27), *Laboratory of Infectious Diseases, National Institute of Allergy and Infectious Diseases, National Institutes of Health, Bethesda, Maryland 20205*

Suzanne Urjil Emerson (79, 95), *Department of Microbiology, University of Virginia School of Medicine, Charlottesville, Virginia 22908*

George Englund (359), *Laboratory of Molecular Genetics, IRP, National Institute of Neurological and Communicative Disorders and Stroke, National Institutes of Health, Bethesda, Maryland 20205*

Linda E. Fisher (513), *Department of Natural Sciences, University of Michigan–Dearborn, Dearborn, Michigan 48128*

Anne Flamand (295, 497), *Laboratoire de Génétique 2, Université de Paris-Sud, 91405 Orsay Cedex, France*

Debra W. Frielle (139), *Department of Microbiology, University of Pittsburgh School of Medicine, Pittsburgh, Pennsylvania 15261*

Robert S. Fujinami (427), *Department of Immunology, Scripps Clinic and Research Foundation, La Jolla, California 92037*

Patricia N. Fultz (489), *Department of Biology, University of California, San Diego, La Jolla, California 92093*

H. P. Ghosh (193), *Department of Biochemistry, McMaster University, Hamilton, Ontario L8N 3Z5, Canada*

Colomba Giorgi (3), *Department of Microbiology, University of Geneva Medical School, 1205 Geneva, Switzerland*

P. Giraudon (505), *Unité de Virologie, INSERM, 69008 Lyon, France*

Gail Goodman-Snitkoff (405), *Department of Microbiology and Immunology, Albany Medical College, Albany, New York 12208*

N. T. Gorman[7] (167), *Division of Virology, Department of Pathology, University of Cambridge, Cambridge CB2 2QQ, United Kingdom*

Todd L. Green (95), *Department of Microbiology, University of Virginia School of Medicine, Charlottesville, Virginia 22908*

Brian W. Grinnell (131), *Department of Microbiology, University of Virginia School of Medicine, Charlottesville, Virginia 22908*

K. C. Gupta (11, 17, 483), *Division of Virology and Molecular Biology, St. Jude Children's Research Hospital, Memphis, Tennessee 38101*

George G. Harmison (35, 469), *Laboratory of Molecular Genetics, IRP, National Institute of Neurological and Communicative Disorders and Stroke, National Institutes of Health, Bethesda, Maryland 20205*

[6]Present address: The Stockholm County Council Central Microbiological Laboratory, S-101 22 Stockholm 1, Sweden.

[7]Present address: College of Veterinary Medicine, Department of Medical Sciences, University of Florida, Gainesville, Florida 32610.

Ronald C. Herman (475), *Virology Laboratory, Center for Laboratories and Research, New York State Department of Health, Albany, New York 12201*

L. E. Hightower (301), *Microbiology Section U-44, The University of Connecticut, Storrs, Connecticut 06268*

E. M. Hoey (55, 333), *Department of Biochemistry, The Queen's University of Belfast, Belfast BT9 7BL, Northern Ireland*

R. Nick Hogan (359, 421), *Laboratory of Molecular Genetics, National Institute of Neurological and Communicative Disorders and Stroke, National Institutes of Health, Bethesda, Maryland 20205*

John J. Holland (489), *Department of Biology, University of California, San Diego, La Jolla, California 92093*

C.-H. Hsu (265), *Division of Virology and Molecular Biology, St. Jude Children's Research Hospital, Memphis, Tennessee 38101*

Alice S. Huang (87), *Department of Microbiology and Molecular Genetics, Harvard Medical School, and Division of Infectious Diseases, The Children's Hospital, Boston, Massachusetts 02115*

Y. T. Huang (365), *Department of Microbiology and Immunology, University of North Carolina, School of Medicine, Chapel Hill, North Carolina 27514*

J. D. Hull (55), *Department of Biochemistry, The Queen's University of Belfast, Belfast BT9 7BL, Northern Ireland*

D. Margaret Hunt (123), *Department of Biochemistry, The University of Mississippi Medical Center, Jackson, Mississippi 39216*

J. Ideler[8] (279), *The Wistar Institute, Philadelphia, Pennsylvania 19104*

Ronald M. Iorio (309), *Department of Molecular Genetics and Microbiology, University of Massachusetts Medical School, Worcester, Massachusetts 01605*

Steven Jacobson (435), *Neuroimmunology Branch, National Institute of Neurological and Communicative Disorders and Stroke, National Institutes of Health, Bethesda, Maryland 20205*

C. Yong Kang (321, 461), *Department of Microbiology and Immunology, University of Ottawa School of Medicine, Ottawa, Ontario K1H 8M5, Canada*

Jack D. Keene (103, 109, 469), *Department of Microbiology and Immunology, Duke University Medical Center, Durham, North Carolina 27710*

D. W. Kingsbury (11, 17, 265, 483), *Division of Virology and Molecular Biology, St. Jude Children's Research Hospital, Memphis, Tennessee 38101*

Akio Kiuchi (41), *Department of Environmental Health Sciences, School of Public Health, University of Alabama in Birmingham, Birmingham, Alabama 35294*

Shinae Kizaka (405), *Department of Microbiology and Immunology, Albany Medical College, Albany, New York 12208*

M. J. Klutch (183), *Division of Virology, Office of Biologics, National Center for Drugs and Biologics, Bethesda, Maryland 20205*

Robert Knobler (489), *Department of Immunology, Scripps Clinic and Research Foundation, La Jolla, California 92037*

[8]Present address: Department of Virology, Agriculture University, Wageningen, The Netherlands.

Daniel Kolakofsky (3), *Department of Microbiology, University of Geneva, 1205 Geneva, Switzerland*
K. Koschel (545), *Institute of Virology and Immunobiology, Würzburg University, D-8700 Würzburg, Federal Republic of Germany*
Delphine Krantz (223), *Department of Microbiology, University of Illinois, Urbana, Illinois 61801*
Michael G. Kurilla (103), *Department of Microbiology and Immunology, Duke University Medical Center, Durham, North Carolina 27710*
Florence Lafay (497), *Laboratoire de Génétique 2, Université de Paris-Sud, 91405 Orsay Cedex, France*
M. Lafon[9] (279, 285), *The Wistar Institute, Philadelphia, Pennsylvania 19104*
Dennis M. Lambert (369), *Department of Molecular Virology, Christ Hospital Institute of Medical Research, Cincinnati, Ohio 45219*
Robert A. Lazzarini (359, 393, 399), *Laboratory of Molecular Genetics, IRP, National Institute of Neurological and Communicative Disorders and Stroke, National Institutes of Health, Bethesda, Maryland 20205*
W. C. Leung (49), *Department of Medicine, University of Alberta, Edmonton, Alberta T6G 2G3, Canada*
G. Lund (49), *Departments of Biochemistry and Medicine, University of Alberta, Edmonton, Alberta T6G 2H7, Canada*
B. W. J. Mahy (167), *Division of Virology, Department of Pathology, University of Cambridge, Addenbrooke's Hospital, Cambridge CB2 2QQ, United Kingdom*
Lorraine L. Marnell (255), *Department of Cellular, Viral and Molecular Biology, University of Utah, Medical Center, Salt Lake City, Utah 84132*
S. J. Martin (55, 333), *Department of Biochemistry, The Queen's University of Belfast, Belfast BT9 7BL, Northern Ireland*
Gustave N. Mbuy (175), *Department of Molecular Virology, Christ Hospital Institute of Medical Research, Cincinnati, Ohio 45219*
Patrick W. McClear (71), *Department of Microbiology and Immunology, Washington University School of Medicine, St. Louis, Missouri 63110*
Marcella A. McClure (71), *Department of Microbiology and Immunology, Washington University School of Medicine, St. Louis, Missouri 63110*
John McCulloch (461), *Department of Microbiology and Immunology, University of Ottawa School of Medicine, Ottawa, Ontario K1H 8M5, Canada*
Henry F. McFarland (435), *Neuroimmunology Branch, National Institute of Neurological and Communicative Disorders and Stroke, National Institutes of Health, Bethesda, Maryland 20205*
D. E. McFarlin (421), *Neuroimmunology Branch, National Institute of Neurological and Communicative Disorders and Stroke, National Institutes of Health, Bethesda, Maryland 20205*
John J. McGowan (215), *Department of Microbiology, Uniformed Services University of the Health Sciences, Bethesda, Maryland 20814*

---

[9]Present address: Institute Pasteur, Service Rage 25, 5724 Paris Cedex 15, France.

James J. McSharry (405), *Department of Microbiology and Immunology, Albany Medical College, Albany, New York 12208*

Ellen Meier (35, 469), *Laboratory of Molecular Genetics, IRP, National Institute of Neurological and Communicative Disorders and Stroke, National Institutes of Health, Bethesda, Maryland 20205*

David C. Merz[10] (443), *Department of Neurology, University of Texas Health Science Center, Houston, Texas 77025*

Chester A. Meyers[11] (359), *Laboratory of Molecular Genetics, IRP, National Institute of Neurological and Communicative Disorders and Stroke, National Institutes of Health, Bethesda, Maryland 20205*

J. B. Milstien (183), *National Center for Drugs and Biologics, Food and Drug Administration, Bethesda, Maryland 20205*

E. M. Morgan (17, 483), *Division of Virology and Molecular Biology, St. Jude Children's Research Hospital, Memphis, Tennessee 38101*

Trudy G. Morrison (161, 201), *Department of Molecular Genetics and Microbiology, University of Massachusetts Medical School, Worcester, Massachusetts 01605*

Sue A. Moyer (153), *Department of Microbiology, Vanderbilt University School of Medicine, Nashville, Tennessee 37232*

P. Münzel (545), *Institute of Virology and Immunobiology, Würzburg University, D-8700 Würzburg, Federal Republic of Germany*

Erling Norrby (351), *Department of Virology, School of Medicine, Karolinska Institute, Stockholm, Sweden*

Ward F. Odenwald (399), *Laboratory of Molecular Genetics, IRP, National Institute of Neurological and Communicative Disorders and Stroke, National Institutes of Health, Bethesda, Maryland 20205*

John R. Ogden (271), *Department of Microbiology, University of Virginia Medical School, Charlottesville, Virginia 22908*

Michael B. A. Oldstone (427, 489), *Department of Immunology, Scripps Clinic and Research Foundation, La Jolla, California 92037*

Ruth Park (461), *Department of Microbiology and Immunology, University of Ottawa School of Medicine, Ottawa, Ontario K1H 8M5, Canada*

R. G. Paterson[12] (115), *MRC Virology Unit, Institute of Virology, University of Glasgow, Glasgow G11 5JR, Scotland*

Jean L. Patterson[13] (3), *Department of Microbiology, University of Geneva, 1205 Geneva, Switzerland*

R. C. Patterson (167), *Division of Virology, Department of Pathology, University of Cambridge, Cambridge CB2 2QQ, United Kingdom*

---

[10]Present address: Department of Medicine, University of Michigan, Ann Arbor, Michigan 48109.

[11]Present address: Laboratory of Chemoprevention, National Cancer Institute, National Institutes of Health, Bethesda, Maryland 20205.

[12]Present address: Department of Biochemistry, Molecular and Cell Biology, Northwestern University, Evanston, Illinois 60201.

[13]Present address: Department of Clinical Veterinary Medicine, Cambridge University, Cambridge CB2 2QQ, United Kingdom.

John T. Patton[14] (147, 393), *Department of Bacteriology and Immunology, University of North Carolina, Chapel Hill, North Carolina 27514*

Mark E. Peeples[15] (315), *Department of Molecular Genetics and Microbiology, University of Massachusetts Medical Center, Worcester, Massachusetts 01605*

Richard W. Peluso (153), *Department of Microbiology, Vanderbilt University School of Medicine, Nashville, Tennessee 37232*

M. Pepin[16] (295), *Centre National d'Etudes sur la Rage, 54220 Malzeville, France*

Jacques Perrault (71), *Washington University, Department of Microbiology and Immunology, School of Medicine, St. Louis, Missouri 63110*

Ernst Peterhans (451), *Institute of Virology, University of Zürich, CH-8057 Zürich, Switzerland*

Helen Piwnica-Worms (109), *Department of Microbiology and Immunology, Duke University Medical Center, Durham, North Carolina 27710*

Marcel W. Pons (369), *Department of Molecular Virology, Christ Hospital Institute of Medical Research, Cincinnati, Ohio 45219*

A. Portner (339, 345), *Division of Virology and Molecular Biology, St. Jude Children's Research Hospital, Memphis, Tennessee 38101*

A. Ramsingh[17] (49), *Department of Medicine, University of Alberta, Edmonton, Alberta T6G 2G3, Canada*

Jessica Randle (247), *Virology and Immunology Section, Queensland Institute of Medical Research, Herston, Brisbane, Queensland 4006, Australia*

G. G. Re (11, 17, 483), *Division of Virology and Molecular Biology, St. Jude Children's Research Hospital, Memphis, Tennessee 38101*

Kevin J. Reagan (387), *The Wistar Institute, Philadelphia, Pennsylvania 19104*

M. E. Reichmann (223), *Department of Microbiology, University of Illinois, Urbana, Illinois 61801*

James Remenick (215), *Department of Microbiology, Uniformed Services University of the Health Sciences, Bethesda, Maryland 20814*

Chris D. Richardson (359, 421), *Laboratory of Molecular Genetics, National Institute of Neurological and Communicative Disorders and Stroke, National Institutes of Health, Bethesda, Maryland 20205*

F. Rickaert[18] (421), *Laboratory of Molecular Genetics, National Institute of Neurological and Communicative Disorders and Stroke, National Institutes of Health, Bethesda, Maryland 20205*

B. K. Rima (55, 333), *Department of Biochemistry, The Queen's University of Belfast, Belfast BT9 7BL, Northern Ireland*

Steven J. Robbins (247), *Virology and Immunology Section, Queensland Institute of Medical Research, Herston, Brisbane, Queensland 4006, Australia*

---

[14]Present address: Department of Biology, University of South Florida, Tampa, Florida 33620.

[15]Present address: Department of Immunology, Rush Medical College, Chicago, Illinois 60612.

[16]Present address: Institut National de la Recherche Agronomique, Station de Pathologie de la Reproduction, 37380 Nouzilly, France.

[17]Present address: Department of Human Genetics, Yale University School of Medicine, New Haven, Connecticut 06510.

[18]Present address: Department of Pathology, Hospital Erasme, B-1070 Brussels, Belgium.

Olga M. Rochovansky (175), *Department of Molecular Virology, Christ Hospital Institute of Medical Research, Cincinnati, Ohio 45219*

Laurent Roux (3), *Department of Microbiology, University of Geneva, 1205 Geneva, Switzerland*

Jayasri Roy (329), *Roche Institute of Molecular Biology, Roche Research Center, Nutley, New Jersey 07110*

P. Roy (41), *Department of Environmental Health Sciences, School of Public Health, University of Alabama in Birmingham, Birmingham, Alabama 35294*

Shmuel Rozenblatt (359), *Laboratory of Molecular Genetics, IRP, National Institute of Neurological and Communicative Disorders and Stroke, National Institutes of Health, Bethesda, Maryland 20205*

S. E. H. Russell (55), *Department of Biochemistry, The Queen's University of Belfast, Belfast BT9 7BL, Northern Ireland*

Angeles Sanchez (63), *Roche Institute of Molecular Biology, Roche Research Center, Nutley, New Jersey 07110*

Masanobu Satake (27), *Laboratory of Infectious Diseases, National Institute of Allergy and Infectious Diseases, National Institutes of Health, Bethesda, Maryland 20205*

Manfred Schubert (35, 393, 469), *Laboratory of Molecular Genetics, IRP, National Institute of Neurological and Communicative Disorders and Stroke, National Institutes of Health, Bethesda, Maryland 20205*

Harlan B. Scott II (223), *Department of Microbiology, University of Illinois, Urbana, Illinois 61801*

D. Scraba (49), *Departments of Biochemistry and Medicine, University of Alberta, Edmonton, Alberta T6G 2H7, Canada*

I. Seif[19] (295), *Université Paris-Sud, 91405 Orsay Cedex, France*

A. S. Seifried (183), *Division of Virology, Office of Biologics, National Center for Drugs and Biologics, Bethesda, Maryland 20205*

Bernard Semadeni (451), *Institute of Virology, University of Zürich, CH-8057 Zürich, Switzerland*

Jeong Sun Seo[20] (461), *Department of Microbiology and Immunology, University of Ottawa School of Medicine, Ottawa, Ontario K1H 8M5, Canada*

Alfred C. Server (443), *Department of Neurology, University of Texas Health Science Center, Houston, Texas 77025*

Hooshmand Sheshberadaran (351), *Department of Virology, School of Medicine, Karolinska Institute, Stockholm, Sweden*

E. J. B. Simpson (333), *Department of Biochemistry, The Queen's University of Belfast, Belfast BT9 7BL, Northern Ireland*

---

[19]Present address: Centre National de la Recherche Scientifique, Laboratoire de Génétique des Virus, 91190 Gif sur Yvette, France.

[20]Present address: Department of Biochemistry, Seoul National University College of Medicine, Seoul, Korea.

Abigail L. Smith (379), *Section of Comparative Medicine and Department of Epidemiology and Public Health, Yale University School of Medicine, New Haven, Connecticut 06510*

C. L. Smith (279, 285), *The Wistar Institute, Philadelphia, Pennsylvania 19104*

G. W. Smith (301), *Microbiology Section, The Biological Sciences Group, The University of Connecticut, Storrs, Connecticut 06268*

Douglas G. Storey (321), *Department of Microbiology and Immunology, University of Ottawa, School of Medicine, Ottawa, Ontario K1H 8M5, Canada*

Donald F. Summers (255), *Department of Cellular, Viral and Molecular Biology, University of Utah Medical Center, Salt Lake City, Utah 84132*

J. F. Szilágyi (115), *MRC Virology Unit, Institute of Virology, University of Glasgow, Glasgow G11 5JR, Scotland*

Volker ter Meulen (233, 521), *Institute of Virology, University of Würzburg, D-8700 Würzburg, Federal Republic of Germany*

George B. Thornton (329), *Roche Institute of Molecular Biology, Roche Research Center, Nutley, New Jersey 07110*

Gregory H. Tignor (379), *Department of Epidemiology, Yale University School of Medicine, New Haven, Connecticut 06510*

Mary W. Treuhaft (537), *Marshfield Medical Foundation, Marshfield, Wisconsin 54449*

J. Tucker (49), *Department of Biochemistry, University of Alberta, Edmonton, Alberta T6G 2H7, Canada*

Christine Tuffereau (497), *Laboratoire de Génétique 2, Université de Paris-Sud, 91405 Orsay Cedex, France*

D. L. J. Tyrrell (49), *Departments of Biochemistry and Medicine, University of Alberta, Edmonton, Alberta T6G 2H7, Canada*

A. Varrichio (285), *The Wistar Institute, Philadelphia, Pennsylvania 19104*

Sundararajan Venkatesan (27), *Laboratory of Infectious Diseases, National Institute of Allergy and Infectious Diseases, National Institutes of Health, Bethesda, Maryland 20205*

Robert R. Wagner (131, 271), *Department of Microbiology, University of Virginia School of Medicine, Charlottesville, Virginia 22908*

Lori J. Ward (201), *Department of Molecular Genetics and Microbiology, University of Massachusetts Medical Center, Worcester, Massachusetts 01605*

M. Neal Waxham (443), *Department of Neurology, University of Texas Health Science Center, Houston, Texas 77025*

Steven L. Wechsler (239, 529), *Department of Molecular Virology, Christ Hospital Institute of Medical Research, Cincinnati, Ohio 45219*

Maurice J. Weidemann (451), *Department of Biochemistry, Faculty of Science, Australian National University, Canberra, ACT, Australia*

Gail W. Wertz (21, 147, 365), *Department of Microbiology and Immunology, University of North Carolina School of Medicine, Chapel Hill, North Carolina 27514*

Patricia Whitaker-Dowling (207), *Department of Microbiology, University of Pittsburgh School of Medicine, Pittsburgh, Pennsylvania 15261*
T. J. Wiktor (279), *The Wistar Institute, Philadelphia, Pennsylvania 19104*
T. F. Wild (505), *Unité de Virologie, INSERM, 69008 Lyon, France*
A. Wilde (161), *Department of Molecular Genetics and Microbiology, University of Massachusetts Medical School, Worcester, Massachusetts 01605*
Margaret M. Willcocks[21] (521), *Institute of Virology, University of Würzburg, D-8700 Würzburg, Federal Republic of Germany*
Paul M. Williams (79), *Department of Microbiology, University of Virginia School of Medicine, Charlottesville, Virginia 22908*
Jeffrey Wilusz (103), *Department of Microbiology and Immunology, Duke University Medical Center, Durham, North Carolina 27710*
C. Winkelkötter (545), *Institute of Virology and Immunobiology, Würzburg University, D-8700 Würzburg, Federal Republic of Germany*
Jerry S. Wolinsky (443), *Department of Neurology, University of Texas Health Science Center, Houston, Texas 77025*
William H. Wunner (279, 285, 387), *The Wistar Institute, Philadelphia, Pennsylvania 19104*
Karen K. Y. Young (239), *Department of Molecular Virology, Christ Hospital Institute of Medical Research, Cincinnati, Ohio 45219*
Julius S. Youngner (139, 207), *Department of Microbiology, University of Pittsburgh School of Medicine, Pittsburgh, Pennsylvania 15261*
G. Zwingelstein (505), *Unité de Virologie, INSERM, 69008 Lyon, France*

---

[21]Present address: Department of Genetics, University of Newcastle, Newcastle upon Tyne, United Kingdom.

Preface

Negative strand viruses include several families of large, enveloped RNA viruses. They are responsible for many important common or exotic, human and animal diseases, including influenza, mumps, measles, rabies, canine distemper, some human encephalitis, and hemorrhagic fevers. Four previous symposia have been held on negative strand viruses. Three were organized by Drs. R. D. Barry and B. W. J. Mahy and were held in Cambridge, England, in 1969, 1973, and 1977. A fourth symposium was held in St. Thomas, Virgin Islands, in 1980, and was organized by Drs. D. H. L. Bishop and R. W. Compans.

Research on the molecular biology of negative strand viruses has continued over the last years at an accelerated pace and was recently reviewed at a fifth symposium held September 11–17, 1983, at Mariner's Inn, Hilton Head, South Carolina. To focus the subject matter, the proceedings of this symposium are being published in two volumes. One volume concerns the three negative strand virus families with segmented RNA genomes: arenaviruses, bunyaviruses, and orthomyxoviruses. This volume contains papers on negative strand virus families with nonsegmented genomes: paramyxoviruses and rhabdoviruses.

Rhabdoviruses and paramyxoviruses, although evidently different in their biological, biochemical, and structural properties, do share certain basic organizational features such as gene product arrangements and transcriptional signals. Thus studies of the two virus families are complementary in nature and advances in the field have developed in parallel. The papers in this volume reflect such developments and illuminate the various stages in the strategy of negative strand virus infections: adsorption, penetration, mRNA transcription, translation, RNA replication, morphogenesis, virus release. Also addressed are questions concerning the biology of virus infection and host response.

We would like to acknowledge the sponsorship of the symposium by the University of Alabama in Birmingham, and financial support provided by the following organizations: Amicon; Biocell Laboratories; Elsevier/North Holland; Hoffmann–La Roche, Inc.; Imperial Chemical Industries; New Brunswick Scientific Company, Inc.; Pitman-Moore, Inc.; Rheem Manufacturing Company, Inc.; Vangard International, Inc.; and Wheaton Instruments.

We also wish to thank all of the participants who contributed to the success of the meeting and the authors who have provided an excellent and timely series of manuscripts. Our special thanks are also due to the following participants who served as chairmen of the scientific sessions: Drs. G. M. Air, M. A. Bratt, J. M. Dalrymple, N. J. Dimmock, S. U. Emerson, A. Flamand, C. Y. Kang, E. D. Kilbourne, D. W. Kingsbury, H.-D. Klenk, D. Kolakofsky, W. G. Laver, R. A. Lazzarini, D. P. Nayak, M. B. A. Oldstone, P. Palese, M. Schubert, and G. W. Wertz.

Finally, we would like to thank Betty Jeffrey and Denice L. Montgomery, both of whom worked cheerfully and extraordinarily hard in organizing the symposium and in preparing these volumes for publication.

D. H. L. BISHOP
R. W. COMPANS

Contents of *Segmented Negative Strand Viruses*

Genome Structure, Transcription, and Genetics

Coding Analyses of Bunyavirus RNA Species
    D. H. L. Bishop, E. Rud, S. Belloncik, H. Akashi, F. Fuller, T. Ihara, Y. Matsuoka, and Y. Eshita
Molecular Cloning of the Bunyamwera Virus Genome
    J. F. Lees, C. R. Pringle, and R. M. Elliott
Hybrid Selection of Viral mRNA using cDNA Clones of Inkoo Virus M Gene
    M. K. Spriggs and M. J. Hewlett
Rift Valley Fever Virus Intracellular RNA: A Functional Analysis
    M. D. Parker, J. F. Smith, and J. M. Dalrymple
Transcription of the S RNA Segment of Germiston Bunyavirus
    M. Bouloy, N. Pardigon, S. Gerbaud, P. Vialat, and M. Girard
The S Segment of Bunyaviruses Codes for Two Complementary RNAs
    G. Abraham and A. K. Pattnaik
Restriction of Subunit Reassortment in the Bunyaviridae
    C. R. Pringle, W. Clark, J. F. Lees, and R. M. Elliott
Genetic Analyses of Pichinde and LCM Arenaviruses: Evidence for a Unique Organization for Pichinde S RNA
    D. D. Auperin, V. Romanowski, and D. H. L. Bishop
Studies on the Molecular Biology of Lymphocytic Choriomeningitis Virus
    P. J. Southern and M. B. A. Oldstone
Sequence Rearrangements in Influenza Virus RNA and Ribonucleoprotein Structure
    J. S. Robertson, P. A. Jennings, J. T. Finch, and G. Winter
Molecular Topography of the Influenza Virus P Protein Complex during Capped RNA-Primed Messenger RNA Synthesis
    J. Braam, I. Ulmanen, and R. M. Krug

In Vivo Transcription and Translation of Defective-Interfering Particle-Specific
RNAs of Influenza Virus
    T. M. Chambers, R. K. Akkina, and D. P. Nayak
Mutants Obtained following Undiluted Passage of an Influenza A
(Fowl Plague) Virus
    C. Scholtissek, K. Müller, R. T. Schwarz, and H.-D. Klenk
Genetic Stability of A/Ann Arbor/6/60 Cold-Mutant (Temperature-Sensitive)
Live Influenza Vaccine Strains
    N. J. Cox and A. P. Kendal

Gene Expression, Protein Synthesis, and Protein Modification

Subgenomic RNA of Arenavirus Pichinde and Its Implication in Regulation
of Viral Gene Expression in Productive Infection and Persistence
    *W.-C. Leung, A. Ramsingh, G. Jing, K. Mong, A. K. Taneja,
    and R. S. Hodges*
Biochemical Characterization of Hantaan Virus
    C. S. Schmaljohn and J. M. Dalrymple
Eukaryotic Expression of Cloned cDNA Coding for Influenza Viral
Glycoproteins using an SV40 Vector: Use of Recombinant DNA Mutants
to Study Structure–Function Relationships
    T. J. Bos, N. L. McQueen, A. R. Davis, and D. P. Nayak
Studies on the Second Protein Encoded by the Neuraminidase Gene
of Influenza B Virus
    M. W. Shaw, O. Haller, R. A. Lamb, and P. W. Choppin
Functional Expression and Mutational Analysis of Influenza Virus Surface
Glycoproteins
    C.-J. Lai, L. Markoff, K. Sekikawa, and J. Hansen
The Nonstructural Gene Segment of Influenza A Virus: Expression
of NS1 Protein in Mammalian Cells; Analysis of a Deletion Mutant
    *M. Krystal, S. Nakada, D. A. Buonagurio, D. C. DeBorde,
    H. F. Maassab, and P. Palese*
Production of a Subunit Vaccine for Rift Valley Fever Virus
    *A. F. Purchio, M. S. Collett, M. Parker, C. Schmaljohn,
    and J. Dalrymple*
Persistent Expression of Influenza Virus Nucleoprotein in Recombinant
DNA-Transfected Mouse Cells
    B.-C. Lin, C.-J. Lai, and M.-F. Law
Expression of Influenza Virus Subgenomic Virion RNAs in Infected Cells
    C. R. Penn and B. W. J. Mahy
Transport of Viral Glycoprotein and Its Modulation by Monensin
    F. V. Alonso-Caplen, Y. Matsuoka, and R. W. Compans

Interfering with Glycoprotein Processing in Influenza and Sindbis
Virus-Infected Cells
   P. A. Romero, R. Datema, and R. T. Schwarz

Viral Proteins: Antigenic and Functional Analyses

Antigenic and Structural Studies on the Glycoproteins of Lymphocytic
Choriomeningitis Virus
   M. J. Buchmeier
Comparative Analysis of Lassa and Lassa-Like Arenavirus Isolates from Africa
   J. P. Gonzalez, M. J. Buchmeier, J. B. McCormick, S. W. Mitchell, L. H.
   Elliott, and M. P. Kiley
Neutralization of Arenaviruses: Reaction of Tacaribe Virus and Variants
with Monoclonal Antibodies
   L. M. Allison, M. Salter, M. J. Buchmeier, H. Lewicki, and C. R. Howard
Antigenic Sites on the G1 Glycoprotein of La Crosse Virus That Are Involved
in Neutralization
   L. Kingsford and L. D. Ishizawa
Influenza Virus Neuraminidase: Structure and Variation
   W. G. Laver, P. M. Colman, C. W. Ward, J. N. Varghese, G. M. Air,
   and R. G. Webster
Pleiotropic Effects of a Single Amino Acid Change on Antigenicity
and Biologic Function of Swine Influenza Virus Hemagglutinin Mutants
   E. D. Kilbourne, G. W. Both, and W. Gerhard
Selection of Receptor Variants from Human and Avian Influenza Isolates
with the H3 Hemagglutinin
   G. N. Rogers, X.-F. Wang, T. J. Pritchett, L. F. Haber, and J. C. Paulson
Antigenic Alterations of the Influenza Virus Hemagglutinin during the
Infectious Cycle
   T. Bächi, J. W. Yewdell, and W. Gerhard
Antigenic Drift of the H1 Subtype Hemagglutinin from 1977 to 1980
   F. L. Raymond, A. J. Caton, G. G. Brownlee, N. J. Cox,
   and A. P. Kendal
The Binding Sites to Monoclonal Antibodies on A/USSR/90/77 (H1N1)
Hemagglutinin
   S. Nakajima, K. Nakajima, and A. P. Kendal
Sequence Changes in the Hemagglutinin of an Enterotropic H3 Influenza Virus
   C. W. Naeve, V. S. Hinshaw, and R. G. Webster
Structure of the Cleavage Site of Hemagglutinins of Pathogenic
and Nonpathogenic H7 Influenza Viruses
   F. X. Bosch and R. Rott

An Influenza Virus Mutant Whose Lesion Results in Temperature-Sensitive Hemagglutination
    J. W. McCauley
The Carbohydrates of the Hemagglutinin of Influenza Virus
    W. Keil, R. Geyer, H. Niemann, J. Dabrowski, and H.-D. Klenk
Heterogeneity of the Membrane (M1) Protein of Influenza Virus
    A. Gregoriades
Antigenic Characterization of Influenza A Matrix Protein with Monoclonal Antibodies
    K. L. van Wyke, J. W. Yewdell, S. M. Michalek, J. R. McGhee, and B. R. Murphy
Evolutionary Conservation of Influenza Nucleoprotein Genes in Host Species
    W. J. Bean
Virus-Specific Polypeptides in Cells Infected with Influenza C, Dhori, and Thogoto Viruses
    J. P. M. Clerx and H. Meier-Ewert

Biology

Mechanism of Persistent LCMV Infection in Mice: Role of LCMV Variants in Suppression of LCMV-Specific Cytotoxic T-Cell Response and Maintenance of the Carrier State
    R. Ahmed, A. Salmi, and M. B. A. Oldstone
The African Arenaviruses Lassa and Mopeia: Biological and Immunochemical Comparisons
    J. C. S. Clegg and G. Lloyd
Replication of Bunyaviruses in a *Xenopus laevis* Cell Line
    G. E. Watret, C. R. Pringle, and R. M. Elliott
Interaction of Neutralized Influenza Virus with Avian and Mammalian Cells
    N. J. Dimmock, H. P. Taylor, and A. S. Carver
Restricted Mobility of Influenza Hemagglutinin on HeLa Cell Plasma Membranes
    S. Basak, R. W. Compans, and M. B. A. Oldstone
Reassembly of Membranous Particles with Influenza Virus Hemagglutinin and Restoration of Biological and Immunological Activities
    Y. Hosaka
Incorporation of M Protein of Influenza Virus into Liposomes: Further Characterization of Liposomal Structures
    D. J. Bucher, R. R. Dourmashkin, and J. A. Greenberg
In Vitro and *in Vivo* Properties of an Influenza A Host Range Virus
    D. C. DeBorde, A. M. Donabedian, S. M. Peters, and H. F. Maassab
Self-Inactivation of Influenza A Virus and Autolysis of the Virus-Infected Cells
    K. Shimizu, K. Aihara, and T. Miyamoto

Use of Avian–Human Reassortant Influenza A Viruses as Live Vaccine Viruses in Man
   B. R. Murphy, A. J. Buckler-White, S.-F. Tian, M. L. Clements,
   W. T. London, and R. M. Chanock

Index

Genome Structure

ANALYSIS OF THE SENDAI VIRUS GENOME BY MOLECULAR CLONING

Benjamin M. Blumberg, Colomba Giorgi,[2]
Peter C. Dowling,[3] Lyle A. Dethlefsen,[4]
Jean L. Patterson, Laurent Roux and
Daniel Kolakofsky

Department of Microbiology
University of Geneva Medical School
1205 Geneva, Switzerland

A special focus of this laboratory is the study of the control mechanisms of negative strand RNA viral transcription and replication. We have demonstrated that the VSV leader region contains at least two signals, a nucleation site for leader chain encapsidation and a genomic transcription termination site, both of which are functionally keyed to the presence of the VSV nucleocapsid (N) protein (1,2,3). Control of transcription and replication in Paramyxoviruses and in Bunyaviruses is likely to be more complicated, as these viruses produce multiple leader RNAs (4,5), and each also codes for nonstructural proteins which have been suggested to function in such a control capacity (6,7,8).

The C protein of Sendai virus is particularly interesting because, although it is barely detectable in the mature virion, it is present in infected cells and can be synthesized *in* *vitro* using mRNA from infected cell cytoplasm (6,9). The C protein has been shown to have a unique peptide map (6,10), and although annealing studies have shown that esentially the entire Sendai genome is transcribed into viral mRNAs (11,12), no viral mRNA corresponding to the small size (22kd) of this protein has been reported. The finding that C protein is among the *in* *vitro* translation products of mRNAs cosedimenting on sucrose gradients with that for the 79kd viral P protein (9) suggested that the C mRNA might be part

---

[1]This work was supported by grants from the Swiss National Science Fund (DK), from the Zyma and Whitehall Foundations, and from the Multiple Sclerosis Society USA (PCD).
[2]On leave from Laboratorio di Virologia, Istituto Superiore di Sanita, Rome.
[3]On leave from New Jersey Medical School, Newark, and VA Medical Center, East Orange, NJ.
[4]On leave from University of Utah Medical Center, Salt Lake City, Utah.

of a larger structure. Since the 15kb coding capacity of Sendai would allow for any of several possibilities - tandem or overlapping cistrons, or even spliced mRNA, however unlikely for this cytoplasmic virus - cloning of the viral genome was essential for determining its organization.

CLONING AND MAPPING OF THE 3´ PROXIMAL THIRD OF THE SENDAI VIRUS GENOME

The molecular cloning of the Sendai virus genome has been described by Dowling et al.(13). Briefly, methylmercury-denatured virion 50S RNA was reverse transcribed using calf thymus pentamers as primers, the second DNA strand was synthesized by boiling at pH8 and incubating with the Klenow fragment of DNA polymerase, and the mixture was treated with S1 nuclease to generate ds-DNA. These viral-specific DNAs were cloned into the PstI site of pBR322 by poly dC/poly dG tailing, and tet^R amp^S transformants were isolated and screened by the method of Grunstein and Hogness (14) using as probes partially alkali digested, radiolabelled Sendai 50S RNA. To order the genome, clones containing viral 3´ end sequences were selected using ^{32}P - *in vitro* virion polymerase RNAs as probes, since these RNAs should favor 3´ end gene products(4,9)- whence the designation of these clones as S(Sendai), L(leader) or N(NP). A series of overlapping clones selected by these probes was provisionally ordered by a combination of dot blotting, primer extension and restriction mapping. The DNA sequence of one clone was found to match the RNA sequence of the genome from position 22 of the 3´ end onwards, thus establishing the orientation of the series. Positive selection or hybrid arrest of translation of the viral proteins using restriction fragments from these clones annealed to CsCl pellet RNA from Sendai virus infected cells then revealed the gene order to be 3´-NP-P+C-M.

Giorgi et al.(15) have determined by DNA sequencing of the P+C region of the genome that P and C are translated in overlapping reading frames from a single mRNA. Here we report the addition to our series of a further clone which contains the termination site of the M gene, as shown in the updated genomic map (Figure 1).

IDENTIFICATION OF A CLONE CONTAINING THE END OF THE M GENE

A rough calculation of the coding capacity required for the NP (59kd), P+C (79kd) and M (35kd) proteins, about 4800 bp, suggested that the end of the M gene would lie beyond the clones mapped in Dowling et al.(13). To find a further clone which would contain this site, duplicates of the Grunstein-

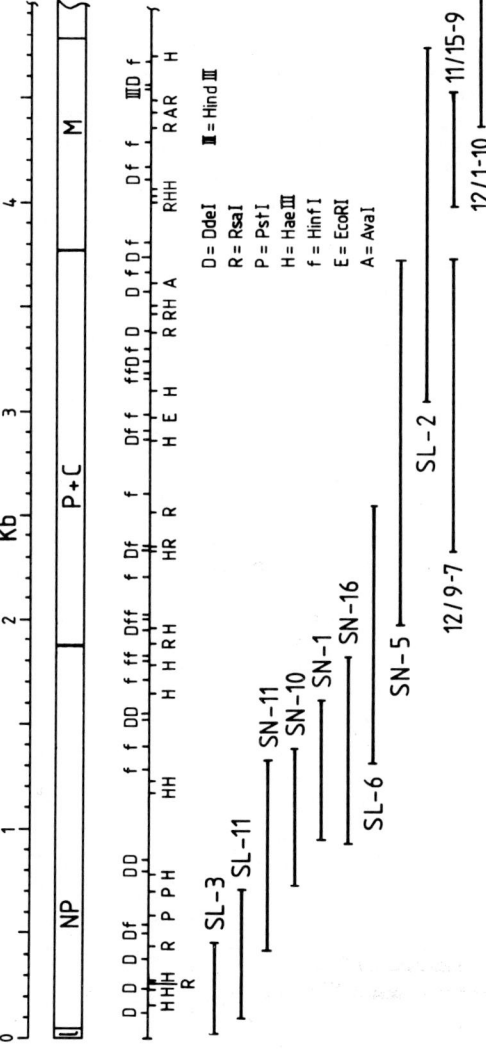

Figure 1. Map of the 3′ proximal third of Sendai virus. Bars denote the clones used to determine the map. The restriction sites shown are sufficient to establish the alignment of the clones, but not all sites have been marked.

Hogness filters used originally to screen for viral-specific clones were therefore re-screened with nick-translated (16) probes made from the excised viral-specific PstI inserts of these clones. Those colonies which were positive to the SL-2 probe and negative to a mixed probe made from the inserts of all clones more 3´ proximal than SL-2 were picked for further analysis, and a preliminary PstI restriction digest was performed on minilysates of these colonies to determine the sizes of the viral-specific inserts in their plasmids (data not shown). Twelve clones containing the largest inserts were then grown in one-liter cultures, and the plasmids were extracted (17) and isolated by centrifugation in CsCl.

These DNAs were then subjected to a combination of dot blot and detailed restriction analysis, as described in Dowling et al.(13). Plasmid DNA (0.5ug) from these and other clones was dotted identically onto four nitrocellulose filters (18), and hybridized in turn, under formamide annealing conditions (19), with nick-translated probes made from their excised PstI inserts. Figure 2 panel A shows the results using a mixed probe specific for 3´ proximal sequences through SN-5, panel B using SL-2 as a probe, and panel C using the clone 11/15-9 (upper right corner of these grids - the numbering system of this series of clones derives from their Grunstein blot number and location). Nine new clones were strongly positive to SL-2. Those at grid positions 3-8 and 5-1 (this clone corresponds to 12/9-7 on the map) were identified as lying to the left side of SL-2 because they were also positive to the mixed probe. A gap between 12/9-7 and the clone at grid position 1-8 (corresponding to 11/15-9 on the map) was indicated since 12/9-7 (grid position 5-1) did not hybridize to the 11/15-9 probe (Figure 2, panel C). The clones were next aligned by digestion with several restriction enyzmes. Two of the ethidium bromide stained gels are shown in Figure 3. Insert bands which are unchanged by PstI digestion are identified as internal fragments of the clone. The clone 11/15-9 apparently contained only internal restriction fragments (e.g. the 135bp Rsa-Rsa band) which were also contained within SL-2, whereas another strongly selected clone (12/1-10 on the map, circled on the grid) clearly contained no internal restriction fragments in common with SL-2. The presence uniquely in 12/1-10 of the large 370 bp Hinf-Hinf fragment further suggested that this clone might lie well beyond the end of SL-2, and thus the map layout would be as shown in Figure 1.

Comparison of the restriction patterns established the probable insert orientations and boundaries of the mapped clones, and a small (~30bp) Hinf-Hinf fragment common to SL-2 and 12/1-10 was eventually found by radiolabelling (not

ANALYSIS OF THE SENDAI VIRUS GENOME

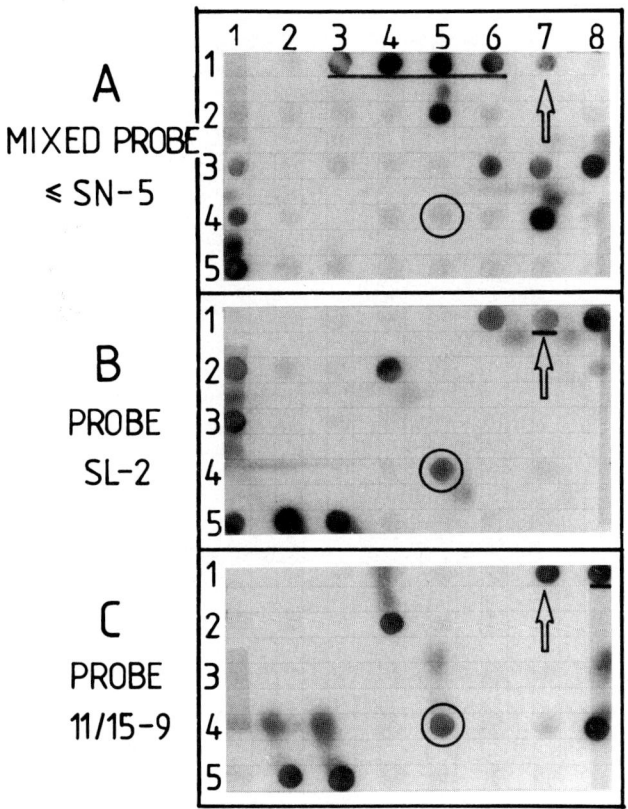

Figure 2. Dot blot analysis of SL-2 related clones. Clones used as probes are underlined. The grid position of SL-2 is marked with an arrow, that of clone 12/1-10 is circled. Control DNAs from E coli B and pBR322 are at grid positions 1-1 and 1-2 (upper left).

shown). Maxam-Gilbert (20) sequencing from the Hind III site common to SL-2 and 12/1-10 has revealed the genome sequence 5´- AUUUAUUCUUUUUGAAUCCC -3´, located just beyond the G-C tail of SL-2. This sequence appears near the end of a sequence of 700 continuous bases in SL-2 and 12/1-10, in which there is a single open reading frame (unpublished, work in progress), and contains the conserved Sendai polyadenylation signal determined by Gupta and Kingsbury (21). It would also code for a mRNA beginning Cap-AGGG---, a sequence also found at the start of the viral NP, P+C and M genes (15,21).

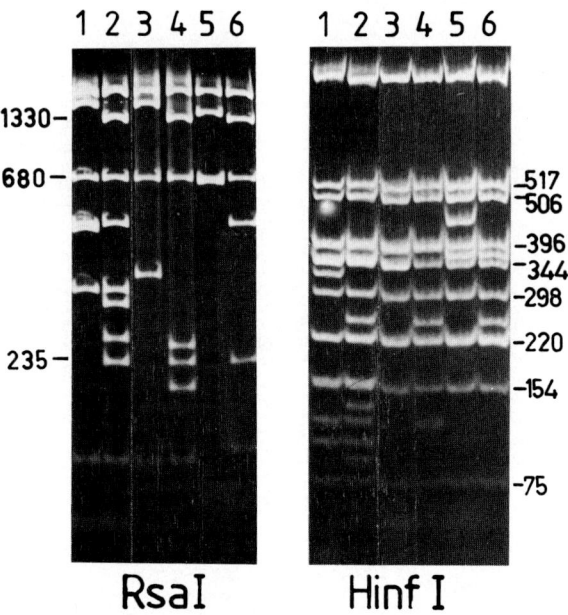

Figure 3. Restriction analysis of SL-2 related clones. Three-microgram aliquots of SL-2 (lanes 1,2), 11/15-9 (lanes 3,4) and 12/1-10 (lanes 5,6) were digested with either RsaI or HinfI alone (odd-numbered lanes), or with these enzymes plus Pst I (even-numbered lanes).

This sequence must thus represent the end of the M gene and the start of the next.

REMARKS

Ordering the Sendai virus genome by the overlapping clone technique may appear cumbersome, but it has the virtue of giving a continuous sequence at intercistronic regions. It furthermore allows us to control for spliced mRNAs by S1 mapping and primer extension studies (15), and restriction fragments for these studies are relatively easy to isolate from the smaller clones. There are pitfalls in this technique: for example, poly G-C tailing may create spurious restriction sites at the ends of a clone. The resulting spurious restriction fragments may then make ordering of the clone difficult. Such a spurious Dde I restriction fragment was actually present in 11/15-9 (data not shown).

We are presently trying to patch over a different sort

of pitfall: the location of the glycoprotein genes. Although the Sendai genome was provisionally mapped some time ago by UV inactivation (22), the order reported (3´NP-F-M-P-HN-L) differs both from that reported (23) for NDV (3´ NP-P-(F+M)-HN-L) and from the map determined in our studies. As discussed by Dowling et al.(13), the assumption of a single entry promoter made in these UV studies may not hold true for Sendai virus. If, for example, a separate internal promoter were responsible for the relatively high molar rate of F protein synthesis in Sendai virus (22,24), this gene could now occupy any position on the genome after M. By means of our cumbersome technique, we have assembled two further clones, accounting for about 1800 bp of new sequence. Hopefully these clones should provide enough information to identify the next gene.

ACKNOWLEDGMENTS

We thank Colette Pasquier, Rosette Bandelier, Pascale Beffy, Joseph Menonna, Barbara Goldschmidt and Jean Adamus for expert technical assistance, Stuart Clarkson, Robert Hipskind, Eric Long, Jack Gorski, Bernard Mach, Claude de Preval, Gary Gray, Albert Schmitz and Peter Bromley for much helpful advice and encouragement.

REFERENCES

1. Blumberg, B.M., Giorgi, C. and Kolakofsky, D. (1983) Cell. 32, 554.
2. Giorgi, C., Blumberg, B.M., and Kolakofsky, D. (1983). J.Virol. 46, 125.
3. Blumberg, B.M., and Kolakofsky, D. (1983) J. Gen. Virol., in press.
4. Leppert, M., Rittenhouse,L., Perrault, J., Summers, D.F., and Kolakofsky, D. (1979). Cell. 18, 735.
5. Patterson, J.L., Cabradilla, C., Holloway, B.P., Obijeski J.K., and Kolakofsky, D. (1983). Cell. 33, 791.
6. Etkind, P.R., Cross, R.K., Lamb, R.A., Merz, D.C., and Choppin, P.W. (1980). Virology. 100, 22.
7. Fuller, F., and Bishop, D.H.L. (1982). J. Virol. 41, 643.
8. Cabradilla, C.D., Holloway, B.P., and Obijeski,J.F. (1983). Virology. 128, 463.
9. Dethlefsen, L., and Kolakofsky, D. (1983). J. Virol. 46, 321.
10. Lamb, R.A., and Choppin, P.W. (1978). Virology. 84, 469.
11. Roux, L. and Kolakofsky, D. (1975). J. Virol. 16, 1426.
12. Amesse, L.L., and Kingsbury, D.W. (1982). Virology. 118,

13. Dowling, P.C., Giorgi, C., Roux, L., Dethlefsen, L.A., Galantowicz, M., Blumberg, B.M., and Kolakofsky, D. (1983). PNAS, in press.
14. Grunstein, M. and Hogness, D. (1975). PNAS. 72,3961.
15. Giorgi, C., Blumberg, B.M. and Kolakofsky, D. (1983). Cell. In press.
16. Rigby, P.W.J., Dieckmann, M., Rhodes, C., and Berg, P. (1977). J. Mol. Biol. 113,237.
17. Birnboim, H.C., and Doly, J. (1979). Nuc. Acids Res. 7, 1513.
18. Kafatos, F.C., Jones, C.W. and Efstradiatis, A. (1977). Nuc. Acids Res. 7, 154.
19. Dawid, I. (1977). Biochim. Biophys. Acta. 477,191.
20. Maxam, A.M., and Gilbert, W. (1980). Methods Enzymol. 65, 499.
21. Gupta, K.C., and Kingsbury, D.W. (1982). Virology. 120,518.
22. Glazier, K., Raghow, R., and Kingsbury, D.W. (1977). J.Virol. 35, 682.
23. Collins, P.L., Hightower, L.E., and Ball, L.A. (1980). J.Virol. 35, 682.
24. Roux, L., and Waldvogel, F. (1982). Cell. 28, 293.

NON-CODING REGULATORY SEQUENCES OF THE SENDAI
VIRUS GENOME[1]

K.C. Gupta, G.G. Re, and D.W. Kingsbury

Division of Virology and Molecular Biology
St. Jude Children's Research Hospital
P.O. Box 318, Memphis, Tennessee 38101

LEADER RNA TEMPLATES

Of the 15,000 bases in the Sendai virus genome, it appears that less than 2% are dedicated to regulatory purposes. At the genome's 3' end, there is a leader RNA template about 50 bases long, where RNA synthesis initiates and where the RNA-synthesizing enzyme complex may be committed to proceeding in the replicative mode, producing a full-length antigenome (1; Re, Morgan, Gupta, and Kingsbury, this volume). Alternatively, the enzyme complex may act as a transcriptase, obeying the sequences that signal termination of leader RNA synthesis and initiation and termination of virus mRNAs. At the 5' end of the genome, there is encoded a homolog of the positive strand leader RNA template (1,2). In complementary form at the 3' end of the antigenome, this negative strand leader RNA template would serve as the initiation site for synthesis of genome replicas; the termination signal for the negative strand leader may permit early abortion of genome replication under conditions unfavorable for extending the chain, such as a temporary unavailability of proteins to encapsidate it (3).

In our studies of the terminal and internal regulatory sequences of the Sendai virus genome (2,4,5), we have seen several homologies with the regulatory sequences of the other main model of the nonsegmented negative strand RNA viruses, vesicular stomatitis virus (VSV). As shown in Fig. 1, the 46 base VSV positive strand leader RNA template and the 51 base Sendai virus positive strand leader RNA tem-

---

[1]Supported by research grants RG 1142 from the National Multiple Sclerosis Society and AI 05343 from the National Institute of Allergy and Infectious Diseases, by Cancer Center Support Grant CA 21765 from the National Cancer Institute, and by ALSAC.

```
Sendai virus:
 - 3'-UGGUUUGUUCUCUUCUUUGCACAUACCUUAUAUAUUACUUCAAUCUGUCCU...
 + 3'-UGGUCUGUUCUCAAAUUCUCUAUAAAUAAGAAAAUUUAAAAGAACAGAAGA...
              |          |          |          |          |
             10         20         30         40         50
VSV:          |          |          |          |
 - 3'-UGCUUCUGUUUGUUUGGUAAUAAUAGUAAUUUUCCGAGUCCUCUUU...
 + 3'-UGCUUCUGGUGUUUUGGUCUAUUUUUUAUUUUUGGUGUUCUCCCAG...
```

Base Composition: A+U (%)

Sendai virus:	−	67
	+	75
VSV:	−	67
	+	63

FIGURE 1. 3'-terminal nucleotide sequences of Sendai virus (Enders strain) and VSV (Indiana) genomes (−) (Re, Morgan, Gupta, and Kingsbury, this volume; 13) and antigenomes (+) (2,14).

plate are rich in the bases A and U (especially U), reflecting homologies (and an evolutionary relationship) between the two viruses, most marked in their first 20 bases (5). Even more striking are the homologies between the sequences of positive and negative strand leader templates within each virus genome (6), providing a common initiation signal recognizable by a single RNA polymerase complex (7).

SEQUENCE SIGNALS AT GENE JUNCTIONS

Conceptually and functionally, we can divide the base sequences at the junctions of Sendai virus genes into three regions: the end (E) of the upstream gene, an unexpressed intergenic (I) sequence, and the sequence that signals the start (S) of the next downstream gene (Fig. 3). The E sequence of each Sendai virus gene comprises a tetranucleotide, 3'-AUUC, followed by $(U)_5$ (4). The tetranucleotide appears to represent the termination signal for transcription and the $(U)_5$ is the signal for initiating 3'-terminal poly(A) addition. Reexamination of our data reveals another consensus, a U residue two bases upstream from 3'-AUUC (Fig. 3). However, despite clear homologies with the VSV genome in the terminating tetranucleotide and oligo(U) sequence, nothing comparable to this penultimate U exists in the VSV genome (8,9).

Upstream gene	Next gene	Sequence
leader	NP	3'...AAA UCCCAGUUUCAUAGG...
NP:48	P(F$_o$)	...GAA UCCCACUUUCAAGUA...
P(F$_o$):45	M	...GAA UCCCACUUUCUUUAA...
M:47	F$_o$(P)	...GAA UCCCUAUUUCAGGGA...
F$_o$(P):44	HN	...GAA UCCCACUUUCACUCC...
HN:46	L	...GGG UCCCACUUACCCUUC...
VSV Consensus		...GA UUGUCNNUAG...

FIGURE 2. Intergenic and mRNA start signals of Sendai virus (Enders strain). Numbers in the first column designate the gene-terminal oligonucleotides sequenced previously (4). Uncertainties in the order of the P and F$_0$ genes are indicated by parentheses.

SEQUENCES OF THE I AND E REGIONS

The following steps were taken to determine the sequences downstream from each gene-terminal E region. First, upstream sequences were determined. mRNAs extracted from infected cells were reverse-transcribed using a radiolabeled phasing primer, 5'-(dT)$_{10}$dCdT. Treatment of the products with restriction endonucleases capable of cleaving single stranded DNA (10) yielded several electrophoretically separable 5'-labeled cDNA fragments. Each fragment had a 5'-proximal sequence identical to one of the E sequences (4). A synthetic oligodeoxynucleotide primer complementary to a sequence slightly upstream from each E sequence was then synthesized. Each primer, labeled at its 5' terminus, was separately used in the reverse transcription of virus genomes, for sequence determination by dideoxynucleotide (11) and chemical cleavage (12) methods.

Next to the (U)$_5$ polyadenylation signal, we found the trinucleotide 3'-GAA in four cases and GGG once (Fig. 2). The 3'-AAA that follows the leader template sequence (Re, Morgan, Gupta and Kingsbury, this volume) also has homology with these trinucleotides. Next, the pentanucleotide 3'-UCCCA was found in five of six Sendai virus genes, a U being substituted for the A in one case. After a single heterogeneous nucleotide, the tetranucleotide 3'-UUUC appeared in five of six genes, an A being substituted for U in the third position of one gene. Thereafter, four of six genes had a consensus A, which may also contribute to the mRNA start signal (Fig. 2).

We believe that the 3'-GAA and its congeners represent the intergenic (I) sequence and that mRNA start (S) signals begin at the next U. This conclusion was supported by

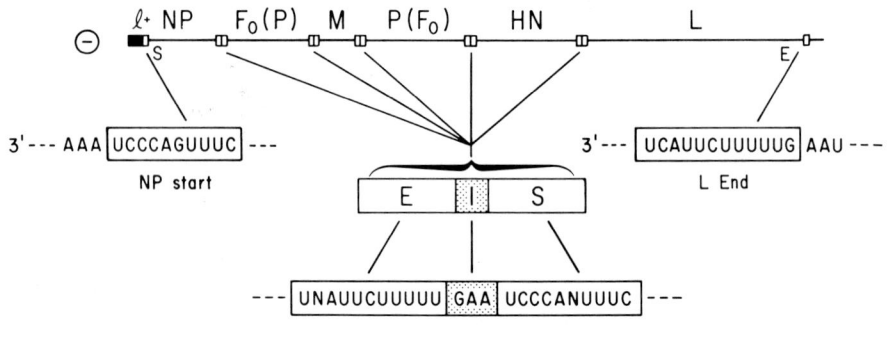

FIGURE 3. Diagram of the Sendai virus genome, the locations of its transcriptional regulatory signals and their nucleotide sequences. E, I and S refer to mRNA end, intergenic and start signals, respectively. Uncertainties in the order of the P and F_0 genes are indicated by parentheses.

sequencing the 5' termini of several viral mRNAs (data not shown; D. Kolakofsky, personal communication). These assignments are clearly analogous to the VSV transcriptional control signals, which have 3'-GA as the I sequence, followed by a consensus pentanucleotide S signal starting with U, then by a heterogeneous dinucleotide and a consensus 3'-UAG (Fig. 2).

THE COMPLETE SET OF SIGNALS

These sequences may represent the entire complement of nucleotides, about 200, that regulate the expression of the remaining 15,000 nucleotides in the virus genome (Fig. 3). The organization of the Sendai virus genome is so similar to that of the VSV genome, not only in gene order, but in the strong homologies between regulatory sequences, that there can be little doubt that these representatives of the paramyxovirus and rhabdovirus families have evolved from a common ancestor (4).

However, a point of divergence between Sendai virus and VSV is in their S signals; although organized in a similar manner, there were few base homologies between them.

Perhaps mRNA initiation (or, more properly, reinitiation) signals do not make stringent demands on the recognition faculties of the viral RNA polymerase. In contrast, strong homologies in the E and I regions of both viruses underscore the special significance of transcription termination for the nonsegmented negative strand RNA viruses. The same template must be used both for transcription and replication and it is essential that the viral RNA-polymerizing enzyme can be instructed to observe those termination signals strictly during transcription and ignore them absolutely during replication. Perhaps this feat is so difficult that alternative solutions have not been readily generated by the evolutionary process.

ACKNOWLEDGMENTS

We appreciate the help of C. Naeve in the synthesis of deoxynucleotide primers and the technical assistance of M.A. Oullette.

REFERENCES

1. Leppert, M., Rittenhouse, L., Perrault, J., Summers, D.F., and Kolakofsky, D. (1979). Cell 18, 735.
2. Re, G.G., Gupta, K.C., and Kingsbury, D.W. (1983). Virology, in press.
3. Blumberg, B.M. and Kolakofsky, D. (1981). J. Virol. 40, 568.
4. Gupta, K.C. and Kingsbury, D.W. (1982). Virology 120, 518.
5. Re, G.G., Gupta, K.C., and Kingsbury, D.W. (1983). J. Virol. 45, 659.
6. Keene, J.D., Schubert, M., Lazzarini, R.A., and Rosenberg, M. (1978). Proc. Natl. Acad. Sci. U.S.A. 75, 3225.
7. Giorgi, C., Blumberg, B., and Kolakofsky, D. (1983). J. Virol. 46, 125.
8. McGeoch, D.J. (1979). Cell 17, 673.
9. Rose, J.K. (1980). Cell 19, 415.
10. Rice, C.M. and Strauss, J.H. (1981). J. Mol. Biol. 150, 315.
11. Sanger, F., Nicklen, S., and Coulson, A.R. (1977). Proc. Natl. Acad. Sci. U.S.A. 74, 5463.
12. Maxam, A.M. and Gilbert, W. (1980). Methods Enzymol. 65, 499.
13. Keene, J.D., Schubert, M., and Lazzarini, R.A. (1980). J. Virol. 33, 789.
14. Yang, F. and Lazzarini, R.A. (1983). J. Virol. 45, 766.

STRUCTURE OF A COMPLETE CLONE OF THE SENDAI VIRUS NP GENE[1]

E.M. Morgan, K.C. Gupta, G.G. Re and D.W. Kingsbury

Division of Virology and Molecular Biology
St. Jude Children's Research Hospital
P.O. Box 318, Memphis, Tennessee 38101

CONSTRUCTION OF THE CLONE

We have been generating molecular clones of Sendai virus genes to determine the primary structures of the proteins they encode and to perform in vitro genetic manipulations (1). Using virus-specific mRNA templates, we obtained cDNA clones representing sequences of the genes for the NP, M and P proteins, but none of the clones contained much more than half of any of those genes (1).

A modified procedure has now provided a full-length clone of the NP gene. As before, the mixture of mRNAs taken from Sendai virus-infected cells was used as template for reverse transcription primed by oligo(dT). However, this time, we selected the largest transcripts by centrifugation in alkaline sucrose gradients (2), tailed them with oligo(dC) and synthesized the second strand with reverse transcriptase primed with oligo(dG)$_{10-12}$ (3). The largest double-stranded DNA molecules were selected in neutral sucrose gradients (2) and inserted into the PstI site of pBR322 (1). NP-specific clones were identified by blot hybridization with radiolabeled DNA from two shorter NP clones (2). One of the colonies contained a plasmid, designated pNP3, with an NP insert about 2000 base pairs long, congruent with the expected size of the NP gene (4).

IDENTIFICATION OF 5' AND 3' GENE TERMINI

Base sequences at the termini of pNP3 were determined both by labelling the PstI sites with [^{32}P] dideoxy ATP and

---

[1] Supported by research grant RG 1142 from the National Multiple Sclerosis Society and AI 05343 from the National Institute of Allergy and Infectious Diseases, by Cancer Center Support Grant CA 21765 from the National Cancer Institute, and by ALSAC.

by labelling the 3' termini of internal restriction sites with the Klenow fragment of Escherichia coli DNA polymerase I. The sequence 5'-AGGGT at the 5' terminus of the clone, followed, after a space of one nucleotide by 5'-AAAG, corresponds to the initiation consensus sequence described for all Sendai virus mRNAs (Gupta, Re and Kingsbury, this volume). Downstream, the sequence corresponded to that of virus genomic RNA (Re, Gupta, Morgan and Kingsbury, this volume), confirming that the gene represented in the pNP3 is NP, which occupies the first position in the genetic map of Sendai virus (4).

At its 3' terminus, the insert contained sequences almost identical to those previously described in a smaller NP clone that embraced the viral mRNA termination and poly(A) initiation signals (1). However, there was a minor discrepancy: the last 3 bases, 5'-AAG, preceding the poly(A) tract, were absent (Fig. 1). This may have occurred because the oligo(dT) primer slipped past the G residue that marks (in complementary form) the end of the mRNA stop signal (5), but it is also possible that the AAG sequence had not been transcribed by the viral RNA polymerase when it made the mRNA template of the cloned DNA. At any rate, coupled with the apparent size of the insert, the terminal sequence data indicate that pNP3 contains the entire coding sequence of the NP protein.

RESTRICTION MAPPING AND SEQUENCING

Fig. 1 also depicts a restriction map of pNP3 and the regions that we have sequenced at least once (indicated by black segments). So far, we have sequence data on more than 1600 nucleotides, leaving less than 200 completely uncharted. The positions of the restriction sites were determined by electrophoretic restriction mapping (6) and confirmed by sequence data.

We have identified four potential initiation codons within the first 310 nucleotides of the gene, at positions 65, 201, 204 and 305. The first AUG appears to initiate the correct NP reading frame, but further sequencing data will be necessary to confirm this.

The availability of a complete NP clone sets the stage for enhancing our understanding of NP structure and for obtaining expression of the gene in either prokaryotic or eukaryotic cells in studies of NP functions. In particular, the hypothesis that NP protein abundance is the regulatory switch between viral RNA replication and transcription (7,8) can be tested.

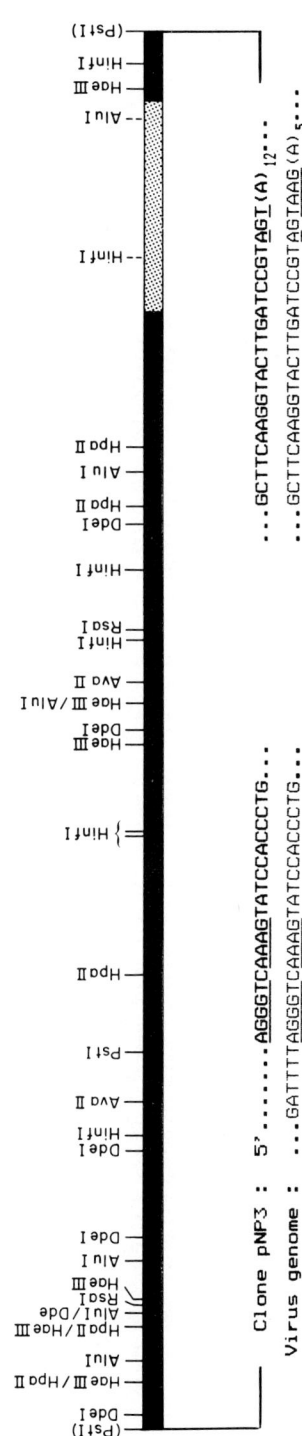

FIGURE 1. Restriction map of clone pNP3. Restriction sites were determined as described in the text. Only sites actually used to obtain nucleotide sequence information are shown. Black zones represent regions that have been sequenced at least once; the stipled area represents a region that has not yet been examined. The 5' and 3' terminal-sequences of the insert and of the corresponding cDNAs deduced from RNA sequence data (Gupta, Re and Kingsbury, this volume) are also shown.

ACKNOWLEDGEMENTS

Karen Rakestraw provided skilled technical assistance.

REFERENCES

1. Gupta, K.C., Morgan, E.M., Kitchingman, G., and Kingsbury, D.W. (1983). J. Gen. Virol., in press.
2. Rothenberg, E. and Baltimore, D. (1976). J. Virol. 17, 168.
3. Land, H., Grez, M., Hauser, H., Lindenmaier, W., and Schütz, G. (1981). Nucleic Acids Res. 9, 2251.
4. Amesse, L.S. and Kingsbury, D.W. (1982). Virology 118, 8.
5. Gupta, K.C. and Kingsbury, D.W. (1982). Virology 120, 518.
6. Smith, H.O. and Birnstiel, M.L. (1976). Nucleic Acids Res. 3, 2387.
7. Kingsbury, D.W. (1974). Med. Microbiol. Immunol. 160, 73.
8. Blumberg, B.M., Leppert, M., and Kolakofsky, D. (1981). Cell 23, 837.

cDNA CLONING, MAPPING AND TRANSLATION OF TEN
RESPIRATORY SYNCYTIAL VIRUS mRNAs[1]

P.L. Collins, L.E. Dickens and G.W. Wertz

Department of Microbiology and Immunology
School of Medicine
University of North Carolina
Chapel Hill, N.C. 27514

Human respiratory syncytial (RS) virus, the prototype of the pneumovirus genus of the paramyxovirus family (1), is an important agent of respiratory tract disease. The RS virus genome is a single negative strand of encapsidated RNA that is transcribed in infected cells into polyadenylated mRNAs (2-6). As described here, we have used molecular cloning techniques to characterize the genome and gene products of RS virus strain A2. Some of this work has been described elsewhere (5).

cDNA cloning. cDNAs were copied from mRNAs from virus-infected HEp-2 cells and cloned using the Escherichia coli plasmid pBR322 (5). Viral-specific clones were catagorized by reciprocal dot blot hybridization and by hybridization to reverse transcripts of viral mRNAs into ten families that lacked detectable cross-hybridization (5, unpublished data).

RNA (Northern) Blots. cDNA clones representing the ten different families were radiolabeled by nick translation and hybridized to RNA blots prepared from intracellular RS viral RNAs.
As shown in Fig. 1, all ten clones hybridized with genomic RNA (RNA 8), confirmation of their viral-specificity. One clone each hybridized with RNAs 4, 5 and 7 (Fig. 1, lanes 4, 5 and 7, respectively), three clones hybridized at the position of RNA gel band 1 (lanes 1a, 1b and 1c), two with RNA 2 (lanes 2a and 2b), and two with RNA 3 (3a and 3b). In hybrid-selected translation experiments (described below), each of these clones selected mRNAs encoding different viral proteins. This demonstrated that each of the ten clones represents a different,

---

[1]This work was supported by PHS Grants AI12464 and AI15134 (NIAID) and by NCI Grants CA09156 and CA19014. P.L.C. received support from PHS Grant AI06956 (NIAID).

unique viral mRNA. Thus, RNA gel band 1 contained three unique mRNAs (designated, in order of increasing size, RNAs 1a, 1b and 1c), and RNA gel bands 2 and 3 each contained two unique mRNAs (mRNAs 2a, 2b, 3a and 3b).

RNAs A-H and 6 each hybridized with two or three different clones (Figs. 1 and 2). This suggested that these RNAs were nonunique di and tricistronic transcripts, an interpretation that was supported fully by mol. wt. comparisons (Fig. 2).

Transcriptional Map. On the premise that the polycistronic RNAs A-H and 6 were generated by readthrough transcription of adjacent genes, their sequence homologies and mol. wts. were interpreted to prepare a viral transcriptional map (Fig. 2). This analysis did not, however, determine the number of viral promoters, the direction of transcription, or the relationship between the two groups of genes in the transcriptional map.

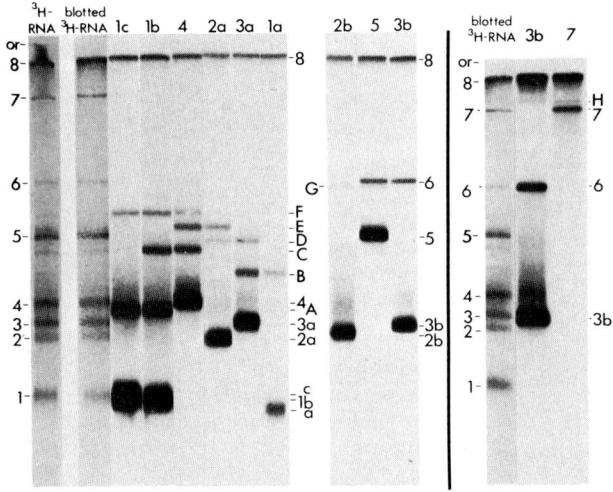

Figure 1. Hybridization of cDNA clones with RNA (Northern) blots. Intracellular (^{3}H) viral RNAs were separated in replicate lanes by electrophoresis in 1.5% agarose-urea gels. One lane was analyzed by fluorography (lane marked "^{3}H RNA") and the remaining lanes were transferred to diazobenzyloxymethyl paper. Two strips were analyzed by fluorography ("blotted ^{3}H RNA"). The remaining strips were analyzed by autoradiography following hybridization with cDNA clones of viral RNAs 1a, 1b, 1c, 2a, 2b, 3a, 3b, 4, 5 and 7 (as indicated over each lane).

Therefore, transcriptional mapping was also performed by exposing virus-infected cells to UV light and quantitating the rates of inactivation of the viral genes. The transcription of surviving genes was monitored by the incorporation of (^{3}H) uridine or (^{32}P) phosphate into RNA in the presence of cycloheximide (to block replication of undamaged genomes) and actinomycin D. Individual transcripts were quantitated by gel electrophoresis and by hybridization to replicate dot blots of the ten viral cDNA clones. In some experiments, cycloheximide was omitted and the replication of surviving genomic RNAs was quantitated by gel electrophoresis. The target mol. wt. for UV-inactivation of genome replication was assumed to be the estimated genome mol. wt., 5.6×10^6. This provided a standard for calculating target mol. wts. for individual genes (Table 1). As shown in Table 1, the target mol. wts. for the ten viral genes increased cumulatively in the order 1c < 1b < 4 2a < 3a < 1a < 2b < 5 < 3b < 7. The target mol. wt. for gene 1c alone was similar to its gene size, suggesting that it alone lay close to a promoter. The simpliest interpretation of these data is that the ten viral genes are transcribed from a single promoter in the sequence 1c -1b -4 -2a -3a -1a -2b -5 -3b -7. In addition, the target size analysis provided evidence that genes 1a and 2b are adjacent in the map; the proximity of these two genes had not been determined by analysis of polycistronic RNAs.

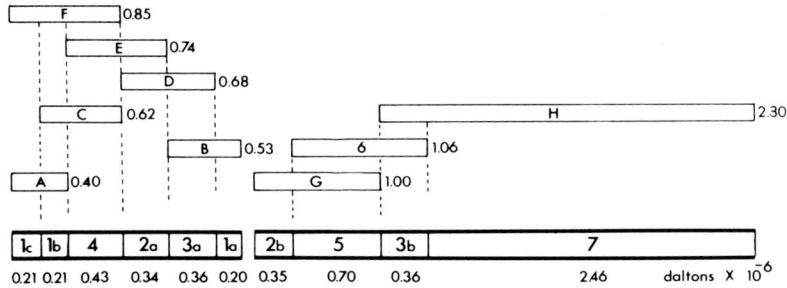

FIGURE 2. Transcriptional map for RS virus. RNAs 1-7 and A-H are drawn overlapping based on sequence homologies determined as shown in Fig. 1. The RNAs are drawn in proportion to their estimated mol. wt. minus the mol. wt. of 125 polyadenylate residues (7). Their adjusted mol. wts. (x 10^{-6}) are indicated.

TABLE 1
UV TARGET SIZES FOR RS VIRAL GENES

GENE	UV TARGET MOL. WT. (x 10^{-6})[a]	GENE MOL. WT. (x 10^{-6})[b]	CUMULATIVE GENE MOL. WT. (x 10^{-6})[c]
1c	0.36[d]	0.21	0.21
1b	0.44[d]	0.21	0.42
4	0.82[d]	0.43	0.85
2a	1.20[d]	0.34	1.19
3a	1.24[d]	0.36	1.55
1a	1.43[d]	0.20	1.75
2b	1.81[d]	0.35	2.1
5	2.69[e]	0.70	2.8
3b	2.97[d]	0.36	3.16
7	6.4[e]	2.46	5.62

[a] Target mol. wt. for an individual gene = (D_0 for RNA replication ÷ D_0 for individual genes) x 5.6 x 10^6. D_0 is the dose of radiation that results in 37% survival.
[b] Calculated by subtracting the mol. wt. of 125 polyadenylate residues (7) from the mRNA mol. wt.
[c] Sum of the gene mol. wt. plus the mol. wt. of preceding genes based on the gene order 1c, 1b, 4, 2a, 3a, 1a, 2b, 5, 3b, 7.
[d] Quantified by dot blot hybridization; average of three experiments.
[e] Quantified by gel electrophoresis; average of two experiments.

Polypeptide Coding Assignments of the RS virus mRNAs. cDNA clones representing nine of the ten unique viral mRNAs were used to purify mRNAs by hybridization-selection (8) for translation in vitro. SDS-gel electrophoresis of the (^{35}S) methionine-labeled translation products (Fig. 3) determined the following polypeptide coding assignments: RNA 1a, 9.5 kilodalton (K) protein (lane 1a); RNA 1b, 11K protein (lane 1b); RNA 1c, 14K protein (lane 1c); RNA 2a, nucleocapsid phosphoprotein P (lane 2a); RNA 2b, 32K and 36K proteins (lane 2b); RNA 3a, matrix protein M (lane 3a); RNA 3b, 24K protein (lane 3b); RNA 4, major nucleocapsid protein N (lane 4); and RNA 5, 59K protein (lane 5). The coding assignment of RNA 7 was not tested, but is assumed by size considerations to be the L protein. The 32K and 36K proteins appeared to be related based on limited digest peptide mapping (not shown). Comparison of the hybrid-selected products with polypeptides extracted

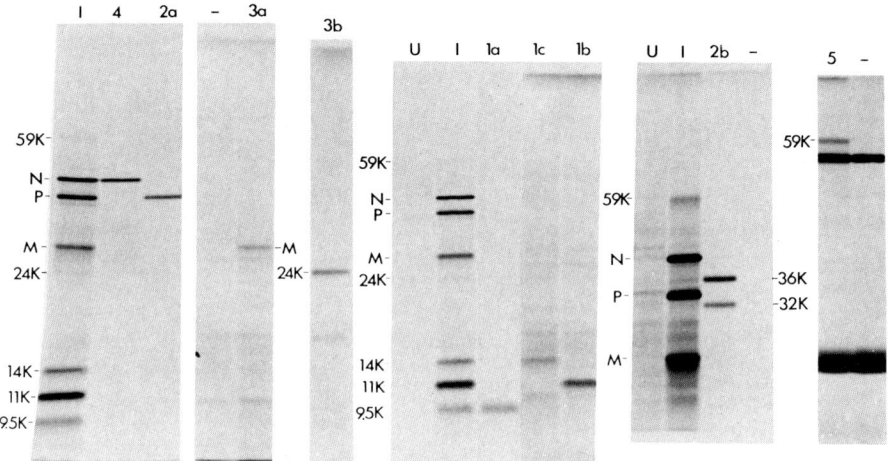

Figure 3. Polypeptide coding assignments of the RS virus mRNAs. mRNAs were purified by hybridization-selection (8) using cloned cDNAs of RNAs 1a, 1b, 1c, 2a, 2b, 3a, 3b, 4 and 5. Wheat germ extracts (lanes 1a, 1b, 1c, 2a, 2b, 3a, 3b and 4) or reticulocyte lysates (lane 5) were programmed with the selected mRNAs (as indicated) or with no added mRNA (lanes marked "-") or with total mRNA from uninfected ("U") and virus-infected ("I") actinomycin D-treated cells. The (^{35}S) methionine-labeled translation products were analyzed by electrophoresis in SDS-polyacrylamide gels. Fluorograms of the fixed, dried gels are shown.

from infected cells by gel electrophoresis and limited digest peptide mapping showed that the 9.5K, 11K, 14K, 24K, M, P and N proteins had authentic counterparts of the same size (not shown). In contrast, the 36K and 59K proteins lacked detectable counterparts of the same size in vivo (not shown) and therefore are candidates to be the unprocessed polypeptide moieties of the viral F and G glycoproteins.

SUMMARY

Cloned cDNAs of viral mRNAs were used as hybridization probes to (i) analyze RNA blots of intracellular viral RNAs, and (ii) purify (hybrid-select) homologous mRNAs for translation in vitro. These analyses demonstrated the existence of ten unique viral mRNAs and identified their sizes and polypeptide coding assignments. The available data indicated that each mRNA encodes a single polypeptide chain. Thus, the mRNA and polypeptide products of ten RS viral genes have been identified.

A transcriptional map for RS virus was determined by two independent methods: (i) analysis of polycistronic readthrough transcripts, and (ii) UV-inactivation studies. Results with both methods supported fully the interpretation that the ten viral genes are transcribed from a single promoter site in the sequence (designated by mRNA number with the encoded protein indicated in parenthesis): 1c (14K), 1b (11K), 4 (N), 2a (P), 3a (M), 1a (9.5K), 2b (36K), 5 (59K), 3b (24K), 7 (L).

REFERENCES

1. Kingsbury, D.W., Bratt, M.A., Choppin, P.W., Hansen, R.P., Hosaka, Y., ter Meulen, V., Norrby, E., Plowright, W., Rott, R., Wunner, W.H. (1978). Intervirology 10, 137.
2. Huang, Y.T. and Wertz, G.W. (1982). J. Virol. 43, 150.
3. Huang, Y.T. and Wertz, G.W. (1983). J. Virol. 46, 667.
4. Lambert, D.W., Pons, M.W., Mbuy, G., and Dorsch-Hasler, K. (1980). J. Virol. 36, 837.
5. Collins, P.L. and Wertz, G.W. (1983). Proc. Natl. Acad. Sci. USA. 80, 3208.
6. Venkatesan, S., Elango, N. and Chanock, R.M. (1983). Proc. Natl. Acad. Sci. USA. 80, 1280.
7. Weiss, S.R. and Bratt, M.A. (1974). J. Virol. 13, 1220.
8. Ricciardi, R.P., Miller, J.S. and Roberts, B.E. (1979). Proc. Natl. Acad. Sci. USA. 76, 4927.

STRUCTURAL ANALYSIS OF HUMAN RESPIRATORY SYNCYTIAL VIRUS GENOME

Sundararajan Venkatesan, Narayansamy Elango,
Masanobu Satake, and Robert M. Chanock

Laboratory of Infectious Diseases
National Institute of Allergy and Infectious Diseases
National Institutes of Health
Bethesda, Maryland 20205

Human respiratory syncytial virus (RS virus), a pleomorphic enveloped RNA virus that replicates in the cytoplasm is a leading causative agent of serious lower respiratory disease among infants and children (1). Although previously classified as a paramyxovirus, it is distinct in its morphology and lack of hemagglutinin and neuraminidase activities. In spite of its importance as a human pathogen in early life, detailed studies relating to the nature of its organization and expression have only recently begun, primarily because of its poor growth in tissue culture and virus instability. We have explored the structure of this virus by use of recombinant DNA techniques, and this report is a summary of some preliminary observations.

The genome of RS virus consists of a single 5000 kDal RNA species of negative polarity. Several workers (2, 3) have identified 7 or 8 virus coded polypeptides consistent with the earlier genetic analysis of ts mutants that identified 7 or 8 nonoverlapping complementation groups (4). Work from our laboratory has identified at least 8 polypeptides that are present in the extracellular virus that have apparent molecular masses of 160, 84, 68, 46, 36, 28, 22 and 18 kDal. The viral origin of these polypeptides was confirmed by specific immunoprecipitation with rabbit RS antiserum. The 84 and 68 kDal proteins are envelope glycoproteins, readily solubilized by nonionic detergents. One of these (84 kDal) probably represents the major envelope glycoprotein (G), the equivalent of the paramyxovirus HN protein. Under reducing conditions of electrophoresis, the 68 kDal glycoprotein is replaced by two proteins of 49 and 16 kDal. This is analogous to the paramyxovirus fusion factor (Fo) that is synthesized as a precursor and subsequently cleaved proteolytically into two subunits (F_1 and F_2) held together by disulfide bonds. The other proteins include a 160 kDal protein (L, probably the viral polymerase by analogy with paramyxoviruses), major nucleocapsid protein (NC, 46 kDal), phosphoprotein (36 kDal), matrix protein (M, 28 kDal), and two nonstructural proteins (NS_1 and NS_2, 22 and 18 kDal). Poly (A) containing RNA from infected cells was translated in vitro yielding proteins corresponding to NC, P, M and the two nonstructural proteins (Fig. 1, Lane 8). Occasionally, a 59 kDal polypeptide was detected. Huang and Wertz have shown that a viral mRNA 2200 bases long is translated into a protein of 59 kDal (5). Whether this represents a precursor of the fusion factor is not clear.

CONSTRUCTION AND CHARACTERIZATION OF cDNA CLONES

In order to analyze the different transcriptional units of RS virus, a cDNA library was constructed in E. coli (HB 101) using plasmid pBR322 as a vector. The strategy used avoided the traditional self-priming reaction for the second strand synthesis and nuclease S_1 treatment to remove the hairpin (6). Discrete sized single stranded (ss) cDNA were used separately for constructing double stranded (ds) cDNAs. A library of 2900 tetracycline resistant and ampicillin sensitive transformants was screened by hybridization to ^{32}P labeled poly (A) RNA from infected cells. Seventy-seven % of the transformants gave a positive reaction and from this group a sublibrary of 75 transformants was selected on the basis of size of the cDNA inserts. Several rounds of hybrid selection of viral mRNAs using recombinant DNA immobilized on nitrocellulose filters and subsequent cell-free translation of selected mRNAs allowed us to identify 4 different classes of recombinants that encoded the RS viral NC, P, M or NS_2 gene (Fig. 1). Certain recombinants hybrid selected mRNA that yielded on translation P protein and a smaller protein comigrating with a viral nonstructural protein NS_1 (Fig. 1, lane 5, bottom arrow). Similarly, another recombinant plasmid hybrid selected RNA that yielded on translation M protein and a smaller protein comigrating with another nonstructural protein (Fig. 1, lane 4). Two dimensional tryptic fingerprinting and partial protease V8 cleavage digestions were utilized to confirm that the translation products were of viral origin. Several independent recombinant plasmids with cDNA inserts adequate in size to accomodate the coding sequence of each of the genes were thus identified (7).

The cDNA library was further screened in duplicate by hybridization with ^{32}P labeled cDNA synthesized in vitro using infected or uninfected cell poly (A) RNAs as templates. Approximately 85% of the recombinants reacted only with cDNA prepared from infected cell RNA. Recombinants hybridizing to the virus specific cDNA were also screened by hybridization with radioactively labeled genomic RNA extracted from intracellular nucleocapsids. The genomic RNA was either 3' end labeled by the RNA ligase reaction or 5' end labeled by the polynucleotide kinase reaction after fragmentation and dephosphorylation.

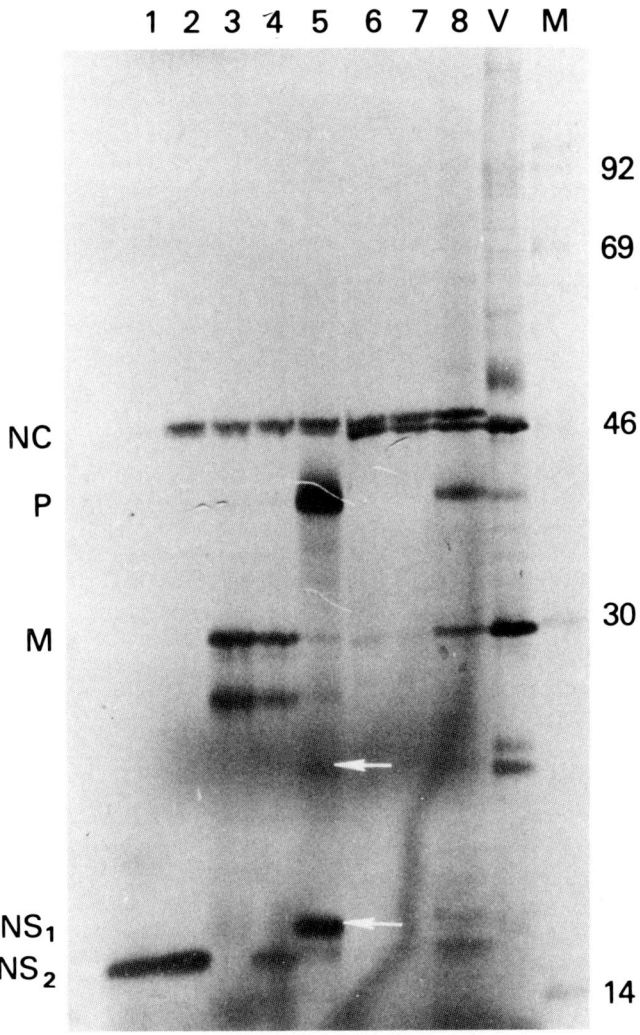

Fig. 1. Cell-free translation products of mRNAs from RS virus-infected cells selected by hybridization to RS virus cDNA plasmids. The plasmids used were $pRSB_8$ (lane 1), $pRSC_6$ (lane 2), $pRSB_7$ (lane 3), $pRSD_3$ (lane 4), $pRSC_1$ (lane 5), $pRSB_{10}$ (lane 6), and $pRSB_{11}$ (lane 7). Translation products of polyadenylylated RNA from infected cells are displayed in lane 8. RS virion polypeptides labeled in vivo with [^{35}S]methionine are shown in lane V.

Subtractive screening eliminated recombinants belonging to the classes described above. The remaining transformants were grown in groups of five. Plasmid DNA was then radioactively labeled by nick translation and used to detect poly (A) RNAs from infected cells. Two transformants (pRS$_4$ and pRSA$_2$) hybridized with poly (A) RNAs 2200 and 1100 bases in length, respectively (Fig. 2). pRS$_4$ had a 1000 bp cDNA insert and hybridized to a mRNA twice its length. We have since identified additional recombinants that contain larger inserts. One of these (pRSA$_{14}$), hybridizes to pRS$_4$, contains 2000 bp of viral sequences and might represent a full-length copy of the 2200 base mRNA. This species of mRNA was shown to encode a polypeptide of 59 kDal (5).

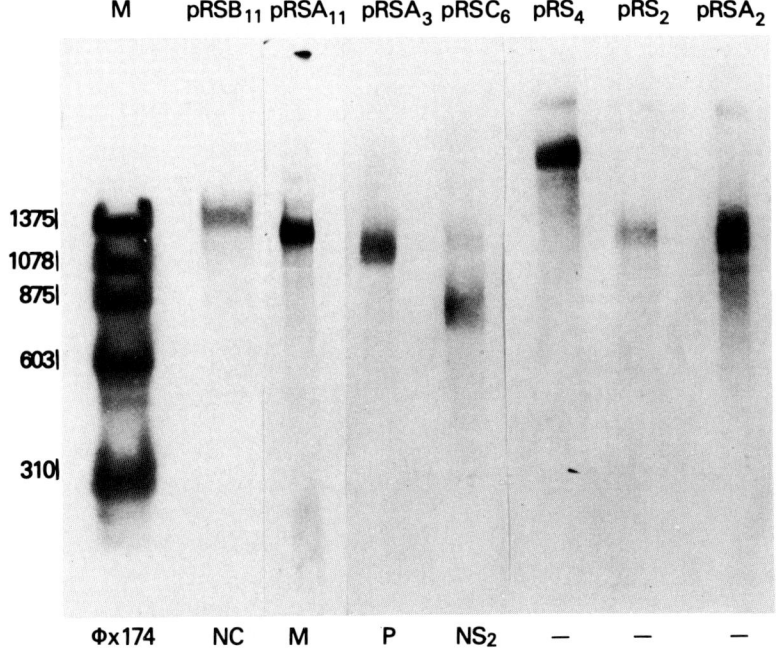

Fig. 2. Hybridization of recombinant plasmids to poly (A) RNA from infected cells. Poly (A) RNA from infected cells was electrophoresed on formaldehyde agarose (1.5%) gels, transferred to nitrocellulose and hybridized to radioactively labeled plasmid DNAs. The symbol (-) denotes that the identity of the respective viral gene product(s) is not known. Molecular weight markers are 5' end labeled Hae III fragments of φx174 Form I DNA (M).

SIZE DETERMINATION OF VIRAL mRNAs

As illustrated in Fig. 2, recombinants encoding the RS viral NC, P, M or NS_2 hybridized to discrete sized poly (A) RNAs from infected cells. Three of these, encoding the NC, P or M gene had cDNA inserts almost identical in size to the RNAs with which they hybridized. However, recombinant ($pRSC_6$), that encoded a viral nonstructural gene, reacted with a poly (A) RNA about 200 bases shorter than the cloned RS viral sequence. All these cDNA inserts had a terminal stretch of A residues derived from the poly (A) tail of the viral mRNAs. By primer extension the recombinant plasmids ($pBRSB_{11}$ and $pRSA_{11}$) encoding the NC and M genes, respectively, were found to lack 6 and 7 nucleotides corresponding to the 5' end their respective mRNA (data not shown).

DNA SEQUENCE ANALYSIS

DNA sequence of the RS viral NC, M and NS_2 gene inserts genes was determined. A 1443 bp DNA sequence of the NC gene insert of plasmid $pRSB_{11}$ had a single open reading frame encoding a protein of 467 amino acids (Fig. 3). The other two reading frames were extensively blocked throughout. The 51 kDal NC protein was relatively rich in basic amino acids but poor in cysteine. There was no clustering of basic amino acid residues within specific domains (8). A recombinant ($pRSA_{11}$) encoding the matrix protein gene had 944 bp of RS virus sequence. There was a single long open reading frame potentially encoding a protein of 256 amino acids (Fig. 3). This 28717 dalton protein was relatively basic and moderately hydrophobic. There were two clusters of hydrophobic amino acids in the C-terminal third of the molecule. These might be involved in the interaction of matrix protein with virus altered plasma membranes. A second open reading frame encoding 75 amino acids and partially overlapping with the C-terminus of the matrix protein was also present (Fig. 3). Interestingly, certain M gene recombinants hybrid selected mRNA that yielded on translation M and a protein comigrating with a viral nonstructural protein (Fig. 1, lane 4). The biological significance of the second open reading frame is not clear at this time. The amino acid sequence of both the viral NC and M did not share homology with the known sequence of other viral capsid and matrix proteins, implying that RS virus has undergone extensive evolutionary divergence.

A recombinant plasmid ($pRSC_6$) encoding a viral nonstructural protein (NS_2) was also selected for sequencing. The viral insert of 1050 bp had two nonoverlapping reading frames capable of encoding proteins of 139 and 140 amino acids, respectively (Fig. 3). Since the molecular weight of the protein obtained by cell-free translation of viral mRNA hybrid selected by this plasmid agreed well with the calculated molecular weight of the proteins encoded by either reading frame, it is not clear whether one or both are used for translation in vivo. Presently we are investigating this question by a combination of 2D gel analysis of the protein(s), primer extension to map the 5' end of the mRNA(s) and hybrid arrested translation of mRNA(s) using restriction fragments that span only one of the two reading frames. The precise gene order of RS virus is not established, although Collins and Wertz (9) have proposed a transcriptional map wherein two viral nonstructural protein genes are 3'-proximal on the genomic template. Also, it is not known whether an untranslated leader RNA is synthesized prior to synthesis of mRNAs as in the case of other negative strand RNA viruses. If the 0 frame encodes NS_2 protein, we may have fortuitously cloned a linked transcript containing this gene and upstream sequences on the genomic template for a second nonstructural protein; conversely, NS_2 protein might be encoded by the +2 frame.

The codon usage of the RS genes showed an inherent bias against CG dinucleotide within the coding region similar to the situation observed with VSV, influenza virus and eukaryotic genomes (10). Each of the viral genes lack the canonical eukaryotic polyadenylation signal AAUAAA upstream of the 3' poly (A) tail. In addition, these genes lack homologous sequences just upstream of the poly (A) tail, unlike VSV and Sendai virus which have a conserved four nucleotide sequence upstream of the poly (A) tail of each transcript (11). The translation initiator codon AUG for each gene is embedded within a canonical PXXAUGG sequence seen around functional eukaryotic initiation sites.

Recombinant $pRSA_3$ contains a P gene cDNA insert that hybridized to a viral mRNA of similar size. Certain recombinants containing this gene hybrid selected viral mRNA(S) that translate P and a protein comigrating with a viral nonstructural protein (NS_1) (Fig. 1). This may be analogous to Sendai virus, where P and C proteins are translated from the same mRNA, either as a tandem nonoverlapping dicistronic RNA or as a mRNA with overlapping reading frames (12). We are presently sequencing cDNA inserts encoding the P gene to determine whether this is the case with RS virus.

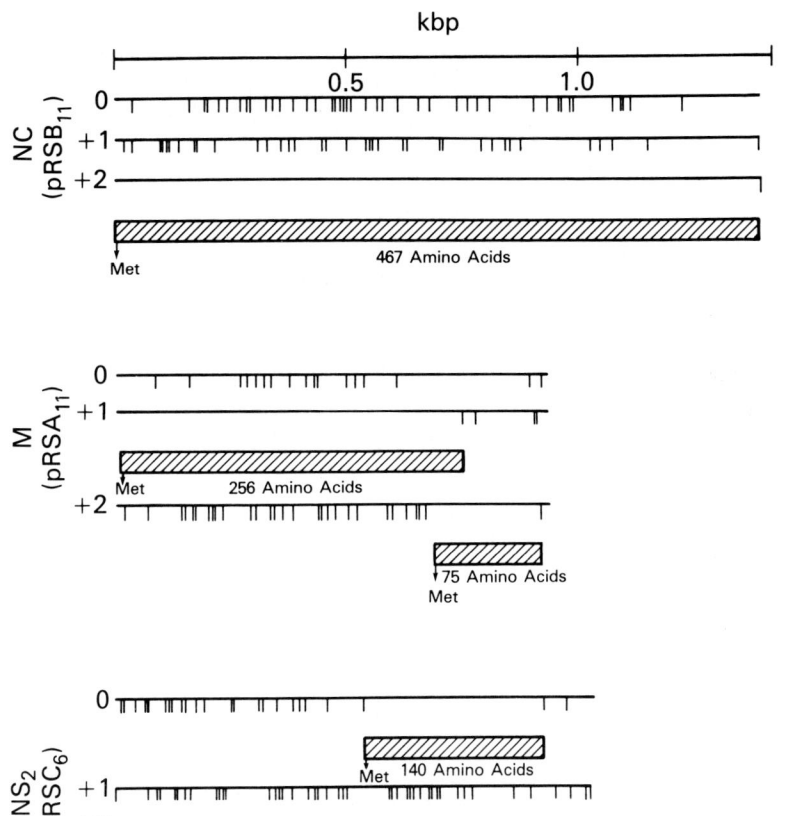

Fig. 3. Schematic illustration of open reading frames within the cDNA inserts of the recombinants encoding RS viral NC, M, or NS_2 gene. The lengths of the cDNA inserts are indicated in kbp. Each plasmid and the gene contained in it are identified on the left. The vertical lines denote stop codons. The hatched rectangles represent the open reading frames. The number of amino acids for each protein is shown below the rectangles.

REFERENCES

1. Chanock, R. M., Kim, H. W., Brandt, C. D., and Parrott, R.H. (1982). In "Viral Infections of Humans: Epidemiology and Control" (Evans, A.S. (ed.), p. 471. Plenum, New York.
2. Bernstein, J. M., and Hruska, J. F. (1981). J. Virol. 38, 278.
3. Wunner, W. H., and Pringle, C. R. (1976). Virology 73, 228.
4. Gimenez, H. B., and Pringle, C. R. (1978). J. Virol. 27, 459.
5. Huang, Y. T. and Wertz, G.W. (1983). J. Virol. 46, 667.
6. Land, H., Gretz, M., Hauser, H., Lindenmaier, W., and Schutz, G. (1981). Nucleic Acids Res. 9, 2251.
7. Venkatesan, S., Elango, N. and Chanock, R. M. (1983). Proc. Natl. Acad. Sci. U.S.A. 80, 1280.
8. Elango, N. and Venkatesan, S. (1983). Nucleic Acids. Res. (In Press).
9. Collins, P. L. and Wertz, G. W. (1983). Proc. Natl. Acad. Sci. U.S.A. 80, 3205.
10. Rose, J. K. and Gallione, C. J. (1981). J. Virol. 39, 519.
11. Gupta, K. C. and Kingsbury, D. W. (1982). Virology 120, 415.
12. Dowling, P. C., Giorgi, C., Roux, L., Dethelfsen, L. A., Galantowicz, M. E., Blumberg, B. M. and Kolakofsky, D. (1983). Proc. Natl. Acad. Sci. U.S.A. 80, 5213.

THE POLYMERASE GENE OF VSV

George G. Harmison, Ellen Meier, and Manfred Schubert

Laboratory of Molecular Genetics, IRP
National Institute of Neurological and Communicative
Disorders and Stroke, NIH
Bethesda, Maryland 20205

The synthesis of translatable messenger RNAs of Vesicular stomatitis virus (VSV) requires a great number of different enzymatic activities. These include specific binding of the polymerase to the ribonucleoprotein template, initiation, polymerization, capping, methylation, polyadenylation and specific termination. In addition, highly conserved signal sequences are recognized by the polymerase along the template.

In vitro transcription experiments demonstrated that, besides the template, only the viral NS and L proteins are required for transcription (1). Both are considered dissimilar subunits of the polymerase complex. Possibly all the functions described above are carried out by the polymerase complex since they can not be separated from it. Exogenous RNAs are not accepted as substrates for any of the modification functions such as capping, methylation and polyadenylation. It, therefore, appears that all of these essential enzymatic activities are not only carried out on the template, but that they also require nascent transcripts as substrates. Because of its giant size, we expect that most of the functions are associated with the L protein.

In this communication, we summarize our nucleotide sequence analysis of the polymerase gene and the deduced amino acid sequence of the L protein. The sequences will be published elsewhere.

CLONING OF THE L GENE

As a first step on our way to dissect the individual functions of the polymerase complex, we have isolated double-stranded cDNA copies of the L gene. VSV genomic RNA (Indiana serotype, Mudd-Summers strain) was reverse transcribed by random priming. After second strand synthesis and oligo (dC) tailing, the DNA fragments were inserted into the oligo (dG)

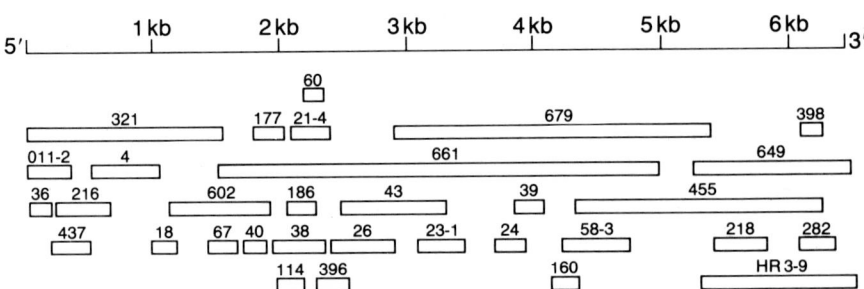

FIGURE 1. Map of 32 cDNA clones of the L gene which were partially or completely used to determine the nucleotide sequence of the L gene. The 5' terminus of VSV genomic RNA is indicated. The 3' end corresponds to the G and L boundary.

tailed Pst 1 restriction site of pBR322. E. coli HB101 cells were transformed and the tetracycline resistant colonies were characterized by filter hybridization. $5'-{}^{32}P$-end labeled RNA fragments of defective interfering particle RNAs such as DI 011, T, T(L) and 611 which lack sequences from the smaller genes (2, Meier et al., this volume) were used as hybridization probes. Approximately 1200 colonies were analyzed and more than 500 cDNA clones of the L gene region were identified and tentatively mapped. More than 50 overlapping cDNA clones covering the entire gene were precisely mapped by hybridization and restriction cuts. Figure 1 shows the map of 32 cDNA clones which were either partially or completely used to determine the nucleotide sequence of the L gene. The sizes of the inserts ranged from 100 bp to 3400 bp.

NUCLEOTIDE SEQUENCE ANALYSIS

Each nucleotide of the L gene was sequenced, on the average, 5 times in plus and minus sense using 3 independently derived cDNA clones (3). This extensive sequence analysis was necessary because more than 20 base differences between various cDNA clones from the same regions were detected. These differences included not only base substitutions, but surprisingly, also single base deletions and insertions, changes which would clearly be lethal to the virus. We are currently investigating whether these biologically very interesting frame shift mutations are present in subpopulations of the VSV genomic RNA or whether they are simply a reflection of the infidelity of the reverse transcriptase.

THE POLYMERASE GENE OF VSV   37

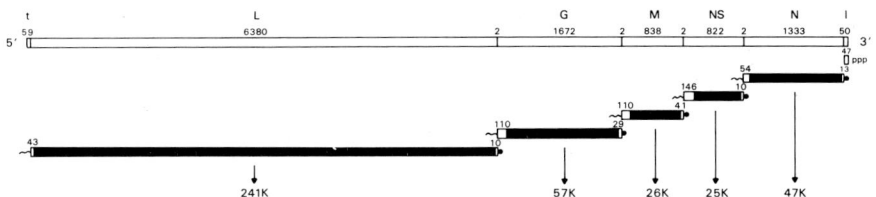

FIGURE 2. The genome of VSV and its transcripts. The sizes of the leader region 1 (6, 7), the N, NS, M, G genes (8, 9), the L gene (this report), the intercistronic regions (4) and the 5' terminal extracistronic trailer region t (5) are indicated. The coding regions of each messenger RNA (closed bars) and the sizes of the terminal untranslated regions are as marked. The calculated molecular weights of the corresponding proteins are listed in kilodaltons.

Our nucleotide sequence analyses showed that the L gene spans 6380 nucleotides, starting at the intercistronic boundary of the G and L genes (position 4724 from the 3' end of the genome) (4) and terminating at its polyadenylation signal at position 60 from the 5' end of the genome (5). These data, together with the published sequences of the N, NS, M and G genes, complete the entire nucleotide sequence of VSV Indiana. The genome consisting of the leader region (6, 7), the 5 genes (8, 9, this report), the four intercistronic regions (4) and the 5' terminal trailer region (5) is 11,162 nucleotides long (Figure 2). The L gene alone comprises 57.2% of the total genome.

TRANSLATION OF L MESSENGER RNA

There is no evidence for post-transcriptional size alterations of VSV messages such as splicing. The ribosome binding site (10) and the polyadenylation site of L messenger RNA (5) have been described earlier. Therefore, the amino acid sequence of the protein can be deduced from the primary structure of the L message. Figure 3 shows the translation of (+) sense L message as well as the corresponding (-) sense genomic RNA in three reading frames. In (+) sense, only three regions were detected which are at least 70 nucleotides in length and contain an AUG start codon. It is obvious that reading frame No. 2 encodes the L protein. Although we found a region in the genomic RNA, B, which has an open reading frame and could potentially code for a peptide with a

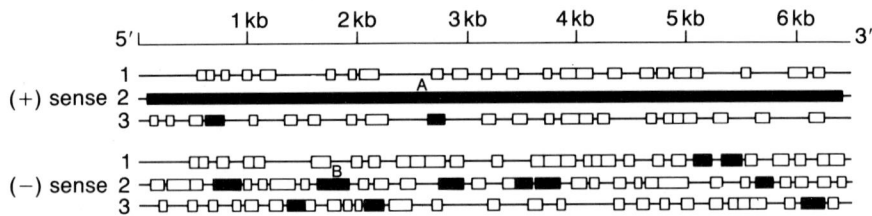

FIGURE 3. Translation of the L messenger RNA (+) and the L gene (-) sequence in all 3 reading frames. Regions without a translational stop codon and at least 70 nucleotides in length are boxed. Closed boxes mark regions containing the translational start codon AUG. A corresponds to the open reading frame for the L protein. B represents the second largest potential coding region.

molecular weight of 10,000 daltons, there is no evidence that it is actually used. Most importantly, (-) sense translatable transcripts from (+) strand templates are highly unlikely.

The L messenger RNA contains a single open reading frame which starts with the first AUG 11 nucleotides from the 5' end of the message -- consistent with the reported ribosome binding site (10) -- and terminates with UAA leaving 43 nucleotides before the poly (A) tail untranslated (Figure 2). The L message contains the least number (53) of noncoding nucleotides of all VSV messages. The total number of encoded amino acids is 2109 which gives the L protein a calculated molecular weight of 241,000 daltons. Its size exceeds earlier estimations which ranged from 160,000 to 230,000 daltons (11).

The total coding region of the VSV genome is 93.9% leaving 6.1% for the untranslated terminal message regions, the leader and the trailer regions. From this coding capacity, the L protein comprises 60.4% compared to 6.3% for the NS protein which is approximately one tenth the size of L.

AMINO ACID COMPOSITION OF THE L PROTEIN

The amino acid composition of the L protein as deduced from the nucleotide sequence of the L message is listed in Table 1. The amounts of basic, acidic, polar and nonpolar amino acids are summarized. The L protein is a basic protein, but less basic than the M protein which contains 17.1% basic and 11.4% acidic amino acids (9). Unlike the NS, M and G proteins (8, 9), L does not exhibit large clusters of amino acids of similar characteristics, i.e., charged,

TABLE 1. Amino acid composition of the VSV L protein. The amino acid sequence of the L protein was deduced from the nucleotide sequence of L messenger RNA.

ala	92		leu	219
arg	124		lys	129
asn	97		met	60
asp	120		phe	93
cys	34		pro	98
gln	70		ser	178
glu	112		thr	122
gly	127		trp	42
his	64		tyr	70
ile	155		val	103
		Total: 2109 aa		
Basic	15.0%		Acidic	11.0%
Polar	31.9%		Nonpolar	42.1%

hydrophobic. Although the overall amounts of basic, acidic, polar and nonpolar amino acids are almost identical to the PB2 protein of influenza virus (12), the relative amounts of the individual amino acids are quite different.

Comparison of the nucleotide sequence and amino acid sequences of L and the P proteins of influenza virus (12, 13), which combined are roughly the size of L, did not reveal any homology. In addition, computer analysis of the sequence to all nucleotide sequences in the Genbank Data Bank in Los Alamos as well as the Protein Sequence Data Base of the National Biomedical Research Foundation in Washington, D.C., did not reveal any significant similarities to other sequences or proteins. The total number of nucleotides screened was about 2×10^6, including the sequences of many proteins which interact with nucleic acids. The L protein has a unique sequence. A comparison to the L proteins of other rhabdoviruses or paramyxoviruses awaits nucleotide sequence information from these viruses.

The VSV L gene represents the first large polymerase gene of a negative strand virus to be cloned and sequenced. This work completes the cloning and sequencing of the VSV Indiana genome. Despite the massive size of the gene, our sequence analysis suggests that only one protein is encoded. The molecular weight of 241,000 daltons makes L a perfect candidate for a multifunctional protein. With the final assembly of the L gene and its insertion into an expression vector, we anticipate to unravel the functional domains of the protein involved in transcription and replication and study its interactions with the nucleocapsid template and the NS protein.

ACKNOWLEDGMENTS

We are indebted to Dr. Robert A. Lazzarini for his constructive comments on this research and to Dr. Lynn Hudson for allowing us access to her VSV cDNA library. We also would like to express our appreciation to Charlene French and David Powell for their continuous support in computer programming and analysis of the L gene sequences. We thank Dr. Jacob V. Maizel and John Owens for comparing the L gene sequence to other sequences.

REFERENCES

1. Emerson, S. U., and Yu, Y. (1975). J. Virol. 15, 1348.
2. Lazzarini, R. A., Keene, J. D., and Schubert, M. (1981). Cell 26, 145.
3. Maxam, A. M., and Gilbert, W. (1977). Proc. Natl. Acad. Sci. (USA) 78, 2090.
4. Rose, J. K. (1980). Cell 19, 415.
5. Schubert, M., Keene, J. D., Herman, R. C., and Lazzarini, R. A. (1980). J. Virol. 34, 550.
6. McGeoch, D. J., and Dolan, A. (1979). Nucl. Acids Res. 6, 3199.
7. Rowlands, D. J. (1979). Proc. Natl. Acad. Sci. (USA) 76, 4793.
8. Gallione, C. J., Greene, J. R., Iverson, L. E., and Rose, J. K. (1981). J. Virol. 39, 529.
9. Rose, J. K., and Gallione, C. J. (1981). J. Virol. 39, 519.
10. Rose, J. K. (1978). Cell 14, 345.
11. Kang, C. Y., and Prevec, L. (1969). J. Virol. 3, 404.
12. Fields, S., and Winter, G. (1982). Cell 28, 303.
13. Winter, G., and Fields, S. (1982). Nucl. Acids Res. 10, 2135.

CLONING AND SEQUENCING OF M mRNA OF SPRING VIREMIA OF CARP VIRUS

Akio Kiuchi and P. Roy

Department of Environmental Health Sciences
School of Public Health
University of Alabama in Birmingham
Birmingham, AL 35294

Spring Viremia of Carp Virus (SVCV) is a fish Rhabdovirus structurally similar to vesicular stomatitis virus (VSV). As shown in Fig. 1 like VSV, SVCV has three major polypeptides, a glycoprotein G, a nucleocapsid protein N and a matrix (or membrane) protein M, and two minor proteins L and NSI. In addition, another minor protein, a second phosphoprotein NS2, which is an easily distinguished viral protein, is also present. It is of interest that the SVCV virion proteins are similar both in type, size and relative abundance to those of VSV, except that (1) the two SVCV NS phosphoproteins are easily resolved, (2) the SVCV G polypeptides are larger than the 67×10^3 VSV G and SVCV M polypeptide is slightly smaller than the 27×10^3 VSV M.

We have shown previously that although these two viruses are serologically distinct, they can synthesize mRNA species with similar 5' termini either <u>in vitro</u> or <u>in vivo</u> (1). Our recent studies on the 3' termini sequence of SVCV genome shows that the first 20 nucleotides are exactly similar to that of VSV Indiana and New Jersey serotypes (Unpublished data).

Since SVCV and VSV were recovered from two diverse sources, fish and mammals respectively, we were interested to know the structure and evolution of the viral membrane protein, particularly, whether there are any homology between their membrane protein M. To approach the question of M protein homology, we have extended our sequence analyses to the mRNA encoding the M protein from SVCV. In order to do this, we isolated a cDNA clone of mRNA encoding the protein. We have compared the predicted amino acid sequences of this M protein with that of VSV M protein (San Juan strain) determined previously by Rose et.al.(2).

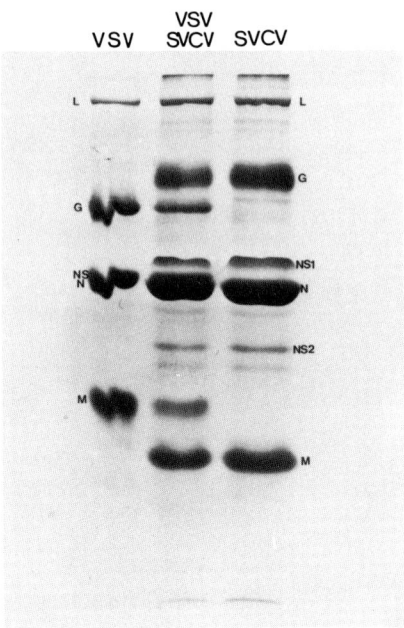

FIGURE 1. The autoradiogram of a polyacrylamide gel showing a comparison of [^{14}C]-leucine labeled VSV proteins and [^{3}H]-leucine labeled SVCV proteins.

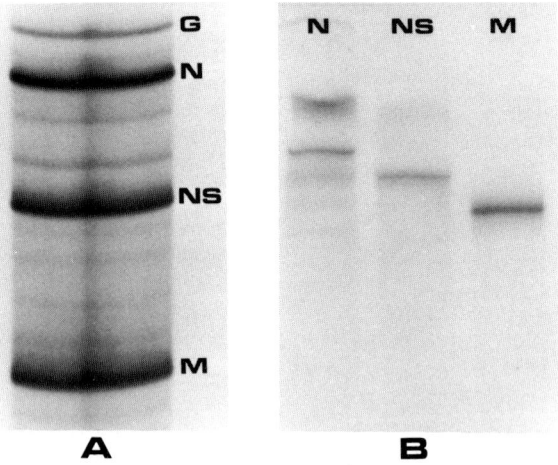

FIGURE 2. Autoradiogram of denatured gel showing cDNAs synthesized from polyA selected (A) total mRNA's of SVCV (B) individual mRNA's of SVCV.

The Strategy for Cloning SVCV mRNAs. To prepare cDNA we have used total mRNA from infected cells. PolyA tailed RNA were selected on oligo(dT) cellulose column and cDNA were synthesized using oligo(dT)$_{12-18}$ primer and reverse transcriptase. After boiling off the RNA template in formamide, the individual first strand cDNA were purified through a denatured gel as shown in Fig. 2A. To confirm the size of each cDNA the individual mRNA, were isolated in agarose gel containing 10mM methyl mercury hydroxide (3) and were used for cDNA synthesis as shown in Fig. 2B. Eluted purified cDNA were then polydA tailed at the 3' end and second strand were synthesized using oligo(dT)$_{12-18}$ primer. Double stranded cDNA were treated with nuclease S1 and repaired with Klenow's fragment and were ligated to Pvu II cut blunt end of pBR322. Hybrid plasmids were used for E. Coli. transformation and ampicillin resistant colonies were selected. Positive clones were screened by colony hybridization techniques using ^{32}P-labeled viral RNA probes.

Identification of the Cloned SVCV mRNAs. In order to identify the cDNA clones of SVCV mRNAs, we use the dot blot hybridization technique of Thomas (4) with slight modifications. Individual mRNA species that were isolated from agarose gel, are spotted (1-3µl) onto dry nitrocellulose paper that has been already saturated with 20 x SSC. The nitrocellulose paper containing the dots were then baked for 2 hours at 80°C under vacuum, then prehybridized for 8 hours at 42°C, and then hybridized for overnight with ^{32}P labeled, Hinf I digested plasmid DNA probe. After washing, the nitrocellulose paper was exposed at -70°C for 1 hour with intensity screen. The results are shown in Fig. 3. This clone was only hybridized to SVCV viral RNA, total mRNA and individual M mRNA. Clearly, this clone contained DNA capable of hybridizing specifically to the M mRNA.

DNA Sequencing. DNA sequencing was carried out by standard techniques using different restriction endonucleases as shown in Fig. 4. The distances and direction sequenced are shown by arrows. All sequence analyses were done by the Maxam & Gilbert method (5).

The Sequence of the Cloned M mRNA. The cloned M cDNA was sequenced. The complete nucleotide sequences of cDNA of M messenger RNA and the predicted amino acid sequences of M protein are shown in Fig. 5. Excluding poly A tail, the entire sequence is 710 bp long, encoding a protein of 223 amino acids. The amino acid composition of SVCV M protein and VSV M protein are shown in Table 1. SVCV M protein has a calculated size of 25,599 daltons while VSV M protein has a molecular weight of 26,064.

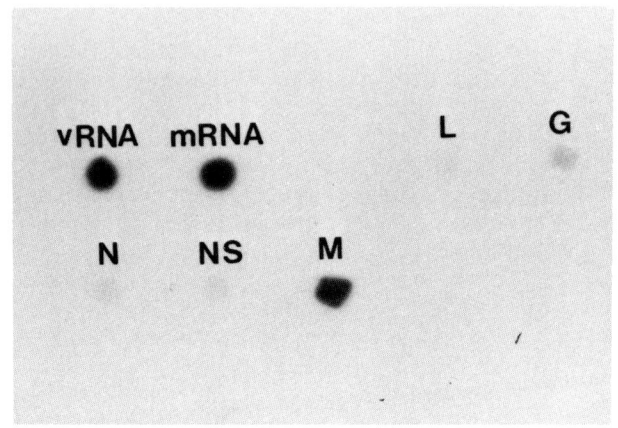

FIGURE 3. Dot blot hybridization of SVCV RNA species with ^{32}P-probes prepared from cDNA clone.

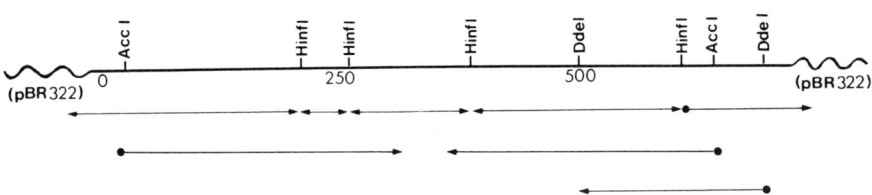

FIGURE 4. The strategy employed to sequence the cloned M mRNA. Sites for restriction enzymes that were used for sequencing are shown. The distances and directions sequenced are shown by arrows.

```
              M  S  T  L  R  K  L  F  G  T  K  K  S  K  G  T
GAACAGACATCATGTCTACTCTAAGAAAGCTCTTTGGAACCAAGAAGTCAAAAGGTACTC
            10        20        30        40        50        60

 P  P  T  Y  E  E  T  L  A  T  A  P  V  L  M  D  T  H  D  T
CTCCCACTTATGAGGAGACACTGGCGACTGCGCCAGTGTTAATGGATACTCATGATACTC
        70        80        90       100       110       120

 H  S  H  S  L  Q  W  M  R  Y  H  V  E  L  D  V  K  L  D  T
ACTCCCACTCACTGCAGTGGATGAGGTATCATGTTGAATTGGACGTAAAATTGGATACAC
       130       140       150       160       170       180

 P  L  K  T  M  S  D  L  L  G  L  L  K  N  W  D  V  D  Y  K
CCTTAAAAACGATGTCGGATCTTCTCGGACTCCTGAAAAATTGGGATGTAGATTACAAAG
       190       200       210       220       230       240

 G  S  R  N  K  R  R  F  Y  R  L  I  M  F  R  C  A  L  E  L
GTTCTAGGAACAAGCGTAGATTCTACAGATTGATCATGTTCCGTTGTGCGTTAGAACTCA
       250       260       270       280       290       300

 K  H  V  S  G  T  Y  S  V  D  G  S  A  L  Y  S  N  K  V  Q
AGCATGTATCGGGAACATACTCTGTTGACGGGTCGGCTTTGTACTCTAACAAGGTGCAAG
       310       320       330       340       350       360

 G  S  C  Y  V  P  H  R  F  G  G  M  P  P  F  K  R  E  I  E
GGAGTTGTTATGTGCCTCATCGATTCGGTCAAATGCCTCCTTTCAAGAGGGAGATCGAGG
       370       380       390       400       410       420

 V  F  R  Y  P  V  H  Q  H  G  Y  N  G  M  V  D  L  R  M  S
TCTTTAGATACCCAGTACATCAACATGGATACAACGGGATGGTAGATCTGAGAATGTCGA
       430       440       450       460       470       480

 I  C  D  L  N  G  E  K  I  G  L  N  L  L  K  E  C  Q  V  A
TCTGTGACCTAAACGGGGAAAAGATAGGCCTCAATTTGTTGAAAGAGTGTCAGGTAGCAC
       490       500       510       520       530       540

 H  P  N  H  F  G  K  Y  L  E  E  V  G  L  E  A  A  C  S  A
ACCCCAACCATTTCCAAAAATATCTGGAAGAAGTGGGGCTGGAAGCAGCCTGTTCAGCCA
       550       560       570       580       590       600

 T  G  E  M  I  L  D  M  T  F  P  M  P  V  D  V  V  P  R  V
CAGGAGAATGGATTCTTGATTGGACATTTCCTATGCCAGTAGACGTGGTGCCTCGCGTTC
       610       620       630       640       650       660

 P  S  L  F  M  G  D  ***
CTTCCCTGTTCATGGGAGATTAAATTGAGATTGATGTTCGCTGAGATATG
       670       680       690       700       710
```

FIGURE 5. The nucleotide and corresponding amino acid sequence of the SVCV M mRNA as deduced from the sequence data of cDNA clone. Nucleotides are numbered below their respective residues. Amino acids are centered over the corresponding nucleotide triplets.

TABLE 1 - The amino acid composition of SVCV M protein and VSV M protein.

Amino acid (abbreviation)	No. of residues in:	
	SVCV[a]	VSV[b]
Ala (A)	8	17
Arg (R)	12	10
Asn (N)	7	6
Asp (D)	13	13
Cys (C)	5	1
Gln (Q)	6	3
Glu (E)	11	13
Gly (G)	15	15
His (H)	10	8
Ile (I)	5	11
Leu (L)	25	17
Lys (K)	15	21
Met (M)	10	11
Phe (F)	9	14
Pro (P)	13	16
Ser (S)	14	18
Thr (T)	13	11
Trp (W)	4	3
Tyr (Y)	10	12
Val (V)	17	9

[a] Total number of residues, 223: Molecular weight. 25,599.
[b] Total number of residues, 229: Molecular weight. 26,064.

The Hydropathic Pattern of SVCV and VSV M Protein. When the hydropathic pattern of M proteins of SVCV and VSV were compared, no significant characteristics were observed. However, as shown in Fig. 6, both proteins have clusters of positive charges in similar positions. Both proteins are rich in basic amino acids and contain highly basic amino terminal domains. When the sequences of the two proteins were compared visually, as shown in Fig. 7, it was observed that there were about 63 identities which comprises of 28% and there were six gaps. Although the standard deviation is not known at the present time, it is quite clear that there is high homology between the two M proteins.

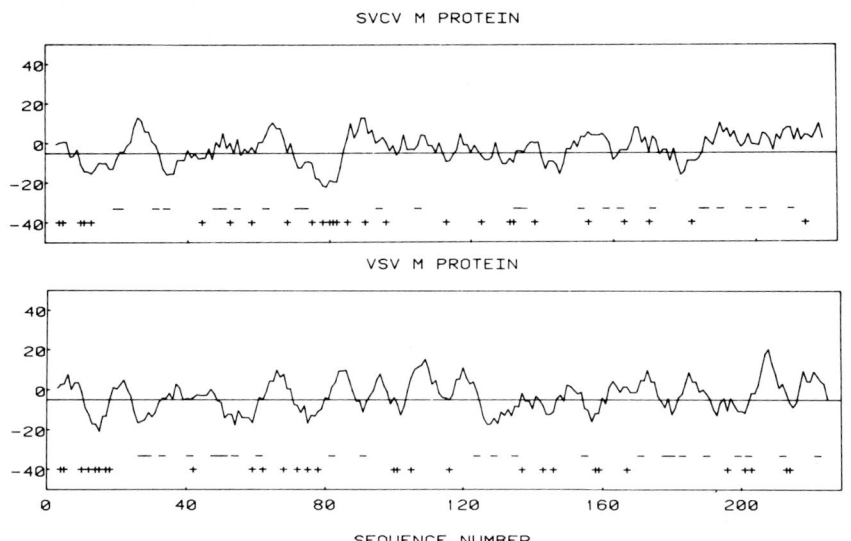

FIGURE 6. The hydropathic pattern of SVCV and VSV M proteins.

```
SVCV  MSTLRKLFGT    KKSKGT    PPTYEETLATAPVL              MDTHDT
VSV   MSSLKKILGKGKKKSKKLGIAPPPYEEDTSMEYAPSAPIDSYFGVDEMDTYDP

SVCV  HSHSLQWMRYHVELDVKLDTPLKTM SDLLGLLKNWDVDYKGSRNKRRFYRLIM
VSV   NQ  PNQLRYEKFFTVKMTVRSNPFTYSDVAAAVSHWDHMYIGMAGKRPFYKILA

SVCV  FRCALELKHVSGTYSVDGSALYSNKVQGSCYVPHRFGQMPFFKREIEVFRYPVHQ
VSV   FLGSSNLKATPAVLADQGQPFYHTHCEGRAYLPHRMGKTPEMLNVPEHFRRFFNI

SVCV  HGYNGMVDLQMSICDLNGEKIGLNLLKECQVAHPNHFQKYLEEVGLEAACSATGE
VSV   GLYKGTIELTMTLYDDESLEAAPMIWDHFNSSKFSDERFKALMFGLIVEKKASGA

SVCV  WILDWTFPMPVDVVPRVPSLFMGD
VSV   WVLDSISHFK
```

FIGURE 7. Comparison between the predicted sequences of VSV M protein and SVCV M protein. The identities between the two sequences are shown by boxed areas.

Rose, et al., (6) have suggested previously both VSV M protein and Influenza virus M protein begins with the sequence Met-Ser-X-Lys. It is notable that SVCV M protein also begins with a similar sequence. From these data it seems that certain amino acid residues are highly conserved in these sequences and although SVCV and VSV were recovered from two diverse sources, it is quite likely that the two viruses might have originated from a common ancestor among the Rhabdoviridae family.

ACKNOWLEDGEMENTS

We thank Dr. David H.L. Bishop for many helpful discussions and suggestions. This work was supported by Grant No. PCM-8104086 from the National Science Foundation.

REFERENCES

1. Gupta, K.C., Bishop, D.H.L. and Roy, P. (1979). J. Virol. 30:735-745.

2. Rose, J.K., and Gallione, C.J. (1981). J. Virol. 39:519-528.

3. Bailey, J.M., and Davidson, N. (1975). Anal. Biochem. 70:75-85.

4. Thomas, P.S. (1980). Proc. Natl. Acad. Sci. USA 77:5201-5205.

5. Maxam, A.M., and Gilbert, W. (1976). Proc. Natl. Acad. Sci. USA 74:560-564.

6. Rose, J.K., Doolittle, R.F., Anilionis, A., Curtis, P.J., and Wunner, W.H. (1982) J. Virol. 43:361-364.

CHARACTERIZATION OF MEASLES VIRUS RNA[1]

J. Tucker, A. Ramsingh, G. Lund, D. Scraba
W. C. Leung and D. L. J. Tyrrell

Departments of Biochemistry and Medicine
University of Alberta
Edmonton, Canada T6G 2H7

MEASLES VIRUS RNA

Measles virus, a paramyxovirus, is a pathogen known to cause acute and persistent infections in man. The measles virus polypeptides have been identified, however, there is considerable variation in the literature reports of the size of the measles virus RNA. Estimations of the molecular weight have been from 4.5×10^6 to 6.4×10^6 using sucrose gradient sedimentation(1), electron microscopy(2), estimation from the structure of nucleocapsids(3), and from agarose or acrylamide gels under denaturing conditions(4,5). In this report we have determined the size of measles virus RNA using both morphological and biochemical techniques.

Virus Growth and Purification. Vero cells were grown in roller bottles (1600 cm^2) and at confluency infected at 1 pfu/cell with the Lec strain of measles virus. The infectious virus was twice plaque purified and used after one passage. Measles virus was purified from the supernatant fluid at 72 hours post-infection as previously described(6). Cell material was retained for nucleocapsid preparation using the following procedure. The cell pellet was washed 2x with PBS and disrupted by the addition of 1% Triton X-100. After a 30 minute incubation, the nuclei were pelleted at 1000 g for 20 minutes. The supernatant was layered onto a discontinuous gradient of 6 ml 25%, 4 ml 30%, and 2 mg 40% CsCl in PBS and centrifuged on an SW27 rotor at 25,000 rpm for 1.5 hours. The nucleocapsid band present in the 30% CsCl region was removed, diluted in PBS and layered on a second gradient composed of 4 ml each of 20%, 25%, 30%, 35%, and 40% CsCl in PBS and centrifuged at 4°C in an SW rotor at 24,000 rpm for 16 hours. The nucleocapsid band was removed, diluted

---

[1] This work was supported by the Medical Research Council of Canada.

with PBS and pelleted at 40,000 g for 30 minutes. The nucleocapsid pellet was resuspended in 1.0 ml of 0.01 M Tris-HCl, pH 8.0, 0.1 M NaCl and 1 mM EDTA prior to RNA extraction.

RNA Extraction. Purified measles virus and nucleocapsid were diluted 1:1 with TNE buffer (0.01 M Tris-HCl, pH 7.5, 0.1 M NaCl, 0.001 M EDTA). Virions were disrupted with 0.1% 2-mercaptoethanol and 1.0% SDS. The suspension was extracted with an equal volume 90% phenol/chloroforum and the organic phase was reextracted with TNE buffer. The pooled aqueous phase was mixed with 2.5 volume ethanol and RNA precipitated at -20°C. The RNA was pelleted at 16,000 g for 30 minutes at 4°C and resuspended in sterile distilled water.

Electron Microscopy of Measles Virus, Nucleocapsid and RNA. Purified nucleocapsid and viral proteins were negatively stained using 1.0% sodium phosphotangstate and examined in a Phillips EM 300 electron microscope. Electron micrographs were made of measles virus RNA molecules after treatment with DMSO and glyoxal(7). Internal standards for the RNA measurements were pBR322 DNA (partially cleaved by Pst 1) and intact mengo virus RNA. Nucleic acid lengths were measured on prints (3x enlarged) using a Hewlett-Packard 9874A digitizer coupled to a Tektronix 4051 graphics computer.

Agarose Gel Elecrophoresis Under Denaturing Conditions. Agarose gel electrophoresis of RNA in 10 mM methylmercury hydroxide was done using the method of Lehrach et al(8) with minor modifications. Horizontal slab gels were run in a BioRad mini electrophoresis apparatus and 1-3 ug of RNA samples used in each well. After electrophoresis at 35mA at room temperature for 3 hours, the gels were stained in 2 ug/ml ethidium bromide and photographed under shortwave UV light.

Agarose gel electrophoresis of RNA denatured in glyoxal and DMSO was performed using the method described by McMaster and Carmichael(9). Gels were either blotted onto nitrocellulose or stained in acridine orange and photographed as previously described.

In Vitro Synthesis of cDNA Probes and RNA Blot Analysis. RNA isolated from purified measles virus preparations was electrophoresed in 1% low melting temperature agarose gel containing 10 mM methylmercury hydroxide. The gels were stained with ethidium bromide and the measles viral RNA band was excised. The RNA was extracted and used as a template for cDNA synthesis. Cellular ribosomal RNA was also used as a template for cDNA synthesis. The cDNA probles were synthesized as described by Leung et al(10). The technique

CHARACTERIZATION OF MEASLES VIRUS RNA

for RNA blot analysis used was similar to that described by Thomas(11).

RESULTS

RNA extracted from purified measles virus was denatured in 10 mM methylmercury hydroxide or in glyoxal and DMSO. A typical gel profile is shown in Figure 1.

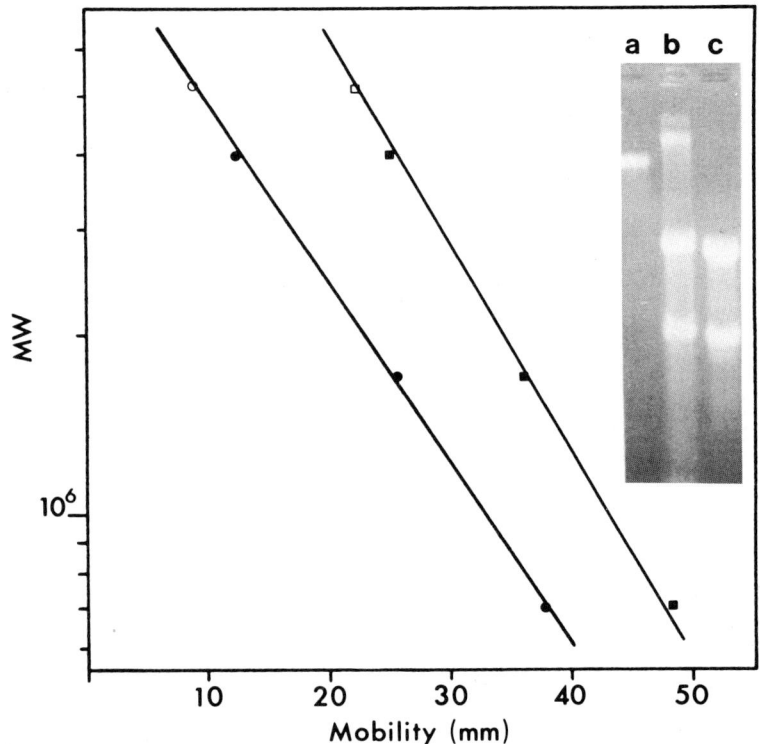

FIGURE 1. Electrophoresis of RNA Under Denaturing Conditions in Agarose Gels. A typical gel of RNA preparations in a methylmercury hydroxide agarose gel (a-Vesicular stomatitis virus RNA; b-Measles virus RNA; and c-Ribosomal RNA). The results of the mobility versus molecular weights of RNA have been plotted for methylmercury hydroxide (●——●) and DMSO and glyoxal (■——■) gels.

The profile shows measles virus RNA with three major RNA bands, two of which co-migrate with host cell ribosomal 28S and 18S RNA. The largest RNA species is believed to represent measles virus RNA. In order to obtain molecular weight estimations of the measles virus RNA, three RNA markers, i.e. vesicular stomatitis virus (VSV) RNA (4.0×10^6), 28S (1.7×10^6) and 18S (7×10^5) ribosomal RNA. The results obtained are shown in Figure 1. The molecular weight of measles viral RNA from these studies was estimated to be 5.2×10^6. RNA extracted from purified measles virus preparations and analyzed in DMSO and glyoxal denaturing gels was also compared to RNA markers. By this method, the molecular weight of measles virus RNA was 5.2×10^6 (Figure 1).

Electron microscopy measurements of measles virus RNA spread after treatment with DMSO and glyoxal showed a good deal of fragmentation, but the longest molecules were 5.1 um. An internal standard of pBR322 DNA (4362 base pairs, 1.47 um in length) was used (Figure 2). Mengo virus spread under the same conditions was 2.5 um in length and has a known molecular weight of 2.56×10^6. Using these internal standards, the measles virus RNA molecular weight determined in this manner was 5.2×10^6. Thus the results by electron microscopic measurements confirmed the results obtained by denaturing agarose gel electrophoresis.

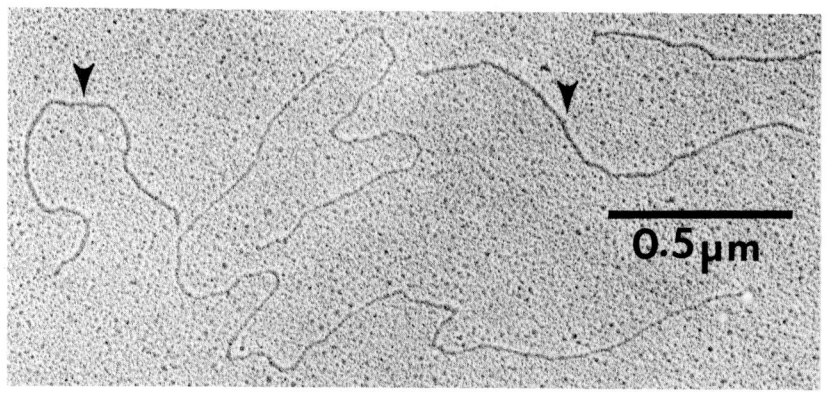

FIGURE 2. Electron Micrograph of Intact Measles Virus RNA. Arrows indicate pBR322 cleaved by Pst I which was used as an internal standard (length 1.47 um). Intact measles RNA with a length of 5.1 um is shown.

Since the 28S and 16S ribosomal RNA species co-purify with measles virus RNA, it is important to eliminate the possibility that large RNA species (5.2×10^6) was not a ribosomal precursor RNA, e.g. the 45S RNA. This possibility was ruled out by the following experiment, RNA extracted from VSV, measles and measles nucleocapsids was denatured in glyoxal and DMSO and electrophoreses in agarose. The gel was blotted onto nitrocellulose paper and probed with radio-labelled cDNA prepared from RNA extracted from uninfected Vero cells. The cDNA probe hybridized to the cellular 28S and 18S RNA species and a few minor bands, but did not hybridize to the RNA isolated from VSV, measles virus, or purified nucleocapsid. When the measles RNA band was excised from methylmercury gels and used for cDNA synthesis, this probe hybridized to measles virus RNA from purified virus and nucleocapsid, but not to VSV RNA or rRNA (data not shown).

CONCLUSION

The molecular weight of measles virus RNA has been difficult to accurately determine. The molecular weight estimations have varied from 4.5×10^6(5) to 6.4×10^6(3). Using either purified virus or nucleocapsid preparations as the source of measles virus RNA, a molecular weight to 5.2×10^6 was determined by both biochemical and morphological techniques. Although some smaller species of RNA were observed on the electron micrographs of measles virus RNA extracted from nucleocapsid, the largest samples had a molecular weight of 5.2×10^6, the same length as the largest species of RNA seen in electron micrographs of RNA extracted from whole virus. Our results are in agreement with the molecular weight estimations recently reported by Dunlap et al(2).

ACKNOWLEDGEMENTS

We appreciate the technical assistance of Ian McRobbie with the cell cultures and virus propagation and Roger Bradley with the electron microscopic studies. We also thank Roberta Wardhaugh for typing this manuscript.

REFERENCES

1. Schluederberg, A. (1977). Biochem. Biophy. Res. Commun. 42, 1012.
2. Dunlap, R.C., Milstein, J.B., and Lundquist, M. (1983). Intervirology 19,169.
3. Nakai, T., Shand, F.L., and Howatson, A.F. (1969). Virology 38,50.
4. Hall, W.W., and Martin, S.J. (1973). J. Gen. Virol. 19, 175.
5. Backo, K., Billeter, M., and ter Muelen, V. (1983). J. Gen. Virol. 64, 1409.
6. Tyrrell, D.L.J., and Norrby E. (1978). J. Gen. Virol. 39, 219.

7. Murant, A.F., Taylor, M., Duncan, G.H., and Raschke, J.H. (1981). J. Gen. Virol. 53, 321.
8. Lehrach, H., Diamond, D., Wozney, J.M., and Boedtker, H. (1977). Biochemistry 16, 4743.
9. McMaster, G.K., and Carmichael, G.G. (1977). Proc. Natl. Acad. Sci. U.S.A. 74, 4835.
10. Leung, W.C. Ramsingh, A., Dimock, K., Rawls, W., Petrovich, J., and Leung, M. (1981). J. Virol. 37, 48.
11. Thomas, P.S. (1980). Proc. Natl. Acad. Sci. U.S.A. 77, 5201.

THE CLONING OF MORBILLIVIRUS
SPECIFIC RNA[1]

J.D. Hull, S.E.H. Russell, E.M. Hoey
B.K. Rima and S.J. Martin

Department of Biochemistry
The Queen's University of Belfast
Belfast BT9 7BL, N. Ireland

Measles virus (MV) and canine distemper virus (CDV), both morbilliviruses have been associated with several degenerative neurological diseases such as subacute sclerosing panencephalitis and old dog encephalitis. These diseases have been suggested to involve a persistent infection by these morbilliviruses. In order to understand the persistent infection fully, it is necessary to identify the factors which are responsible for initiation and maintenance. Several models have been proposed to explain the phenomenon in cell culture such as the selection of virus mutants, the modulation of matrix protein synthesis and the selection of defective interfering particles. Few studies have been made of the molecular basis of persistence due to the low levels of gene expression in the persistently infected cell. However, virus gene specific radiolabelled cDNA probes could overcome this problem allowing the quantification of changes in gene expression during the initiation and course of a persistent infection. Furthermore, control of gene expression at the transcriptional and translational level could be distinguished.

THE PURIFICATION AND CHARACTERISATION OF MORBILLIVIRUS MESSENGER RNA

The most abundant and readily obtainable morbillivirus specific RNA species in the infected cells are the mRNA's. Thus, they were the template of choice for cDNA synthesis. The mRNA's of the morbilliviruses are probably monocistronic and have been shown to posses polyadenosine (polyA) tails (1).

---

[1]This work was supported by the Multiple Sclerosis Society and the Medical Research Council (U.K.).

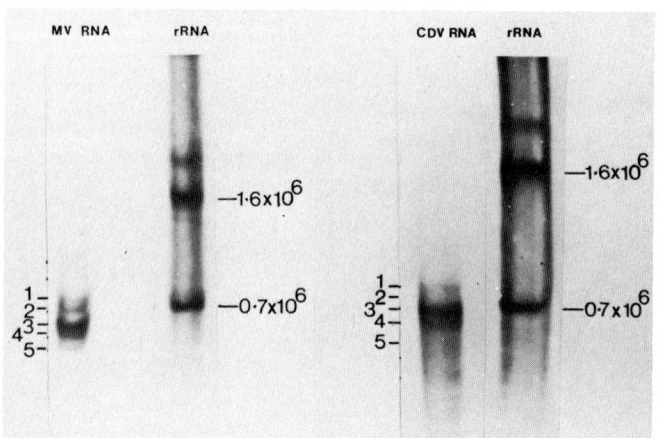

FIGURE 1. MV and CDV infected cell polyA$^+$ RNA fractionated in the presence of CH_3HgOH.

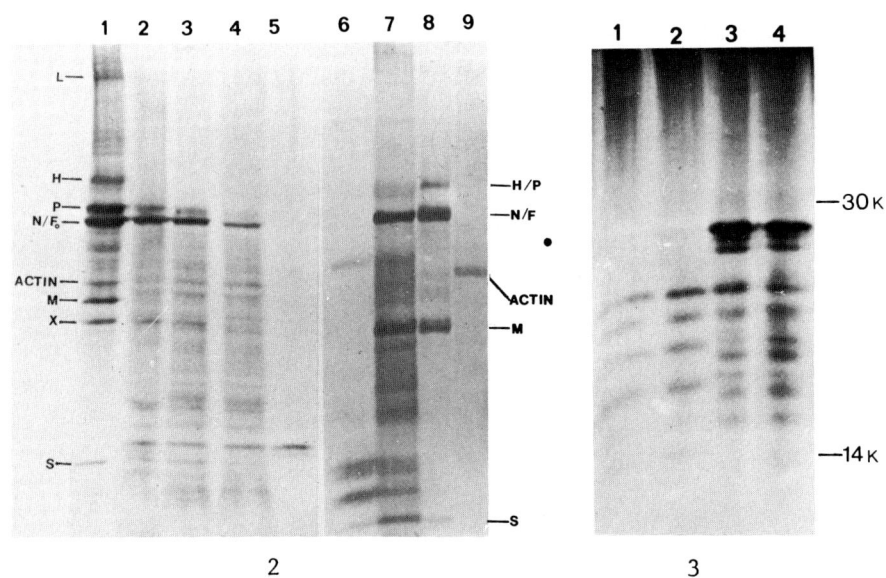

FIGURE 2 (left). The products of in vitro translation of MV and CDV polyA$^+$ RNA. Lane (1) MV in vivo proteins; (2) proteins directed by RNA harvested at 21 hpi; (3) 19 hpi; (4) 17 hpi; (5) and (6) zero RNA control; (7) CDV in vitro and (8) CDV in vivo proteins.

FIGURE 3 (right). The products of limited proteolytic cleavage of MV X (lanes 1,2) and P (lanes 3,4) proteins with V8 protease synthesized in vivo (1,3) and in vitro (2,4).

They code for seven major viral proteins: L (Large; mol.wt. 180K), H (Haemagglutinin; 80K), P (Polymerase; 70K), N (Nucleocapsid; 60K), F (Fusion; 60K), M (Matrix; 40K) and S (Small; 20K) (reviewed in 2). Figure 1 shows the principal polyA containing RNA species synthesized in MV and CDV infected Vero cells during the latter part of a lytic infection, in the presence of sufficient actinomycin D to inhibit >90% of host cell RNA synthesis. The total, infected cell RNA was obtained by the method of Glisin and coworkers (3), enriched for polyA RNA by oligo-dT-cellulose affinity chromatography and fractionated in agarose gels in the presence of methylmercury (II) hydroxide. These gels indicate the presence of five major RNA species in the approximate mol. wt. range of 0.7-1.0×10^6 in MV infected cells and five species in CDV infected cells, which do not comigrate with the MV mRNA's, but are in the same mol. wt. range. The mol. wts. of the MV mRNA species are similar to those reported (4). However, as has been noted (5), the maximum coding capacity of these molecules is in excess of that required to produce the MV specific proteins. Similar investigations of the labelled polyA$^+$ RNA of mock infected cells show a different pattern of bands with mol. wts. in the range of 0.3-1.2×10^6, with a specific activity five times less than the infected cell polyA$^+$ RNA.

To characterize more fully the virus specific mRNA species in the total polyA$^+$ RNA population the ability of this RNA to direct the synthesis of proteins in a cell free translation system was examined (Fig. 2). MV infected cell polyA$^+$ RNA directed the synthesis of six major proteins having approx. mol. wts. of 70K, 60K, 45K, 37K and 20K. These have been identified by limited proteolysis with V8 protease as P, N, actin, M, X and S respectively, where X is considered to be a derivative of P (earlier also identified in vivo), as it shares several major peptides with the P protein (Fig. 3). CDV infected cell polyA$^+$ RNA directed the synthesis of four major proteins which have been identified as the P, N, M and S proteins. In the case of MV the cell free translation of M specific mRNA was highly variable between preparations. This was not the case for CDV. Neither CDV nor MV infected cells polyA$^+$ RNA directed the synthesis of L protein, nor have nonglycosylated forms of the H or F proteins been detected. However, in both systems virus specific proteins appear to be the major products in in vitro translation of infected cell mRNA populations.

THE SYNTHESIS AND CLONING OF MORBILLIVIRUS cDNA

Single stranded cDNA (sscDNA) was synthesized by reverse transcription of oligo-dT primed infected cell polyA$^+$ RNA. The average yield of sscDNA per µg of RNA template was 100 ng. By sizing on alkaline sucrose gradients or in denaturing gel systems, the sscDNA was found to be a heterogeneous population of molecules ranging from 200->1100 nucleotides in length. This material was cloned using the procedure described by Cann and coworkers (6) whereby sscDNA/mRNA hybrids, immediately after reverse transcription, were directly extended at their 3' termini with an oligo-dC tail using terminal transferase in the presence of Co^{++}. The tailed hybrids were reannealed to PstI cut pAT153 plasmid which had its 3' termini extended with an oligo-dG tail of similar length to the dC tails of the hybrids (30-40 nucleotides). The putative recombinant plasmids were then used to transform Escherichia coli K12 (HB 101 or ED8767) (7). Putative insert carrying bacteria were selected using the insertional inactivation of the pAT153 ampicillin resistance gene as indicator. The proportion of transformed colonies with an inactivated ampicillin gene varied from 60-25% depending upon the extend of the initial PstI digestion of the vector. Colony hybridisation (8) with sscDNA probes (specific activity approx. 4.5 x 10^8 cpm/µg), reverse transcribed from uninfected Vero cell or morbillivirus infected cell polyA$^+$ RNA, selected out those colonies that carried infection specific sequences. Of those bacteria which yielded a reproducible signal approximately one quarter were infection specific. About 60% of these plasmids contained inserts greater than 300 base pairs (bp) which could be released by restriction with PstI.

THE CHARACTERIZATION OF MORBILLIVIRUS SPECIFIC CLONES

Of 25 MV infection specific clones, 5 have been shown to carry cDNA inserts greater than 600 bp (pMV1: 1kbp; pMV11: 1kbp - both with internal PstI sites -; pMV5: 800 bp; pMV14: 750 bp; pMV9: 600 bp). Of the 15 CDV infection specific clones examined 3 have large PstI releasable inserts (pCDV2: 1450 bp; pCDV6: 800 bp; pCDV7: 400 bp). When clones pMV1-6 and pCDV2,6,7 were challenged with a cDNA probe, prepared by reverse transcription of mumps virus infected Vero cell polyA$^+$ RNA, no signal was obtained which suggested that they are MV or CDV virus specific and not derived from paramyxovirus induced cell specific mRNA's. Cross hybridisation studies are presently underway to establish sequence

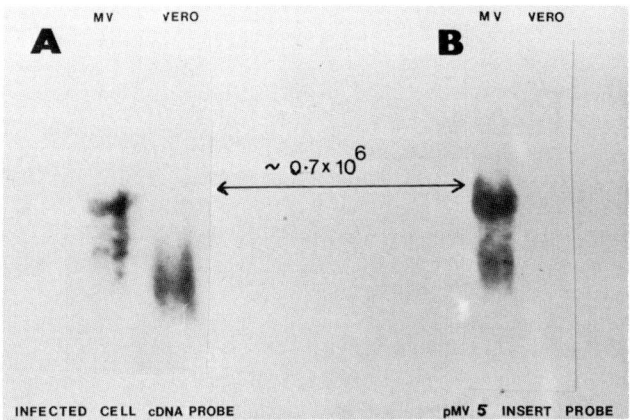

Figure 4. MV and Vero cell RNA immobilized on DBM paper and probed with (a) total MV infected cell cDNA probe (b) nick translated pMV5 insert DNA probe.

homologies between the clone banks in our possession derived from two strains of mumps virus, MV and two plaque morphology mutants of CDV.

Probing the 25 MV specific clones with nick-translated purified insert of pMV5, lit up two other related clones, pMV2 and pMV3, both of which carry small inserts (200-300 bp). Further work of this type to group the remaining 22 clones is currently in progress.

To demonstrate that the clones were complementary to viral mRNA's, polyA$^+$ RNA immobilized on DBM paper (9) after northern blots has been probed. Figure 4A shows the results obtained when MV infected cells polyA$^+$ RNA and Vero cell polyA$^+$ RNA were challenged with a MV infected cell cDNA probe. Both types of RNA yielded a signal. However, the hybridisation patterns were quite distinct. This demonstrated that both types of RNA had been successfully transferred and immobilized and also, that there appeared to be extra RNA species in the infected cell RNA population. Figure 4B shows the results when this RNA was challenged with a nick-translated pMV5 insert probe. A band was apparent which corresponded closely to the position of band 3 (Fig. 1). No hybridisation to Vero cell RNA could be detected (even at very long exposure times) at the stringency conditions used (0.1 x SSC at 65°C). To position the hybridisation signals with respect to MV mRNA bands (Fig. 1) a ^{3}H-uridine labelled infected cell polyA$^+$ RNA sample was fractionated in the

in the original gel before northern blotting. This was cut off and fluorographed separately. The band pattern and separation was similar to Fig. 1. However, the electrophoretic separation differences between various mRNA species appeared to be too small to allow an identification to be made of the band to which the insert of pMV5 hybridised. Similar experiments with pCDV2 probes yielded the same type of interpretation difficulties.

Therefore, this communication reports the isolation of morbillivirus specific mRNA and the preparation of a number of MV and CDV infection specific cDNA clones. The evidence obtained to date suggests strongly that these clones are virus specific. Further characterisation of these clones is currently being carried out by hybrid arrest or hybrid selected translation experiments.

ACKNOWLEDGEMENTS

We would like to thank Ms C. Lyons for assistance with the cell culture work involved in this project and Ms D. Ingram for typing the manuscript.

REFERENCES

1. Hall, W.W., and ter Meulen, V. (1977). J. gen. Virol. 35, 497.
2. Rima, B.K. (1983). J. gen. Virol. 64, 1205.
3. Glisin, V., Crkvenjakov, R., and Byus, C. (1974). Biochemistry 13, 2633.
4. Rozenblatt, S., Gesang, C., Lavie, V., and Neumann, T.S. (1982). J. Virol. 42, 790.
5. Gupta, K.C., Morgan, E.M., Kitchingman, G., and Kingsbury, D.W. (1983). J. gen. Virol. 64, 1679.
6. Cann, A.J., Stanway, G., Hauptmann, R., Minor, P.D., Schild, G.C., Clarke, L.D., Mountford, R.C., and Almond, J.W. (1983). Nucleic Acids Res. 11, 1267.
7. Dagert, M. and Ehrlich, S.D. (1979). Gene 6, 23.
8. Grunstein, M., and Hogness, D. (1975). Proc. natn. Acad. Sci. U.S.A. 72, 3961.
9. Alwine, J.C., Kemp, D.J., and Starke, G.R. (1977). Proc. natn. Acad. Sci. U.S.A. 74, 5350.

Transcription and Replication

SPECIFICITY OF INTERACTION OF L, NS, AND M PROTEINS WITH N-RNA COMPLEX OF VESICULAR STOMATITIS VIRUS IN A HETEROLOGOUS IN VITRO RECONSTITUTION SYSTEM

Bishnu P. De, Angeles Sanchez, and Amiya K. Banerjee

Roche Institute of Molecular Biology
Roche Research Center
Nutley, New Jersey 07110

The transcribing nucleocapsid of vesicular stomatitis virus (VSV) consists of the N protein-RNA complex associated with two dissociable proteins: L and NS (1). An efficient reconstitution system for in vitro transcription was developed by Emerson and Wagner (2), where addition of dissociated L and NS proteins to the purified N-RNA complex resulted in faithful transcription of the genome RNA. Transcription products included the leader RNA and five 5' capped, 3'-polyadenylated mRNA species coding for the five VSV structural proteins (3). Although both L and NS are needed for transcription (4), the precise role(s) of each protein in this process is not yet established.

In order to study specificity of interaction of the L and NS proteins with the N-RNA complex, in the present studies we have used two serotypes of VSV i.e., Indiana (VSV_{IND}) and New Jersey (VSV_{NJ}). Although these two viruses are serologically distinct, previous studies have shown evidence of relatedness based on cross-reactive group-specific antigen (ribonucleoprotein core, RNP) (5). Moreover, it was shown that L protein of VSV_{NJ} could be utilized in vivo in the transcription of a defective particle obtained from a heat-resistant mutant of VSV_{IND} (HR-DI) which lacks the L protein gene (6). Also of interest are the observations that wild-type VSV_{NJ} can complement the transcription defect of a VSV_{IND} Group I mutant (7); VSV-NJ interferes with the replication of VSV_{IND} (8) and HR-DI IND can cause heterotypic interference with VSV_{NJ} in coinfected cells (9). These results clearly indicate that interaction of L and NS proteins with heterologous N-complex probably occurs in vivo despite the fact that corresponding heterologous template in a reconstitution experiment failed to complement the serotype-specific proteins to restore transcription in vitro (10). The N genes of VSV_{IND} and VSV_{NJ} share considerable sequence homology (D. Rhodes, unpublished observations) in addition to coding for a common leader RNA (11) and 5'-terminal hexanucleotide on each of their mRNAs (12). It was, therefore, of interest to reexamine the specificity of interaction of L and NS proteins with heterologous N-RNA complex.

HETEROLOGOUS RECONSTITUTION SYSTEM.

The L and NS proteins of VSV_{IND} (Mudd-Summers strain) and VSV_{NJ} (Ogden strain) were purified from virions disrupted with detergent in the presence of 0.4 M NaCl. The RNP containing the L and NS proteins were removed from the G and M proteins by centrifugation (13). The L and NS proteins were then dissociated from the RNP in 0.8 M NaCl and further fractionated by chromatography on phosphocellulose column as described by Emerson and Yu (4). As shown in Fig. 1, (#1) purified NS protein (Indiana) free from L protein was recovered as flow through while the L protein that eluted at 1M NaCl contained only a trace amount of NS protein (#2). The band migrating slightly faster than the NS protein in both fractions was found to be N protein by Western blot using N-specific antibody. The L and NS proteins of VSV_{NJ} were purified similarly and had an elution profile the same as shown in Fig. 1. The N-RNA complexes of both the serotypes were prepared as described previously (13). In a typical reconstitution system, the NS protein (#1) was transcriptionally inactive. The #2 fraction retained some activity but addition of #1 stimulated transcription by 3 to 4-fold.

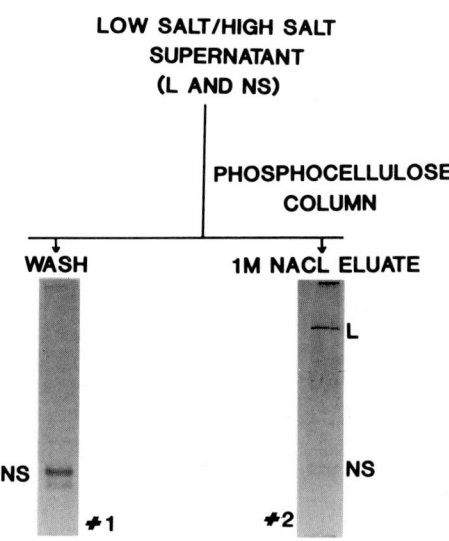

Fig. 1. Separation of L and NS proteins of VSV. Purified virus of both the serotypes were disrupted with Triton X-100 and L and NS

proteins were purified as described earlier (13). The proteins were further separated by chromatography on a phosphocellulose column (4). The wash (#1) and 1M NaCl eluate (#2) were analyzed by polyacrylamide gel electrophoresis and proteins were stained by silver reagent.

Fig. 2 shows the specificity of interaction of L and NS proteins of VSV_{NJ} with N-RNA complex of VSV_{IND} based on their ability to synthesize RNA in vitro. The RNA products were analyzed by polyacrylamide gel electrophoresis. The L and NS proteins of VSV_{NJ} were active as shown by their ability to synthesize RNA with homologous N-RNA complex (Lane A). In contrast, the proteins failed to synthesize mRNA and leader RNA when added to N-RNA complex of VSV_{IND} (Lane B). In sharp contrast, when L and NS proteins of VSV_{NJ} were added together with purified NS_{IND} (#1, Fig. 1) RNA synthesis ensued as evidenced by the synthesis of leader RNA and mRNA which stayed at the origin (Lane D). Virtually no RNA product was synthesized when NS_{IND} was added alone to N-RNA_{IND} (Lane C). These results clearly indicate that L and NS_{NJ} can replace the L_{IND} function, and subsequent addition of NS_{IND} was required for mRNA synthesis.

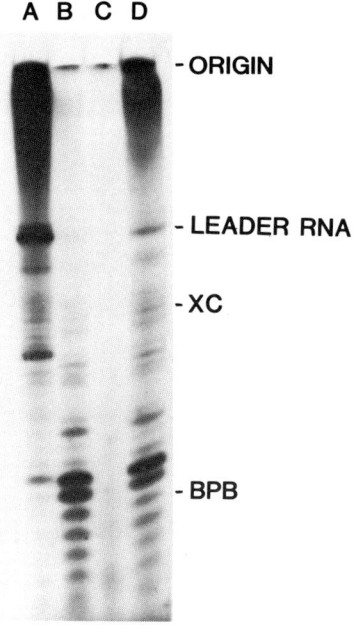

Fig. 2. RNA synthesis by in vitro reconstituted system. Purified fractions #1 and #2 (Fig. 1) were added to N-RNA complex under RNA synthesis conditions, and products were analyzed by electrophoresis in a 20% polyacrylamide gel followed by autoradiography (13). (A) $(L+NS)_{NJ}$ + N-RNA_{NJ}; (B) $(L+NS)_{NJ}$ + N-RNA_{IND}; (C) NS_{IND} + N-RNA_{IND}; (D) $(L + NS)_{NJ}$ + NS_{IND} + N-

RNA_{IND}. Positions of leader RNA, xylene cyanol (XC) and bromophenol blue (BPB) dyes are shown.

Fig. 3 Lane B shows that in a partial reaction (14) containing only ATP and α-^{32}P CTP, $(L+NS)_{NJ}$ were capable of synthesizing AC (oligo I) and AAC and AACA (oligo II) on the N-RNA_{IND}, i.e., similar to the homologous components (Lane A). In contrast, NS_{IND} failed to synthesize significant oligomers when reacted with its homologous template (Lane C). Addition of NS_{IND} together with $(L+NS)_{NJ}$ to the N-RNA_{IND} (Lane D) did not increase oligonucleotide synthesis. (Compare Lanes B & D). From the results shown in Figs. 2 and 3, it can be concluded that L_{NJ} could bind to N-RNA_{IND} and initiate RNA chains but homologous NS_{IND} could interact with L_{NJ} for subsequent chain elongation and continued RNA synthesis.

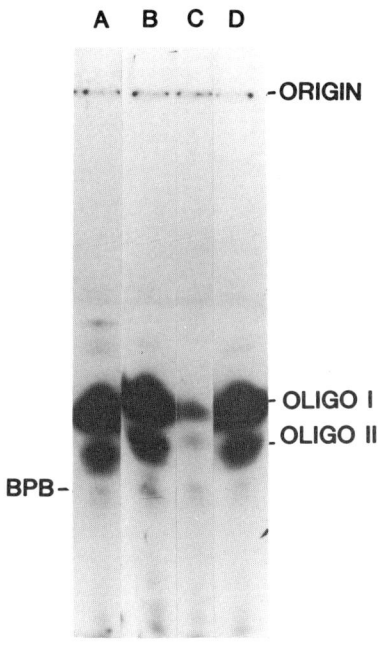

Fig. 3. Initiation of RNA synthesis in the reconstituted system. Synthesis of oligonucleotides AC (oligo I) and AAC, AACA (oligoII) was carried out in the presence of ATP and α-^{32}P CTP and products analysed by polyacrylamide gel electrophoresis (14). (A) $(L+NS)_{NJ}$ + N-RNA_{NJ}; (B) $(L+NS)_{NJ}$ + N-RNA_{IND}; (C) NS_{IND} + N-RNA_{IND}; (D) $(L+NS)_{NJ}$ + NS_{IND} + N-RNA_{IND}.

In contrast to the above results, when N-RNA_{NJ} was used in a similar reconstitution sysem with $(L+NS)_{IND}$, different results were obtained. $(L+NS)_{IND}$ was able to bind to N-RNA_{NJ} and initiate RNA chains similar to that shown in Fig. 3, but NS_{NJ} was unable to synthesize RNA when added to the mixture (data not shown). These results indicated that NS_{NJ} lacks the ability to bind to L_{IND} to synthesize RNA chains, whereas as shown above, NS_{IND} can specifically interact with L_{NJ}. In order to further confirm this specific interaction between L_{NJ} and NS_{IND}, we studied the capacity of NS_{IND} to replace NS_{NJ} in a homologous reconstitution system. The results of such an experiment is shown in Fig. 4. The $(L+NS)_{NJ}$ (#1, Fig. 1) in a homologous reconstitution system was capable of synthesizing mRNA (Lane A), and addition of homologous NS_{NJ} (#2) stimulated mRNA synthesis by 3-fold (lane B). When NS_{NJ} was replaced by NS_{IND}, identical stimulation of mRNA synthesis occurred (lane C). These results demonstrate that NS_{IND} could specifically interact with L_{NJ} to synthesize mRNA. Thus, NS_{IND} and NS_{NJ} share common reactive sites that are involved in interaction with L_{NJ}, but interaction of NS_{IND} with L_{IND} is mediated by a different site not shared by NS_{NJ}.

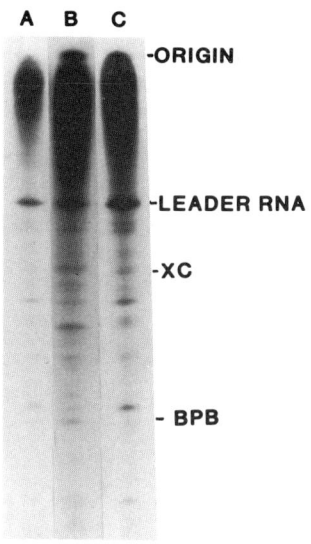

Fig. 4. Effect of NS_{IND} in VSV_{NJ} reconstituted system. RNA synthesis was carried out using template $N-RNA_{NJ}$ and (A) L_{NJ}(#2); (B) L_{NJ}(#2) + NS_{NJ} (#1); (C) L_{NJ}(#2) + NS_{IND}; products analysed by electrophoresis in a 20% polyacrylamide gel.

SPECIFICITY OF INTERACTION OF M PROTEIN WITH HETEROLOGOUS RNP.

It has been previously shown that the matrix protein (M) interacts strongly with RNA at low ionic condition with virtual cessation of RNA synthesis in vitro (13). To study whether this interaction was serotype specific, we purified a mixture of G and M proteins from VSV_{IND} and VSV_{NJ} and studied their effect on RNA synthesis in heterologous systems. As shown in Table 1, M_{IND} was capable of inhibiting transcription efficiently (90%) when added to RNP_{NJ}, although at a slightly higher concentration compared to homologous inhibition. Similarly, M_{NJ} interacted with RNP_{IND} although at a lower efficiency (60%). These results indicate that the M protein of both the serotypes share homologous regions that are specifically involved in binding to the RNP and inhibiting transcription.

TABLE 1

Inhibition of Transcription in Vitro by M Protein

	M_{IND}[1] (µg)			M_{NJ} (µg)	
	10	30	60	30	60
RNP_{IND}	80%	97%	N.D	48%	76%
RNP_{NJ}	N.D	76%	90%	61%	75%

[1] Fraction containing M and G proteins were prepared from purified VSV_{IND} and VSV_{NJ} (13); they correspond to M_{IND} and M_{NJ}, respectively. Indicated amounts were added to purified RNP fraction (10 µg) of both serotypes and inhibition of RNA synthesis in vitro was assayed as described (14). N.D; not determined.

In order to further confirm these findings, we raised monoclonal antibodies against purified M_{IND} and M_{NJ} using the procedure described previously (15). A total of 72 and 31 hybridomas for M_{IND} and M_{NJ}, respectively, were obtained. Using a micro ELISA for M protein antibody, it was found that 22 of 72 M_{IND} hybridomas reacted specifically with M_{IND} and 50 with both M_{IND} and M_{NJ}. Only one of 31 M_{NJ} hybridomas was specific for M_{NJ} and 30 reacted with both M_{IND} and M_{NJ}. A typical micro ELISA for M proteins of both the serotypes is shown in Table 2. These results indicate that M_{IND} and M_{NJ} must share common epitopes which are involved in specific interactions with the N protein.

Table 2

Micro ELISA for M Protein Antibody[1]
(O.D at 410 mµ)

M_{IND} Antibody Hybridoma #	Antigen		M_{NJ} Antibody Hybridoma #	Antigen	
	M_{IND}	M_{NJ}		M_{IND}	M_{NJ}
35	1.27	0.13	26	0.08	0.64
44	1.22	1.25	19	1.24	1.22
60	0.19	0.81	16	0.41	0.04

[1] M and G proteins were isolated from virions and M protein was further purified by Con A-Sepharose column and used to raise monoclonal antibodies according to the method described previously (15). Supernatant fluids from different hybridomas were used for micro ELISA using purified M protein as antigen and β-galactosidase conjugated anti-mouse IgG as the screening reagent.

Conclusion. Although VSV_{IND} and VSV_{NJ} are serologically distinct, considerable homology exists between the N genes, leader RNAs and some specific domains on the genome RNAs. The above results, in addition, indicate that the interaction of L and NS proteins with heterologous template is also quite specific. We have shown that L_{NJ} could complement L_{IND} only in the presence of NS_{IND} in the in vitro reconstitution system. Thus, it seems that NS_{IND} has specific domains that recognize L_{NJ}. In contrast NS_{NJ} lacks specificity to interact with L_{IND}. These results would clarify some of the previous observations, such as replication of $HR-DI_{IND}$ in the presence of VSV_{NJ} (6) and complementation of VSV_{IND} group I mutant with VSV_{NJ} (7). Nucleotide sequence determinations of $L_{IND,NJ}$ and NS_{NJ} genes would help locate regions of homology within the corresponding genes. The above results also give insight into the possible role of L and NS proteins in the transcription process. The L protein appears to be involved in the initiation process, while NS protein mediates chain elongation and continued RNA synthesis. The precise mode by which L and NS interact with each other and with the template N-RNA remains to study. The M protein of both serotypes appears to possess common regions that interact with heterologous RNPs. The complete nucleotide sequence of M_{NJ} gene and further studies with monoclonal antibodies should determine which regions in the M proteins are involved in that interaction.

REFERENCES

1. Wagner, R.R. (1975). In "Reproduction of Rhabdoviruses" (H. Fraenkel-Conrat and R.R. Wagner, eds.). p.1, Vol.4, Plenum Publishing Corp. New York.
2. Emerson, S.U., and Wagner, R.R. (1973). J. Virol. 12, 1325.
3. Abraham, G., and Banerjee, A.K. (1976). Virology 71, 230.
4. Emerson, S.U., and Yu, YH. (1975). J. Virol. 15, 1348.
5. Cartwright, B., and Brown, F. (1972). J. Gen. Virol. 16, 391.
6. Chow, J.M., Schnitzlein, W.M., and Reichmann, M.E. (1977). Virology 77, 579.
7. Repik, P., Flamand, A., and Bishop, D.H.L. (1976). J. Virol. 20, 157.
8. Legault, D., Takayesu, D., and L. Prevec (1977). J. Gen. Virol. 35,53.
9. Prevec, L., and Kang, C-Y. (1970). Nature (London) 228, 25.
10. Bishop, D.H.L., Emerson, S.U., and Flamand, A. (1974). J. Virol. 14, 139.
11. Colonno, R.J. and Banerjee, A.K. (1978). Nucl. Acids. Res. 5, 4165.
12. Franze-Fernandez, M.T., and Banerjee, A.K. (1978). J. Virol. 26, 179.
13. De, B.P., Thornton, G.W., Luk, D., and Banerjee, A.K. (1982). Proc. Nat. Acad. Sci. (U.S.A.) Vol. 17, 7137.
14. Chanda, P., and Banerjee, A.K. (1981). J. Virol. 39, 93.
15. De, B.P., Tahara, S., and Banerjee, A.K. (1982). Virology, 122, 510.
16. Voller, A., Bartlett, A., and Bidwell, D.E. (1978). J. Clinical Pathology, 31, 507.

DOES MODIFICATION OF THE TEMPLATE N PROTEIN PLAY A ROLE IN REGULATION OF VSV RNA SYNTHESIS?[1]

Jacques Perrault, Patrick W. McClear, Gail M. Clinton,[2] and Marcella A. McClure

Department of Microbiology and Immunology
Washington University School of Medicine
St. Louis, Missouri 63110

INTRODUCTION

Vesicular stomatitis virus (VSV) RNA synthesis, as well as that of other negative-strand RNA viruses takes place on ribonucleoprotein (RNP) templates, and is regulated via specific termination events (1). What controls termination is crucial for understanding the overall regulation of virus growth in lytic and persistent infections. VSV transcription templates contain six major termination sites corresponding to the ends of the leader gene and the five virus structural genes. To achieve the first step in replication, termination at each of these sites must be suppressed to allow genome-size plus strand synthesis. This switch is dependent on new viral protein synthesis in vivo (2). However, the coupling of replication to translation can be circumvented in vitro using purified virus in the presence of base analogues (3,4,5). Kolakofsky and colleagues have recently provided some evidence for a model whereby suppression of termination in vivo may be achieved by binding of soluble N protein to nascent leader RNA chains (6,7). The initiation of the helical RNP assembly process is thus viewed as being sufficient to prevent termination at the leader RNA sites. Our recent studies with a unique class of VSV mutants, called pol R, suggest that antitermination at leader sites may be governed by modification of template-associated N protein subunits (8,9,10). We shall briefly summarize our results here and discuss a model incorporating these ideas.

---

[1]This work was supported by a research grant (AI14365) and a Research Career Development Award from the NIH to J.P. and a NSF grant (PCM-8208155) to G.M.C.
[2]Department of Biochemistry, Louisiana State University Medical Center, New Orleans, Louisiana 70112.

PROPERTIES OF POL R VSV MUTANTS

We have documented several changes in the nature of transcripts synthesized in vitro by purified pol R VSV particles as compared to wild-type virus. Most important among these are 1) a large increase (~7X) in readthrough transcripts spanning the plus strand leader-N gene junction (10), 2) a quantitatively similar increase in readthrough of the minus strand leader RNA termination site (measured on DI templates) (8,10), 3) an increased resistance to inhibition by replacement of ATP by AMP-PNP during initiation of synthesis (Perrault and McClear, in preparation), and 4) a decreased sensitivity to virion M protein inhibition (McClear and Perrault, unpublished). Other minor changes include 1) an approximately 60% decrease in the accumulation of plus strand leader RNAs (9,10), 2) a two-fold increase in initiation at the 3' end of the standard template relative to total RNA synthesized (10), and 3) a significant increase in the size and heterogeneity of poly(A) tails on otherwise normal mRNAs which are also abundantly synthesized (9). The importance of the readthrough property of pol R VSV is enhanced by the fact that this phenomenon is specific for leader RNAs since no such increase was detected for the junctions between structural genes (9). It should also be emphasized that the phenotype of pol R VSV represents a major change in behaviour for the virus polymerase at least in vitro. All productive VSV RNA synthesis apparently begins at the 3' end of the templates (11), and mRNAs are produced in decreasing amounts as a function of the distance from this 3' end. The leader RNA termination sites, only 46-48 nucleotides from the entry sites, are therefore the most frequent termination signals encountered by the polymerase.

Wild-type virus also reads through the leader RNA termination sites in vitro under standard conditions, but the frequency of such readthrough (~12%) versus pol R1 VSV (~83%) clearly suggests a shift from a transcription mode to a replication mode in the latter (10). At the present time, we do not understand why most pol R in vitro readthrough transcripts terminate heterogeneously only 300 to 800 bases from the 3' terminus, instead of continuing synthesis until the end of their templates. This perhaps indicates that a second function is needed for sustained elongation beyond the critical leader junction. This second function would then presumably be operative when elongation is carried out in the presence of base analogues since full-length plus strands are synthesized in vitro under these conditions by wild type virus (3).

The properties of the two independently isolated pol R1 and pol R2 viruses are qualitatively similar but show distinct quantitative differences (Perrault, unpublished). Although these mutants were obtained following cycles of heat inactivation of virus lysates, they are not heat resistant compared to wild-type. In vivo, their growth is somewhat restricted (~10-50% of wild type yields). Viral RNAs in mutant-infected cells show characteristics similar to those seen in vitro but less dramatically so. We have recently detected unencapsidated pol R readthrough transcripts in vivo and are currently examining their properties (Perrault, unpublished).

From our point of view, the most unexpected aspect of the pol R mutants concerns the mapping of the mutation responsible for this interesting phenotype. In vivo construction of chimeric DI composed of wild-type RNA template and pol R1 proteins, or mutant template RNA and wild-type proteins, allowed us to test whether the readhtrough phenotype was due to sequence changes at the sites of leader RNA termination, or due to changes in a protein(s) involved in termination. The answer was clearly a protein change (8,10). We then attempted in vitro reconstitution of polymerase activity using purified RNP templates (standard and DI) and solubilized polymerase proteins (L+NS). Again, the answer was clear. Readthrough of plus strand or minus strand leader termination sites partitioned with the highly purified RNP template fractions (10). These results very strongly implicate the N protein of pol R viruses in the readthrough activity.

Despite the limitations inherent in studying non-conditional mutants, we were able to independently confirm that the properties of the pol R mutants are due to an N protein alteration. We found a qualitatively and quantitatively similar isoelectric charge shift in the N protein of both pol R mutants (10). Additional comparisons with wild-type virus revealed only two other protein changes, one in M for pol R1, and one in NS for pol R2. These results indicate that the N protein charge shift common for both independent mutant isolates is in all likelihood responsible for most or all of phenotypic changes seen in pol R virus.

TEMPLATE MODIFICATION MODEL OF VSV RNA SYNTHESIS

The specificity and magnitude of the changes in termination of RNA synthesis seen in pol R viruses argue that an important element controlling the balance between transcription and replication has been altered. Whether N protein alone, or N protein in concert with other viral and/or host determinants, controls this switch is not yet clear. What our studies point out, however, is that assembled VSV RNP templates, containing only N protein can carry the information necessary to direct the polymerase

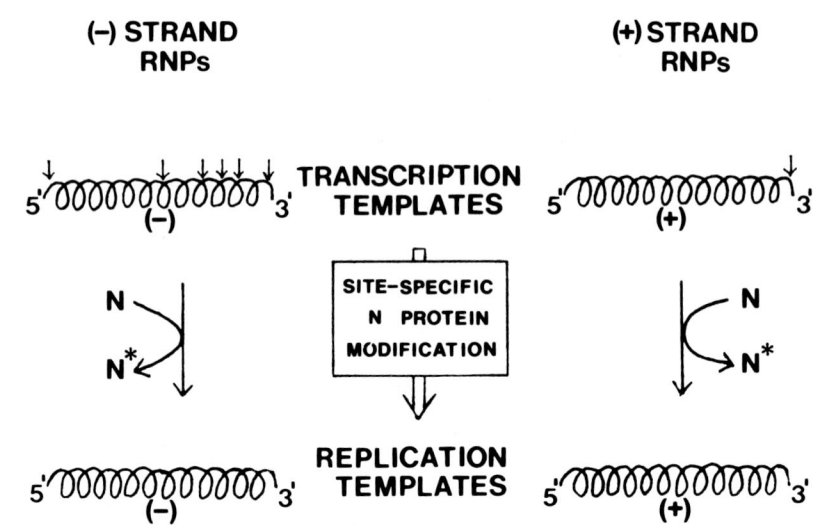

FIGURE 1. A model for the control of VSV transcription and replication by site-specific N protein modification. The six termination sites on minus strand RNP templates (arrows) correspond to the ends of the leader gene and the five structural genes. Only one termination site, that giving rise to the minus strand leader RNA transcript, is found on the plus strand standard RNP template (also found on both plus and minus strand DI RNP templates). We propose that one or more N protein subunits at these sites are specifically modified (N>N*) in vivo before the viral polymerase can replicate or readthrough the templates.

towards a transcription mode (termination) or a replication mode (antitermination). The shift towards the replication mode in the pol R mutant RNPs, solely as a result of an N protein mutation, in turn suggests that the simplest way to carry out this switch is by modification of the template protein. We have therefore recently proposed that the VSV RNP transcription templates, normally found in virions (both standard and DI), are modified in vivo to replication templates whose structure allows readthrough of the leader termination sites (10). We further suggested that this change may involve a post-translational modification of the N protein (see figure 1). This proposal, of course, does not rule out a need for additional factors to achieve successful and sustained replication in vivo such as a pool of soluble N protein capable of initiating assembly of nascent replication products. It may be that concurrent assembly is normally needed to suppress termination sites beyond that of the leader gene on the minus strand genome template.

How can we explain the pol R mutant phenotype in the context of this model? The difference in readthrough activity between wild-type and mutant virus is not so much a qualitative difference but a quantitative difference (~12% vs 83% readthrough). This suggests that both types of viruses contain the two postulated forms of RNPs in different ratios. The amino acid change in the pol R mutant N protein could either affect the rate and/or extent to which the conversion of transcribing RNPs to replicating RNPs occurs in vivo, or affect the assembly mechanism which selects transcription-type RNPs for packaging in virions. It should be noted here that structural differences between virion and cellular RNPs, based solely on N-N or N-RNA interactions, have been documented before (12).

Our studies have not yet identified what kind of N protein modification could be involved. The multiple isoelectric species of N protein we and other have observed in wild-type VSV (10,13) strongly suggest that this protein is post-translationally modified. However, all of the four subspecies we detected were shifted to the anode in the pol R viruses, and the relative difference in charge between these species within wild-type or mutant viruses remained the same (10). So, the putative charged amino acid substitution in the mutant N protein does not appear to affect this particular post-translational modification, at least superficially. Some other modification may therefore be involved in regulating the activity of the templates. One possibility is that the switch from transcription to replication RNPs invokes changes in only a few of the 1500-2000 N protein subunits on each template. This idea of site-specific modification of the template (see figure 1) transpired from results of our in vitro reconstitution experiments. As mentioned previously, readthrough products synthesized on pol R1 standard RNP templates appear to terminate non-specifically, i.e., at a large number of sites ~300 to 800 bases from the 3' end. Curiously, this very heterogeneous pattern of termination is reproduced (band for band on acrylamide gels) whether we examine transcripts from unfractionated pol R1 standard virus, or reconstituted reactions with either homologous pol R1 solubilized enzyme or heterologous wild-type solubilized enzyme (10). Thus, the pol R1 RNP standard templates behave as if the termination sites are already "tagged" in the assembled virion (the same phenomenon is also seen with pol R1 DI templates and to a more limited extent with wild type standard and DI templates). Such behaviour could be explained by modifications of N protein subunits at selected sites along the template which act as termination signals. The abnormal pol R mutant N protein might then allow "illegitimate" or incorrect modification of sites on its RNP templates.

Lastly, we have not yet addressed the question of which entity might be responsible for N protein modifications. It seems likely to us that this function is carried out normally by a virally-coded polypeptide. The aforementioned observation that full-size plus strands can be synthesized in the absence of new viral protein synthesis with purified virions in the presence of base analogues, suggests that this is the case and that the modification can also be carried out in vitro under special conditions. The NS phosphoprotein and the M protein would seem to be logical candidates for the modifying function since they are known to bind directly to the RNP template and, in the case of NS, at sites close to both plus and minus leader RNA termination sites (14,15,16). Perhaps, the modification can also be carried out or complemented by host cell factors under different conditions. A role for host cell factors in VSV replication is well established (17) and a direct or indirect effect in the reaction responsible for the switch from transcription to replication is a distinct possibility.

ACKNOWLEDGMENTS

We thank Joanne Shuttleworth and Aileen Torres for excellent technical assistance.

REFERENCES

1. Ball, L.A., and Wertz, G.W. (1981). Cell 26, 143-144.
2. Wertz, G.W., and Levine, M. (1973). J. Virol. 12, 253-264.
3. Testa, D., Chanda, P.K., and Banerjee, A.K. (1980). Proc. Natl. Acad. Sci. USA 77, 294-298.
4. Chanda, P.K., Yong Kang, C., and Banerjee, A.K. (1980). Proc. Natl. Acad. Sci. USA 77, 3927-3931.
5. Chanda, P.K., Roy, J., and Banerjee, A.K. (1983). Virology 129, 225-229.
6. Leppert, M., Rittenhouse, L., Perrault, J., Summers, D.F., and Kolakofsky, D. (1979). Cell 18, 735-748.

7. Blumberg, B.M., Giorgi, C., and Kolakofsky, D. (1983). Cell 2, 559-567.
8. Perrault, J., Lane, J.L., and McClure, M.A. (1980). In "Animal Virus Genetics", ICN-UCLA Symposia on Molecular and Cellular Biology, Vol. XVIII (B. Fields, R. Jaenisch, and C.F. Fox, eds.), pp. 379-390. Academic Press, New York.
9. Perrault, J., Lane, J.L., and McClure, M.A. (1981). In "The Replication of Negative Strand Viruses" (D.H.L. Bishop and R.W. Compans, eds.) pp. 829-836. Elsevier North Holland, New York.
10. Perrault, J., Clinton, G.M., and McClure, M.A. (1983). Cell, in press.
11. Emerson, S.V. (1982). Cell 31, 635-642.
12. Naeve, C.W., Kolakofsky, C.M., and Summers, D.F. (1980). J. Virol. 33, 856-865.
13. Hsu, C.-H., and Kingsbury, D.W. (1982). J. Virol. 42, 342-345.
14. Keene, J.D., Thornton, B.J., and Emerson, S.U. (1981). Proc. Natl. Acad. Sci. USA 78, 6191-6195.
15. Isaac, C.L., and Keene, J.D. (1982). J. of Virol. 43, 241-249.
16. Pinney, D.F., and Emerson, S.U. (1982). J. of Virol. 42, 897-904.
17. Pringle, C.R. (1978). Cell 15, 597-606.

BINDING STUDIES OF NS1 AND NS2 OF VESICULAR STOMATITIS VIRUS[1]

Paul M. Williams and Suzanne Urjil Emerson

Department of Microbiology
University of Virginia School of Medicine
Charlottesville, Virginia 22908 U.S.A.

INTRODUCTION

Vesicular stomatitis virions (VSV) are composed of five virally coded proteins: N, NS, M, G, and L. Of these, N, NS, and L are all needed for transcription (1). The template for transcription contains the virion's negative strand RNA tightly encapsidated by N protein. This template, when combined with column-purified L and NS in the presence of nucleotides, $MgCl_2$ and D.T.T., is capable of in vitro transcription.

The NS protein is a phosphoprotein which exists in several different phosphorylated forms. DEAE-cellulose chromatography used by Kingsford and Emerson (2) separates virion NS into two species called "NS1" and "NS2." The NS2 is more highly phosphorylated than NS1. Each of these NS species when run on urea-SDS polyacrylamide gels separates into two closely migrating bands (TOP and BOTTOM) which differ in the extent of phosphorylation. The NS BOTTOM is the more highly phosphorylated species. Urea-acetic acid gels and isoelectric focusing gels also separate species of NS differing in phosphorylation (3,4).

VSV is thought to encode proteins capable of functions necessary for its own replication and transcription including initiation, elongation, capping, and methylation enzymes. Since there are only five proteins to supply all

---

[1] This work was supported by Public Health Service Grant 5R01A111722-10 from the National Institute of Allergy and Infectious Diseases. P.M.W. was supported by Public Health Service training grant 5T32-CA09109 from the National Cancer Institute.

of these functions, the assumption is made that some if not all of the proteins play a multifunctional role in the life cycle of the virus. Because the NS protein exists in different phosphorylated states, it is a good candidate for such a multifunctional protein.

Kingsford and Emerson (2) have examined transcriptional activity of NS1 and NS2 and found only NS2 is active in in vitro transcription. Kingsbury (5) employed bacterial alkaline phosphatase to dephosphorylate NS on virion nucleocapsids. These treated nucleocapsids were found to lack transcriptional activity. Several approaches follow which further examine this correlation between the state of phosphorylation of NS and its activity in transcription.

RESULTS

L protein is unable to bind to template in the absence of NS protein (6), but it has not been shown whether NS1, NS2, or both are required for L binding. Two simple explanations for the inability of NS1 to function in a reconstituted transcription system are that NS1 does not bind to template or is unable to promote the binding of L protein. Therefore, VSV virions were examined to determine if both NS1 and NS2 are bound to the nucleocapsid core. The virions were solubilized in low salt and detergent to remove G and M proteins and any other proteins which were unbound or loosely associated with nucleocapsids. The nucleocapsids were separated from the free proteins by centrifugation. These purified nucleocapsids were then treated with high salt, 3.5 M urea, and detergent to release L and NS. Again free proteins were separated from the template by centrifugation. These freed proteins were analyzed for NS1 and NS2 content (see Fig. 1). As is seen, both NS1 and NS2 are associated with nucleocapsid cores purified from virions.

Next, virions were solubilized with high salt, 3.5 M urea, and detergent, and NS1 and NS2 were purified by the method of Kingsford and Emerson (2). Each protein was then assayed for its ability to bind to purified template (N and RNA) (see Fig. 2). Both NS1 and NS2 rebind to template. The rebinding of both NS species reached saturation at approximately the same level. This level of saturation is very close to that of whole NS (which is a mixture of NS1 and NS2) on intact virions (7).

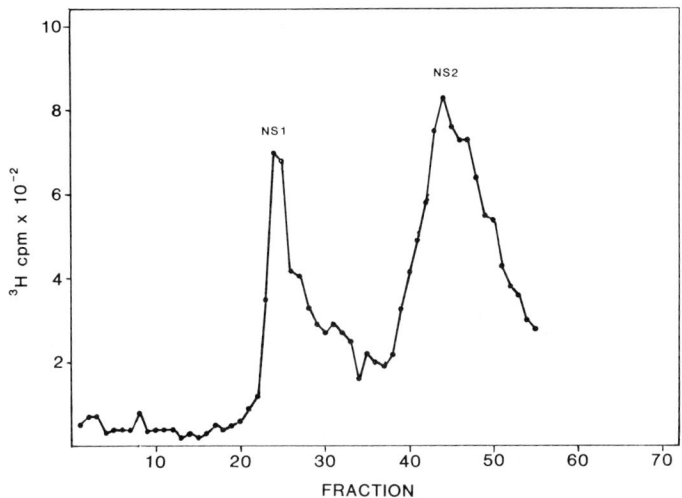

Figure 1. DEAE cellulose chromatography of NS protein. VS virions were solubilized in 0.25 M NaCl, 1% Triton X-100, 10 mM Tris 7.2, and 5 mM D.T.T. Nucleocapsids were separated from solubilized proteins by centrifugation in a SW50.1 rotor for 3 hours at 40K rpm. The nucleocapsids were spun onto a 200 µl pad of 100% glycerol overlaid by 1 ml of 20% glycerol. The nucleocapsid band was removed along with the 100% glycerol and diluted to 12.5 mls with 10 mM Tris 7.2. This was added to 12.5 mls of high-salt, urea-solubilizing solution (1.44 M NaCl, 7 M urea, 1% Triton X-100, 20% glycerol, 20 mM Tris 7.2, and 10 mM D.T.T.). Again the mixture was spun in a SW50.1 rotor for 3 hours at 40K rpm. This time the supernatant was pooled and dialyzed overnight against 0.1 M NaCl column wash (0.1 M NaCl, 0.2% Triton, 25% glycerol, 0.2 mM D.T.T., and 10 mM Tris 7.2). The DEAE column chromatography was performed as published by Kingsford and Emerson (2). 0.5 ml fractions were collected, and 50 µl fractions were analyzed by liquid scintillation spectrophotometry.

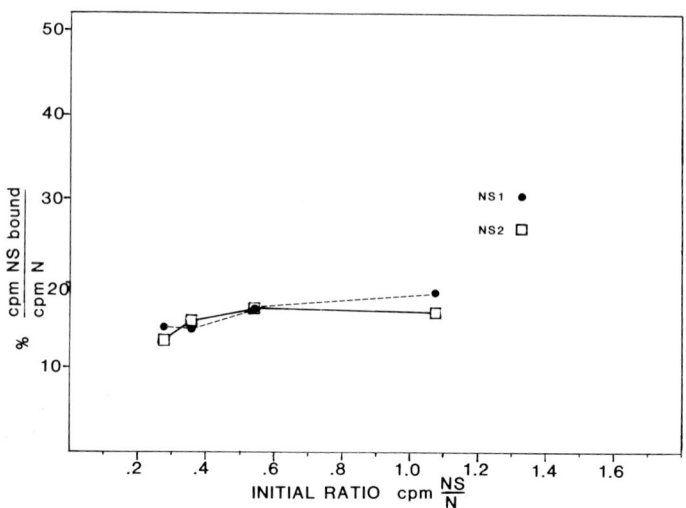

Figure 2. NS1 and NS2 were purified as outlined by Kingsford (2). Peak fractions were pooled and dialyzed against 0.1 M NaCl column wash. Purified template was prepared as described by Mellon and Emerson (7). An equal number of ^{3}H leucine counts of NS1 or NS2 were added to a reaction mix (yielding a final concentration of 0.057 M NaCl, 2% Triton, and 10 mM Tris 7.2). To this were added decreasing amounts of template. Rebinding was allowed to occur for 30 minutes at 31°C. The reaction was then layered onto 200 µl of 30% glycerol overlaid on 20 µl of 100% glycerol. This was spun for 2 hours at 40K rpm in a SW50.1 rotor. The supernatant containing unbound proteins and the 30% glycerol were pulled and discarded. The template plus rebound NS was recovered from the 100% glycerol pad and was pooled and diluted with PAGE sample buffer without glycerol and run on 12.5% polyacrylamide gels (see Kingsford and Emerson [2]). The gels were fixed and stained. The NS and N bands were cut out and eluted with NCS tissue solubilizer from Amersham and quantitated by liquid scintillation spectrophotometry.

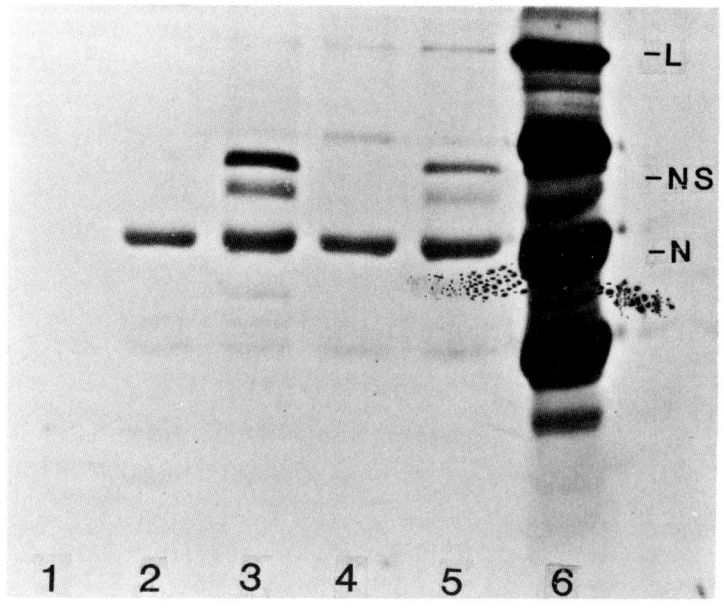

Figure 3. Autoradiograph of PAGE of rebound L with NS1 or NS2 and template. Rebinding reactions were set up as in Figure 2 but the final salt concentration was 0.3 M NaCl. After allowing rebinding to proceed for 1 hour on ice, the reaction was spun as in Figure 2 (the 30% glycerol contained 0.3 M NaCl). The templates and rebound proteins were dialyzed against 10 mM Tris 7.4. The samples were lyophilized to dryness and resuspended in sample buffer and run on 12.5% SDS polyacrylamide gels. The gels were fixed, stained, and enhanced before drying and autoradiography. Lane 1 is a reaction of L alone. Lane 2 is L + template. Lane 3 is whole virion NS (NS1 and NS2) + L + template. Lane 4 is NS1 + L + template. Lane 5 is NS2 + L + template. Lane 6 is VSV. All proteins are labeled with S^{35} methionine.

Since both NS1 and NS2 are bound to nucleocapsids in virions and both can rebind to purified template, templates with either NS1 or NS2 bound to them were examined to see whether one or both species were required for L protein binding. Reconstitutions were set up with purified L, NS, and templates. Rebinding of L occurred whether NS1 or NS2 was used (see Fig. 3).

SUMMARY

The state of phosphorylation of a protein can affect protein charge and/or conformation and thus alter protein-protein or protein-nucleic acid interactions. The difference in phosphorylation of NS1 and NS2 could affect charge and/or conformation of NS and result in NS2 being the only transcriptionally active form _in vitro_. We have attempted to find the reasons why NS2 is active in transcription and NS1 is not.

We have shown that both NS1 and NS2 can rebind to template. They both rebind to the same level of saturation as NS in intact virions. Also we found NS1 and NS2 both can promote the binding of L. Therefore, the functional difference between NS1 and NS2 does not reflect the inability of NS1 to interact with either the template or L protein.

Footprinting experiments from Jack Keene's lab showed NS protein interacts with the virion RNA between bases 16 and 30 from the 3' end (8). Sue Emerson has shown that the entry site for the transcriptase is at the 3' end of the virion. These data suggest that NS interaction specifically with these bases might constitute an active promoter region. We therefore are attempting to footprint NS1 and NS2 bound to templates to see if NS2 but not NS1 interacts with these same bases.

In vitro data indicate that the template is the limiting transcription factor in virions. In other words, an excess of polymerase components (L and NS) are packaged in virions. We can postulate that, although NS1 and NS2 are bound to templates in virions and both can rebind to templates _in vitro_, this binding in the case of NS1 is a nonspecific, transcriptionally inactive packaging phenomenon. Perhaps only NS2 is capable of specific and transcriptionally active binding.

ACKNOWLEDGMENTS

We thank Todd Green, Deborah Pinney, and Karen Williamson for their helpful discussion.

REFERENCES

1. Emerson, S.U., and Yu, Y.-H. (1975). Journal of Virology 15, 1348-1356.
2. Kingsford, L., and Emerson, S.U. (1980). Journal of Virology 33, 1097-1105.
3. Clinton, G.M., Burge, B.W., and Huang, A. (1978). Journal of Virology 27, 340-346.
4. Hsu, C.-H., and Kingsburg, D.W. (1982). Journal of Virology 42, 342-345.
5. Kingsbury, D.W., and Hsu, C.-H. (1981). In "The Replication of Negative Strand Viruses" (D.H.L. Bishop and R.W. Compans, eds.), pp. 821-827.
6. Mellon, M.G., and Emerson, S.U. (1978). Journal of Virology 27, 560-567.
7. Pinney, D.F., and Emerson, S.U. (1982). Journal of Virology 42, 897-904.
8. Keene, J.D., Thornton, B.J., and Emerson, S.U. (1981). Proc. Natl. Acad. Sci. 78, 6191.

DYAD-SYMMETRY IN VSV RNA MAY DETERMINE THE
AVAILABILITY OF THE RNA-POLYMERASE BINDING SITE.[1]

Anneka C. Ehrnst[2] and Alice S. Huang

Department of Microbiology and Molecular Genetics
Harvard Medical School
and
Division of Infectious Diseases
Children's Hospital Medical Center
Boston, Massachusetts 02115

VSV RNA SYNTHESIS

After infection of cells, the VSV polymerase initiates RNA synthesis at the 3' end of the genomic negative strand. This may result either in replication of RNA or in transcription. Subsequently the polymerase generates more negative strands by interaction with the 3' end of the complementary positive strand. In addition to the specific polymerase binding sites near the termini (1, 2), sequences further into the genome are thought to be important for the efficiency of replication (3, 4). This is based upon studies on the replication of defective interfering (DI) particles, where different sequence rearrangements result in their amplification at the expense of standard virus (5).

The nature of polymerase-RNA interactions during the replication of VSV is unknown. Aside from sequence specificity, secondary structures may be implicated. Possible stable or quasi-stable secondary structures at the 3' ends of the positive and the negative RNA strands were sought by using computer analysis with special reference to the presentation of the polymerase binding site. The secondary structures are discussed in relation to possible binding of the N, NS and M proteins of VSV which modulate polymerase activity during RNA synthesis. The possible secondary RNA structures at the 3' ends of three related defective interfering particles are also presented. Although our interpretations are based mainly on

---

[1] Supported by a research grant NIH AI 16625. AE was supported by the National Multiple Sclerosis Society (No. FG531-A-1) and by the Fogarty International Center, (NIH FO5 TWO 2968-0151), as well as by the Swedish Medical Research Council, the Swedish Society for Medical Sciences, the Swedish Multiple Sclerosis Society, and the Karolinska Institutet.
[2] Present address: The National Bacteriological Laboratory S-105 21 Stockholm, Sweden.

theoretical considerations, such sequence analysis points to the necessity for quantitative binding studies between VSV RNA and proteins.

SOURCES OF VSV SEQUENCES

All RNA sequences are of the Indiana serotype of VSV determined by McGeoch and Dolan (8) and by Schubert et al. (9). Sequences from defective interfering (DI) particles were from Schubert et al. (3) as confirmed by Rao and Huang (4).

These DI sequences relate to standard viral RNA sequences in the following way. The first 60 nucleotides at the 3' end of the negative strand of DI 011 are identical to the 3' end of the positive strand of standard VSV. DI 0.45 (also known as DI-T(L)) and DI 611 have shorter 3'-end sequences (48 nucleotides) that are identical to the 3' ends of the positive strands. DI-T contains even a shorter portion of complementary sequences since it has only 45 nucleotides at its 3' end that are identical to sequences on the positive strand. All of the DI RNAs retain a portion of the L gene.

COMPUTER ANALYSIS

The SEQ program from the Molgen group at Stanford (10) was used for the computer analysis. Parameters were initially set to maximize the yield of dyad-symmetries, or stems, and then narrowed to yield the most significant ones based on estimates of the stability of these secondary structures according to Tinoco et al. (11). The most stable stem and loop structures give the lowest free energy values, ΔG, in kilocalories per mole.

SECONDARY STRUCTURES AT THE 3' END OF THE NEGATIVE STRAND RNA OF STANDARD VSV.

Since secondary structure might influence replication, possible stem and loop structures were looked for at the 3' end of VSV RNA. Figure 1 shows the secondary structures that might form with the first 77 nucleotides at the 3' end of the minus strand. Significant secondary structures were not obtained with only the first 50 nucleotides. There were two principal hairpin structures _a_ and _b_. Each contained loops of 12 nucleotides. The one with the longest stem had two possible stem endings, _a_ and _a'_. Structure _a_ ran from nucleotide 10 through nucleotide 72 while structure _a'_ started with the first nucleotide. The _a_ structures were slightly more stable than those of _b_.

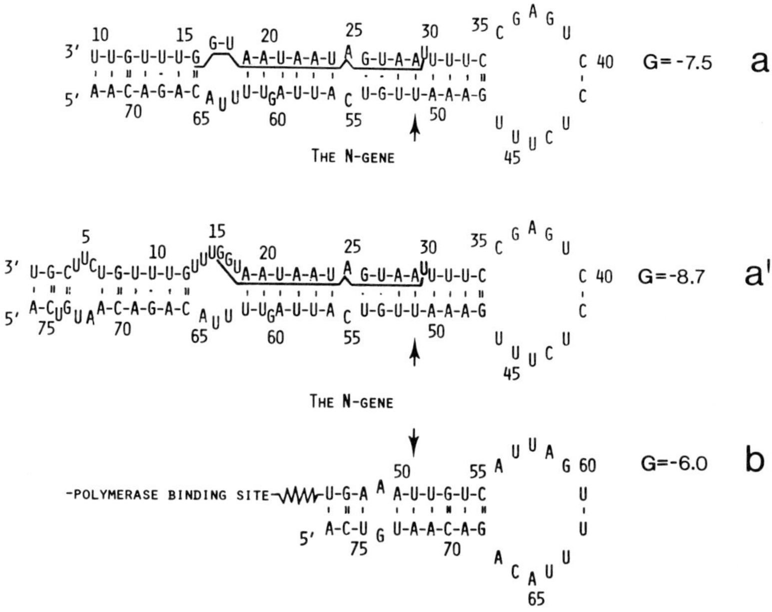

DYAD-SYMMETRY IN VSV RNA   89

```
        10       15   G-U  20          25          30      35  C G A G U
    3' U-U-G-U-U-U-G      A-A-U-A-A-U    G-U-A-A U-U-U-C              C 40     G=-7.5    a
        | | | | | | |  "  | | | | | | |  | | | |  - | | | |  | | | |  |
    5' A-A-C-A-G-A-C  A  U U-U G A-U-U-A  C U-G-U-A-A-A-G              C
        70         65  U   60         55 ↑       50                 U U C U
                                                                      45
                        THE N-GENE
```

```
         5          15                                                G A G
              10         U U G G           20          25         30     35 C    U
    3' U-G-C U U C-G-U-U-U-G U      A-A-U-A-A-U    G-U-A-A U-U-U-C              C 40     G=-8.7    a'
        |  " |"  |||||| "    |||||||| -|||| ||||
    5' A-C G U A A-C-A-G-A-C A U U-U G A-U-U-A  C U-G-U-A-A-A-G                  C
        75     70         65    60          55 ↑          50                 U U C U
                                                                                45
                        THE N-GENE
                           ↓
                                                        50         55  A U U A G 60          G=-6.0    b
    -POLYMERASE BINDING SITE-WWW-U-G-A A A-U-U-G-U-C                        U
                                | " | |   | | " | |                         U
                           5' A-C-U G U-A-A-C-A-G                          A
                              75         70                                 C A U
                                                                             65
```

Figure 1. Possible secondary structures formed by the 3' end of the negative strand RNA of standard VSV. The polymerase binding site (1) is indicated by a straight line in a and a'.

All three structures failed to place the polymerase binding site in a hairpin loop. Rather, the site, as indicated by a dark line, was enclosed within stems of a and a', or free, as in b. Note that the hexamer 3'UAGUUU 5' was also part of the stem structure in a and a' but that the configuration in structure b presented the hexamer in the loop.

SECONDARY STRUCTURES AT THE 3' END OF THE POSITIVE STRAND RNA OF STANDARD VSV.

Figure 2 shows three mutually exclusive hairpin loop formations c, d, and e formed by the first 55 nucleotides at the 3' end of the positive strand. Two of them, c and e, contained the polymerase binding site in a loops of 33 and 25 nucleotides, respectively. The smallest hairpin loop, d, had the least significant G value. Here the polymerase binding site was free and not in the loop or stem. Note that the loop closings contained at least one G-C pair. The hexamer 3' UAG-UUU 5' was free as a single strand in structures c and d but only partially in e.

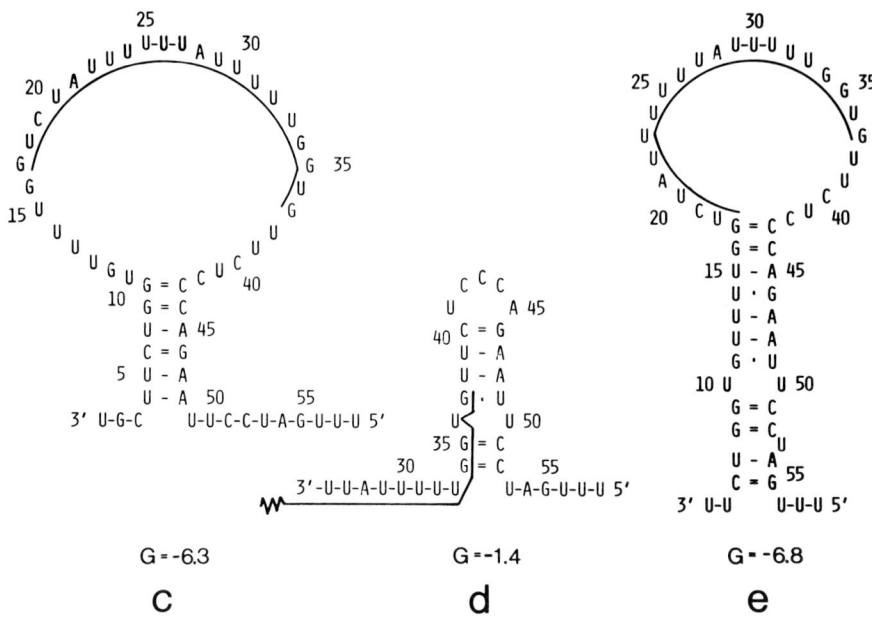

Figure 2. Mutually exclusive possible secondary structures formed by the 3' end of the positive strand of standard VSV. The polymerase binding site (2) is indicated by a straight line.

SECONDARY STRUCTURES OF THE 3' END OF THE NEGATIVE STRAND RNA OF DI PARTICLES.

Secondary structures at the 3' ends of the RNA of VSV DI particles were similar to those of the 3' end of the positive strand RNA of standard VSV, except for variations introduced by shorter sequence identities. The RNA of DI 011 at the 3' end, identical for 60 nucleotides to the 3' end of the positive strand, formed the same secondary structures as those shown in Figure 2. The RNAs of DI 611 and DI 0.45 formed the same size loops, containing 33 or 25 nucleotides, as the positive strand of standard VSV but their stems differed (Figure 3). The RNA of DI-T had the least stable structure (Figure 4). These low, negative G values may be improved if sequences internal to the complementary termini were included in the analysis. Unfortunately, such sequences have yet to be defined.

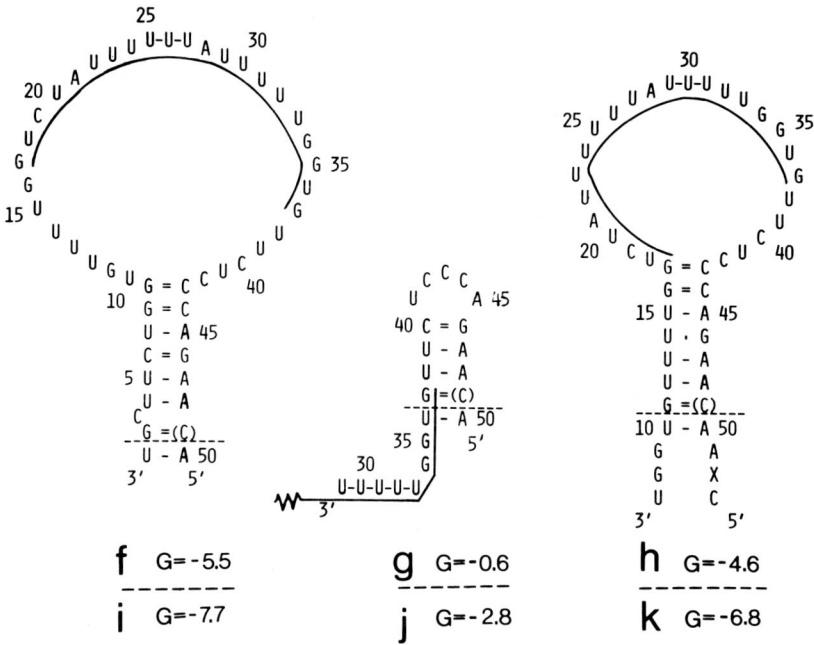

Figure 3. Possible secondary structure of the 3' end of the negative strand RNA of DI 0.45 and DI 611 particles. Structures f, g, and h are those above the respective dotted lines and are derived from DI 0.45. Structures i, j, and k extend beyond the dotted lines and are derived from DI 611.

Despite such instability, all of the RNA termini of DI particles closed their loops with at least one G-C bond similar to the loops formed by the positive strand of standard VSV. The small hairpin loop structure, m, might form, depending on what X represents. Structure i, interestingly, started from the first nucleotide and had the greatest free energy of formation among the DI RNA sequences and was comparable in stability to the secondary structures formed at the 3' end of the genomic negative strand. Despite the differences in the G values between DI RNAs and positive strand VSV RNA, the large stable loops, containing the polymerase binding sites, were common to both.

THE INFLUENCE OF ALTERED SEQUENCES ON SECONDARY STRUCTURE.

The influence of two known site mutations (12) on the secondary structures presented here was considered. At position 37, A is exchanged with U, and at position 48, A is

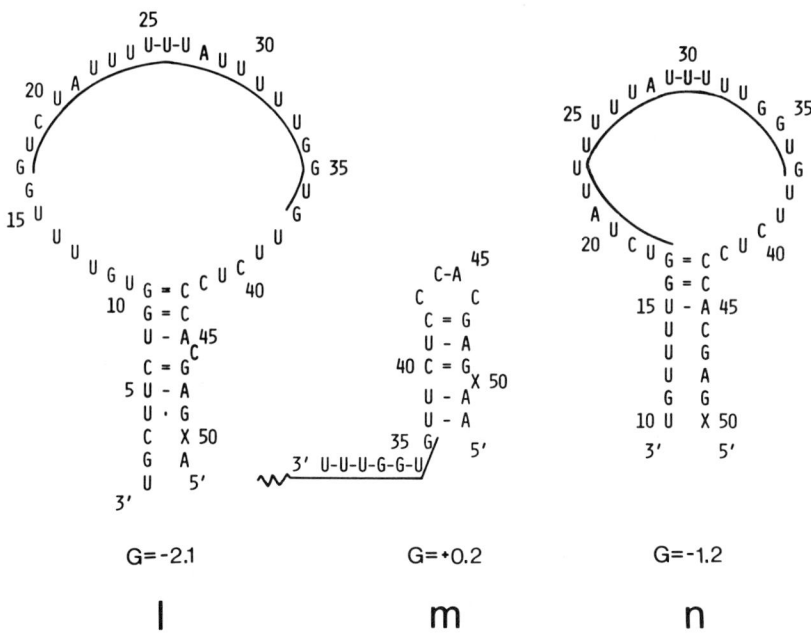

Figure 4. Possible secondary structures of the 3' end of the negative strand RNA of DI-T particles.

exchanged with C at the 3' end of the negative strand of VSV Indiana. Placing these in Figure 2 showed that the first of the mutations was located in the loop of a and a' and, thus, had no influence on the stability of the secondary structure. The mutation at position 48, however, prevented closing of the hairpin loop of structure a and a' (11, 13). Structure b remained principally unaltered ($\Delta G = -5.2$).

INTERPRETATION OF DATA.

The most stable structures of the negative strand "closed up" the polymerase binding site in a stem. In contrast the structures of the positive strand presented the polymerase binding site in an "open" hairpin loop. Therefore, these analyses support the hypothesis that there may be differential affinities in the binding of polymerase by the negative strand and its complement, resulting in the asymmetrical synthesis of plus and minus strands during VSV replication (5). Although one cannot be sure that stem structures actually prevent polymerase attachment or, alternatively, that the presentation of polymerase binding sites in loops facilitates polymerase binding, such assumptions have been suggested for the interaction of the tobacco mosaic virus RNA with its coat protein (15). Also, Peattie et al. (16) identified a "bulged" double helix in an RNA-protein contact site of Escherichia coli 5S RNA and found that this bulging appeared to be conserved throughout evolution. One of the stems of DI 0.45 and DI 611, f and i, looks very similar both in length and location to this 5S RNA ribosomal binding site.

Previously, Schubert et al. (9), using the first 125 nucleotides at the 5' end of the negative strand of VSV RNA, presented a complex flower-like secondary structure. The complementary 3' end of the positive strand could be deduced to form similar symmetrical complex structures. Unlike the "open" polymerase binding sites proposed by our studies, the complex structure formed with 125 nucleotides places the polymerase binding site in a closed stem.

For stable secondary structures of naked RNA it has been estimated that the free energy, ΔG, should be less than -15 kilo calories per mole (10). However, it is conceivable that less stable structures may alternate in a fluctuating manner, whereby signals in a non-coding region may be turned on and off. With proteins, the RNA might remain stably in one form. Such a control mechanism based on alternating secondary structures is analogous to the control of transcription termination at the attenuator of the tryptophan operon on Escherichia coli where mutually exclusive, stable base-paired structures have been identified (17, 18).

The major structural proteins that bind to VSV RNA are the RNA-binding, N protein, the large, L polymerase protein and the NS phosphoprotein. N protein is necessary for both template activity (19) and for continued genomic synthesis (20, 21). Recently, a temperature-sensitive N mutant showed that the lesion fails to affect the replication of DI RNA in the same way as standard genomic RNA (7). Therefore, both the binding to the RNA and interaction with polymerase must be considered when N protein functions are examined. A complication of such regulation of RNA synthesis is the attenuating roles of the M and the G proteins (14, 22, 23, 24).

To determine whether or not there is any validity for the structures presented here, in vitro RNA synthetic systems should be used where binding constants of the proteins and competition between synthetic oligonucleotides can be measured.

REFERENCES

1. Keene, J.D., Thornton, B.J., and Emerson, S.U. (1981). Proc. Natl. Acad. Sci. USA 78, 6191.

2. Isaac, C.L., and Keene, J.D. (1982). J. Virol. 43, 241.
3. Schubert, M., Keene, J.D., and Lazzarini, R. A. (1979). Cell 18, 749.
4. Rao, D.D., and Huang, A.S. (1982). J. Virol 41, 210.
5. Huang, A.S., Little, S.P., Oldstone, M.B.A., and Rao, D.D. (1978). In "Persistent Viruses" (G. Todaro, J. Stevens, and C.F. Fox, eds.), p. 399, Academic Press, N.Y.
6. Keene, J.D., Rosenberg, M., and Lazzarini, R.A. (1977). Proc. Natl. Acad. Sci. USA 74, 1353.
7. Rao, D.D., and Huang, A.S. (1980). J. Virol. 36, 756.
8. McGeoch, D.J., and Dolan, A. (1979). Nucleic Acids Res. 6, 3199.
9. Schubert, M., Keene, J.D., Herman, R.C., and Lazzarini, R.A. (1980). J. Virol. 34, 550.
10. Clayton, P., Friedland, P., Kedes, L., and Brutlay, D. (1981). "SEQ-sequence analysis system" Copyright 1981 by the Board of Trustees, Stanford Univeristy, California.
11. Tinoco, I., Jr., Bover, P.N., Dengler, B., Levine, M.D., Uhlenbeck, O.C., Crothers, M.D., and Gralla, J. (1973). Nature New Biol. 246, 40.
12. Rowlands, D., Grabau, E., Spindler, K., Jones, C., Semler, B., and Holland, J. (1980). Cell 19, 871.
13. Tinoco, I., Jr., Uhlenbeck, O.C., and Levine, M.D. (1971). Nature 230, 362.
14. McGeoch, D.J., Dolan, A., and Pringle, C.R. (1980). J. Virol. 33, 69.
15. Zimmern, D. (1977). Cell 11, 463.
16. Peattie, D.A., Douthwaite, S., Garrett, R.A., and Noller, H.F. (1981). Proc. Natl. Acad. Sci. USA 78, 7331.
17. Lee, F., and Yanofsky C. (1977). Proc. Natl. Acad. Sci. USA 74, 4365.
18. Stroymowski, I., and Yanofsky, C. (1982). Proc. Natl. Acad. Sci. USA 79, 2181.
19. Bishop, D.H.L., and Roy, P. (1971). J. Mol. Biol. 57, 513.
20. Perlman, S.M., and Huang, A.S. (1973). J. Virol. 12, 1395.
21. Knipe, D.M., Lodish, H.F., and Baltimore, D. (1977). J. Virol. 21, 1140.
22. Perrault, J., and Kingsbury, D.T. (1974). Nature 248, 45.
23. Clinton, G.M., Little, S.P., Hagen, F.S., and Huang, A.S. (1978). Cell 15, 1455.
24. Pinney, D.F., and Emerson, S.U. (1982). J. Virol. 42, 897.

THE EFFECT OF THE β-γ BOND OF ATP ON TRANSCRIPTION OF VESICULAR STOMATITIS VIRUS[1]

Todd L. Green and Suzanne Urjil Emerson

Department of Microbiology
University of Virginia School of Medicine
Charlottesville, Virginia 22908 U.S.A.

In vitro transcription of vesicular stomatitis virus (VSV) requires much higher concentrations of ATP than of the other three nucleotides -- 500 µM compared to 22 µM for GTP and 33 µM for CTP and UTP (1). Although it is not known why so much ATP is needed, an examination of VSV transcription suggests a number of steps where the nucleotide could be involved (2,3). Transcription by the viral-encoded polymerase is sequential and polar in the gene order 3'-leader-N-NS-M-G-L-5'. Leader RNA is a 47-nucleotide-long transcript which initiates with A and contains a high proportion of A residues, but is not polyadenylated. The monocistronic mRNAs coded for by the N through L genes all initiate with an A residue which is capped and methylated, and they are polyadenylated at their 3' end. The polymerase enters at the 3' end of the genome (4), and transcription occurs by either a stop-start or processing mechanism (3). Since leader begins with the sequence (p)ppACG (5), it has been suggested that the high K_m for ATP reflects initiation with an A residue. This interpretation is based on in vitro studies where pre-incubation of virus with CTP and high concentrations of ATP allows subsequent RNA synthesis at low concentrations of ATP (1). Imido ATP, which has a non-hydrolyzable β-γ phosphate bond (6,7), inhibits in vitro transcription as assayed by incorporation of labeled nucleotides into acid-insoluble material. However, if the virus is pre-incubated with ATP and CTP, then some RNA synthesis occurs when imido ATP is substituted for ATP.

---

[1]This work was supported by Public Health Service grant 5R01AI11722-10 from the National Institute of Allergy and Infectious Diseases. T.L.G. was supported by Public Health Service training grant 5T32-CA09109 from the National Cancer Institute.

From this it was concluded cleavage of the β-γ bond of ATP was required for initiation at the 3' end of the viral genome.

We have carried out <u>in vitro</u> transcription experiments with imido analogs. Figure 1 compares the results of transcription products of solubilized standard and DI virions synthesized with ATP, imido ATP, or imido GTP. Transcription of standard virions with ATP generates leader RNA, the 5'-N gene oligonucleotides described previously (8), and high-molecular-weight RNA indicative of messenger synthesis. However, when imido ATP is substituted for ATP, only leader RNA and the N oligonucleotides appear to be synthesized. A longer exposure of the same gel shows the presence of larger transcripts, which could be due to leader-N readthrough (9,10) or N mRNA synthesis from polymerase already at the promoter. Transcription with imido GTP results in a normal pattern of RNA products though in reduced amounts. This indicates that although the imido analogs are not recognized efficiently by the transcriptase, inhibition of mRNA synthesis is specific for the ATP analog and is not due solely to the presence of the imido group. DI virions synthesize mainly the 46-nucleotide-long DI product when either ATP, imido ATP, or imido GTP is included, as would be expected if imido ATP preferentially affects large RNA synthesis. We also see inhibition of mRNA transcription if ATP is used in very low concentrations, i.e., 25 to 50 μM (Figure 2).

Total incorporation of $[\alpha^{32}P]$-UTP into acid-precipitable RNA using imido ATP is 5% of that with ATP. This correlates with the inhibition of mRNA synthesis seen in the gel. The analog has less of an effect on DI product synthesis, about 30% of the incorporation of ATP, perhaps because only a small product is being made. Imido GTP is incorporated about five times more efficiently than imido ATP by the B virus (30% of ATP). Since DI virions use imido GTP two to three times better than B virions (60-75% of ATP), this may be a reflection on the need to cap mRNAs in the standard virus transcription reaction.

Next we directly demonstrated that imido ATP could initiate transcripts. An <u>in vitro</u> transcription reaction lacking UTP and GTP, but containing ATP and CTP, will generate large amounts of (p)ppAC representing the 5' end of leader RNA or DI particle product (4). Calf intestinal alkaline phosphatase will remove the 5' terminal phosphates from the ATP-initiated dimer but the imido group will

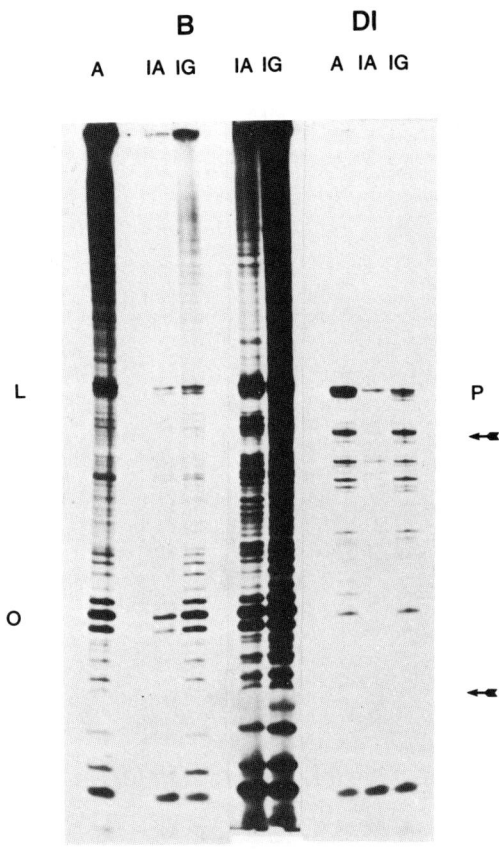

Figure 1. Autoradiograph of a 20% polyacrylamide gel with 7 M urea showing the [$\alpha^{32}P$]-UTP-labeled transcripts synthesized in a standard in vitro reaction (9) by wt (lanes A-E) and DI (lanes F-H) virions. RNA synthesized for 3 h at 31°C was phenol-extracted, ethanol-precipitated, and lyophilized. Samples were resuspended in 7 M urea with xylene cyanol and bromophenol blue as dyes before electrophoresis. Lanes D and E are the same as lanes B and C, respectively, but the autoradiograph was exposed for 7 times as long. A and F (1 mM ATP); B, D, and G (1 mM imido ATP); C, E, and H (1 mM imido GTP). The location of leader (L), DI product (P), the 11- to 14-nucleotide-long 5' N-gene oligonucleotides (O), and the dyes (arrows) are noted.

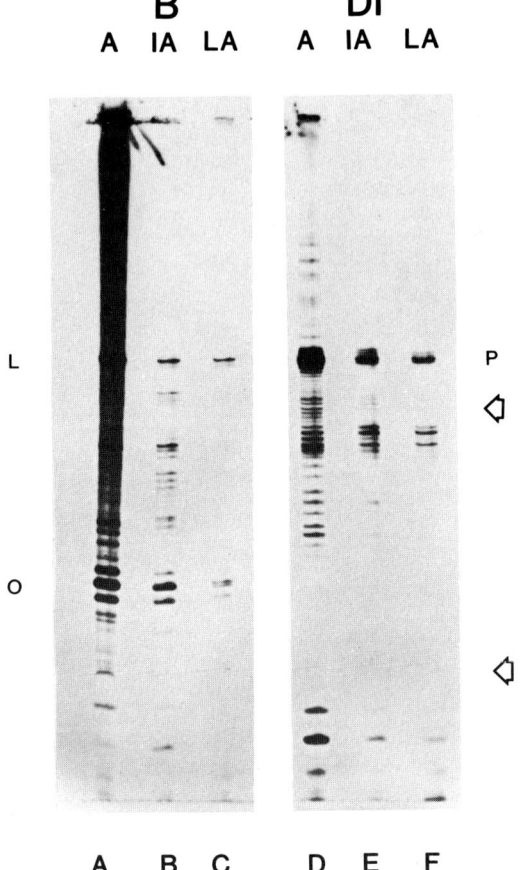

Figure 2. Autoradiograph of a 20% polyacrylamide gel with 7 M urea, showing the $[\alpha^{32}P]$-UTP-labeled transcripts synthesized in a standard in vitro reaction by wt (lanes A-C) and DI (lanes D-F) virions. RNA synthesized for 3 h at $31°C$ was phenol-extracted, ethanol-precipitated, and lyophilized. Samples were resuspended in 7 M urea with xylene cyanol and bromophenol blue before electrophoresis. A and D (1 mM ATP); B and E (1 mM imido ATP); C and F (33 μM ATP). Leader (L), DI product (P), 5' N-gene oligonucleotides (O), and dyes (arrows) are noted.

protect an imido-initiated transcript from phosphatase digestion, so the two AC dimers can be distinguished by gel electrophoresis. Figure 3 indicates that imido ATP does initiate transcription of leader and the DI product. There is some ATP contamination in imido ATP, but it is less than 0.1% (6). Transcripts containing all four nucleotides were also initiated with imido ATP (data not shown).

The results presented demonstrate that both initiation and elongation of transcripts occur with imido ATP, but under our conditions a restricted set of transcripts is synthesized. This is in apparent conflict with earlier results (1), which stated that imido ATP could elongate but not initiate transcripts. However, different methods were used to look at the products.

In view of the high K_m for ATP, there are several explanations for our results with the ATP and GTP analogs. First, the polymerase protein NS is a phosphoprotein, and its function is affected by the extent of phosphorylation (11,12). Since imido ATP is not a substrate for protein kinases, inhibition of mRNA synthesis may reflect a requirement for NS phosphorylation prior to mRNA synthesis. Alternatively, cleavage of the β-γ bond of ATP causing a conformational change in the transcriptase may be required for synthesis of the messengers. Transcription of the 5'-N gene oligonucleotides indicates imido ATP can be used for internal initiations once the transcriptase is bound to the beginning of a gene. There are four nucleotides separating the leader and N genes (13); if these nucleotides are not normally transcribed, ATP hydrolysis may be needed to translocate the transcriptase from the end of the leader gene to the start of the N gene. A third possibility is that because there are so many A residues as well as long stretches of the nucleotide in leader and the mRNAs, the polymerase has an increased chance of falling off the template if imido ATP is substituted for ATP. Therefore it is less likely transcription of the N mRNA would be completed and the transcripts would not appear in large enough numbers to be seen on gels. Polyadenylation would probably also be inhibited, and if that is necessary before the next gene, NS, is transcribed, incorporation would be drastically reduced.

The results with imido GTP show that the DI particle uses the analog more efficiently than the B particle. This may be due to a requirement for cleavage of GTP in order to cap the mRNAs, or to the fact that there are more G residues

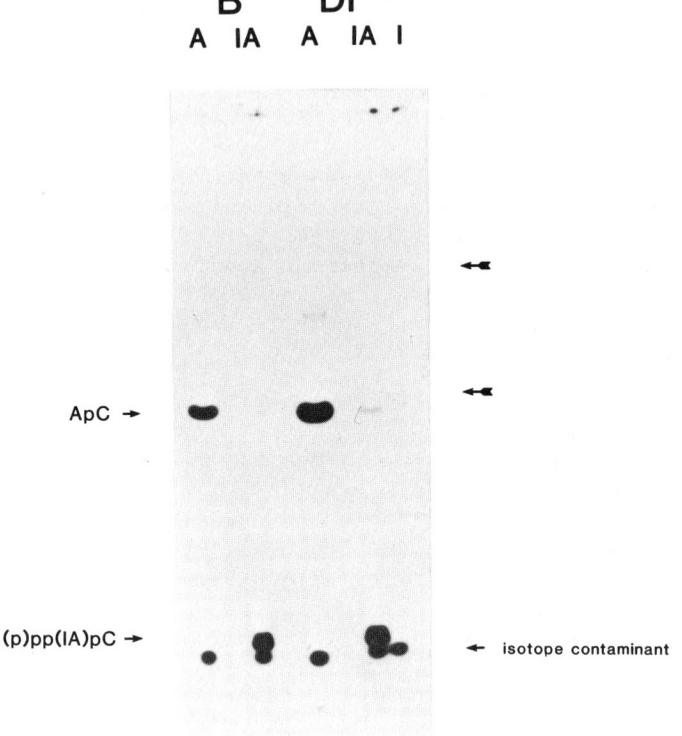

Figure 3. Autoradiograph of a 20% polyacrylamide gel showing AC dimers synthesized by B (lanes A and B) and DI (lanes C and D) virions in the absence of UTP and GTP (5). The [α^{32}P]-CTP-labeled products synthesized at 31°C for 3 h were phenol-extracted, ethanol-precipitated, and treated with calf intestinal alkaline phosphatase (0.52 U in 0.1 M Tris [pH 8.0]) for 60 min at 37°C. 7 M urea was added before electrophoresis on gels containing no urea. Bromophenol blue and xylene cyanol (arrows) were run in lanes with no samples. A and C (1 mM ATP); B and D (1 mM imido ATP); E (isotope control, no virus).

in the standard virus transcripts. We do not know. Because low ATP gives the same results as with imido ATP, it suggests high ATP concentrations are needed to transcribe the larger RNAs. But again, whether the need for ATP in large amounts is due to a specific step in the transcription process which requires cleavage of the β–γ bond, or to the presence of so many A residues in the transcripts, is not clear. It is possible that a combination of steps -- initiation, elongation, translocation, phosphorylation -- is the reason for the high K_m for ATP. At present we cannot distinguish among these possibilities.

REFERENCES

1. Testa, D., and Banerjee, A.K. (1979). J. Biol. Chem. 254, 2053.
2. Ball, L.A., and Wertz, G.W. (1981). Cell 26, 143.
3. Banerjee, A.K., Abraham, G., and Colonno, R.J. (1977). J. Gen. Virol. 34,1.
4. Emerson, S.U. (1982). Cell 31, 635.
5. Colonno, R.J., and Banerjee, A.K. (1976). Cell 8, 197.
6. Penningroth, S.M., Olehnik, K., and Cheung, A. (1980). J. Biol. Chem. 255, 9545.
8. Yount, R.G., Babcock, D., Ballantyne, W., and Ojala, D. (1971). Biochemistry 10, 2484.
9. Chinchar, V.G., Amesse, L.S., and Portner, A. (1982). Biochem. Biophys. Res. Commun. 105, 1296.
10. Testa, D., Chanda, P.K., and Banerjee, A.K. (1980). Proc. Natl. Acad. Sci. USA 77, 294.
11. Kingsford, L., and Emerson, S.U. (1980). J. Virol. 33, 1097.
12. Sokol, D., and Clark, H.F. (1973). Virology 52, 246.

INTERACTIONS BETWEEN CELLULAR LA PROTEIN AND LEADER RNAs

Jack D. Keene, Michael G. Kurilla, Jeffrey Wilusz,
and Jasemine C. Chambers

Department of Microbiology and Immunology
Duke University Medical Center
Durham, North Carolina 27710

INTRODUCTION

Antisera from patients with systemic lupus erythematosus have been useful in the isolation and identification of cellular ribonucleoproteins.[1] In general, these autoantibodies react with determinants on cell proteins or ribonucleoproteins that are associated with small nuclear or cytoplasmic RNAs. One lupus antigen that is bound to cell RNAs of known function is the La specificity. La is a 45Kd molecular weight phosphoprotein[2] that is bound to precursor transcripts of RNA polymerase III such as transfer RNA, 4.5S, 5S, 7-2, 7S, and others.[3,4] In addition, antisera from lupus patients with the La specificity also react with small viral RNPs from adenovirus (VA) and Epstein-Barr virus (EBERs) infected cells.[5] These viral RNAs are also products of cell RNA polymerase III and tend to be more stably bound to La protein.[6] One characteristic of all of these short RNA polymerase III precursor RNAs is that they tend to have additional 3' terminal uridylate residues that may influence binding of La protein.

We have recently reported that the plus strand[7] and minus strand[8] leader RNAs of vesicular stomatitis virus also react with La antisera during the infectious cycle. The extent of La protein binding to leader RNAs was at least 50% in these studies whereas less than 1% of the RNA polymerase III precursor transcripts are normally detected in association with La protein.[9] Thus, leader RNAs of VSV differ from other La bound small RNAs in that they are not products of RNA polymerase III and that large amounts of the leader RNAs present in infected cells are bound to La protein.

Steitz and coworkers have proposed that La protein is a transcription factor for RNA polymerase III, binds to precursor RNAs and is released from the mature form of the RNA.[10] It seems unlikely that VSV uses La protein as an indispensable transcription factor *per se* because leader RNA synthesis is highly efficient *in vitro*. La protein has not been detected in VSV particles and leader RNA synthesized *in vitro* can not be precipitated by La antisera (unpublished observations). Although the association of leader RNAs with La protein in VSV infected cells may be fortuitous, specificity of the association is evident in that some shorter species of the VSV (Indiana) minus strand leader RNA are not associated with La protein.[8] Thus, as few as 8 additional bases at the 3' end of minus strand leader RNA result in a dramatic increase in the binding of La protein. In this paper, we describe recent evidence that generalizes the association of La protein with leader RNAs of VSV (New Jersey) and rabies virus, and we propose a role for La protein in the replicative cycle of rhabdoviruses.

RESULTS AND DISCUSSION

Detailed studies of the association of the cellular La protein and the VSV N protein with plus strand leader RNA of VSV (Indiana) have indicated a transition from La to N protein binding as the infection progresses.[7] The La protein bound leader RNA accumulated linearly for 4 hours, plateaued until about 6 hours, and then declined. The N protein bound leader RNA, on the other hand, accumulated gradually at early times in the infection but increased dramatically by 4-6 hours in an inverse relationship to La bound leader RNA. By 16 hours post infection, all of the leader RNA detected was bound to N protein and none was detected in association with La protein. This finding suggested a role for La protein in VSV replication.[7]

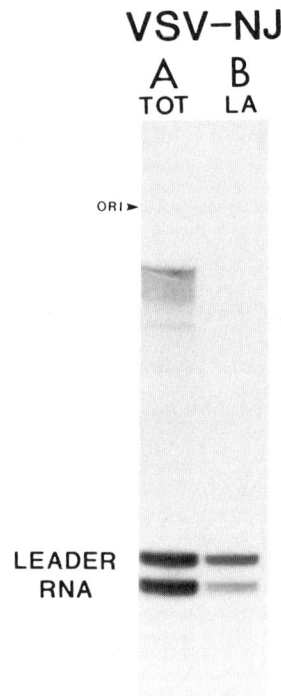

Figure 1. Immunoprecipitation of the leader RNA of the New Jersey serotype of VSV with anti-La specific serum. Lane A, probing of the total (+) leader RNA from extracts of infected BHK cells; lane B, probing of (+) leader RNA in infected BHK cell extracts after immunoprecipitation with lupus antisera of the La specificity.

We have investigated whether the New Jersey serotype of VSV shows similar interactions between the La protein and (+) leader RNA. Figure 1 shows that the plus strand leader RNA of VSV-NJ can be immunoprecipitated from infected BHK cells with La antisera.

The sequence of the VSV-NJ (+) leader RNA is similar to the Indiana serotype at the 5' end but differs significantly in the 3' half of the molecule (figure 2). We have also recently demonstrated the association of La protein with rabies virus (+) leader RNA (Kurilla, Holloway, Cabradilla, and Keene, unpublished). In these studies, a significant portion of the (+) leader RNA of rabies virus was immunoprecipitated with anti-La sera midway through the infectious cycle. Figure 2 shows a comparison of sequences in the 3' halves of the (+) leader RNAs of VSV-IND, VSV-NJ, and rabies virus. Although a consensus sequence is not readily evident, certain features are characteristic in these regions. In the 3' terminal 20 nucleotides there is a purine-rich region followed by a few terminal uridylate residues. Terminal uridylate residues have been implicated previously in the binding of La protein.[6] Probing of leader RNAs, although indirect, has shown that RNAs terminating in 3' adenylates are also bound to La protein in infected cells.[7] The presence of U residues appears to enhance binding, however.

$VSV_{INDIANA}$ ...C A U U A A A A G G C U C A G G A G A A A C U (U U)-OH

$VSV_{NEW\ JERSEY}$ ...A A U U A U U U G G C C U A G A G G G A G C U (U U)-OH

RABIES ...A G C G U C A A U U G C A A A G C A A A A A U G U-OH

Figure 2. Comparison of the 3' ends of the (+) leader RNAs of VSV-IND, VSV-NJ, and rabies virus. Predicted sites of interaction of La protein reside in the 3' half of the leader RNAs.

Figure 3 shows nucleotide sequences at the 3' end of the minus strand leader RNAs of VSV-IND and VSV-NJ. The regions encompassing the polyadenylation site of the L mRNA and the 5' end of the genomic RNAs of these viruses are structurally different.[11] Furthermore, the nature of the interaction with La protein differs between the two VSV serotypes. The Indiana serotype synthesizes a minus strand leader RNA of about 46 nucleotides in length that is the major species in infected cells.[8] The La bound form of the VSV-IND minus strand leader RNA is eight nucleotides longer and is a less abundant species in infected cells (Figure 3). The minus strand leader RNA of VSV-NJ is about 46 nucleotides long and this major species is bound to La protein (Wilusz and Keene, unpublished). Thus, the Indiana serotype utilizes the entire noncoding region of the template (up to the polyadenylylation site of the L gene) for the production of minus strand leader RNAs, while the New Jersey serotype appears to contain an additional 57 nucleotides of noncoding function that is not involved in minus strand leader RNA synthesis or La protein binding. These findings demonstrate the specificity of La protein binding as shown in figure 3 and argue for a general role of La protein in the VSV replication cycle.

INDIANA 5' pppACGAAGACCACAAAACCAGAUAAAAAAUAAAAACCACAAGAGGGUC<u>UUAAGGAU</u>-OH

INDIANA 5' pppACGAAGACCACAAAACCAGAUAAAAAAUAAAAACCACAAGAGGGUC-OH

NEW JERSEY 5' pppACGAAGACAAAAAAACCAUUCCUAAUACAAAAGGCCAAAAGAAGGU-OH

Figure 3. Minus strand leader RNAs of VSV-IND and VSV-NJ showing the 3' region involved in binding La protein. The La bound minus strand leader RNA of VSV-IND is 8 nucleotides longer (underlined sequence) than the major form of minus strand leader RNA. The minus strand leader RNA of VSV-NJ is 46 nucleotides long and fully bound to La protein.

The scheme shown in figure 4 depicts the binding of the cellular La protein near the 3' end of the (+) leader RNA early in the infectious cycle. As the infection proceeds, viral nucleocapsid protein accumulates and binds at the 5' end of the (+) leader RNA.[12] A crossover point eventually occurs at 4 to 6 hours post infection where the accumulation of La bound (+) leader RNA ceases but the binding of N protein continues. We propose that La protein binds leader RNA in the infected cell and participates in the release of the transcript from the template. In this sense La protein may function in the maintenance of the transcription mode <u>in vivo</u>. When sufficient N protein is synthesized to initiate the process of encapsidation, the N protein molecules nucleate at the 5' end of the leader RNAs and displace or prevent binding of molecules of La protein. Thus, nucleocapsid assembly and the synthesis of full length replicative strands may be controlled and perhaps fine tuned by interplay between molecules of the cellular La and the viral N protein. This model also allows the N protein to maintain a high affinity for the VSV RNA and still raise the threshold level of N protein necessary to induce replication. Thus the feedback of N protein to inhibit its own synthesis (by inhibiting transcription of its mRNA) is delayed without the need to reduce the affinity of N protein for VSV RNA. This would allow a large pool of N protein to be available in vivo.

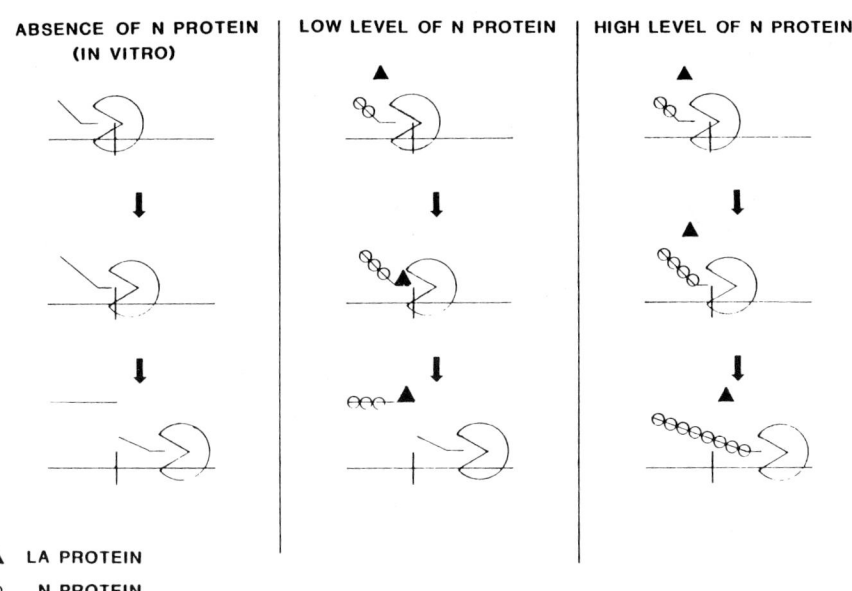

Figure 4. The proposed role of La protein in the replication of negative strand RNA viruses. The genomic template is indicated by the horizontal line with the leader gene/N gene junction marked with a vertical line. The large circle represents the RNA polymerase. In the absence of N protein, the polymerase will terminate at the end of the leader gene and reinitiate the N gene mRNA. In the presence of low levels of N protein, the La protein binds to the 3' end of the leader RNA transcript and prevents N protein from suppressing termination. The transcript is then released with La protein bound. In the presence of high levels of N protein, La protein is either displaced or prevented from binding to the transcript and termination is suppressed resulting in replication.

REFERENCES

1. Lerner, M.R. and Steitz, J.A. (1979): Proc. Nat. Acad. Sci. (USA) 76, 5495-5499.
2. Francoeur, A.M. and Matthews, M.B. (1982): Proc. Nat. Acad. Sci (USA) 79, 6772-6776.
3. Lerner, M.R. and Steitz, J.A. (1981): Cell 25, 298-300
4. Chambers, J.C., Kurilla, M.G., and Keene, J.D. (1983): J. Biol. Chem.: (in press).
5. Rosa, M.D., Gottlieb, E., Lerner, M.R., and Steitz, J.A. (1981): Mol. Cell. Biol. 1, 785-796.
6. Lerner, M.R., Andrews, N.C., Miller, G., and Steitz, J.A. (1981): Proc. Nat. Acad. Sci. (USA) 78, 805-809.
7. Kurilla, M.G. and Keene, J.D. (1983): Cell: (in press).
8. Wilusz, J., Kurilla, M.G., and Keene, J.D. (1983): Proc. Nat. Acad. Sci. (USA): (in press).
9. Rinke, J. and Steitz, J.A. (1982): Cell 29, 149-159.
10. Keene, J.D., Schubert, M. and Lazzarini, R.A. (1980): J. Virol. 33, 789-794.
11. Keene, J.D., Piwnica-Worms, H., and Isaac, C.L. (1981): In "The Replication of Negative Strand Viruses," D.H.L. Bishop and R.W. Compans, eds., Elsevier North Holland, Inc., New York.
12. Blumberg, B.M., Giorgi, C., and Kolakofsky, D. (1983): Cell 32, 559-567.

EFFECTS OF CELL EXTRACTS ON TRANSCRIPTION
BY VIRION AND INTRACELLULAR NUCLEOCAPSIDS
OF VESICULAR STOMATITIS VIRUS

Helen Piwnica-Worms and Jack D. Keene

Department of Microbiology and Immunology
Duke University Medical Center
Durham, North Carolina 27710

INTRODUCTION

Several small RNA species are synthesized *in vitro* by the endogenous transcriptase of vesicular stomatitis virus. These small RNAs can be divided into two groups. The first group consists of RNAs that are triphosphated at their 5' termini and show uv inactivation kinetics and times of appearance similar to the leader RNA. These RNAs provided the basis for the multiple-entry model of VSV transcription proposed by Testa et al.[1] The second group consists of small capped RNA species that have uv inactivation kinetics and time courses of appearance similar to messenger RNA. These characteristics of the latter species are consistent with a single-entry model for VSV transcription.[2] Controversy remains concerning the role of these small RNAs in VSV transcription. For various reasons it has been difficult to detect them *in vivo*. As one approach to determine whether they are synthesized *in vivo*, we have prepared infected cell extracts to use in *in vitro* transcription reactions. Under these conditions, we found that the small capped and triphosphated RNAs were not synthesized. The leader RNA was the only small RNA species detected. Furthermore, when uninfected cell extracts were added to *in vitro* reactions under conditions which normally produce large quantities of the small RNAs, only the leader RNA was synthesized. These results suggest that the small RNAs represent aberrant products of *in vitro* transcription reactions rather than actual precursors to mature messenger RNAs as suggested previously.[1]

RESULTS

Using purified viral or intracellular transcriptive complexes (RNP cores) of VSV, we have reported previously the identity of several small RNA species synthesized *in vitro* by the genomic RNA.[2] Figure 1A shows the results of a time course obtained from synchronized *in vitro* transcription reactions utilizing purified virion RNP cores. Species 1 was identified as the leader RNA, species 2 and 7 correspond to the 5' terminal sequences of the NS mRNA, species 3,4,5, and 8 correspond to the 5' terminal sequences of the N mRNA and species 6 originates from the leader gene.

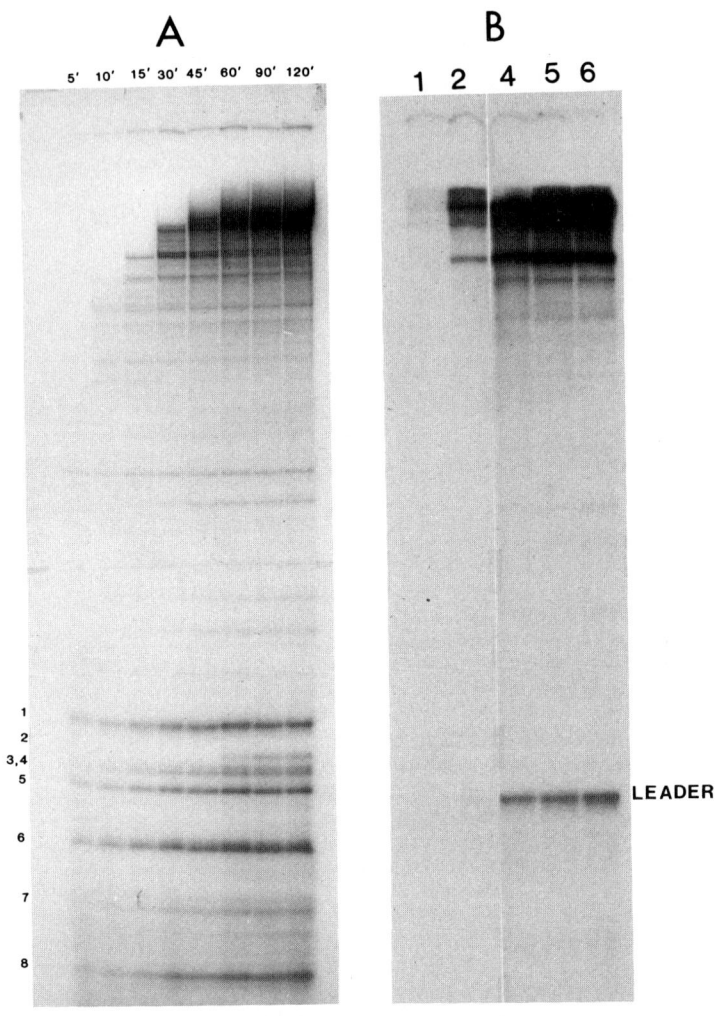

FIGURE 1 (A) Time course of appearance of small RNA species in a synchronized *in vitro* transcription reaction utilizing purified virion RNP cores. Transcription reactions were synchronized by the addition of $MgCl_2$ or purified RNP cores. At times indicated (min. of incubation) samples were removed and anaylzed on a 5% acrylamide gel. (B) Time course of appearance of RNA species *in vitro* utilizing infected cell extracts. BHK spinner cells were infected with VSV at an m.o.i. of 10. At times indicated (hours post infection), cells were lysed and extracts prepared. Transcription reactions utilizing infected cell extracts were incubated for one hour at 30°C. Samples were analyzed on 5% acrylamide gels.

EFFECTS OF CELL EXTRACTS

In order to study the synthesis of these small RNAs in vivo, we examined transcripts synthesized in in vitro reactions using infected cell extracts prepared at different times after infection of BHK spinner cells with VSV. As shown in figure 1B, the pattern of small RNAs synthesized by infected cytoplasmic extracts is very different from that observed when purified RNP cores were used (figure 1A). The leader RNA is the only small RNA observed in infected cell extracts. Messenger RNA migrating in the upper portion of the gels is similar under both sets of transcription conditions, however, and suggests that degradation was minimal under these conditions.

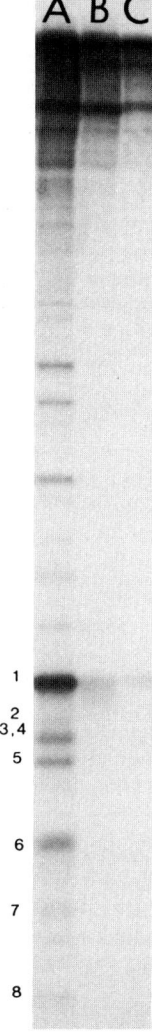

FIGURE 2. RNA transcripts synthesized in vitro using: purified intracellular RNP cores in the absence of cell extracts (A), infected cell extracts isolated 4 hours post infection (B), or purified virion RNP cores in the presence of cytoplasmic cell extracts (C).

In order to determine if the difference in the pattern of small RNA synthesis was due to a structural difference between virion and intracellular transcriptive complexes, we substituted purified intracellular RNP cores for virion RNP cores in _in vitro_ transcription reactions. As shown in figure 2A, the pattern of small RNA synthesis was identical to that seen with purified virion RNP cores (figure 1). This suggests that the differences in small RNA synthesis are not due to structural differences between the two templates.

When whole cell or cytoplasmic cell extracts were added to _in vitro_ reactions utilizing purified virion or intracellular RNP cores, however, the pattern of small RNA synthesis changed (figure 2B). Although synthesized in reduced amounts, the leader RNA was the only small RNA detected under these conditions. Nuclear extracts and other proteins such as bovine serum albumin did not have this effect (data not shown). Thus, there are factors present in the cytoplasm of cells that influence the initiation or termination of VSV transcripts. Whether these factors interact with the template, the RNA products or the polymerase is unknown.

In order to futher characterize the factors responsible for this activity, heat inactivation studies were performed. Cytoplasmic extracts were untreated (figure 3A), heated at $65°$C for 1 min.(figure 3C) or heated at $90°C$ for 1 min.(fig. 3D) and added to _in vitro_ transcription reactions utilizing purified virion RNP cores. In the presence of untreated cytoplasmic extracts the leader RNA was the only small RNA species synthesized (figure 3B). When the extract was heated prior to transcription, however, the ability to suppress the synthesis of small RNAs was reduced as shown by the appearance of species 3 to 8.

The sensitivity of the cell factors to heat inactivation suggests that they may be proteinaceous in nature. In order to determine whether the factors responsible for the suppression of small RNA synthesis could stably associate with transcription complexes, purified virion RNP cores were preincubated in the presence of cytoplasmic cell extracts and then repurified by centrifugation over discontinuous glycerol gradients as described previously. The pattern of small RNAs synthesized by the repurified virion cores (fig. 3E) resembled that seen normally with purified cores in the absence of cell extracts (fig. 3A). When intracellular cores were similarly isolated from infected cell extracts and incubated in standard _in vitro_ transcription reactions in the absence of added extract, small RNA species 1 through 8 were synthesized (fig. 3G). This is in contrast to the pattern seen when total infected cell extracts were analyzed (fig. 3F). Thus, the factors responsible for the suppression do not irreversibly modify the template nor do they remain stably bound during the nondenaturing isolation procedures used to prepare RNP cores. The activity is precipitable in the presence of 80% ammonium sulfate, however, and is stable to dialysis at $4°C$ overnight (data not shown). To rule out nonspecific protein effects as being responsible for the suppression activity, bovine serum albumin and BHK cell nuclear extracts were added to purified viral or RNP cores, at concentrations identical to those used with cell extracts. These proteins were not able to substitute for cytoplasmic extracts and the small RNAs synthesized were identical to those synthesized in the absence of cell extracts. Thus, the effects appear to be specific for factors in the cytoplasm of host cells.

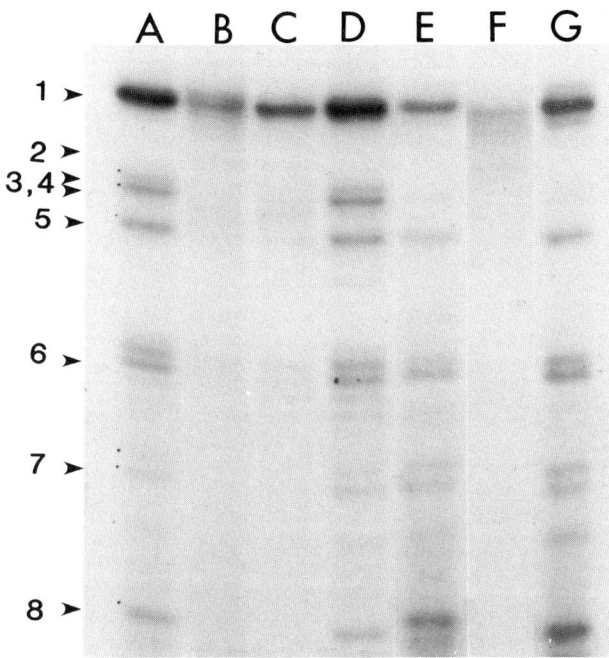

FIGURE 3. Small RNA transcripts synthesized in vitro using (A) purified virion RNP cores in the absence of cell extracts; (B) purified virion RNP cores in the presence of cell extracts; (C) purified RNP cores in the presence of cell extract heated at 60 C for 1 min; (D) purified RNP cores in the presence of cell extracts heated at 90°C for 1 min; (E) repurified virion RNP cores that were previously incubated with cell extracts at 37°C for 10 min. and reisolated by centrifugation and transcribed in the absence of cell extracts; (F) infected cell extracts prepared at 4 hr post infection; (G) intracellular RNP cores purified from extracts prepared 4 hr post infection and transcribed in the absence of cell extract.

DISCUSSION

Although it is known that VSV transcription proceeds in a linear and sequential fashion,[3,4,5] the exact mechanism of synthesis remains unclear. One unresolved issue is whether the viral polymerase must enter and initiate transcription at precisely the 3' terminus of the template (single-entry model)[6] or whether it can also enter internally at each of the intergenic sites (multiple-entry model).[1]

The multiple-entry model was based on the finding of small triphosphated RNAs which are initiatied internally at the N and NS intergenic sites during in vitro transcription reactions. This model proposes that the short triphosphated RNAs are synthesized simultaneously at the onset of transcription and are then sequentially extended into mRNA. Whether these small triphosphated RNAs represent precursors to mature mRNA or are abortive products of the in vitro system remains unresolved. Much evidence has accumulated recently, however, to suggest that the small triphosphated RNAs may not function as precursors to mRNA. Lazzarini and coworkers,[7] have demonstrated that the triphosphated RNAs do not remain template associated, are metabolically stable and cannot be chased into capped mRNA. We and others[2,8] have shown that the 5' portions of the N and NS mRNAs accumulated sequentially during in vitro transcription suggesting that the polymerase is not solely dependent upon the prior synthesis of the short triphosphated RNAs in order to transcribe each mRNA start. The synthesis of the small triphosphated RNAs probably reflects recycling by polymerase molecules which are in excess and bound to the template at the intergenic sites during the assembly and release of the virion. Furthermore, transcription studies performed with the DI particle, LT-2,[9] and reconstitution studies[10] with the nondefective template of VSV, suggest that there is a single entry site on the VSV genome.

In this paper, we present evidence which suggests that the small RNAs are not synthesized in extracts of infected cells. In addition, there are factors present in the cytoplasm of host cells which suppress the synthesis of these small RNA species. Whether this suppression occurs at the level of initiation or termination is unknown. Taken together, these studies provide strong evidence against the suggestion that the short triphosphated RNAs serve as mRNA precursors and are supportive of a single-entry model for VSV transcription.

REFERENCES

1. Testa, D., Chanda, P.K. and Banerjee, A.K. (1980): Cell 21: 267-275.
2. Piwnica-Worms, H. and Keene, J.D. (1983): Virology 125:206-218.
3. Ball, L.A. and White, C.N. (1976): Proc. Natl. Acad. Sci.(USA) 73: 442-446.
4. Abraham, G. and Banerjee, A.K. (1976): Proc. Natl. Acad. Sci.(USA) 73: 1504-1508.
5. Iverson, L.E. and Rose, J.K.(1981): Cell 23: 477-484.
6. Banerjee, A.K., Abraham, G. and Colonno, R.J. (1977): J. Gen. Virol. 34: 1-8.
7. Lazzarini, R.A., Chien, I., Yang, F. and Keene, J.D. (1982): J. Gen. Virol. 58:429-441.
8. Iverson, L.E. and Rose, J.K. (1982): J. Virol. 44: 356-365.
9. Keene, J.D., Chien, I. and Lazzarini, R.A. (1981): Proc. Natl. Acad. Sci.(USA) 78: 2090-2094.
10. Emerson, S. (1983): Cell 31:635-642.

SYNTHESIS OF THE VARIOUS RNA SPECIES IN CELLS INFECTED WITH THE TEMPERATURE-SENSITIVE MUTANTS OF VSV NEW JERSEY

J.F. Szilágyi, R.G. Paterson[1] and C. Cunningham

M.R.C. Virology Unit, Institute of Virology, University of Glasgow, Glasgow, G11 5JR, Scotland

TEMPERATURE-SENSITIVE MUTATIONS AFFECTING RNA SYNTHESIS OF VSV NEW JERSEY

Temperature-sensitive mutants of vesicular stomatitis virus New Jersey serotype (VSV New Jersey) were classified into six non-overlapping complementation groups (1). On the basis of the total amounts of RNA synthesised in infected BHK cells at the permissive (31°C) and restrictive (39°C) temperatures, mutants of complementation groups A, B and F were classified as "RNA negative", mutants of groups C and D as "RNA positive", whereas in the case of group E mutants two (tsE1 and tsE3) were RNA negative while the third (tsE2) was RNA positive (1).

Viral RNA synthesis in infected cells is presumed to take place in three separate stages: transcription of the mRNA species followed by synthesis of the replicative intermediate, and finally synthesis of the virion RNA (2). Activity of the virion-associated RNA transcriptase was assayed in vitro at 31°C and 39°C in order to determine at which stage RNA synthesis was inhibited in the RNA negative mutants (3). Three mutants, tsB1, tsE1 and tsF1, were found to possess a temperature-sensitive transcriptase. Since they belong to three different complementation groups, this suggests the involvement of at least three polypeptides in the transcription process.

---

[1] R.G. Paterson was a recipient of a Medical Research Council grant for training in research methods.

Furthermore, since the transcriptase activity of other mutants of complementation groups E and F (tsE2, tsE3 and tsF2) was not inhibited the mutated polypeptides of these groups are presumably multifunctional, taking part both in transcription and replication and in the case of the "RNA positive" mutant tsE2 possibly also in virus maturation. Since transcription of the group A mutants was not inhibited in the in vitro assay they were presumed to be defective in some stage of replication.

High resolution DATD cross-linked polyacrylamide gel electrophoresis and Cleveland's partial digestion analysis established that ts mutation affected polypeptide N of the group A mutants (4) and polypeptide NS in the group E mutants (5). Dissociation-reconstitution experiments confirmed that NS is the mutated polypeptide in group E and showed that L is the mutated polypeptide in group B (6, 7). As yet it has not proved possible to determine which polypeptide of the group F mutants is affected by the mutation.

DETAILED ANALYSIS OF THE RNA SYNTHESISED IN INFECTED CELLS AT THE PERMISSIVE AND RESTRICTIVE TEMPERATURES

To study the RNA synthesis by wild type VSV New Jersey and its temperature-sensitive mutants BHK cells were infected with purified virions (50 pfu per cell) and were incubated in the presence of 10 ug/ml Actinomycin D at the permissive and restrictive temperatures. After 6 hours of incubation RNA species were extracted by hot-phenol, separated by agarose-urea gel electrophoresis and the [^{3}H]-uridine labelled RNAs were visualised by fluorography.

RNA Synthesis by Wild Type VSV New Jersey. In BHK cells infected with wild type VSV New Jersey (Figures 1 and 3) the messenger RNAs for the five viral polypeptides (L, G, N, NS, M) are the predominant labelled RNA species. Besides these there are also present four minor RNA species (a, b, c and d). Three of these, a, c and d, are also synthesised during in vitro RNA synthesis using purified virus or transcribing nucleoprotein (TNP) complexes and for this reason we believe that they

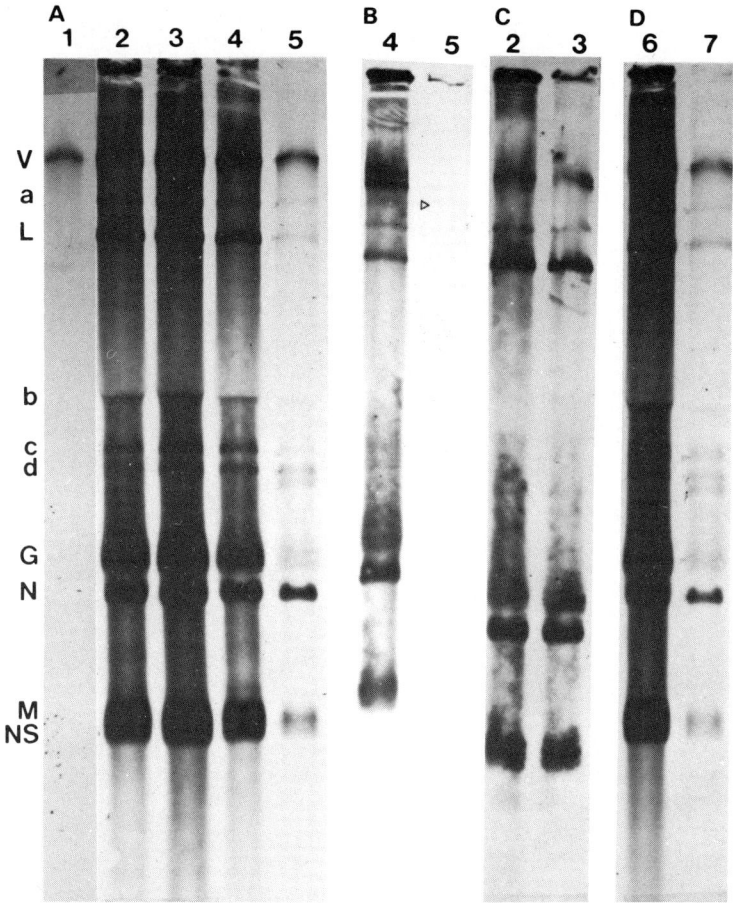

FIGURE 1. RNA species synthesised either in BHK cells infected with wild type VSV New Jersey (Panel A, tracks 2 and 3) and with the temperature-sensitive mutants tsB1 (Panel A, tracks 4 and 5), tsC1 (Panel C, tracks 2 and 3) and tsD1 (Panel D, tracks 6 and 7), or in CE cells infected with the mutant tsB1 (Panel B, tracks 4 and 5). The first track of each pair contains the RNA species synthesised at 31°C and the second track contains those synthesised at 39°C. Purified virion RNA (Panel A, track 1) was used as marker. The positions of the virion RNA (V), messenger RNAs L, G, N, NS and M (for the corresponding polypeptides) and the minor RNA species a, b, c, and d are indicated.

are also products of the transcription.

It was not possible to analyse in detail the various RNA species in the region of the virion RNA, since the technique used did not resolve the high molecular weight RNAs into discrete bands.

RNA Synthesis by the Groups B, C and D Mutants.
Temperature-sensitive mutants of complementation groups B, C and D (tsB1, tsC1 and tsD1) synthesised large amounts of RNA at the restrictive temperature in BHK cells (Fig. 1, Panels A, C and D). In the case of tsC1 similar amounts of RNA were synthesised both at 39^oC and 31^oC. Thus both transcription and replication occurred at the restrictive temperature.

Cells infected with tsB1 or tsD1 synthesised the five mRNA species, the minor components and also high molecular weight RNAs with electrophoretic mobilities similar to the virion RNA. Although the amounts of RNAs synthesised at 39^oC were less than at 31^oC the results suggest that both mutants were capable of transcribing and replicating their RNA at the restrictive temperature.

The results obtained with tsD1 are in good agreement with earlier observations but those obtained with tsB1 were unexpected since in vitro transcription by this mutant is inhibited at 39^oC. To study this anomaly we infected chick embryo (CE) cells with this mutant and found that RNA synthesis was completely inhibited at 39^oC (Fig. 1, Panel B).

Thus these results confirm that mutants of complementation groups C and D are "late" mutants. In the case of the group B mutant they also indicate the involvement of a host factor, present in BHK but not in CE cells, which in some way helps polypeptide L to overcome its heat sensitivity and to participate in transcription and perhaps also in replication. In this respect this mutant appears to have some similarity to the temperature-sensitive host range mutant tdCE3.

RNA Synthesis by the Group A Mutants.
Mutant tsA1 and tsA4 synthesised very little RNA at 39^oC (Fig. 2), indicating that transcription by the virion-associated RNA transcriptase was restricted in the infected cells. These results appear to contradict the in vitro transcriptase assays where the amounts of RNA synthesised by these mutants at 39^oC were similar to those synthesised by the wild

SYNTHESIS OF THE VARIOUS RNA SPECIES

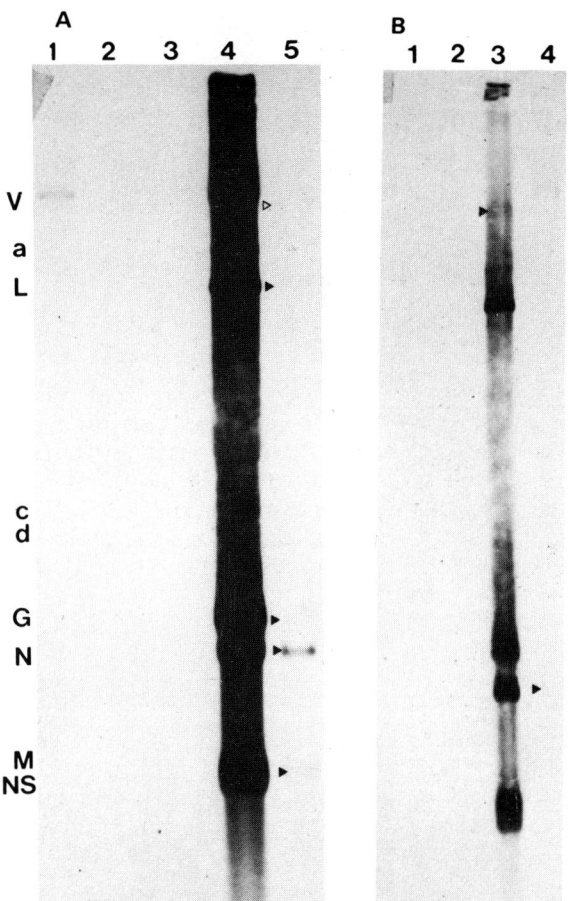

FIGURE 2. RNA species synthesised in BHK cells infected with the temperature-sensitive mutants tsA1 (Panel A, tracks 4 and 5) and tsA4 (Panel B, tracks 3 and 4). The first tracks in both pairs contain the RNAs synthesised at 31°C, and the second contain the RNA synthesised at 39°C. Mock-infected BHK cells incubated at the two temperatures (Panel A, tracks 2 and 3; Panel B, tracks 1 and 2) were used for comparison. Purified virion RNA (Panel A, track 1) was used as marker. The positions of the virion RNA (V), messenger RNAs L, G, N, NS and M (for the corresponding polypeptides) and the minor RNA species a, c, and d are indicated.

type virus (3). Furthermore, the RNA species synthesised by these viruses in vitro at 39°C were similar to those synthesised by wild type virus, suggesting that the ts mutation affected replication rather than transcription. It is difficult to reconcile these two sets of results. Since the ts mutation affected the N polypeptide of these mutants it is possible that the transcriptase synthesises RNA at 39°C once it has complexed with the virion RNA-polypeptide N template. However, on reaching the end of the template, it is no longer able to recombine with the template at 39°C presumably because the configuration of the N polypeptide has been altered at this temperature.

RNA Synthesis of the Groups E And F Mutants.

RNA synthesis at the restrictive temperature in cells infected with the group E mutants (Fig. 3, Panels A and B) varied according to the mutant used for the infection. In cells infected with tsE1 no viral RNA synthesis could be detected whereas in cells infected with tsE2 and tsE3 viral messenger RNA species were observed. These results are in good agreement with the earlier findings and provide further evidence that tsE1 is probably a transcriptase negative mutant, tsE3 a transcriptase positive but replicase negative mutant and tsE2 a transcriptase positive and perhaps also replicase positive mutant. Since NS is the mutated polypeptide in this complementation group these results provide further evidence that this polypeptide plays more than one role in virus development.

In cells infected with the group F mutants (Fig. 3, Panel C) RNA synthesis appeared to be inhibited in the case of tsF1 but messenger RNA synthesis was observed in the case of tsF2. These results are also in agreement with earlier findings and show that the mutated polypeptide in complementation group F is also likely to be multifunctional.

CONCLUSION

Analysis of the RNA species synthesised in tissue culture cells infected with the ts mutants of VSV New Jersey provided further evidence that in group E and F the mutated polypeptides are involved

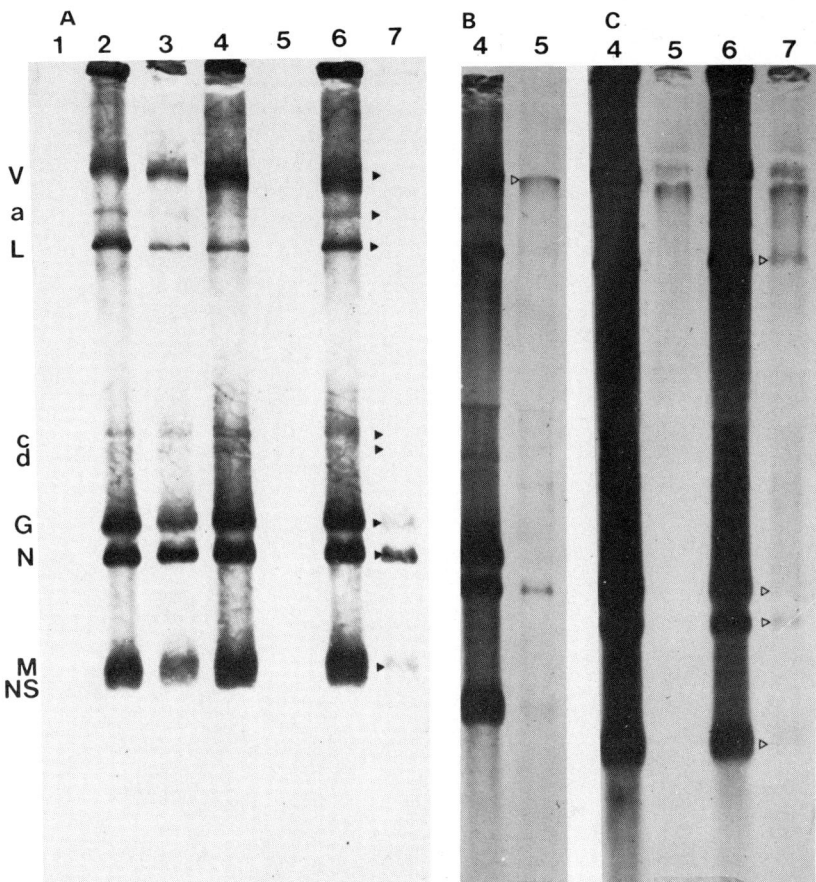

FIGURE 3. RNA species synthesised in BHK cells by the temperature-sensitive mutants of group E (Panels A and B) and group F (Panel C). The cells were infected either with wild type VSV New Jersey (Panel A, tracks 2 and 3), or with mutants tsE1 (Panel A, tracks 4 and 5), tsE2 (Panel A, tracks 6 and 7), tsE3 (Panel B, tracks 4 and 5), tsF1 (Panel C, tracks 4 and 5) and tsF2 (Panel C, tracks 6 and 7). Purified virion RNA (Panel A, track 1) was used as marker. The positions of the virion RNA (V), messenger RNAs L, G, N, NS and M (for the corresponding polypeptides) and the minor RNA species a, c, and d are indicated.

both in transcription and replication, and that mutants in complementation groups C and D were defective in a late function, presumably in assembly. However, in the case of the group A mutants analysis of the RNA species in infected cells suggests that mutation might have affected transcription due to a mutation in polypeptide N. In the group B mutants the involvement of a host factor in transciption was indicated. These new findings may help us towards the better understanding of VSV replication, a process only partially understood at present.

ACKNOWLEDGEMENTS

We thank C.R. Pringle for providing the temperature-sensitive mutants and J.T. Poyner for helping with the preparation of the manuscript.

REFERENCES

1. Pringle, C.R., Duncan, I.B., and Stevenson, M. (1971). J. Virol. 8, 836.
2. Flamand, A., and Bishop, D.H.L. (1974). J. Mol. Biol. 87, 31.
3. Szilágyi, J.F., and Pringle, C.R. (1979). J. Virol. 30, 692.
4. Cunningham, C., and Szilágyi, J.F. In preparation.
5. Evans, D., Pringle, C.R., and Szilágyi, J.F. (1979). J. Virol. 31, 325.
6. Ongrádi, J., Cunningham, C., and Szilágyi, J.F. (1983). Submitted for publication.
7. Ongrádi, J., Cunningham, C., and Szilágyi, J.F. In preparation.

CHARACTERIZATION OF A MUTANT OF VESICULAR STOMATITIS VIRUS WITH AN ABERRANT IN VITRO POLYADENYLATION ACTIVITY[1]

D. Margaret Hunt

Department of Biochemistry
The University of Mississippi Medical Center
Jackson, Mississippi 39216

The genome of the rhabdovirus vesicular stomatitis virus (VSV) is a non-segmented, negative-sense RNA of molecular weight $3.6-4.2 \times 10^6$ (1). Purified virions contain an RNA-dependent RNA polymerase (transcriptase) (2) which, under appropriate conditions, transcribes the genome in vitro, producing monocistronic mRNAs which are capped, methylated and polyadenylated (3). Three of the five viral proteins are required for transcription: the template is genomic RNA complexed tightly with N protein, and L and NS proteins are also necessary for transcription (4). A fourth virally coded protein, M, is not essential for transcription (4) but can modulate the process (5-9). It is not known whether transcription requires, or is significantly affected by, host components which may be packaged in virions.

We have previously reported (10) that our stock of tsG16(I), a temperature-sensitive (ts) mutant of VSV (Indiana) (11), overproduces polyadenylic acid (poly A) in an in vitro transcription system at all temperatures tested (27°C, 31°C and 35°C) and that the mRNAs made in vitro by this virus contain abnormally long tracts of poly(A). It appears that the 5' end of tsG16(I) RNA made in vitro at 31°C is normally modified since it is methylated to almost the same extent as wild type (wt) product RNA, contains the same two methylated cap structures as wt RNA in a similar ratio and is translated as well (per pmol of UMP residues) as wt RNA in an in vitro reticulocyte cell-free translation system (10).

---

[1]This work was supported by Public Health Service grant AI 18201 from the National Institute of Allergy and Infectious Diseases.

One question which arises is which component of tsG16(I) virions is associated with this abnormal polyadenylation activity. TsG16(I) is known to have an L protein defect which is apparently correlated with the thermolability of transcription in vitro by this virus (12). However, as mentioned above, experiments to determine whether aberrant polyadenylation by this mutant is a temperature-dependent phenomenon suggested that it was not. As yet, we have not been able to isolate revertants of our stocks of tsG16 to determine whether the abnormal polyadenylation phenotype is correlated with the temperature-sensitivity of this mutant. The approach used, therefore, in the studies reported here, has been to try to identify the moiety responsible for the poly(A) phenotype of tsG16(I) by separating mutant or wild type virions into various components and by reconstituting the system using homologous or heterologous components.

RECONSTITUTION EXPERIMENTS USING TEMPLATE-AND ENZYME-CONTAINING FRACTIONS

Emerson and Wagner (13) have shown that virions treated with Triton X-100 in the presence of 0.72 M NaCl can be fractionated into a sedimentable ribonucleoprotein template fraction and a non-sedimentable enzyme-containing fraction. Individually, these fractions possess little transcriptive activity but they exhibit considerable transcriptive activity when recombined. To determine whether the poly(A) phenotype of tsG16(I) was associated with the template- or enzyme-containing fraction of the virus, virus was grown on BHK-21 cells and highly purified virions (10) of either mutant or wt VSV Indiana (San Juan strain) were treated with an equal volume of 2X-Triton high salt solubilizer (2X-HSS; 1.44 M NaCl-18.7% glycerol-3.74% Triton-X100-1.2 mM dithiothreitol and 10 mM Tris-hydrochloride, pH 7.4) and separated into pellet (template-containing) and supernatant (enzyme-containing) fractions by ultracentrifugation as previously described (14). The fractions were assayed after homologous or heterologous reconstitution to determine their poly(A) phenotype. Virions which were solubilized at the same time but which were not fractionated were used as controls. Since tsG16(I) synthesizes such a large excess of poly(A) compared to wt virus in an in vitro transcription system, a convenient assay for the aberrant polyadenylation phenotype of tsG16(I) was to use transcription reactions (10) which contained [^{3}H]ATP, [α-^{32}P]UTP, CTP and GTP, and to compare the molar amounts of AMP and UMP incorporated into ice-cold trichloroacetic acid insoluble material. The mutant and wt polyadenylation phenotypes can readily be distinguished by this assay ((10) and Table 1).

TABLE 1

AMP AND UMP INCORPORATION BY UNFRACTIONATED WT AND TSG16(I) VIRIONS OR BY VIRIONS AFTER DISSOCIATION AND RECONSTITUTION FOLLOWING TREATMENT WITH 2X-HSS[a]

Virion Components	Incorporation in 120 min at 31°C[b] pmoles UMP	pmoles AMP	Ratio AMP/UMP
Unfractionated wt	90.2	151.0	1.67
wtP + wt SN	52.9	75.4	1.43
Unfractionated ts	232.9	708.6	3.04
tsP + ts SN	70.6	215.1	3.05
tsP + wt SN	55.7	90.6	1.63
wtP + ts SN	76.4	270.4	3.54

[a] Purified virions were treated with an equal volume of 2X-HSS and separated into pellet (P) and supernatant (SN) fractions by ultracentrifugation as described previously (14). Solubilized, unfractionated virions and P and SN fractions (before or after recombination) were assayed in transcription reactions containing [α-^{32}P]UTP, [^{3}H]ATP, CTP, GTP and S-adenosyl methionine (10); incorporation of each isotope into ice-cold trichloroacetic acid precipitable material was determined as described previously (10).

[b] Values are not corrected for incorporation by P and SN fractions assayed separately. Neither SN fraction showed any detectable incorporation. The UMP incorporation by wtP and tsP fractions was 10.4 pmoles and 5.2 pmoles, respectively, and the incorporation of AMP was 15.9 and 23.9 pmoles, respectively.

As can be seen from Table 1, when tsG16(I) or wt virions were separated into pellet and supernatant fractions by the method described above and then recombined homologously, the parental phenotype was recovered. When the pellet and supernatant fractions were recombined heterologously, the AMP/UMP incorporation phenotype of the ts pellet plus wt supernatant recombination was wt whereas the phenotype for recombined wt pellet plus ts supernatant fractions was that of the mutant. It appeared, therefore, that the aberrant in vitro polyadenylation by tsG16(I) was not due to an altered polyadenylation signal on the template but was due to an altered component in the supernatant fraction.

Similar results were obtained when wt VSV Indiana (Glasgow strain), the parental wt from which tsG16(I) was derived (11), was used (data not shown). However, pellet fractions prepared by the above method from our stocks of this wt strain retained considerable transcriptase activity and we, therefore, have used the San Juan strain since it has been shown that transcriptase fractions and purified L and NS fractions are functionally interchangeable between the Glasgow mutants and the San Juan wild type virus (12,14).

EXPERIMENTS USING TRANSCRIBING NUCLEOCAPSIDS

The supernatant fractions prepared after treatment of the virus with an equal volume of 2X-HSS and ultracentrifugation contain L, NS, G and M proteins (13). To determine whether M or G proteins could be responsible for the abnormal polyadenylation phenotype of tsG16(I), we prepared transcribing nucleocapids (TNCs) which retained L and NS proteins on the N protein-RNA template but from which M and G proteins had been removed. Highly purified virions suspended in 15% glycerol-10 mM Tris-hydrochloride, pH 7.4 were treated with an equal volume of 0.5 M solubilizer (0.5 M NaCl-3.74% Triton X-100-20% glycerol-1.2 mM dithiothreitol-4.3 mM Tris-hydrochloride, pH 8.0) and subjected to gel filtration on agarose A-50m (BioRad Laboratories, Richmond, California) in 0.25 M solubilizer (1 volume of 0.5 M solubilizer plus one volume 10 mM Tris-hydrochloride, pH 8.0). TNCs prepared by this procedure were virtually free of M and G proteins (Fig. 1), but retained almost all of the L and NS proteins (D. M. Hunt, unpublished data) and most of the original transcribing activity of the virus (Table 2).

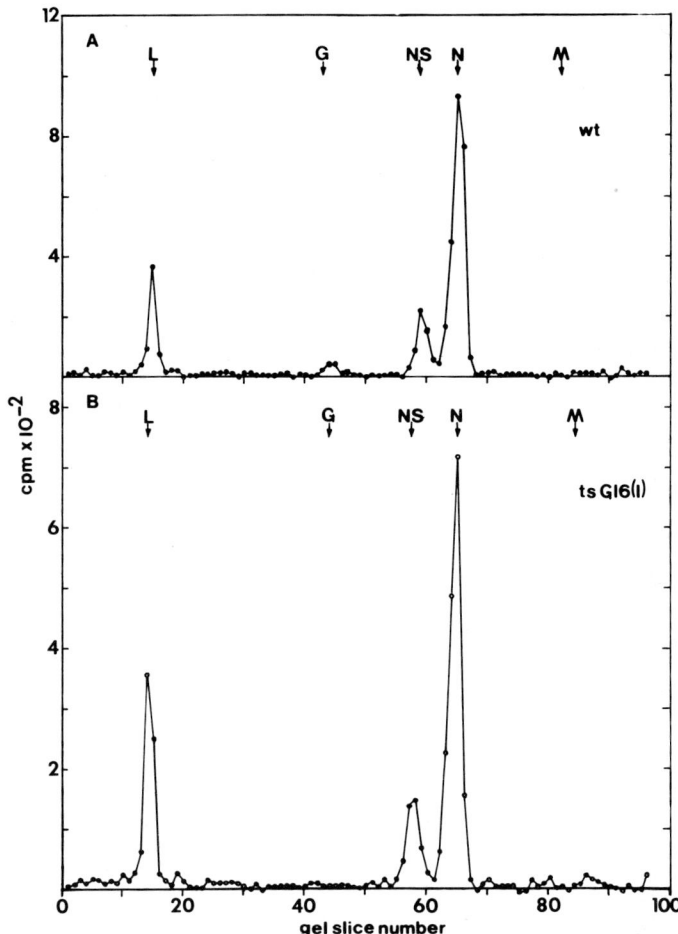

FIGURE 1. Proteins present in transcribing nucleocapsid (TNC) fractions. [^{3}H] leucine labeled virions were treated with an equal volume of 0.5 M solubilizer and subjected to gel filtration on agarose A-50m columns equilibrated with 0.25 M solubilizer (see text). Nucleocapsid containing fractions were collected, precipitated with trichloroacetic acid and prepared for polyacrylamide gel electrophoresis (13). Samples were electrophoresed on 12.5% polyacrylamide-SDS slab gels (10) with marker virion proteins in adjacent lanes. Gels were dried, sliced, digested with NCS (Amersham Corp.): water (9:1) for 3.5 h at 55°C and counted by liquid scintillation spectrometry after addition of a xylene-based fluor. A, wt TNCs; B, tsG16(I) TNCs.

TABLE 2

AMP AND UMP INCORPORATION BY UNFRACTIONATED WT AND TSG16(I) VIRIONS OR BY TRANSCRIBING NUCLEOCAPSIDS (TNCs) PREPARED BY GEL FILTRATION[a]

Virion Components	Ratio pmoles AMP incorporated / pmoles UMP incorporated	% recovery[b] of activity TNCs/virus
Unfractionated wt	2.26	
wt TNCs	2.29	91
Unfractionated ts	4.24	
Ts TNCs	4.42	101

[a] $[^3H]$-leucine labeled purified virions were treated with an equal volume of 0.5 M solubilizer, and TNCs prepared by gel filtration on agarose A50m in 0.25 M solubilizer (see text). Fractions were assayed as described in Table 1. Incorporations were measured after 120 min at 31°C.

[b] Recoveries are calculated by comparing the UMP incorporation by virions and TNCs and correcting for the relative protein concentrations (determined by comparison of $[^3H]$ leucine cpm in each fraction) and the loss of M and G proteins from the TNC (determined by polyacrylamide gel electrophoresis).

The polyadenylation phenotype of such TNCs, as measured by the ratio of pmoles AMP/UMP incorporated, corresponded to that of the parent virus (Table 2). The results in Figure 1 and Table 2 are taken from the same experiment so that a direct comparison can be made. It appeared, therefore, that M and G proteins were probably not directly responsible for the aberrant polyadenylation by tsG16(I) since virtually all of these proteins were removed (Figure 1), but almost all of the transcriptase activity and of the abnormal polyadenylation activity of tsG16(I) was recovered.

DISCUSSION

From separation-reconstitution experiments (Table 1) in which virions were treated with 0.72 M NaCl prior to fractionation, it appeared that the moiety directly responsible for abnormal polyadenylation by tsG16(I) was not the ribonucleoprotein template but, rather, a component of the supernatant fraction, which contained L, NS, G and M proteins. However, since removal of almost all of the G and M proteins by gel filtration in the presence of 0.25 M NaCl-solubilizer (Figure 1) resulted in little loss of either transcriptive activity or the aberrant polyadenylation activity (Table 2) it seemed unlikely that G or M were directly responsible for the polyadenylation phenotype of tsG16(I). Experiments are currently being performed in an attempt to determine whether the abnormal component in tsG16(I) is L protein, NS protein, or perhaps due to an abnormality in the spectrum of host proteins packaged due to a mutation in a virion component.

ACKNOWLEDGMENTS

The technical assistance of Edith F. Smith and Mark Britt is gratefully acknowledged. We thank Drs. C. R. Pringle and R. R. Wagner for VSV isolates.

REFERENCES

1. Wagner, R.R. (1975). In "Comprehensive Virology" (H. Fraenkel-Conrat and R. R. Wagner, eds.), Vol. 4, p. 1. Plenum Publishing Corp., New York.
2. Baltimore, D., Huang, A. S., and Stampfer, M. (1970). Proc. Natl. Acad. Sci. U.S.A. 70, 572.
3. Banerjee, A. K., Abraham, G., and Colonno, R. J. (1977). J. Gen. Virol. 34, 1.
4. Emerson, S. U. (1976). Curr. Top. Microbiol. Immunol. 73, 1.

5. Martinet, C., Combard, A., Printz-Ane, C., and Printz, P. (1979). J. Virol. 29, 134.
6. Carroll, A. R., and Wagner, R. R. (1979). J. Virol. 29, 134.
7. Clinton, G. M., Little, S. P., Hagen, F. S., and Huang, A. S. (1978). Cell. 15, 1455.
8. Pinney, D. F., and Emerson, S. U. (1982). J. Virol. 42, 897.
9. De, B. P., Thornton, G. B., Luk, D., and Banerjee, A. K. (1982). Proc. Natl. Acad. Sci. U.S.A. 79, 7137.
10. Hunt, D. M. (1983). J. Virol. 46, 788.
11. Pringle, C. R. (1970). J. Virol. 5, 559.
12. Hunt, D. M., Emerson, S. U., and Wagner, R. R. (1976). J. Virol. 18, 596.
13. Emerson, S. U., and Wagner, R. R. (1972). J. Virol. 10, 297.
14. Hunt, D. M., and Wagner, R. R. (1974). J. Virol. 13, 28.

ROLE OF THE VIRAL LEADER RNA
IN THE INHIBITION OF TRANSCRIPTION
BY VESICULAR STOMATITIS VIRUS[1]

Brian W. Grinnell and R.R. Wagner

Department of Microbiology
University of Virginia Medical School
Charlottesville, Virginia 22908

Following infection of many cell types by vesicular stomatitis virus (VSV), a rapid decrease in cellular RNA synthesis is observed (1-4). Studies with UV-inactivated virus demonstrated that only a small portion of the 3' end of the genome need be transcribed to obtain this inhibitory activity (5). Because the VSV genome was shown to be transcribed sequentially from its 3' end (6, 7), it was suggested that the first viral transcript, the leader RNA, was the inhibitor molecule. McGowan et al. (8) demonstrated that purified plus-strand leader RNA could inhibit DNA-dependent transcription in a soluble HeLa cell extract and studies by Kurilla et al. (9) showed that the leader RNA was present in the nucleus of infected cells.

In recent studies, we examined the ability of the VSV_{NJ} and VSV_{Ind} serotypes to inhibit transcription and provided additional evidence for the role of the leader RNA (10). We also demonstrated that the VSV_{NJ} leader RNA was a better inhibitor of transcription than the VSV_{Ind} leader RNA. Additional studies (Grinnell and Wagner, unpublished) have defined the sequence and structure of the leader RNA involved in the inhibition of transcription. In this paper we provide a summary of these data.

INHIBITION OF CELLULAR RNA SYNTHESIS IN VSV INFECTED CELLS AND QUANTITATION OF LEADER RNA PRESENT

The effects of VSV_{NJ} and VSV_{Ind} on cellular RNA synthesis were compared by pulse labeling infected cells with tri-

---

[1]This work was supported by PHS grant AI-1112 from the NIAID, grant MV-9E from the ACS and grant PCM-88-00494 from the NSF. BWG is a postdoctoral trainee supported by PHS Training Grant CA-9109 from the NCI and NRSA fellowship 1F32A106894-01 from NIAID.

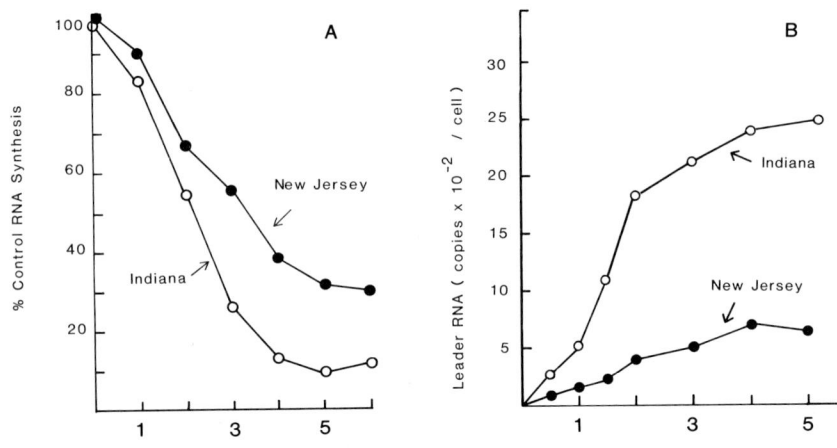

FIGURE 1. (A) Inhibition of cellular RNA synthesis and (B) kinetics of leader RNA production by VSV_{NJ} and VSV_{Ind} in infected cells. MPC-11 cells were infected with the viruses and were pulse-labeled with [^{3}H]uridine for 10 min at hourly intervals after infection. The amount of cell RNA synthesis was determined, relative to mock-infected cells, by the amount of acid precipitable radioactivity in each sample. To detect the plus-strand leader RNA, cell RNA samples were hybridized with a 42S VSV genomic RNA that had been 3'-end-labeled with [^{32}P]pCp (11), followed by digestion with pancreatic and T1 ribonucleases. Double-stranded products were separated on 10% polyacrylamide gels and the amount of leader RNA was determined from the number of Cerenkov counts hybridized and the specific activity of the probe.

tiated uridine at various times after infection and measuring the amount of acid precipitable incorporation. As shown in Figure 1A, both serotypes inhibited cellular RNA synthesis although VSV_{Ind} did so more rapidly and to a greater extent. This difference in inhibitory activity between the two serotypes was independent of the multiplicity of infection or the cell type used.

To determine if the difference in the level of inhibition observed in Figure 1A might relate to differences in the amount of intracellular leader RNA, we examined the kinetics of leader RNA production. In both VSV_{NJ} and VSV_{Ind}-infected cells, the amount of intracellular leader RNA increased steadily with time (Figure 1B). The level of leader in VSV_{Ind}-infected cells reached ~2400 copies/cell by 5h after infection, but in VSV_{NJ}-infected cells, only reached ~625 copies/cell, or approximately four times less.

PRODUCTION OF LEADER RNA IN VIVO BY UV-INACTIVATED VIRUS

The difference in the level of cellular RNA synthesis inhibition obtained with VSV_{NJ} and VSV_{Ind} (Figure 1A) was not as drastic as the four-fold difference in the amount of the leader RNAs detected in these cells (Figure 1B). While this suggested that the VSV_{NJ} leader RNA might be a more efficient inhibitor of transcription, such an assessment would best be made by comparing levels of leader RNA in cells at the same level of RNA synthesis inhibition. To this end, cells were infected with virus that had been irradiated with increasing UV-doses and after 5h, the levels of transcription inhibition and intracellular leader RNA were determined. At increasing UV doses, the ability of VSV to inhibit transcription and to produce leader RNA decreased. For both of these functions, the UV inactivation curves had essentially the same slope (not shown) and even at a dose of 20,000 ergs/mm², the infecting virus still inhibited transcription and produced leader RNA. In contrast, the 37% (1/e) survival dose <u>in vivo</u> for the N protein, whose gene is the first to be transcribed after the leader gene, is ~400 ergs/mm² (5) and at a dose of 600 to 700 ergs/mm² is not detectable.

From the UV inactivation curves, we determined the amount of leader RNA present at fixed levels of cellular RNA synthesis inhibition. At all levels of inhibition, there was approximately five times more leader RNA present in VSV_{Ind}-infected cells than in those infected with VSV_{NJ}. Thus, although much less VSV_{NJ} leader RNA is produced in infected cells (Fig. 1B), this virus may effectively shut off cellular RNA synthesis by means of a more efficient inhibitor than that of VSV_{Ind}.

INHIBITION OF TRANSCRIPTION IN VITRO BY THE LEADER RNAS

The apparent difference in inhibitory efficiencies of the two leader RNAs in infected cells was also observed when we directly compared their abilities to inhibit DNA-dependent RNA synthesis in the <u>in vitro</u> HeLa cell transcription system described by Manley et al. (12). In this system, we utilized two DNA templates, pBR322 recombinants of the adenovirus-2 late promoter (LP), transcribed by RNA polymerase II, and the adenovirus-associated (VA) gene, transcribed by RNA polymerase III. The effect of increasing concentrations of purified VSV_{NJ} and VSV_{Ind} leader RNA on the transcription of both templates is shown in Figure 2. At nearly all concentrations tested, the New Jersey leader was more effective in inhib-

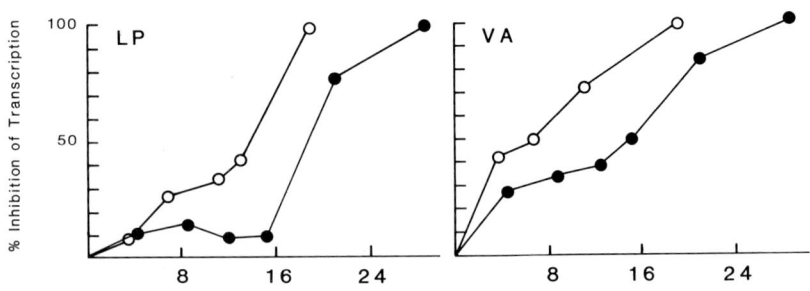

FIGURE 2. Dose response to the inhibition of transcription by the New Jersey (O---O) and Indiana (O---O) leader RNAs. pBR322 recombinants of the Ad2 late promoter (LP) and associated RNA (VA) genes (37ng/µl of reaction, 9 x 10^{-8}M) were incubated in a soluble HeLa cell extract alone or with increasing concentrations of purified leader RNA. The extent of inhibition was determined by dividing the amount of transcription of each template in reactions containing leader by the amount in reactions with no leader added. The amount of transcription was determined by Cerenkov counting of the specific transcripts sliced from gels.

iting $\underline{in}$ $\underline{vitro}$ transcription of both the LP and VA genes. In the range tested, it appeared that the minimal concentration of Indiana leader needed to inhibit the LP gene was much higher, approximately 16ng/µl of reaction compared to ~4ng/µl of reaction for the New Jersey leader. However, increasing the concentrations of VSV_{Ind} leader above 16ng/µl resulted in rapid inhibition of transcription within a relatively narrow range of concentration. In contrast, the inhibition of VA gene transcription by the VSV_{Ind} leader occurred at lower doses. Also, the VSV_{NJ} leader was more effective in inhibiting the VA gene than was the Indiana leader. However, both leader RNAs appeared to be more effective in their ability to inhibit transcription by RNA polymerase III (VA gene) than by polymerase II (LP gene).

EFFECT OF DOUBLE-STRANDED REGIONS IN THE LEADER RNA ON THE INHIBITION OF TRANSCRIPTION

A comparison of the RNA sequences of the VSV_{NJ} and VSV_{Ind} leader RNAs (Figure 3) shows that they are quite similar; the ten nucleotide differences are highlighted by underlining. Four of the nucleotide differences are in an AU-rich region (nucleotides 18-34) suggested to resemble the

```
                           10        20        30        40
                           |         |         |         |
                                 •••••••• ••         •••
VSV_Ind  Leader RNA  5'ppACGAAGACAAACAAACCAUUAUUAUCAUUAAAAGGCUCAGGAGAAAAC-OH  3'

                                ••••••••••  •  ••  •••  ••
VSV_NJ   Leader RNA  5'ppACGAAGACAAAAAAACCAUUAUUACAAUUAAUUGGCCUAGAGGGAAAC-OH  3'
```

FIGURE 3. Comparison of the nucleotide sequence and sites of RNase V1 cleavage in the leader RNAs of VSV$_{Ind}$ and VSV$_{NJ}$. The unique regions of VSV$_{Ind}$ leader are highlighted by underlining. RNase V1 cleavage sites (dots) were determined by analyzing partial digests of 3' end-labeled leader RNA on RNA sequencing gels. Sequence data obtained from Colonno and Banerjee (16, 17).

TATA homology and five are in the purine-rich 3' sequences, which McGowan et al. (8) suggested to be similar to RNA polymerase recognition sequences. We determined that both leader RNAs had secondary structure, which encompassed the AU-rich region. This was accomplished by digesting end-labeled leader RNAs with the double-strand specific RNase V1 (13) and separating the partial digestion products on a 20% polyacrylamide sequencing gel (data not shown). The locations of major cleavage sites are indicated in Figure 3. Using the method of Tinoco et al. (14), we have predicted the stabilities of several stem-loop structures; the VSV$_{NJ}$ leader RNA appears to be capable of forming a more stable secondary structure than the VSV$_{Ind}$ leader RNA (i.e. -4.5 vs. -0.6 kcals for VSV$_{NJ}$ and VSV$_{Ind}$, respectively).

We also examined the effect of eliminating the secondary structure of the leader RNAs on their ability to inhibit transcription of the VA and LP templates <u>in vitro</u>. The intercalating agent, proflavin, has been used to perturb secondary structure (15), and leader RNA heat denatured in the presence of 1µM proflavin contained no double-stranded regions detectable by RNase V1 cleavage. Therefore, we measured the level of transcription inhibition by the leader RNAs in the presence and absence of proflavin. At a concentration of 1µM, proflavin itself did not significantly reduce the level of VA or LP gene transcripts. In the presence of this agent, there was a slight reduction in the inhibitory activity of the VSV$_{Ind}$ leader RNA but a significant reduction in the inhibitory activity of the VSV$_{NJ}$ leader RNA (Table 1). After proflavin treatment, the levels of inhibition by VSV$_{NJ}$ leader RNA approximated the levels obtained with an equimolar amount of VSV$_{Ind}$ leader RNA. However, a significant level of inhibition remained. Thus, secondary structure may be impor-

TABLE 1
INHIBITORY EFFECT OF LEADER RNAS ON IN VITRO TRANSCRIPTION OF THE AD 2 LATE PROMOTER (LP) AND VA GENES IN THE PRESENCE OF PROFLAVIN

Leader RNA added[a] (amount)	Proflavin (1µM)	% inhibition of RNA transcribed on template VA	LP
VSV_{Ind} (16ng/µl)	–	50	38
VSV_{Ind} (16ng/µl)	+	43	36
VSV_{NJ} (15.5ng/µl)	–	87	74
VSV_{NJ} (15.5ng/µl)	+	44	42

[a]The amount of leader used represents a molar concentration of approximately $1 \times 10^{-6} M$. The amount of each template (37ng/µl) represents a concentration of $1 \times 10^{-8} M$ for the VA gene and $9 \times 10^{-9} M$ for the LP gene.

tant in the efficiency of transcription inhibition particularly by the VSV_{NJ} leader RNA.

VSV SEQUENCES INVOLVED IN IN VITRO TRANSCRIPTION INHIBITION

We examined the possibility that a recombinant cDNA of the leader RNA might inhibit polymerase II and III-directed transcription and would allow for a more detailed analysis of the sequences involved. The recombinant plasmid, originally a gift from R. Lazzarini, National Institutes of Health, contained approximately the first 390 nucleotides of the 3' end of the VSV_{Ind} genome, i.e. leader RNA sequences and part of the N gene. The cDNA insert was isolated by Pst-1 cleavage from the pBR322 vector (VSV sequences plus GC tails) and was then digested with the restriction endonuclease DdeI to produce three fragments: DdeIC, containing most of the leader RNA sequence (GC tails and leader nucleotides 1 to 36); DdeIB, containing the remaining leader sequence and 5' end of the N gene mRNA sequence; and DdeIA, containing only N gene mRNA sequence and GC tails. For each fragment, we isolated the DNA strand corresponding to the VSV plus-sense and examined its ability to inhibit both VA and LP gene transcription. The DdeIC leader fragment exhibited significant inhibitory activity (Figure 4). We found no significant effect of the DdeIB or DdeIA fragment or poly [dG-C] on transcription at levels ten times those needed to obtain inhibition by

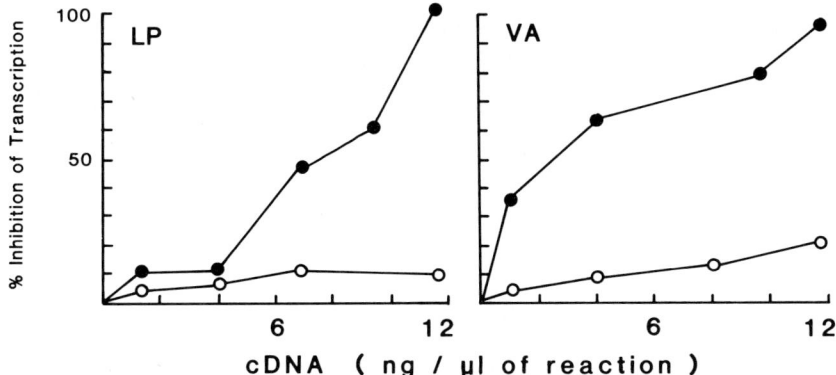

FIGURE 4. Dose response to the inhibition of transcription of the Ad2 LP and VA genes by the plus strands of the DdeIC fragment (leader sequences; O---O) and DdeIA fragment (N gene sequences; (O---O) of a cDNA clone of the 3' end of the VSV genome. The extent of transcription inhibition of each template was determined as in Figure 2.

the DdeIC fragment. In fact, less cDNA leader was required to inhibit transcription than was required for the leader RNA itself (Figure 2). This probably is because the cDNA is less sensitive to nuclease degradation during incubation in the HeLa cell extract than is the RNA. It is interesting that the dose-response patterns of the DdeIC fragment for the inhibition of both the VA and LP genes (Figure 4) were similar to those of purified leader RNA (Figure 2), i.e. inhibition of the VA gene transcription occurred at lower concentrations and reached higher levels than observed for the inhibition of LP gene transcription. This similarity in dose response patterns suggests that the cDNA leader serves as an appropriate model for the leader RNA.

The DdeIC fragment contains the AU-rich (AT) or TATA-like sequences that also appear to be important in the efficiency of transcription inhibition. To further assess the importance of these sequences, we examined the effect of two synthetic oligodeoxynucleotides on transcription of the VA and LP genes: one containing the AT-rich sequences (Oligo I: ATTATAATCATTA) and the other, sequences near the 3' end of the leader (Oligo II: AGGCUCAGGAGA). The AT-rich Oligo I molecule markedly reduced the levels of both VA and LP gene transcription. However, the Oligo II molecule exhibited little inhibitory activity even at very high concentrations (data not shown).

Leader RNA appears to play a dominant role in the inhibition of cellular transcription in VSV-infected cells.

Using a reconstituted HeLa cell transcription system, we have determined that the AU(AT)-rich region of the leader RNA or cDNA sequence alone is capable of inhibiting transcription directed by both RNA polymerase II and III and that secondary structure in this region is important. The leader RNA of VSV may prove to be a useful model for studying regulation by small nuclear RNAs of cellular origin.

ACKNOWLEDGEMENTS

We thank Ernst-L. Winnacker, Institute of Biochemistry, University of Munich for synthesizing oligodeoxynucleotides.

REFERENCES

1. Baxt, B., and Bablanin, R. (1976). Virology 72, 383.
2. Weck, P.K., and Wagner, R.R. (1978). J. Virol. 25, 770.
3. Weck, P.K., and Wagner, R.R. (1979). J. Virol. 30, 410.
4. Yaoi, Y., Mitsui, H., and Amano, M. (1970). J. Gen. Virol. 8, 165.
5. Weck, P.K., Carroll, A.R., Shattuck, D.M., and Wagner, R.R. (1979). J. Virol. 30, 746.
6. Abraham, G., and Banerjee, A.K. (1976). Proc. Natl. Acad. Sci. USA 73, 1504.
7. Ball, L.A., and White, C.N. (1976). Proc. Natl. Acad. Sci. USA 73, 442.
8. McGowan, J.J., Emerson, S.U., and Wagner, R.R. (1982). Cell 28, 325.
9. Kurilla, M.G., Piwnica-Worm, H., and Keene, J.D. (1982). Proc. Natl. Acad. Sci. USA 79, 5240.
10. Grinnell, B.W., and Wagner, R.R. (1983). J. Virol. 48, In press.
11. Leppert, M., Rittenhouse, L., Perrault, J., Summers, D.F., and Kolakofsky, D. (1979). Cell 18, 735.
12. Manley, J.L., Fire, A., Cano, A., Sharp, P.A., and Gefter, M.L. (1980). Proc. Natl. Acad. Sci. USA 77, 3855.
13. Favocova, O.O., Fasiolo, F., Keith, G., Vassilenko, S.K., and Ebel, J.P. (1981). Biochem. 20, 1006.
14. Tinoco, I., Borer, P.N., Dengler, B., Levin, M.D., Uhlenbeck, O.C., Crothers, D.M., and Gralla, J. (1973). Nature New Biol. 246, 40.
15. Hay, N., Skolnik-David, H., and Aloni, Y. (1982). Cell 29, 183.
16. Colonno, R.J., and Banerjee, A.K. (1978). Cell 15, 93.
17. Colonno, R.J., and Banerjee, A.K. (1978). Nucl. Acids. Res. 51, 4165.

TEMPERATURE SENSITIVE MUTANTS OF VSV INTERFERE WITH THE
GROWTH OF WILD-TYPE VIRUS AT THE LEVEL OF RNA SYNTHESIS[1]

Debra W. Frielle and Julius S. Youngner

Department of Microbiology
School of Medicine
University of Pittsburgh
Pittsburgh, PA 15261

Temperature-sensitive (ts) mutants of a number of viruses are capable of interfering with the replication of wild-type virus (1). Youngner and Quagliana (2) demonstrated that ts mutants of VSV interfere with the replication of wild-type VSV at both the permissive and nonpermissive temperatures. Since the ts mutants form plaques which are much smaller than those formed by wild-type VSV (wt-VSV), the progeny of a mixed infection can be analyzed with respect to the relative yields of ts and wt-VSV. By such analysis, it was found that the growth of some ts mutants is enhanced during a mixed infection at the nonpermissive temperature. Therefore, these ts mutants not only interfere with the replication of wt-VSV, but are "rescued" during a mixed infection.

To extend these observations, we have examined in greater detail the level at which ts mutants of VSV interfere with the replication of wild-type virus. Mixed infections with wt-VSV and 2 ts RNA$^-$ mutants were studied.

RNA SYNTHESIS DURING MIXED INFECTION

Cumulative RNA synthesis was measured in actinomycin D treated BHK cells which were either singly infected with wt-VSV, ts G 11, or ts G 41, or were doubly infected with wt-VSV and a ts mutant. Infected cells were incubated at 39.5°C; virus yield and the incorporation of ^{3}H-uridine were measured (Figure 1 and Table 1).

When cells are infected with wt-VSV and ts G 11 at a multiplicity of infection (MOI) of 1 and 10, respectively, the yield of wt-VSV is reduced by greater than 90% relative to an infection by wt-VSV alone. The yield of ts G 11 reflects a

---

[1]This work was supported by research grant (AI-06264) from the NIAID, NIH.

TABLE 1

VIRUS PRODUCED DURING MIXED INFECTION AT 39.5°C WITH wt-VSV AND ts MUTANTS

Cells infected with:	Ratio of yields Mixed infection/single infection	
	wt-VSV[b]	ts Mutant[b]
wt-VSV x ts G 11 (1 x 10)[a]	0.07	5800
wt-VSV x ts G 11 (10 x 10)	0.45	14000
wt-VSV x ts G 41 (1 x 10)	0.0002	1.2
wt-VSV x ts G 41 (10 x 10)	0.001	1.0

[a] Numbers in parentheses are the multiplicities of infection of wt-VSV and the ts mutant: (wt-VSV x ts virus).

[b] Culture medium was harvested at 8 hours after infection. Plaque-forming virus was assayed in chicken embryo monolayers at 34°C.

rescue of greater than 5000-fold (Table 1). However, cumulative RNA synthesis is restricted in such doubly-infected cells (Figure 1, Panel A). There is approximately an 8-fold reduction of RNA synthesis in doubly infected cells relative to cells infected with wt-VSV alone. When an MOI of 10 of each virus is used, interference and rescue are again reflected in the yields of each virus. Cumulative RNA synthesis is also reduced under these conditions, although to a lesser extent (Figure 1, Panel B).

In contrast, during a mixed infection of wt-VSV (MOI = 1 or 10) and ts G 41 (MOI = 10), the yield of wt-VSV is decreased by at least 3 logs and there is no rescue of ts G 41 (Table 1). The synthesis of viral RNA is almost completely inhibited (Figure 1, Panels C and D).

In a double infection with wt-VSV and ts G 11 the level of RNA synthesis represents a combination of the RNA synthesized by wt-VSV and "rescued" RNA synthesis by ts G 11.

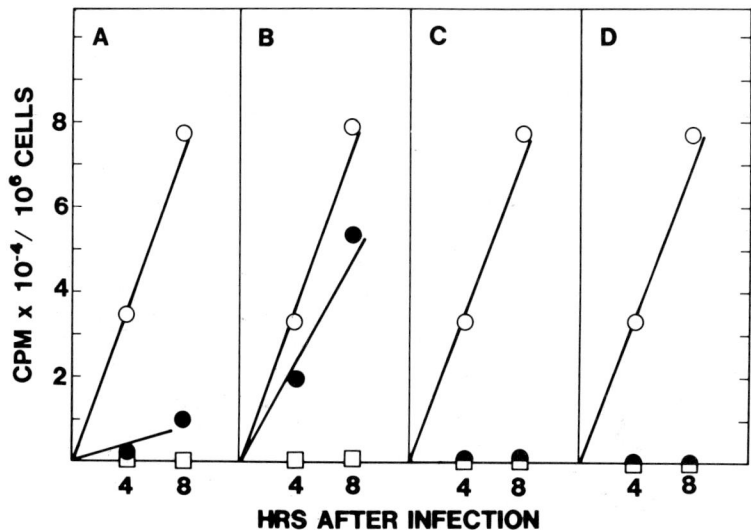

FIGURE 1. RNA synthesis in BHK cells infected with wt-VSV and ts mutants. Cells were singly infected with wt-VSV (O—O), a ts mutant (□—□), or doubly infected with wt-VSV and a ts mutant (●—●). Cumulative RNA synthesis at 39.5°C was measured at 4 and 8 hours post-infection. Panel A, wt-VSV x ts G 11 (MOI = 1 x 10); Panel B, wt-VSV x ts G 11 (MOI 10 x 10); Panel C, wt-VSV x ts G 41 (MOI 1 x 10); Panel D, wt-VSV x ts G 41 (MOI 10 x 10).

On the other hand, in a mixed infection with wt-VSV and ts G 41, there is neither rescue nor viral RNA synthesis. For this reason, ts G 41 was used in the majority of mixed infections described in this report.

Co-infection with wt-VSV and either ts G 11 or ts G 41 did not reduce primary transcription compared to wt-VSV alone (data not shown). Therefore, the inhibition of wt-VSV RNA synthesis during a mixed infection is not due to interference at the level of primary transcription.

WHAT SPECIES OF RNA ARE SYNTHESIZED DURING A MIXED INFECTION WITH wt-VSV AND ts MUTANTS?

Cells were singly infected with wt-VSV, ts G 11 or ts G 41, or were doubly infected with wt-VSV and a ts mutant at an MOI of 10 for each virus. RNA was extracted after an 8 hour incubation at 39.5°C in the presence of ^{3}H-uridine and subjected to electrophoresis in 1% agarose gels. For analytical purposes an equal amount of TCA precipitable radioactivity was applied to the gel for each virus combination. Therefore, the RNA patterns presented in Figure 2 do not represent the amount of RNA synthesized, but show the relative amounts of the individual RNA species synthesized.

It is clear from the fluorogram (Figure 2) that all species of mRNA, as well as 42S virion RNA, are synthesized during a single infection with wt-VSV and during mixed infections with wt-VSV and either ts G 11 or ts G 41 (compare lanes A, C, and E). This is not unexpected during co-infection with ts G 11. In this case, although growth of wt-VSV is reduced, the rescue of the ts mutant clearly requires synthesis of all viral RNAs. In a mixed infection with wt-VSV and ts G 41, on the other hand, there is no rescue of the interfering mutant, yet all viral RNAs are synthesized. Interference is not due to a decreased level of synthesis of one or several species of RNA; all viral RNAs are synthesized in near normal proportions, although at much reduced levels.

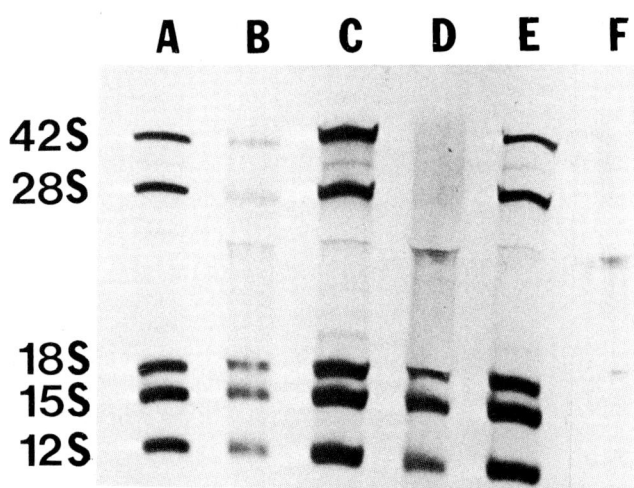

FIGURE 2. RNA synthesis during single and mixed infection. BHK cells were infected with either wt-VSV (Lane A), ts G 11 (Lane B), ts G 41 (Lane D), or were doubly infected with wt-VSV and ts G 11 (Lane C) or wt-VSV and ts G 41 (Lane E). Lane F shows RNA synthesis in mock-infected cells.

EFFECT OF RELATIVE MULTIPLICITIES OF wt-VSV AND ts G 41 ON RNA SYNTHESIS

Cells were singly infected with ts G 41 at an MOI of 10, wt-VSV at multiplicities of 100, 50, 10, 5, and 1, or were doubly infected at those multiplicities (Figure 3). Cumulative RNA synthesis was measured over a 10 hour incubation at 39.5°C. While the synthesis of RNA was inhibited in all double infections relative to RNA synthesis by wt-VSV alone, the degree of inhibition was dependent on the ratio of the MOIs of wt-VSV and ts G 41. When wt-VSV was used at an MOI of 100 and ts G 41 at an MOI of 10, cumulative RNA synthesis was reduced to 67% of the value for cells singly infected with wt-VSV at an MOI of 100. With decreasing multiplicity of wild-type virus, more inhibition was observed; when the ts virus was present in 10-fold excess, RNA synthesis was less than 2% of the control value.

An obvious interpretation of these results is that the degree of inhibition of RNA synthesis is a reflection of the proportion of cells which are actually infected by both a

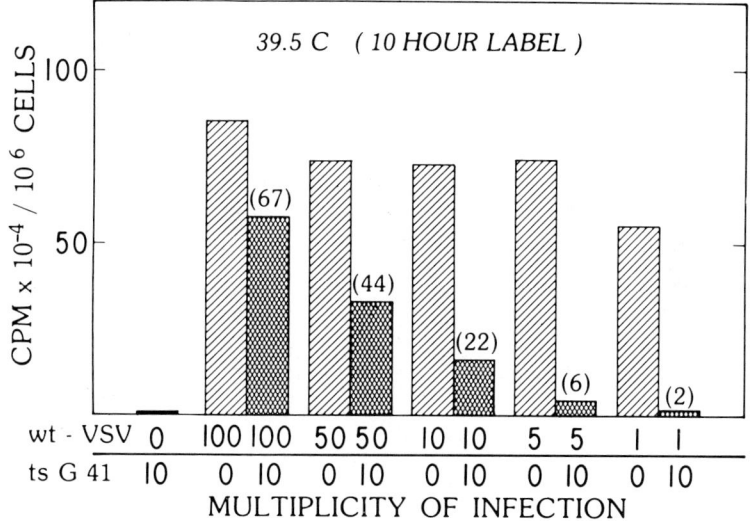

FIGURE 3. The effect of relative multiplicities of infection of wt-VSV and ts G 41 on RNA synthesis in mixed infections. Cells were infected at the indicated multiplicities of infection. Cumulative RNA synthesis was measured over a 10 hour incubation at 39.5°C.

wild-type and a ts virus. Only when the multiplicity of ts virus is greater than the multiplicity of wt virus will all cells infected by wt virus also be infected with a ts virus. Synthesis of viral RNA would be inhibited under these conditions of infection. The failure of Spindler and Holland (3) to observe interference in a mixed infection with ts G 11 and wt-VSV when both viruses were used at a multiplicity of 2.5 is difficult to understand. Perhaps at those MOIs a significant fraction of the cells may not have been doubly infected.

Interference was also examined under conditions where infection with the ts mutant was at hourly intervals after infection with wt-VSV. The ability of the ts mutant to interfere with wt-VSV replication decreased with increasing time between infections (data not shown). Further experiments demonstrated that this loss of activity was due, at least in part, to superinfection exclusion; the ts virus was unable to infect cells which had been previously infected with wt-VSV (4).

SENSITIVITY OF INTERFERENCE TO UV-IRRADIATION

In order to determine if the ability of ts mutants to interfere with wt-VSV replication is dependent on gene expression by the mutant, an interference experiment was carried out with UV-irradiated ts G 41. The ts mutant was diluted in saline to yield an MOI of 10 for infection. The diluted virus was irradiated with a UV lamp (15W) at a distance of 50 cm for the times indicated, mixed with wt-VSV (MOI = 1), and used to infect cells. Under these conditions of irradiation, the titer of ts G 41 was reduced 10-fold by a 14 second exposure (Figure 4). Infected cells were labeled with ^{3}H-uridine for 8 hours at 39.5°C. Duplicate cell cultures were used to assess viral yield. Virus yields and RNA synthesis are expressed as a percentage of the values obtained during an infection with wt-VSV alone (Figure 4).

With unirradiated ts G 41, there is a 95% inhibition of wt-VSV replication, as reflected by yield and RNA synthesis. With increasing times of irradiation, ts G 41 gradually loses the ability to interfere with both yield of wt-VSV and the synthesis of RNA. After 30 seconds of irradiation, the titer of ts G 41 is reduced by 100-fold, and RNA synthesis and wild-type yield are restored to approximately 20% of control levels. By 75 seconds of irradiation, the infectivity of ts G 41 is reduced by at least 5 logs, reducing the effective MOI to $\sim$ 0.001. At this point, the yield and RNA synthesis in mixed infections are identical to those in cells infected with wt-VSV alone. These results suggest that some gene expression by the ts mutant is essential.

DISCUSSION

We have examined the level at which ts mutants of VSV interfere with the replication of wt-VSV. Ts G 41 was chosen for study because, unlike some other interfering mutants, this mutant is not rescued during a mixed infection at the non-permissive temperature. The synthesis of viral RNA is inhibited during a mixed infection with ts G 41; under the appropriate conditions of infection, RNA synthesis is inhibited by greater than 98%. However, primary transcription

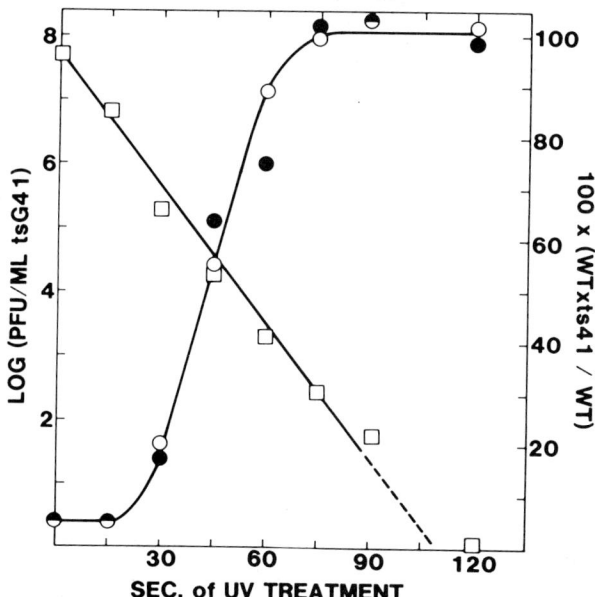

FIGURE 4. The effect of UV-irradiation of ts G 41 on its ability to interfere with wt-VSV. Ts G 41 was UV-irradiated for the times shown (□-□). Cells were infected with wt-VSV and ts G 41 which had been irradiated for increasing times. Cumulative RNA synthesis (O——O) and yield of wt-VSV (●——●) were measured at 8 hours after infection. Results are expressed as percentage of the values obtained with wt-VSV alone.

is not affected. All species of mRNA and 42S genome RNA are synthesized during the mixed infection. UV-irradiation of the ts mutant results in a progressive loss of its ability to interfere with RNA synthesis by wt-VSV and the production of infectious virus.

The ability of RNA⁻, ts mutants of NDV to interfere with the replication of wild-type NDV has also been reported to involve inhibition of RNA synthesis (5). When primary chicken embryo cells are doubly infected with wt-NDV and a ts mutant, RNA synthesis at the non-permissive temperature is inhibited by 50-90% relative to infection with wt-NDV alone. Inhibition of RNA synthesis is also observed in cells doubly infected with wild-type Semliki Forest virus and a ts mutant (6). It is possible that a common mechanism, namely, a block of RNA synthesis, may be responsible for interference by ts mutants with the growth of RNA viruses.

ACKNOWLEDGEMENTS

The authors thank Gregory S. Buzard for his assistance with the preliminary phase of this work. We also thank Dr. T. K. Frey for assistance with RNA gel electrophoresis and Dr. P. A. Whitaker-Dowling for investigating superinfection exclusion.

REFERENCES

1. Youngner, J. S., and Preble, O. T. (1980). In Comprehensive Virology, Vol. 16 (H. Fraenkel-Conrat and R. R. Wagner, eds.) p. 73, Plenum Press, New York.
2. Youngner, J. S., and Quagliana, D. O. (1976). J. Virol. 19, 102.
3. Spindler, K. R., and Holland, J. J. (1982). J. gen. Virol. 62, 363.
4. Whitaker-Dowling, P. A., Youngner, J. S., Widnell, C. C., and Wilcox, D. K. Virology, in press.
5. Preble, O. T., and Youngner, J. S. (1973). J. Virol. 12, 472.
6. Keranen, S. (1977) Virology 80, 1.

ROLE OF VESICULAR STOMATITIS VIRUS PROTEINS IN RNA REPLICATION[1]

John T. Patton[2], Nancy L. Davis, and Gail W. Wertz

Department of Microbiology and Immunology
Medical School
University of North Carolina
Chapel Hill, North Carolina 27514

VSV RNA REPLICATION IN VITRO

Replication of the genome of vesicular stomatitis virus (VSV) is dependent upon protein synthesis (1,2). It is presently unknown which viral proteins are actually required to support this process. It is also unknown which of the VSV proteins, if any, act to regulate RNA replication. In order to more fully characterize VSV RNA replication in terms of the role of the viral proteins in promoting and regulating genome-length RNA synthesis, we have developed an in vitro system that supports the RNA replication and nucleocapsid assembly of a defective interfering (DI) particle of VSV as a function of protein synthesis (3,4). This system, consisting of intracellular DI nucleocapsid templates and a micrococcal nuclease-treated rabbit reticulocyte lysate, replicates VSV RNA when protein synthesis is programmed by the addition of viral mRNA.

In the experiments described here, we have used nucleocapsid templates of the VSV DI particle, DI-T, which contains only the 5' 25% of the genome of wild type virus (5,6). This was done to produce a system in which RNA replication could be studied in the absence of any mRNA synthesis. The DI nucleocapsids are not able to synthesize viral mRNAs because they lack complete information for all the structural genes. In the absence of protein synthesis, the DI templates synthesize only a 46 base long leader RNA (3,7,8). Since mRNAs are not made by DI nucleocapsids, protein synthesis and, hence, RNA replication in this system can be controlled by the

---

[1]This work was supported by Public Health Service grants AI12464 and AI15134 from the National Institute of Allergy and Infectious Disease and grant CA19014 from the National Cancer Institute.

[2]Present address: Department of Biology, University of South Florida, Tampa, Florida 33620.

addition of exogenous mRNA. Therefore, by adding individual VSV mRNAs to program the expression of individual viral proteins, it was possible to investigate the requirement for each of the viral proteins in VSV RNA replication. Further, by adding varying amounts of individual VSV mRNAs to reactions synthesizing genomic RNA, it was possible to examine the ability of each of the viral proteins to regulate VSV RNA replication.

PROTEIN REQUIREMENT FOR RNA REPLICATION

The protein requirement for synthesis of genome-length RNA was examined by programming the in vitro system with individual hybrid-selected N, NS, or M mRNAs (4) to determine if N, NS, or M proteins, when translated individually in the presence of enzymatically active DI nucleocapsids, could promote the synthesis of genome-length RNA. Synthesis of viral RNA and proteins was examined simultaneously in each reaction by double-labeling with [^{3}H]-UTP and [^{35}S]-methionine. Fig. 1A shows the ^{3}H-labeled RNA products and Fig. 1B shows the ^{35}S-labeled protein products. The results demonstrated that synthesis of N protein alone was sufficient to promote the synthesis of both positive- and negative-strand genome-length DI RNA. Neither NS or M protein by themselves were able to support the synthesis of genome-length DI RNA.

EFFECT OF M AND NS PROTEINS ON RNA REPLICATION

The results described above showed that N protein alone was required for the synthesis of genome-length RNA in vitro. The following experiments were designed to determine whether the synthesis of a viral protein, specifically NS or M protein, in addition to N protein affected the level of genome-length RNA synthesis.

Various studies have indicated that NS and M proteins may play a role in regulating VSV RNA synthesis. For NS protein, several different phosphorylated forms are known to exist (9,10). It has been shown that different forms of NS protein differ in their ability to support VSV transcription in vitro (9). Also, several groups have shown that purified M protein when added in large amounts to in vitro transcription reactions will inhibit mRNA production (11,12,13,14). Although both NS and M proteins are able to affect transcription, the ability of these proteins to affect VSV RNA replication has not been studied.

NS Protein Inhibits RNA Replication. To examine the effect of NS and M proteins on VSV RNA replication, varying amounts of hybrid-selected NS and M mRNA were added

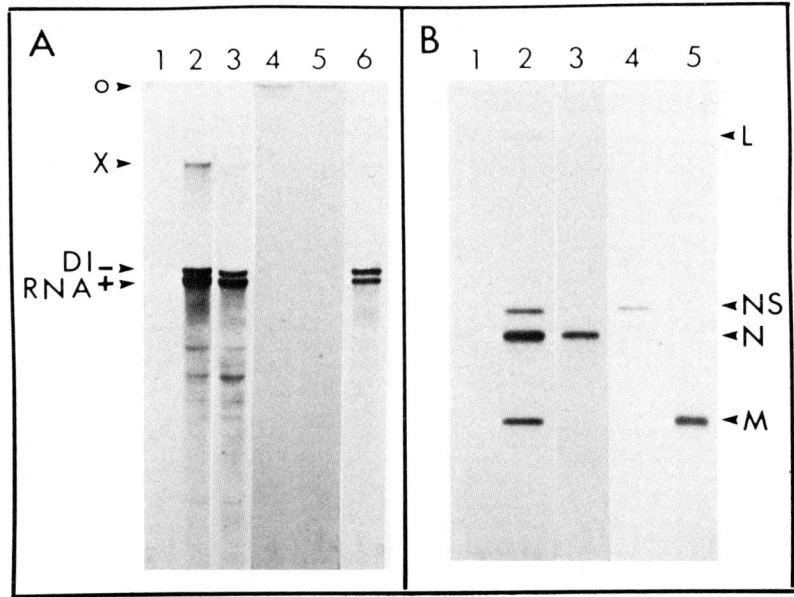

FIGURE 1. In vitro genome-length RNA and protein synthesis in reactions containing individual hybrid-selected viral mRNAs. Reaction mixtures consisted of a nuclease-treated rabbit reticulocyte lysate, buffers, salts, nucleoside triphosphates, amino acids, [^{3}H] UTP, [^{35}S] methionine, purified intracellular DI nucleocapsids and the appropriate viral mRNA (3). Individual mRNAs were purified by hybridization to cDNA immobilized on nitrocellulose filters (4). Incubation of reaction mixtures for 3 hr at 30°C was followed by electrophoresis of deproteinized RNA in 6M urea-agarose gels (Panel A) and by PAGE analysis of proteins (Panel B). (1) No added mRNA, (2) 0.74 µg of total polyadenylated VSV RNA, (3) 5 µl of hybrid-selected N mRNA, (4) 4 µl of hybrid-selected NS mRNA, (5) 4 µl of hybrid-selected M mRNA, (6) in vivo-labeled DI intracellular nucleocapsid template RNA, including a larger DI genome (X).

individually to the replication system along with a constant amount of N mRNA and intracellular DI nucleocapsids. The RNA and protein products are shown in Fig. 2. As above, the reaction programmed with only N mRNA produced both positive- and negative-strand genome-length RNA (Fig. 2A, lane 5). Reactions programmed with an increasing amount of NS mRNA and a constant amount of N mRNA produced an increasing amount of NS

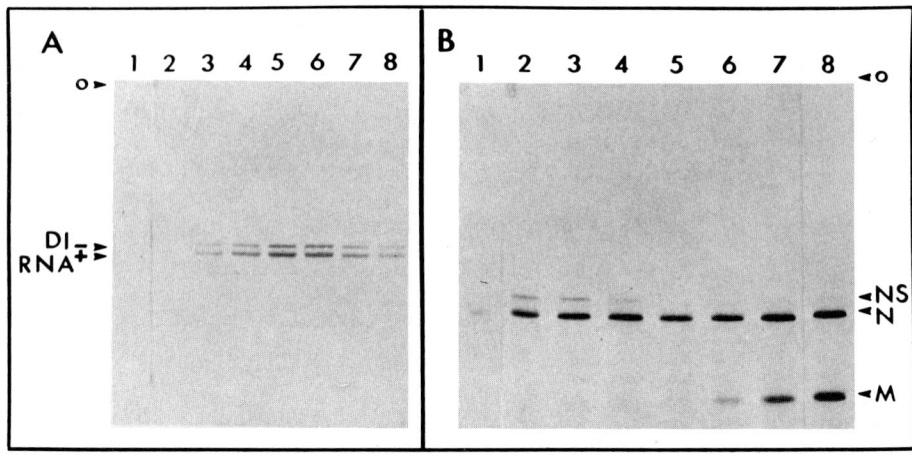

FIGURE 2. Effect of addition of hybrid-selected NS mRNA or M mRNA to reactions containing N mRNA. Reaction mixtures (see Figure 1) contained constant levels of purified intracellular DI nucleocapsids and hybrid-selected N mRNA, except for the control to which no mRNA was added. Varying amounts of either purified NS mRNA or purified M mRNA were added with the N mRNA. RNA (A) and protein (B) products were analyzed as for Figure 1. (1) No added mRNA, (2-8) 3 µl N mRNA plus the following additions, (2) 3 µl NS mRNA, (3) 1.5 µl NS mRNA, (4) 0.5 µl NS mRNA, (5) N mRNA alone, (6) 0.5 µl M mRNA, (7) 1.5 µl M mRNA, (8) 3 µl of M mRNA.

protein relative to a constant amount of N protein (Fig. 2, Panel B). As the molar ratio of newly made NS:N protein increased, the amount of DI genome-length RNA synthesized decreased. In this experiment, at a molar ratio of newly made NS:N protein of 2:1, DI genome-length RNA synthesis was completely inhibited. The molar ratio of 2:5 gave 50% inhibition of RNA replication.

M Protein Inhibits RNA Replication. The ability of M protein to influence RNA replication was measured by adding increasing amounts of purified M mRNA to reactions containing a constant amount of N mRNA and DI intracellular nucleocapsids. The results showed that, similar to NS protein, as the molar ratio of newly made M:N protein increased the amount of RNA replication decreased. However, whereas RNA replication was blocked at a molar ratio of newly made NS:N protein of

2:1, at a molar ratio of newly made M:N protein of 2:1, replication was inhibited only by approximately 50%.

<u>NS and M Proteins Inhibit Replication in the Presence of All Viral Proteins</u>. To test the possibility that increasing amounts of NS or M protein relative to N protein may not be inhibitory to replication when all five newly synthesized viral proteins are present, variable amounts of NS and M mRNAs were added to reactions containing constant amounts of total VSV mRNA and intracellular DI nucleocapsids. The results showed that as the amount of newly made NS or M protein relative to N protein increased, the amount of VSV DI RNA replication decreased (data not shown). Thus, newly made NS and M protein, whether in the presence of other newly synthesized viral proteins or only newly synthesized N protein, inhibited replication.

In order to determine whether any mRNA when added to the replication system in increasing amounts relative to N mRNA would nonspecifically inhibit RNA replication, varying amounts of rabbit globin mRNA were added to reactions containing a constant amount of total VSV mRNA and DI intracellular nucleocapsids. No difference was observed in RNA replication between reactions containing no globin mRNA and those containing differing amounts of this message (data not shown).

M protein has an isoelectric point (pI) of approximately 9.0 (13). To test whether the ability of this protein to inhibit RNA replication was solely the result of its high pI, cytochrome C, a protein with a pI of 9.1, was added in increasing amounts to reactions programmed with total viral mRNA and intracellular DI nucleocapsids. The addition of cytochrome C at molar concentrations approximating those produced <u>in vitro</u> by M mRNA did not affect replication (data not shown). This indicated that the high pI of M protein could not be the only mechanism by which this protein inhibited replication.

ROLE OF N, NS, AND M PROTEINS IN RNA REPLICATION

We have examined the protein synthesis requirement for VSV RNA replication and the ability of newly made NS and M proteins to influence this process using a cell-free replication system. Our results demonstrated that only N protein was required to promote genome-length RNA synthesis by intracellular DI nucleocapsids. Newly made NS and M proteins both inhibited VSV RNA replication <u>in vitro</u>. We can only suggest purposes for these inhibitory effects. It is possible that one function of NS protein in the virus life cycle is to modulate the balance between RNA replication and transcription, thereby preventing the depletion of the N protein pool available for RNA replication. During budding, a region may be formed at the junction of the nucleocapsid and the plasma membrane which has a high M:N ratio. It is here that M protein may act to inhibit replication to prepare the nucleocapsid for release from the cell.

In summary, we have used an in vitro system to show that N protein alone is sufficient to fulfill the protein synthesis requirement for VSV RNA replication. In addition, we have examined the effect of newly made NS and M proteins on RNA replication and have found that although not required for progeny RNA synthesis, these proteins may play important roles in the replication of VSV RNA in the infected cell.

REFERENCES

1. Perlman, S. M., and Huang, A. S. (1973). J. Virol. 12, 1395.
2. Wertz, G. W., and Levine, M. (1973). J. Virol. 12, 253.
3. Wertz, G. W. (1983). J. Virol. 46, 513.
4. Patton, J. T., Davis, N. D., and Wertz, G. W. (1983). J. Virol. (submitted).
5. Leamnson, R., and Reichman, M. E. (1974). J. Mol. Biol. 85, 551.
6. Stamminger, G., and Lazzarini, R. (1979). Cell 3, 85.
7. Emerson, S., Dierks, P., and Parsons, J. (1977). J. Virol. 23, 708.
8. Schubert, M., Keene, J., Lazzarini, R., and Emerson, S. (1978). Cell 15, 103.
9. Kingsford, L., and Emerson, S. U. (1980). J. Virol. 33, 1097.
10. Clinton, G. M., Burge, B. W., Huang, A. S. (1978). J. Virol. 27, 340.
11. Clinton, G. M., Little, S. P., Hager, F. S., and Huang, A. S. (1978). Cell 15, 1455.
12. Wilson, T., and Lenard, J. (1981). Biochem. 20, 1349.
13. Carroll, A. R., and Wagner, R. R. (1979). J. Virol. 29, 134.
14. Pinney, D. F., and Emerson, S. U. (1982). J. Virol. 42, 897.

VESICULAR STOMATITIS VIRUS PROTEINS REQUIRED FOR THE IN VITRO REPLICATION OF DEFECTIVE INTERFERING PARTICLE GENOME RNA[1]

Richard W. Peluso and Sue A. Moyer

Department of Microbiology
Vanderbilt University School of Medicine
Nashville, Tennessee 37232

    We have previously described a procedure which yields extracts of vesicular stomatitis virus (VSV) infected cells which will support the in vitro RNA replication of both wild type VSV and its defective interfering (DI) particle, MS-T (1). Infected cell monolayers are permeabilized by incubation for 60 seconds with lysolecithin after which they are scraped from the dish and disrupted by pipetting. Following the removal of nuclei by centrifugation a cell-free extract is obtained. We have shown that this extract will support the replication and encapsidation of wild type 42S RNA and DI 19S RNA in vitro for 90 minutes. We have also shown that initiation of RNA replication occurs in vitro in these extracts by adding purified detergent-disrupted DI particles to extracts derived from cells infected with wild type VSV alone and demonstrating the synthesis and encapsidation of 19S DI genome RNA.
    Replication of viral RNA in the system described here utilizes soluble proteins which are present in the infected cells when extracts are made. We have shown that extracts can be separated by centrifugation to yield one fraction containing viral nucleocapsids and a second fraction consisting of soluble proteins. These two fractions can be combined to reconstitute the replication of viral RNA, and the soluble protein fraction alone can support the initiation of RNA replication. We have, therefore, turned our attention to an investigation of the soluble protein

---

[1] This work was supported by grant R14594 from NIAID, a Faculty Research Award from the American Cancer Society (to S.A.M.) and a training grant from the National Cancer Institute.

TABLE 1
IN VITRO ASSEMBLY OF MS-T NUCLEOCAPSIDS WITH
[^{3}H]LEUCINE-LABELED VSV INFECTED CELL SOLUBLE PROTEIN

	Nucleocapsid Proteins (cpm)[b]		
Reaction Conditions[a]	L	NS	N
− Replication	117	1149	1154
+ Replication	417	1405	8429

[a]VSV infected BHK cells were labeled with [^{3}H]leucine from 3.5 to 4.5 hr postinfection. A cytoplasmic cell extract was prepared in the absence of ribonucleoside triphosphates and the [^{3}H]-labeled soluble protein fraction was isolated as described previously (1). MS-T intracellular nucleocapsids were isolated from VSV + MS-T infected cells at 4.5 hr postinfection (1). Samples of MS-T nucleocapsids were incubated for 90 min at 30°C with the [^{3}H]leucine-labeled VSV soluble protein fraction in the absence or presence of ribonucleoside triphosphates and an energy regenerating system, which prevents or allows RNA synthesis, respectively.

[b]The products were treated with ribonuclease to degrade mRNAs and the nucleocapsids were purified by sucrose gradient centrifugation (2). The nucleocapsid associated [^{3}H]leucine-labeled proteins were quantitated by scintillation counting of individual proteins following their separation by SDS-polyacrylamide gel electrophoresis and detection by fluorography (3).

preparation from infected cells in order to more clearly define the proteins necessary to support RNA replication.

VIRAL PROTEINS ASSOCIATED WITH MS-T NUCLEOCAPSIDS REPLICATING IN VITRO

To determine which viral proteins are assembled into nucleocapsids during in vitro replication, [^{3}H]leucine-labeled infected cell soluble proteins were mixed with MS-T nucleocapsids from infected cells and incubated in the presence or absence of ribonucleoside triphosphates, conditions which support or prevent in vitro replication, respectively. The nucleocapsid products were then treated with ribonuclease and purified on sucrose gradients. Proteins associated with these nucleocapsids were separated by gel electrophoresis and quantitated (Table 1). In the absence of RNA replication, the apparent exchange of all three nucleocapsid proteins, including the N protein, with the template nucleocapsid occurs. When replication of viral RNA is allowed to proceed, there is a 4-fold increase in the amount of L protein associated with the nucleocapsids, and a 7-fold increase in the amount of N protein associated with the nucleocapsids. The amount of NS protein barely increases over the control level.

MONOCLONAL ANTIBODIES AS PROBES FOR VIRAL RNA REPLICATION

Monoclonal antibodies have been used to study the components of the soluble protein fraction of infected cells which are necessary for in vitro RNA replication. The addition of an anti-M monoclonal antibody to an unfractionated in vitro replication reaction has no effect on either transcription or replication. Each of six different anti-N monoclonal antibodies, however, inhibits both replication and transcription by greater than 90% when added to in vitro reactions (data not shown). The anti-N antibodies apparently bind to the N protein on the nucleocapsid template and inhibit all RNA synthesis.

As an alternative approach, soluble protein preparations were depleted of N protein by incubation with anti-N antibodies, followed by removal of antigen antibody complexes by binding to S. aureus (Cowan strain). The remaining soluble protein, depleted of N protein, is unable to support the in vitro replication of MS-T nucleocapsid RNA (data not shown), thus providing evidence that N protein is required for VSV RNA replication.

These monoclonal antibodies were used to immunoprecipitate [^{3}H]leucine-labeled infected cell soluble protein preparations in the absence of detergents (Figure 1). Each of the six anti-N monoclonal antibodies (lanes B through G) precipitate not only the N protein, but they also precipitate about 10% of the NS protein present in the preparations. When detergents are present during the incubation of antibody with antigen, only the N protein is immunoprecipitated (data not shown). These data suggest that a detergent-sensitive complex between the N and NS proteins exists in the soluble protein fraction of infected cells. This protein complex may be required for efficient in vitro RNA replication.

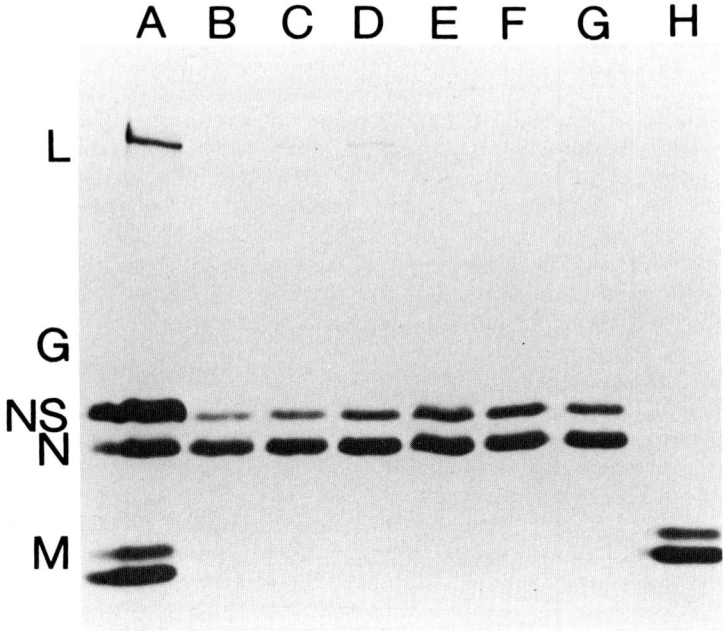

FIGURE 1. Immunoprecipitation of proteins from the soluble protein fraction of VSV infected cells with monoclonal antibodies to VSV proteins. Identical samples of an [^{3}H]leucine-labeled soluble protein fraction from VSV infected BHK cells prepared in a standard replication reaction mixture in the absence of detergents were incubated with antibodies for 2 hr at 4°C. The monoclonal antibodies (ascites fluid) to the VSV N and M proteins were generously donated by Dr. D. Lyles (4). Dr. A.K. Banerjee provided αN-SS (5). The antigen-antibody complexes were collected by binding to S. aureus (Cowan strain), eluted and analyzed by SDS polyacrylamide gel electrophoresis and fluorography (3). Proteins immunoprecipitated by polyclonal αVSV serum (A); αN-SS (B), αN-A3A4 (C); αN-10G4 (D); αN-21F9 (E), αN-19B10 (F); αN-8C11F10(G); and αM-23H12 (H).

COMPONENTS OF THE INFECTED CELL SOLUBLE PROTEIN PREPARATION ACTIVE IN MS-T RNA REPLICATION IN VITRO

To examine possible forms of the VSV N protein functional in in vitro RNA replication and nucleocapsid assembly, in view of our results with the anti-N monoclonal antibodies, we analyzed [^{3}H]leucine-labeled infected cell soluble proteins by centrifugation on 5-20% glycerol gradients. Fractions were collected and immunoprecipitated with anti-VSV serum and the proteins were analyzed by polyacrylamide gel electrophoresis (Figure 2A). The N protein sediments in two broad peaks: as a low molecular weight form in fractions 2, 3 and 4, representing about 60% of the total N protein, and as a high molecular weight form in fractions 8 through 10 with a molecular weight greater than 300,000 based on sedimentation of the L protein ($\sim$200,000 MW) in fraction 6. Note that little if any NS protein sediments with the high molecular weight form of the N protein. Fractions 2, 3 and 4 from the top of the glycerol gradient were also immunoprecipitated with an anti-N monoclonal antibody (Figure 2B). As is seen with unfractionated soluble protein, the anti-N monoclonal antibody precipitates, in addition to N protein, a portion of the NS protein present in the samples, suggesting that an N-NS complex persists in these gradient fractions.

Unlabeled infected cell soluble protein was analyzed on a parallel gradient and selected fractions containing N protein were assayed for their ability to support the replication of MS-T nucleocapsid RNA in vitro as shown in Figure 3. As a control, lane A shows significant synthesis of nucleocapsid associated MS-T genome RNA after the addition of unfractionated infected cell soluble protein to intracellular MS-T nucleocapsids. Proteins in gradient fraction 2 support some MS-T 19S RNA synthesis (lane B); however, the proteins in fraction 3, which contain the peak of the lower molecular weight N protein and of the putative N-NS protein complex (Figure 2), show replication activity close to control levels (lane C). By contrast, low levels of RNA replication occur when the proteins in fractions 8-10 of the gradient are assayed (lanes D-F, respectively). These latter fractions contain the higher molecular weight form of the N protein, representing 40% of the total N protein on the gradient, with only trace amounts of the NS protein. Based on these results, we suggest that a complex consisting of the N and NS proteins exists in VSV infected cells, and that this complex may be the substrate for viral RNA replication. The higher molecular species of the N protein is a poor substrate for viral RNA replication,

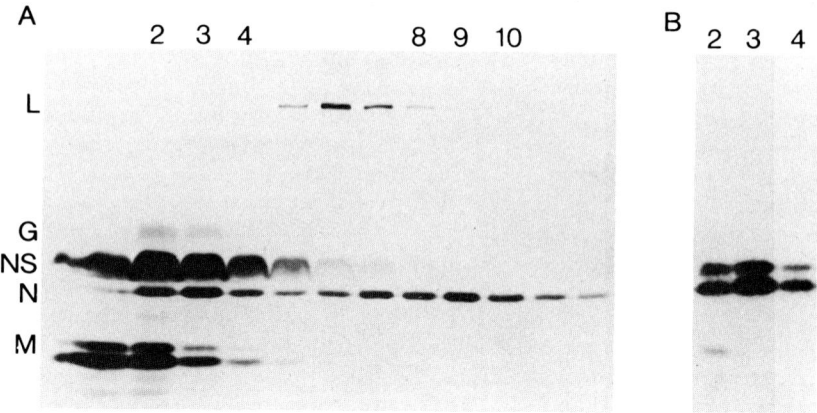

FIGURE 2. Fractionation of the soluble protein portion of VSV infected cells by glycerol gradient centrifugation. VSV infected BHK cells were labeled with [^{3}H]leucine from 3.5-4.5 hr postinfection and a cell extract was prepared in the replication reaction mixture as described (1). The extract was clarified by centrifugation in an SW65 rotor at 50,000 rpm for 75 min at 4°C. The resulting soluble protein fraction was analyzed by sedimentation on a 5-20% (w/v) glycerol gradient in buffer containing 0.2 M NH_4Cl, 0.1 M HEPES, pH 7.4, 7mM KCl, 1 mM DTT in an SW41 rotor at 32,000 rpm for 22 hr at 4°C. One ml fractions were collected from the top of the gradient, viral proteins in the indicated fractions were immunoprecipitated with anti-VSV serum (A) or anti-N monoclonal antibody (B) and analyzed by SDS-polyacrylamide gel electrophoresis. Sedimentation is from left to right.

perhaps due to its state of aggregation and/or the lack of the NS protein.

SUMMARY

The experiments described here utilize monoclonal antibodies to VSV proteins in conjunction with the reconstitution of components of an in vitro replication system to study the mechanism of VSV RNA replication. Based on these data we propose that the essential components for VSV replication include the nucleocapsid template containing the RNA-N complex and the L and NS proteins and, in addition, a putative N:NS protein complex.

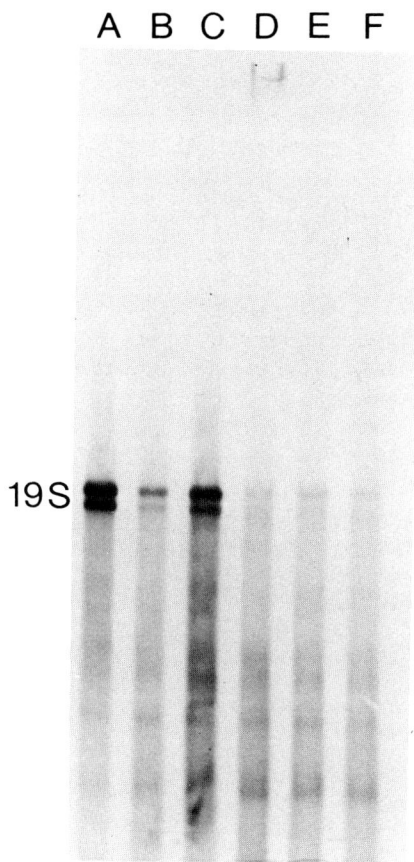

FIGURE 3. Ability of glycerol gradient fractionated VSV soluble protein to support the replication of MS-T RNA. An unlabeled infected cell soluble protein fraction was isolated and separated by glycerol gradient centrifugation as described in Figure 2. Samples of selected fractions were assayed for their ability to support the in vitro replication of MS-T nucleocapsids from infected cells by incubation for 90 min at 30°C in the presence of [^{3}H]UTP. The nucleocapsid associated product RNA was purified by banding on CsCl gradients. The RNA was extracted from the nucleocapsid peak and analyzed by electrophoresis on 1.5% agarose gels in 6M urea containing 25 mM Na citrate, pH 3.5 as described previously (1). MS-T 19S product RNA synthesized with the addition of: unfractionated VSV soluble protein (A), or gradient fractions 2 (B), 3 (C), 8 (D), 9 (E), and 10 (F).

In preliminary experiments we have additional evidence from column chromatography that the N and NS proteins do form a stable complex in vivo. We suggest that it is the N:NS protein complex, rather than N protein alone, which serves as the substrate for N protein binding to nascent replicative RNA. It appears that the NS protein is dissociated upon the binding of N protein to the RNA, since we can detect no coordinate binding of NS and N proteins during in vitro RNA replication and nucleocapsid assembly (Table 1). In accord with the model of Kolakofsky and coworkers (6, 7) the N protein bound to nascent RNA then acts as an antiterminator to allow the polymerase to read through the intercistronic boundaries to synthesize and encapsidate genome RNA.

ACKNOWLEDGEMENT

The authors would like to thank Drs. D. Lyles and A.K. Banerjee for the generous donation of the VSV monoclonal antibodies.

REFERENCES

1. Peluso, R.W., and Moyer, S.A. (1983). Proc. Natl. Acad. Sci. USA. 80, 3198.
2. Moyer, S.A., and Gatchell, S.H. (1979). Virology 92, 168.
3. Horikami, S.M., and Moyer, S.A. (1982). Proc. Natl. Acad. Sci. USA. 79, 7694.
4. LeFrancois, L., and Lyles, D.S. (1982). Virology 121, 157.
5. De, B.P., Tahara, S.M., and Banerjee, A.K. (1982). Virology 122, 510.
6. Blumberg, B.M., Leppert, M., and Kolakofsky, D. (1981). Cell 23, 837.
7. Blumberg, B.M., Giorgi, C., and Kolakofsky, D. (1983). Cell 32, 559.

CHARACTERIZATION OF POLYCISTRONIC TRANSCRIPTS IN NEWCASTLE DISEASE VIRUS-INFECTED CELLS[1]

A. Wilde and T. Morrison

Department of Molecular Genetics and Microbiology
University of Massachusetts Medical School
Worcester, Massachusetts 01605

Newcastle Disease Virus (NDV), a paramyxovirus, is a negative-stranded RNA virus whose genome is transcribed into six monocistronic messenger RNA's (1,2). These RNA's sediment on sucrose gradients at 18S and 35S (3,4). The 18S mRNA's encode the NDV membrane (M) protein, phosphoprotein (P), nucleocapsid protein (NP), fusion protein (F), and hemagglutinin-neuraminidase (HN) protein (5). The 35S mRNA encodes the L protein (6). Figure 1 shows total intracellular NDV RNA resolved on a formaldehyde-agarose gel. The 18S RNA which contains five species (5) resolves into four bands in this gel system. The 35S RNA contains one species (1,2).

Using the nuclease inhibitor nucleoside vanadyl complexes (7), we have detected five new species of intracellular RNA which sediment at 22S to 28S on sucrose gradients. These transcripts contain poly A and represent 15 to 25% of the total intracellular poly A-containing RNA. The molecular weights of these species range from 0.96×10^6 to 1.57×10^6 daltons (Figure 1).

The coding capacity of the NDV genome (mol. weight $5.2 \times 10^6 - 5.7 \times 10^6$ daltons) (8) is accounted for by the six well characterized monocistronic messenger RNA's. Thus it was likely that the 22-28S RNA contained sequences present in the monocistronic RNA's (9, 10). To characterize the specific sequences present in each of the five 22S-28S RNA species, cDNA clones which contained sequences of each of the NDV genes were used as probes on Northern blots of total NDV intracellular RNA. Clones were derived by reverse transcription of NDV mRNA using standard methods. The gene from which each clone is derived was identified by hybrid selected translation (11).

---

1. This work was supported by NIH Grant AI-13847.

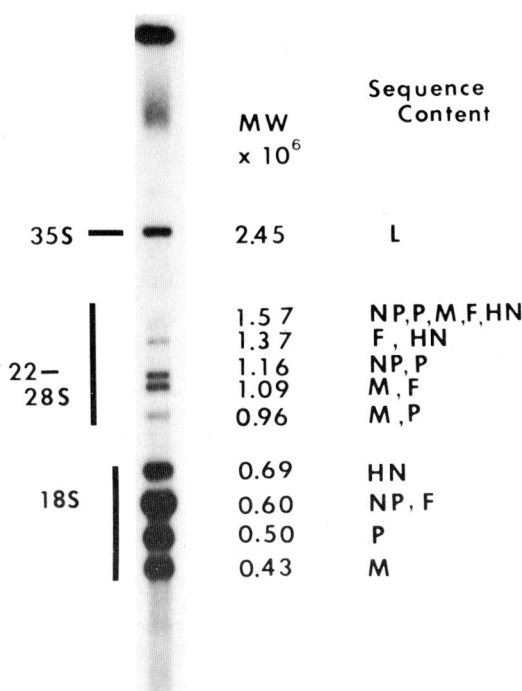

Figure 1: Total intracellular RNA resolved on a 1% formaldehyde gel.

By Northern blot analysis, we found that the 0.96×10^6 dalton species contains M and P gene sequences. The 1.09×10^6 dalton species contains M and F gene sequences while the 1.16×10^6 dalton species contains P and NP gene sequences. The 1.37×10^6 dalton species contains F and HN gene sequences. Thus these four species each contain sequences from two genes (see Figure 1). They likely represent transcription across two adjacent genes. The fifth 22-28S species is resolved as a wide band with a molecular weight of from 1.57 to 1.61×10^6 daltons. This RNA contains sequences of NP, P, M, F and HN genes. The most likely explanation for this species is that it contains a mixture of two or three different RNA's, each of which represents transcription across three adjacent genes. Thus we feel that the 22S-28S RNA's represent transcription across adjacent genes to generate polycistronic transcripts, either dicistronic or tricistronic. And, in fact, the sizes of these transcripts correspond closely to the sizes predicted by the sums of two or three monocistronic transcripts (Table I).

By ordering these species according to overlapping sequences (Figure 2), it is possible to establish a map order for the NDV genome of NP,P,M,F,HN. This order confirms the UV transcription map of Collins et al (12).

TABLE 1

SIZES OF TRANSCRIPTS GENERATED BY TRANSCRIPTION ACROSS ADJACENT GENES x10^6 DALTONS

Gene combinations	Predicted size[a]	Measured size[b]	Predicted size minus Poly A[c]	Measured size minus Poly A[d]
NP-P	1.10	1.16	1.04	1.12
P-M	0.93	0.96	0.87	0.92
M-Fo	1.03	1.09	0.97	1.03
Fo-HN	1.29	1.37	1.22	1.36
HN-L	3.14	NS		
NP-P-M	1.53	1.56	1.44	1.52
P-M-F	1.53	1.58	1.44	1.52
M-F-HN	1.72	NS		
F-HN-L	3.74	NS		
F-HN-Y[e]	–	1.61[e]	–	1.52

[a] Predicted size determined by adding the molecular weights of the monocistronic mRNA's shown in Figure 1.

[b] Sizes measured on Northern blots.

[c] Predicted size determined by adding the molecular weights of the monocistronic mRNA's without Poly A as determined from Figure 3.

[d] Sizes measured from Figure 3.

[e] An undefined space (Y) adjacent to HN must be postulated to account for this species.

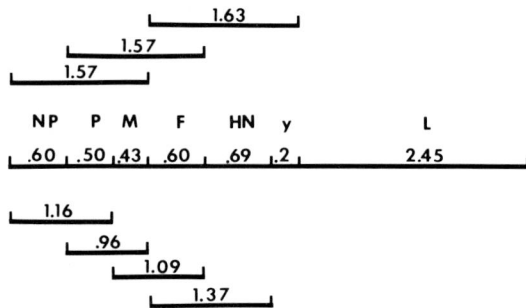

Figure 2: Map order of NDV Generated by overlapping sequences.

Polycistronic transcripts have also been found by Herman et al (13) in in vitro vesicular stomatitis virus transcription reactions, as well as in infected cells. These polycistronic transcripts contain long stretches of internal poly A at the intercistronic boundaries. We, therefore, asked if the NDV polycistronic transcripts contain such sequences. That these transcripts are slightly larger than their predicted size (Table 1), might lead one to suspect the presence of long stretches internal poly A.

In order to detect internal poly A, we hybridized NDV RNA to oligo dT and then digested the hybrid structures with RNase H. Any double-stranded structure should be susceptible to digestion. If there is internal poly A, then the 22S to 28S RNA should disappear after digestion. Such a digestion does reduce the size of the 18S RNA species slightly consistent with the removal of 3' poly A. In addition, the sizes of the 22S-28S RNA species are slightly reduced, consistent with removal of 3' poly A sequences (Figure 3). However, these species remain intact. Thus by this procedure, no internal poly A sequences are detected. Thus the mechanisms responsible for the generation of these species may reflect transcription events different from those responsible for VSV polycistronic transcripts.

CHARACTERIZATION OF POLYCISTRONIC TRANSCRIPTS

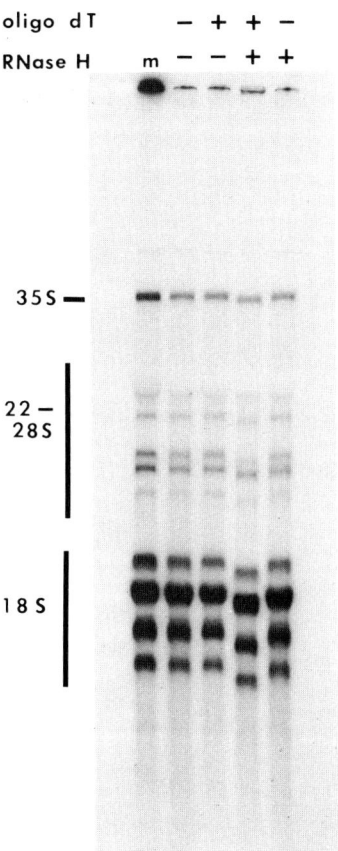

Figure 3: Digestion of NDV RNA with RNase H in the presence of Oligo dT. RNA digested with RNase H in the presence of Oligo dT is indicated by + control digestions containing no RNase H or Oligo dT, or neither are indicated at top of the figure.
m = marker RNA not subjected to reaction conditions.

The details of the transcription mechanisms of paramyxoviruses have not been extensively explored. By analogy with VSV, two mechanisms are possible: processing of a long transcript or sequential start-stop synthesis of each mRNA. The existence of polycistronic transcripts containing no internal poly A is compatible with either model. If a hypothetical nuclease were inefficient in attacking cleavage sites in between cistrons, then various combinations of polycistronic transcripts would be found.

Alternatively, if the NDV polymerase is inefficient in the termination of synthesis at the end of each cistron, then a certain percentage of transcripts would occur in a polycistronic form.

REFERENCES

1. Weiss, S.R., and Bratt, M.A. (1976). J. of Virol. 18, 316.
2. Collins, P.L., Hightower, L.E., and Ball, L.A. (1978). J. of Virol. 28, 324.
3. Bratt, M.A., and Robinson, W.S. (1967). J. Mol. Biol. 23,1.
4. Weiss, S.A., and Bratt, M.A. (1974). J. of Virol. 13, 1220.
5. Collins, P.L., Wertz, G.W., Ball, L.A., and Hightower, L.E. (1982). J. of Virol. 43, 1024.
6. Morrison, T.G., Weiss, S.A., Hightower, L.E., Spanier-Collins, B., and Bratt, M.A. (1975); in vitro transcription and translation of viral genomes (A.L. Haenni and G. Baud, eds.), Inserm, Paris.
7. Berge, S.L., and Birkenmeier, C.S. (1979). Biochem. 19, 5143.
8. Kolakofsky, D.E., de la Tour, B., and Delius, H. (1974). J. of Virol. 13, 261.
9. Varich, N.L., Lukashevich, I.S., and Kaverin, N.V. (1979). Acta Virol. 23, 273.
10. Varich, N.L., Lukashevich, I.S., and Kaverin, N.V. (1979). Acta Virol. 23, 341.
11. Smith, D.F., Searle, P.F., and Williams, J.G. (1979). Nucl. Acids Res. 6, 487.
12. Collins, P.L., Hightower, L.E., and Ball, L.A. (1980). J. of Virol. 35, 682.
13. Herman, R.C., Adler, S., Lazzarini, R.A., Colonno, R.J., Banerjee, A.K., and Westphal, H. (1978). Cell 15, 587.

MOLECULAR STUDIES ON CANINE DISTEMPER
VIRUS REPLICATION[1]

T. Barrett, N.T. Gorman[2], R.C. Patterson[3] and B.W.J. Mahy

Division of Virology
University of Cambridge
New Addenbrooke's Hospital
Hills Road
Cambridge CB2 2QQ

INTRODUCTION

The morbilliviruses, a subgroup of the paramyxoviridae which includes measles virus, canine distemper virus (CDV) and bovine rinderpest virus, cause acute febrile illness in their hosts. Measles virus and CDV are also known to establish persistent infections involving the central nervous system in some cases; measles is the cause of subacute sclerosing panencephalitis (SSPE) in man and CDV causes old dog encephalitis, but the mechanism whereby persistence is established is not understood (1). Molecular differences have been shown in the replication of viruses isolated from SSPE and old dog encephalitis brains. Sera from SSPE patients lack antibodies against the virus M protein while sera from dogs with similar demyelinating diseases contain antibodies to all canine distemper virus proteins (2). The establishment of persistence may nevertheless involve similar genetic mutations, and the induction of defective interfering (DI) RNAs in both viruses. Differential protein expression may occur in addition to this. In this communication we describe the virus-specific RNA species that can be observed in Vero cells infected lytically and persistently with CDV and provide evidence of defective RNA production in the persistently infected cells.

---

[1]The project was supported by grant No. G80/0433/OCA from the British Medical Research Council.

[2]College of Veterinary Medicine, Department of Medical Sciences, University of Florida, Gainesville 32610.

[3]Department of Clinical Veterinary Medicine, Cambridge.

VIRUS-SPECIFIC RNAs PRODUCED IN LYTIC INFECTIONS

For lytic virus infections we used Vero cells infected at a multiplicity of 0.1 p.f.u./cell with CDV (Onderstepoort strain). RNA was labelled in vivo (in both infected and uninfected cells) using ^{32}P-orthophosphate. Nuclear and cytoplasmic extracts were prepared by treating the cells with 1% Nonidet p40 and 0.2% DOC (3). Purified RNA was then fractionated into polyadenylated and nonpolyadenylated RNA by oligo(dT) cellulose chromatography (4). Analysis of RNA by polyacrylamide gel electrophoresis showed only one high molecular weight virus-specific RNA band in the polyadenylated RNA. To improve separation, the RNA was subjected to electrophoresis under completely denaturing conditions in 1.5% agarose gels containing 2M formaldehyde (5). The RNA was then blotted to nitrocellulose paper (6) and autoradiographed. This method greatly improved the separation of the higher molecular weight RNAs and a series of seven virus-specific RNAs could be detected in the polyadenylated RNA. When RNA from similar cells was labelled in the presence of actinomycin D (10 µg/ml) no labelled polyadenylated RNA could be detected in the uninfected cells, while all the virus-specific RNAs remained insensitive to the drug (Figure 1). No virus-specific RNA could be detected in the nuclear fractions.

The most abundant virus messenger RNA species (RNA band 6) has been cloned in a bacterial plasmid, and found to code for the most abundant virus structural protein, the nucleoprotein (Barrett and Mahy, submitted). When infected cells were labelled with ^{32}P at various times post infection (Figure 2) no virus-specific mRNA could be detected up to 12 h.p.i. In more detailed studies on the time of appearance of virus mRNAs (data not shown) virus-specific RNA was detected from 16 h.p.i. and was maximal at 24 h.p.i. It can be seen from Figure 2 that while all host cell polyadenylated RNA is sensitive to 10 µg/ml actinomycin D, some host cell non-polyadenylated RNAs are resistant to inhibition by the drug. These host RNAs are reduced significantly as infection proceeds, probably due to the cytocidal effect of the virus. However, in the absence of actinomycin D the host cells are still capable of significant host mRNA synthesis at 24 h.p.i. (compare tracks 3 and 4, Figure 2).

VIRUS-SPECIFIC RNAs PRODUCED IN PERSISTENT INFECTIONS

We have established a persistent CDV infection by multiple undiluted passages of virus in Vero cells and culturing the surviving cells. The persistently infected cells were

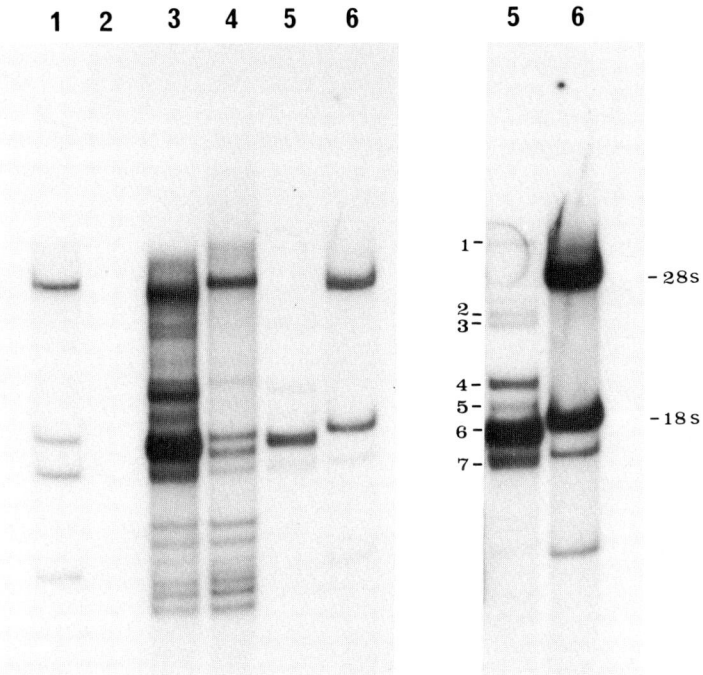

FIGURE 1. Analysis of ^{32}P-labelled polyadenylated (A+) and non-polyadenylated (A-) RNA from uninfected and CDV infected Vero cells. Cells were phosphate-starved at 24 h.p.i. for 3 hr and then labelled for 3 hr with ^{32}P-orthophosphate (2 mCi/15 cm dish). Cytoplasmic and nuclear fractions were prepared by treating the cells with 1% Nonidet P40 and 0.2% deoxycholate (3). RNA was fractionated into (A-) and (A+) species by oligo(dT) cellulose chromatography (4). RNA samples were denatured in 50% formamide, 2M formaldehyde and 1 x MOPS (20 mM MOPS, 5 mM sodium acetate, 1 mM EDTA, pH 7.0) for 5 min at 60% and separated by electrophoresis on a 1.5% agarose (HGT) gel containing 2M formaldehyde and 1 x MOPS at 50 mA for 2 hr (5). The gel was soaked in 20 x SSC and blotted to nitrocellulose (6) before exposure to X-ray film. Where indicated actinomycin D (AMD) was added (10 µg/ml) during both the phosphate starvation and labelling periods. A longer exposure of tracks 5 and 6 is shown on the right. The positions of the AMD-insensitive RNAs (1-7) and the 28S and 18S ribosomal RNAs are indicated.

1, Uninfected (A-) RNA; 2, uninfected (A+) RNA+ AMD; 3, infected (A+) RNA; 4, uninfected (A+) RNA; 5, infected (A+) RNA + AMD; 6, infected (A-) RNA.

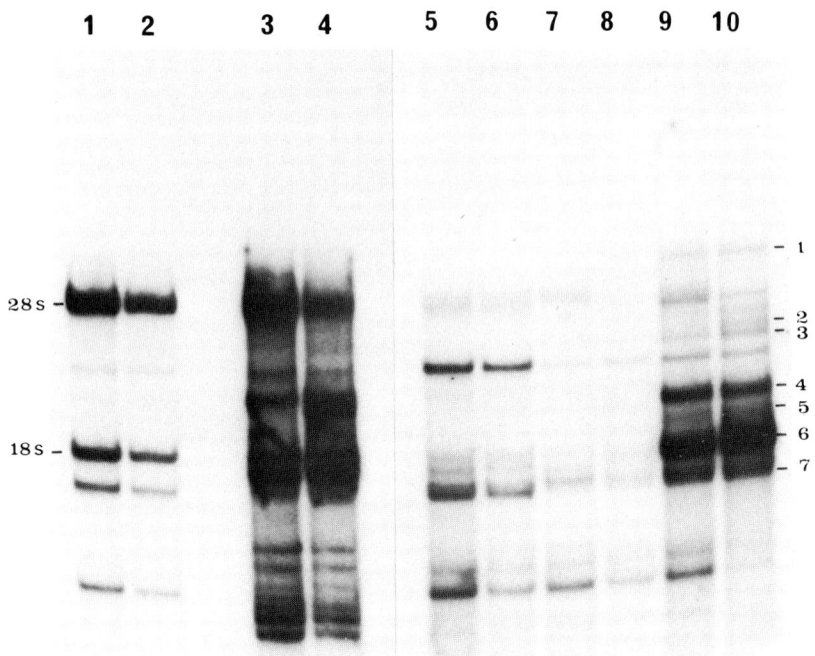

FIGURE 2. Analysis of ^{32}P-labelled RNA isolated from the cytoplasm of uninfected and CDV infected Vero cells at various times post infection. RNA was labelled <u>in vivo</u> with ^{32}P and separated by electrophoresis in 1.5% agarose-formaldehyde gels as detailed in Figure 1. Where indicated AMD was present during the 3 hr phosphate starvation period and during the 3 hr labelling period at 10 μg/ml. The time post infection indicates the time at which phosphate starvation was begun. The positions of the 28S and 18S ribosomal RNA markers and the virus-specific mRNAs (1-7) are indicated at the sides of the figure.

Tracks 1 and 2, (A-) RNA; 3 and 4, (A+) from uninfected and infected cells, respectively. Tracks 5-10 represent total cytoplasmic RNA labelled in the presence of AMD. 5, uninfected cell RNA; 6-10, infected cell RNA labelled at 4, 8, 12, 20 and 24 h.p.i., respectively.

resistant to superinfection and over 90% showed virus specific antigens on immunofluorescence. Electronmicrographs showed typical virus nucleocapsids in the cytoplasm of persistently infected cells (Figure 3). Most of the infectious virus remained cell-associated and was not released into the medium. In contrast, lytically infected cells release most of the

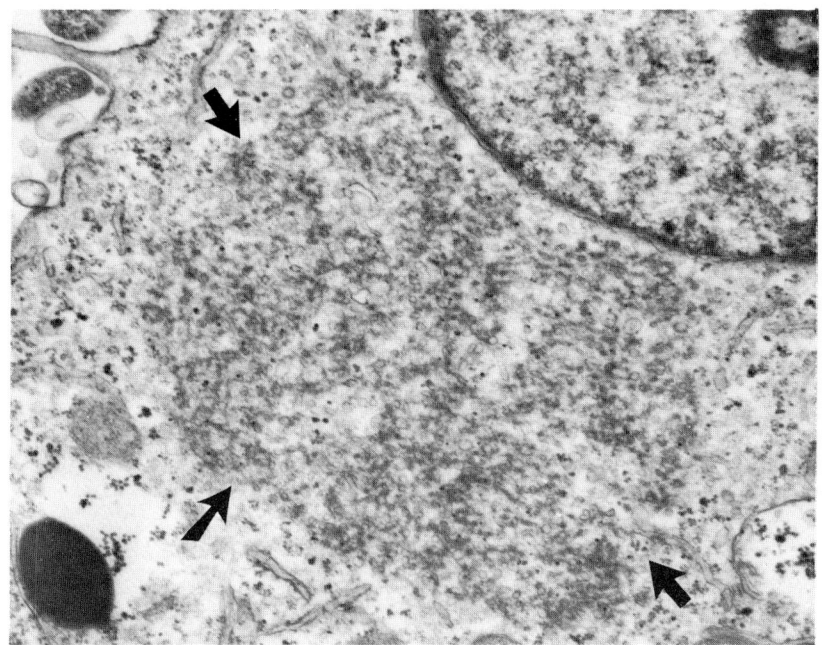

FIGURE 3. Electronmicrograph of a persistently infected Vero cell. Arrows show the position of virus nucleocapsids. Part of the nucleus can be seen in the top right-hand corner. Magnification x 21,600.

infectious virus into the medium and the cell monolayers display gross cytopathic effects by 24 h.p.i.

Figure 4 shows the RNA species detected in persistently infected cells labelled in the presence of actinomycin D (10 μg/ml). Four virus-specific bands were observed which were resistant to DNase$_1$ digestion. Only one RNA (RNA 3) corresponded in size to a virus-specific mRNA detected in lytic infections i.e. to mRNA 6, the mRNA for the virus nucleoprotein (Barrett and Mahy, submitted). The other three RNAs did not correspond to any virus-specific mRNA nor to any host cell RNA resistant to actinomycin D.

To determine whether or not RNA 3 from persistently infected cells was the virus nucleoprotein mRNA, unlabelled RNA was hybridised with ^{32}P-labelled plasmid DNA containing a CDV nucleoprotein-specific insert. This DNA hybridised with two DNase-resistant bands, one of which (number 2) co-migrated exactly with nucleocapsid mRNA from lytically infected cells (Figure 4). The second band of higher molecular weight did not correspond in molecular weight with any of the RNAs

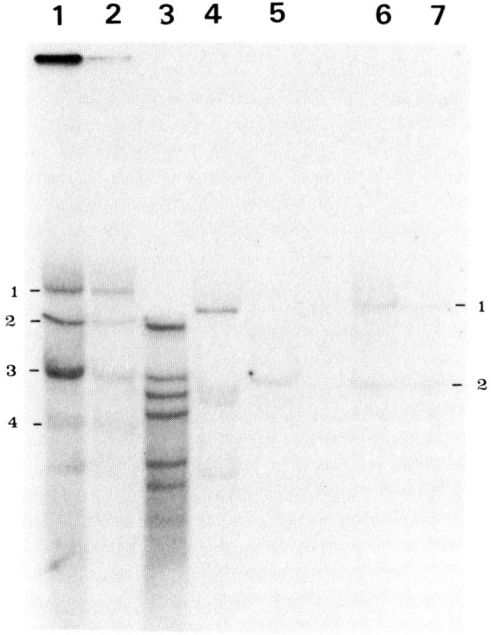

FIGURE 4. Analysis of RNA from Vero cells persistently infected with CDV. The cells were phosphate starved for 3 hr prior to labelling with ^{32}P orthophosphate (2 mCi/15 cm dish). AMD (10 μg/ml) was present during the 3 hr starvation period and during the 3 hr labelling period. Total nucleic acid was extracted from the cells by digesting the cell monolayers with 5 ml pronase (0.5 mg/ml) in 50 mM Tris HCl, pH 7.5, 25 mM Na Cl, 5 mM EDTA, 0.25% SDS at 37° for 1 hr. Immediately after addition of the pronase the viscous cell extract was passed several times through a syringe needle to shear the DNA. Nucleic acid was then purified by phenol:chloroform (1:1) extraction. In some cases the DNA was digested with $DNase_1$ (50 μg) in 2.0 ml 100 mM NaCl, 15 mM $MgCl_2$, 100 mM Tris HCl, pH 7.4, for 5 min at room temperature. RNA was then purified by a second phenol:chloroform extraction. Unlabelled nucleic acid from persistently infected cells was also run on the same gel (tracks 6 and 7). These tracks were cut from the nitrocellulose sheet and hybridised (6) with ^{32}P nick-translated plasmid DNA containing a virus nucleoprotein gene-specific insert (Barrett and Mahy, submitted). The most abundant virus-specific mRNA detected in lytic infections (track 5), which codes for the virus nucleoprotein, can be seen to co-migrate with RNA3 from persistent infections (tracks 1 and 2). The nucleoprotein clone (224) hybridises with an RNA of the same size in (tracks 6 and 7) continued.

observed in either lytically or persistently infected cells labelled in vivo. It is not clear if this band represents a true virus-specific RNA or is an artefact due to aggregation, since a similar high molecular weight band hybridised when the nucleoprotein clone was hybridised to (A+) RNA from lytically infected cells. No hybridisation was detected with (A+) RNA from uninfected cells, however. Table 1 lists the molecular weights of the virus-specific RNAs detected in lytic and persistent virus-infected cells.

TABLE 1
MOLECULAR WEIGHTS OF VIRUS-SPECIFIC RNAs DETECTED IN LYTIC AND PERSISTENT INFECTION OF CANINE DISTEMPER VIRUS

RNA no.	1	2	3	4	5	6	7
Lytic	.96	.78	.76	.64	.58	.50	.43
RNA No.	1	-	2	-	-	3	4
Persistent	.84	-	.70	-	-	.50	.39

RNAs were labelled with ^{32}P-orthophosphate in the presence of 10 µg/ml actinomycin D. Molecular weights ($\times 10^{-6}$) were determined from their migration in agarose-formaldehyde gels by comparison with influenza A virus (fowl plague) RNA markers of known nucleotide composition (7).

CONCLUSIONS

RNA transcribed in lytic and persistent infections with CDV was analysed by electrophoresis in denaturing agarose-formaldehyde gels. Seven virus-specific polyadenylated RNAs (i.e. mRNAs) could be detected from 16 h.p.i. and synthesis

---

Figure 4 continued

Track 1, persistently infected cell RNA; 2, as 1 but treated with DNase$_1$; influenza A (fowl plague) RNA markers; 4, total cytoplasmic RNA from uninfected cells labelled in the presence of AMD; 5, total cytoplasmic RNA from lytically infected cells (24 h.p.i.) labelled in the presence of AMD; 6, unlabelled nucleic acid from persistently infected cells hybridised with ^{32}P-clone 224 DNA; 7, as 6 but treated with DNase$_1$ before electrophoresis.

was maximal at 24 h.p.i. No virus-specific RNA synthesis could be detected in the nuclei of infected cells, and it is probable that the cytoplasm is the site of all virus RNA synthesis. Four virus-specific RNAs could be detected in persistently infected cells, only one of which corresponded in molecular weight to a virus-specific RNA, the nucleoprotein mRNA from lytic infections. The identity of this RNA in persistent infections was confirmed using a plasmid DNA containing a CDV nucleoprotein gene insert as a hybridisation probe. The function of the other virus-specific RNAs in persistently infected cells is unknown. However, aberrant RNAs of similar size have previously been detected in Vero cells persistently infected with SSPE (8). The generation of these unusual subgenomic RNAs in CDV persistent infection is under investigation.

ACKNOWLEDGMENTS

We would like to thank Mr. A.W. Dunne and Miss G.C. Clive for excellent technical assistance.

REFERENCES

1. ter Meulen, V. and Carter, M.J. 1982) in "Virus Persistence", eds. B.W.J. Mahy, A.C. Minson and G.K. Darby, Cambridge University Press, p.97.
2. Hall, W.W., Imagawa, D.T. and Choppin, P.W. (1979) Virology 98, 283.
3. Briedis, D.J., Conti, G., Munn, E.A. and Mahy, B.W.J. (1981) Virology 111, 154.
4. Glass, S.E., McGeoch, D. and Barry, R.D. (1975) J. Virol. 16, 1435.
5. Inglis, M.M. and Darby, G. (1981) Nucl. Acids Res. 9, 5569.
6. Thomas, P.S. (1980) Proc. Natl. Acad. Sci. U.S.A. 77, 5201.
7. McCauley, J. and Mahy, B.W.J. (1983) Biochem. J. 211, 281.
8. Kratsch, V., Hall, W.W., Nagashima, K. and ter Meulen, V. (1977) J. Med. Virol. 1, 139.

RNA-DEPENDENT RNA POLYMERASE ASSOCIATED WITH RESPIRATORY SYNCYTIAL VIRUS

Gustave N. Mbuy and Olga M. Rochovansky

Department of Molecular Virology
The Christ Hospital Institute of Medical Research
Cincinnati, Ohio 45219

Respiratory syncytial virus (RSV), initially isolated by Chanock et al. (1) is a large pleomorphic, enveloped virus which resembles paramyxoviruses in general morphology and size (2-4). The absence of detectable hemagglutinating and neuraminidase activities in RSV envelope glycoproteins has resulted in the classification of this virus in a separate genus, Pneumovirus, within the Paramyxoviridae family (5). RSV contains a negative-strand RNA genome of approximately 5×10^6 daltons (6, 7) that has the capacity to code for at least 9 distinct complementary mRNAs (6-8). The formation of viral mRNAs in infected cells is the first biosynthetic reaction in the virus replication cycle, suggesting that the virus contains a RNA-dependent RNA polymerase (transcriptase). The purpose of the present investigation was to demonstrate whether a transcriptase is associated with purified RS virions and to determine the optimal conditions for its expression in vitro.

IN VITRO REACTION

Conditions for Activity. Human RS virus (Long strain) was used throughout this study. Virus was grown and purified as previously described (6) and further concentrated by sedimentation onto a 60% sucrose cushion. Sucrose, an inhibitor of transcriptase activity, was removed by chromatography of the virus on a short column of Sephadex G-100 equilibrated with STE (0.1 M NaCl, 0.01 M Tris-HCl, pH 7.9, 1 mM EDTA).

Results in Table 1 indicate that a transcriptase was present in purified preparations of RSV and define the conditions required for its in vitro activity. The data were obtained with [^{3}H]UTP as radioactive substrate and activity was measured as the incorporation of label into acid-precipitable product. Optimum activity was obtained at pH 7.9 and 30°. The reaction required both mono- and divalent cations. Sodium or potassium ions were equally effective at concentrations of

TABLE 1
OPTIMUM CONDITIONS FOR RSV IN VITRO TRANSCRIPTION[a]

Conditions	Fresh virus[b] pmols incorp.	Aged virus[b] pmols incorp.
Complete	12.3 (100)[c]	69.8 (100)[c]
-lysolecithin	4.7 (38)	44.6 (64)
$-MgCl_2$; $+MnCl_2$	3.2 (26)	19.5 (28)
-CTP, GTP, ATP	9.6 (78)	51.8 (74)
+DTT (4mM)	1.8 (15)	67.4 (97)
+RNase (50 µg)	1.6 (13)	10.4 (15)
+DNase (50 µg)	11.1 (90)	65.6 (94)
+Act. D (10 µg)	11.8 (96)	66.3 (95)

[a] Assays in an incubation volume of 0.25 ml contained: Tris-HCl, pH 7.9, 60 mM; GTP and CTP, 0.83 mM; ATP, 1.66 mM; NaCl, 120 mM; $MgCl_2$, 8 mM; lysolecithin, 0.04%. When substituted for $MgCl_2$, the $MnCl_2$ concentration was 2 mM. The labeled substrate, [^{3}H]UTP, was present at a concentration of 0.10 mM and contained 200 cpm per pmol. Incubations were for 60 min. at 30°.

[b] Virus concentration was 9.2 µg of viral protein assayed shortly after preparation (fresh) of after 6 weeks of storage (aged).

[c] Values in parentheses represent percentages of the complete reaction.

120 mM. Magnesium appeared to be the divalent cation of choice at an optimum concentration of 8 mM. Substitution of magnesium by manganese ions at concentrations of 2 mM or higher resulted in marked inhibition of activity. The reaction was not sensitive to DNase or actinomycin D but was greatly inhibited by the addition of RNase.

Although the conditions were in general similar to those reported for other negative-strand RNA viruses, the RSV system did show some anomalies. Dependence of the reaction on detergent was incomplete. This was most probably due to the virus purification procedure which includes a sonication step to remove contaminating cellular vesicles (6). It is likely that sonication partially disrupted some virus particles allowing transcriptase activity to be expressed in the absence of detergent. The activity was also not greatly depressed by the omission of unlabeled nucleoside triphosphates. Similar results were obtained when CTP or ATP were used as labeled substrates (data not shown). There is precedent for both lack of dependence on detergent (9, 10) and nucleotides (9, 11, 12) in other viral systems.

A more unusual finding was made when the effect of the reducing agent, dithiothreitol (DTT), on the RSV transcriptase was tested. In contrast to results obtained with transcriptases of other negative-strand RNA viruses (10, 12, 13) which showed at least partial dependence on DTT, the activity associated with fresh preparations of RSV was inhibited 85% by 4 mM DTT. Lower concentrations of DTT (1 to 3 mM) were also found to inhibit the reaction substantially (at least 75%).

Another surprising finding was made when the activity of aged virus was tested. It was observed that transcriptase activity of virus stored at 4° for several weeks increased 5 to 6-fold (Table 1). This was a consistent finding made with all virus preparations. In contrast to fresh virus, the activity of aged virus was neither inhibited nor stimulated by DTT. It was also observed that dependence of the reaction on detergent was decreased. Dependence on nucleotides, mono- and divalent cations as well as sensitivity to RNase of both fresh and aged virus preparations were entirely similar. The reasons for the increased activity of aged virus and for the different responses to DTT and detergent are not presently understood.

Dependence of Activity on Virus Concentration and Time. Figure 1 shows that transcriptase activity of aged virus was not linearly dependent on virus concentration. Increasing viral protein from 1.84 to 3.68 μg resulted in only a 1.47-fold increase in RNA formation. A pattern of progressively lower amounts of product synthesis over the expected was observed when viral protein was increased to 51.52 μg, suggesting the presence of an inhibitor of transcription. Inhibition was most evident when fresh virus was tested. A 10-fold increase in viral protein (1.67 to 16.7 μg) resulted in only a 1.2-fold increase in product formation. Moreover, 20 to 30-fold increases in concentration resulted in inhibition of RNA synthesis (data not shown). These results suggest that the putative inhibitor was partially inactivated during storage of the virus.

Dependence of the activity of aged virus on reaction time is shown in Figure 2. The rate of the reaction was linear for the first 30 min. and then decreased. Synthesis continued at a decreasing rate for 5 hours. Beyond 5 hours the amount of acid-precipitable product slowly declined (data not shown) suggesting breakdown probably caused by RNase contamination of the system. With the exception that less product was formed, fresh virus showed essentially the same time course (data not shown).

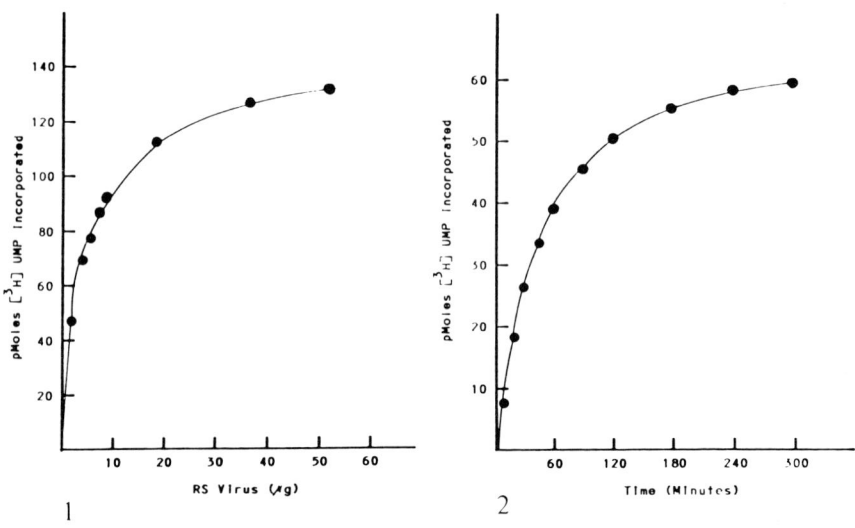

FIGURE 1 (Left). Dependence of transcriptase activity on virus concentration. Reaction volumes of 0.25 ml containing from 1.84 to 51.52 μg of viral protein were prepared (see legend to Table 1). After a 2 hour incubation, RNA was acid-precipitated, filtered and radioactivity determined.

FIGURE 2 (Right). Dependence of transcriptase activity on incubation time. Reaction volumes of 1.5 ml containing 55.2 μg of viral protein were prepared (see legend to Table 1). Reactions were started (0 time) by the addition of ^{3}H-UTP (200 cpm/pmol). At the indicated times 100 μl aliquots were withdrawn, added to 0.2 ml of stopping mixture, and RNA acid-precipitated, filtered, and radioactivity determined.

CHARACTERISTICS OF SYNTHESIZED RNA

Polyadenylation. In order to determine whether in vitro synthesized RNA had the properties expected for viral mRNAs, several characteristics of the RNA were examined. First, polyadenylation of the product was analyzed by affinity chromatography. In vitro formed RNA labeled with [^{3}H]UMP was phenol-extracted and applied to [oligo-dT] cellulose columns. In the several experiments carried out about 50% of the product was bound by the columns, indicating the presence of poly A tracts. After elution of the bound RNA and rechromatography on a second affinity column, about 97% of the RNA was again bound, confirming that 50% of the total synthesized RNA was polyadenylated (poly A+).

Sedimentation. In vitro synthesized RNA was separated into poly A+ and poly A- RNA by chromatography on [oligo-dT] cellulose. Each species of RNA was then sedimented on linear sucrose gradients. As shown in Figure 3A, about 40% of the poly A+ RNA sedimented at 13-25S, approximately the size expected for some RSV mRNAs. The remaining RNA was smaller (4-7S). All of the product was at least 99% sensitive to RNase digestion. Poly A- RNAs gave essentially the same profile with the exception that a small peak, representing about 10% of the total, sedimented at 30-35S (Figure 3, Right). This faster sedimenting RNA was relatively resistant to RNase, suggesting the presence of transcriptive intermediates that were not dissociated from template RNA.

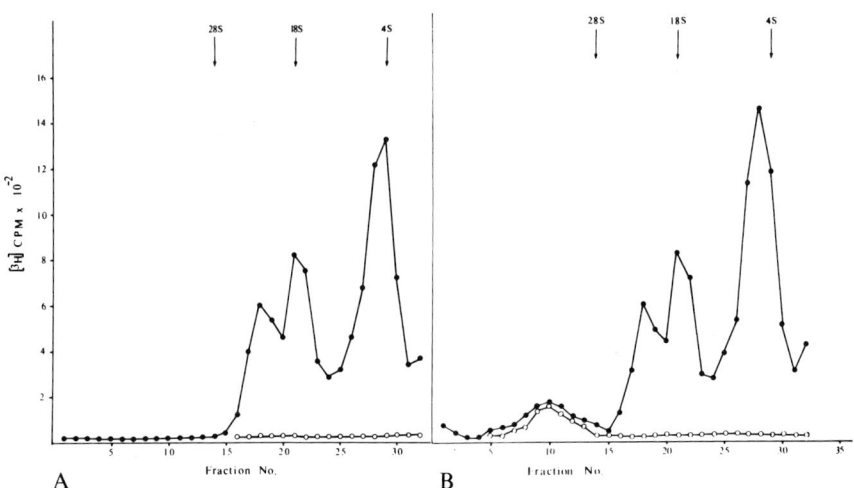

FIGURE 3. Sedimentation analysis of in vitro synthesized RNA. [^{3}H]UMP-labeled product formed in a 2.5 ml reaction volume (see legend to Table 1) containing 92 µg of viral protein and incubated for 2 hours was separated into poly A+ and poly A- RNAs by [oligo-dT] cellulose chromatography. After ethanol precipitation the RNAs were dissolved in H$_2$O, heat denatured and centrifuged on 32 ml sucrose gradients (5-20% sucrose w/v in STE, pH 4.9). Centrifugation was at 20,000 rpm, 4° for 17 hours in a Beckman SW28 rotor. Size markers, 4, 18 and 28S RNAs, were centrifuged on a separate gradient. Fractions of 1 ml were divided into equal aliquots. One set was digested with RNases A and T$_1$ (1 and 10 µg/ml, respectively) for 30 min. at 37°. Aliquots were then acid-precipitated, filtered and radioactivity determined. A (Left): poly A+ RNA, ●-●, no RNase, O-O , RNase treated; B (Right): poly A- RNA.

TABLE 2
ANNEALING OF UNLABELED 50S VIRION RNA TO
^{3}H-URIDINE LABELED IN VITRO POLY A+ RNAs

Annealing Mixture (μl)[a]	cpm[b] No RNase	With RNase	% Resistant
poly A+ RNA (5)	809	473	58
poly A+ RNA (10)	799	479	60
poly A+ RNA (15)	736	478	65
Self-annealing	806	16	2

[a] Varying amounts (μl in parenthesis) of unlabeled genomic 50S RNA extracted from purified virions were mixed with constant amounts of [^{3}H]UMP labeled poly A+ RNA (1800 cpm) synthesized in vitro. RNA samples in 0.1 ml H$_2$O were denatured by heating at 100° for 1 minute and cooled by immersing in ice water. Samples were made 2 X SSC and 50% formamide and annealed as described (6). Each sample was then divided into two equal portions, one set of which was digested with a mixture of RNases A and T$_1$ (1 and 10 μg/ml, respectively) at 37°C for 30 minutes. Acid precipitable radioactivity of each sample was then determined.

[b] Values represent the average of two experiments.

Annealing. Complementarity of in vitro synthesized RNA to RSV genomic 50S RNA was investigated by hybridization. Constant amounts of poly A+ RNA were annealed to increasing amounts of 50S RNA. Results in Table 2 show that self-annealing of poly A+ RNA resulted in only a 2% resistance to RNase digestion indicating that the product was virtually free of contaminating 50S RNA. In contrast, annealing in the presence of RSV 50S RNA increased RNase resistance to 65%. Failure to obtain a value closer to 100% may have been due to insufficient amounts of 50S RNA or to the presence of substantial amounts of small RNAs in the product.

CONCLUSION

The data presented here strongly suggest that RSV contains RNA-dependent and poly A polymerase activities for synthesis of viral mRNAs. Definitive proof for this conclusion depends on obtaining an in vitro system free of contaminating RNase and a demonstration that in vitro formed poly A+ RNAs are of the same size as in vivo synthesized RSV mRNAs and functional in cell-free translation. Studies to obtain this evidence are currently in progress.

ACKNOWLEDGMENTS

We wish to thank Drs. D. Lambert and M. Pons for their helpful discussions.

REFERENCES

1. Chanock, R.M., Roizman, B., and Myers, R. (1957). Am. J. Hyg. 66, 281.
2. Armstrong, J.A., Pereira, H.G., and Valentine, R.C. (1962) Nature. 196, 1179.
3. Bloth, B., Espmark, A., Norrby, E., and Gard, S. (1963). Arch. Gesamte Virusforsch. 13, 582.
4. Joncas, J., Berthiaume, L., and Pavilanis, V. (1969) Virology. 38, 493.
5. Kingsbury, D.W., Bratt, M.A., Choppin, P.W., Hanson, R. P., Hosaka, V., ter Meulen, V., Norrby, E., Plowright, W., Rott, R., and Wunner, W.H. (1978) Intervirology. 10, 137.
6. Lambert, D.M., Pons, M.W., Mbuy, G.N., and Dorsch-Hasler, K. (1980). J. Virol. 36, 837.
7. Huang, Y.T., and Wertz, G.W. (1982) J Virol. 43, 150.
8. Collins, P.L., and Wertz, G.W. (1983) Proc. Natl. Acad. Sci. USA 80, 3208.
9. Bernard, J.P., and Northrop, R.L. (1974) J. Virol. 14, 183.
10. Seifried, A.S., Albrecht, P., and Milstien, J.B. (1978) J. Virol. 25, 781.
11. Robinson, W.S. (1971) J. Virol. 8, 81.
12. Stone, H.O., Portner, A., and Kingsbury, D.W. (1971) J. Virol. 8, 174.
13. Huang, A.S., Baltimore, D. and Bratt, M.A. (1971) J. Virol. 7, 389.

THE EFFECTS OF INTERFERON ON MEASLES VIRUS RNA SYNTHESIS

J. B. Milstien, A. S. Seifried, M. J. Klutch, and B. Bhatia

Division of Virology,
Office of Biologics, NCDB
Food and Drug Administration
Bethesda, Maryland 20205

We have been interested in the effect of interferon (IFN) on measles RNA synthesis in several different systems. Earlier studies examining the effect of IFN on negative strand viruses focused on action at the translational or post-translational level (1-5); the studies described here examine effects at the transcriptional level.

MEASLES INFECTED LYMPHOCYTES

One of the cell-virus systems studied here was human peripheral blood mononuclear cells infected with measles virus. Earlier work suggested that this infection could be induced to change from persistent to productive when IFN-alpha was removed (6). We have confirmed those data, and have further examined the effects of IFN-alpha on accumulation of measles virus specific RNA in these cells.

Table 1 shows that there is a major effect on the accumulation of 50s RNA when IFN is present, correlating with the loss of virus production and a change to a persistent infection. Thus, when the infection becomes productive, after addition of phytohemagglutinin (PHA) (exp. 3) or addition of anti-IFN-alpha, (exp. 8-10), cells show fluorescence with measles positive serum, and 50s measles specific RNA accumulates. Under these growth conditions, IFN concentration in the cell media (columns 4 and 5) is reduced to 50 to 625 U/ml compared to 3000 U/ml. In the presence of anti-measles serum (exp. 4) FA is negative since virus cannot bud from cells in the presence of antibody (Ab); however, 50s RNA does accumulate in the cells and IFN is reduced. The control (exp. 5), using measles-negative serum, resembles the persistent infection (exp. 2). Moreover, the addition of exogenous IFN can overcome the effect of PHA (exp. 6) and, to some extent, that of anti-measles Ab (exp. 7). In contrast, there seemed

TABLE 1
EFFECTS OF TREATMENTS WHICH AFFECT INTERFERON LEVELS
ON MEASLES VIRUS REPLICATION IN HUMAN LYMPHOCYTES[a]

Treatment	FA[b]	50s RNA	IFN After Treatment	IFN At Harvest
1. Uninfected	−	−	125	Less than 5
2. Infected	±	−	3125	1250
3. ___, + PHA[c]	+	+	625	125
4. ___, + Anti-measles Ab after 3 hours	−	+	25	625
5. ___, + Measles-neg Serum after 3 hours	−	−	1000	1000
6. ___, + PHA + 1000 U/ml IFN-beta day 4	−	−	1000	250
7. ___, + Anti-measles Ab after 3 hours + 1000 U/ml IFN-beta day 4	±	+	125	3125
8. ___, + Anti-IFN after 3 hours	+	+	250	125
9. ___, + Anti-IFN after 3 hours + PHA day 4	+	+	50	625
10. ___, + Anti-IFN + PHA both on day 4	±	+	625	125

[a] Exogenously added IFN was interferon-beta or interferon-alpha, which gave comparable results. Anti-interferon was directed against interferon-alpha.
[b] Immunofluorescence, unfixed cells.
[c] Phytohemagglutinin.

to be no qualitative effect on measles mRNA synthesis, though its quantity was decreased following IFN treatment (not shown).

However, because the IFN-treated lymphocytes seem to have additional low molecular weight measles RNA instead of 50s, it is possible that these observations reflect increased breakdown of measles 50s RNA rather than lack of synthesis. This would be consistent with what is known of the antiviral action of IFN, and its action in induction of a ribonuclease (7-11). Studies using monoclonal antibodies specific for components of the measles nucleocapsid suggested that measles nucleocapsids also do not accumulate in these cells (11).

MEASLES INFECTED VERO CELLS

In an attempt to generalize these findings to a simpler system, we examined the effect of IFN on measles virus RNA synthesis in Vero cells, a cell type which produces only low levels of endogenous IFN. Interferon-beta was used for all Vero cell studies. Added IFN at more than 25 U/ml caused a substantial decrease in released virus amounting to 3 to 5 logarithmic units. However, there was no decrease in 50s RNA synthesis when IFN treatment was used (12). Thus the effect of IFN in this system was not analogous to that in lymphocytes.

An examination of mRNAs from Vero cells showed that there was actually an increase in ^{3}H-uridine-labeled mRNA in the presence of IFN especially in mRNA of 16s or less. However, when this mRNA was used to direct in vitro translation, as shown in Figure 1, it was not translated into measles specific proteins at a level commensurate with its concentration, except for a polypeptide coelectrophoresing with measles M protein. If in fact this mRNA represents measles specific RNA, why is it not translated?

To answer this, we first examined polysomes from infected cells and found that after treatment with 250 U/ml of IFN the polysome size was smaller, 151s compared to 175s. Ribosome binding of mRNA after incubation in an in vitro translation assay for various times and separation in high salt sucrose gradients showed, in addition, that polysome formation, which was complete using untreated mRNAs, was not begun in IFN-treated RNAs after 5 minutes incubation. However, binding did occur within a short period of additional incubation. Thus, the kinetics of polysome formation differs in IFN-treated and untreated cells.

TABLE 2
THE EFFECT OF INTERFERON ON CAPPING AND
TRANSLATION OF MEASLES mRNA

	mRNA	Translation	Cap
NO IFN	22s	NP, P, F_1, some M	ND[a]
	16-20s	NP, P, F_1, some M	0, II
	12s	no translation	ND
+ IFN	22s	no translation	ND
	16-20s	no translation	no cap
	12s	M, some NP	I, II

[a]ND = not done.

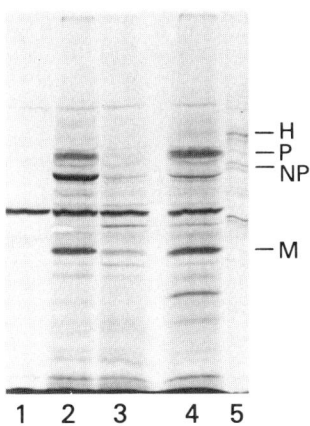

FIGURE 1. In vitro translation of measles mRNA from untreated and IFN-treated Vero cells. Products of an in vitro reticulocyte translation were autoradiographed after separation on a 10% acrylamide gel run in Laemmli buffer (13). Lane 1. Endogenous translation, no added mRNA. Lane 2. Translation of measles mRNA from untreated Vero cells. Lane 3. Translation of measles mRNA from IFN-treated Vero cells (250 units/ml). Lane 4. Translation of control measles mRNA. Lane 5. Measles polypeptides labeled in Vero cell culture.

Attempts to characterize the mRNAs for differences with or without IFN treatment were substantially negative. There was no substantial difference in percentage of polyadenylation, or in methylation and capping. Examination of the cap by DEAE-cellulose ion exchange column chromatography after alkaline phosphatase treatment revealed no aberrant structures at the 5' end as a result of IFN treatment. There was, however, a small difference in the percentage of mRNAs having 5' terminal cap I and cap II structures. When mRNAs were fractionated according to size and examined for the presence of cap and for the ability to direct translation in the reticulocyte lysate system, the smaller mRNAs existing after IFN treatment appeared to have caps and be translatable, while in untreated cells it was the 16-20s mRNAs which were capped and translatable as seen in Table 2. Thus, with IFN treatment of infected cells it appears that the functional mRNAs which accumulate are smaller than those in untreated cells.

IN VITRO STUDIES

We next explored whether this difference was a property of the in vitro polymerase or, alternatively, a posttranscriptional modification of mRNA. To examine this, we studied in vitro RNA synthesis using purified measles virus nucleocapsids. We first examined total in vitro polymerase activity in measles virus infected Vero cells as a function of exogenously added IFN. Cells were treated with IFN 24 hours before infection, and virus and viral nucleocapsids were harvested from the post-nuclear supernatant, centrifuged on isopycnic sucrose-D_2O gradients, and assayed for polymerase activity across the gradient. As Figure 2 shows, the total amount of polymerase activity is about the same with and without IFN treatment; however, after IFN treatment a larger proportion of the activity is associated with nucleocapsids than with complete virions, a reflection of a decrease of intact infectious virions following IFN treatment. When the products of in vitro transcription were analyzed on sucrose velocity gradients, no difference was seen with or without IFN treatment. Similarly, IFN addition to the in vitro RNA polymerase reaction has no effect on product yield or characteristics. When 2.8 nM to 2.8 uM of oligo (2'-5') A, a product which can induce the effect of interferon in intact cells, was added to the in vitro transcription reaction, no effect was seen on the size of the RNA products or the extent of polyadenylation.

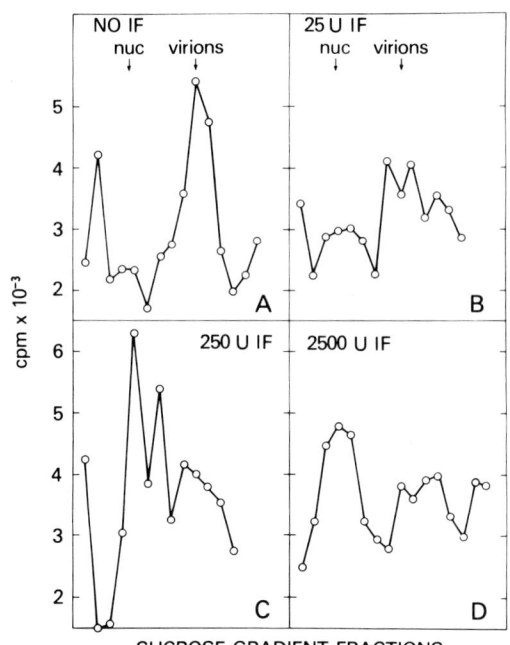

FIGURE 2. The effect of interferon on transcriptase activity of virions. Virions harvested from measles infected Vero cells pretreated with (A) 0, (B) 25 units/ml, (C) 250 units/ml, and (D) 2500 units/ml IFN-beta, were purified on 15-65% sucrose in D_2O gradients, and gradient fractions were assayed for polymerase activity by incorporation of radioactivity (14).

OVERALL EFFECTS OF INTERFERON

Thus, we conclude that the effect of IFN on measles RNA synthesis is probably not at the level of transcription. Instead, it appears to occur posttranscriptionally, perhaps involving RNA stability. In IFN-treated measles infected Vero cells, those mRNAs containing methylated caps at the 5' end appear to be a smaller size than those in untreated infected cells. In IFN-treated cells, the larger mRNAs, of the size of normal measles mRNAs, appear defective in their caps, in their ability to bind to ribosomes, and in their ability to be translated. There is an increased amount of 12s mRNA, which contains a cap and is translatable in vitro to a product coelectrophoresing with measles M protein. Increased RNA breakdown mediated by an induced endonuclease is consistent with known IFN effects (6,7,8,9), but to our knowledge, differential susceptibility of mRNAs to the endonuclease has not been reported. Our data do not support the notion of aberrant ribosome binding due to aberrant cap structure after IFN treatment; rather, when the cap is absent, there appears to be no ribosome binding, while when the cap is present, it appears to be comparable to that found in a normal infection, and ribosome binding and translation occur.

In the lymphocyte system, on the other hand, the RNA species which is unstable is the genomic RNA; mRNAs are made and are translatable into measles protein in vitro and in infected cells. It is our interpretation that the inability to accumulate 50s genomic RNA is a reflection of inhibition of virus budding in lymphocytes when IFN-alpha is induced, which in turn leads to disassembly and breakdown of viral nucleocapsids.

We have examined RNA synthesis in three different measles RNA synthesizing systems and we have found that in vitro, there is no effect of interferon or of interferon induced products, such as oligo A, suggesting that association with cellular components may be needed to produce an effect. In Vero cells, where little endogenous interferon is produced, we have seen that IFN results in a preponderance of lower molecular weight mRNAs which appear to be nonuniformly capped. This results in a differential effect on translatability of the message, probably due to methylation and capping. In lymphocytes, which do not normally support measles infection, the presence of interferon interferes with the accumulation of 50s genomic RNA.

REFERENCES

1. Baxt, B., Sonnabend, J. A., and Bablanian, R. (1977). J. Gen. Virol. 35, 325-334.
2. Marcus, P. I., Engelhart, D. L., Hunt, J. M., and Sekellick, M. J. (1971). Science 174, 593-598.
3. Vilcek, J., Gresser, I., and Merigan, T. C. (eds.). (1980). In "Regulatory Functions of Interferons; Part VII. Molecular Mechanisms of Interferon Action," Annals of the New York Academy of Sciences, 350, p. 432-521.
4. Yau, P. M. P., Godefroy-Colburn, T., Binge, C. H., Ramabhadran, T. U., and Thach, R. E. (1978). J. Virol. 27, 648-658.
5. Maheshwari, R. K., Jay, F. T., and Friedman, R. M. (1980). Science 207, 540-541.

6. Jacobson, S., and McFarland, H. F. (1982). J. Gen. Virol. 63, 351-357.
7. Nilsen, T. W., Maroney, P. A., and Baglioni, C. (1982). J. Virol. 42, 1039-1045.
8. Nilsen, T. W., Maroney, P. A., and Baglioni, C. (1981). J. Biol. Chem. 256, 7806-7811.
9. Repik, P., Flamand, A., and Bishop, D. H. L. (1974). J. Virol. 14, 1169-1178.
10. Johnson, M. I., Zoon, K. C., Friedman, R. M., DeClercz, E., and Torrence, P. F. (1980). Biochem. Biolog. Res. Commun. 97, 375-383.
11. Milstien, J. B., Klutch, M. J., Bhatia, B., and Lucas, C. J. (1983, in press). In "The Biology of the Interferon System." (H. Schellekens, ed.).
12. Milstien, J. B., and Seifried, A. S. (1982). Abstract of the ASM Annual Meeting, Atlanta, Georgia, 253.
13. Laemmli, U. K. (1970). Nature (London) 227, 680-685.
14. Milstien, J. B., and Seifried, A. S. (1981). In "The Replication of Negative Strand Viruses." (D. L. Bishop and R. W. Compans, eds) p. 485-491. Elsevier-North Holland.

Gene Expression, Protein Synthesis, and Protein Modification

CONSTRUCTION AND EXPRESSION OF A CHIMERIC GENE OF
GLYCOPROTEIN G AND MATRIX PROTEIN M OF VESICULAR
STOMATITIS VIRUS[1]

J. Capone[2] and H.P. Ghosh

Department of Biochemistry
McMaster University
Hamilton, Ontario, Canada, L8N 3Z5

The information necessary for the synthesis, transport, and modifications of secretory and membrane proteins is believed to be contained in discrete regions or domains of the polypeptide which are decoded by specialized mechanisms in the cell (1). These domains, or 'topogenic sequences' (2) which can be a transient or permanent feature of the polypeptide, include the signal sequence which initiates the translocation of proteins into and across specific membranes, the stop transfer sequence which in the case of transmembrane proteins serves to interrupt the signal sequence initiated translocation and insures a proper asymmetric orientation of the polypeptide and sorting sequences which serve to target the newly synthesized polypeptide along selected subcellular pathways to its site of function. The glycoprotein G of vesicular stomatitis virus contains a transient amino terminal hydrophobic signal sequence which is involved in translocation of the polypeptide across the endoplasmic reticulum as well as a hydrophobic COOH-terminal domain which anchors the polypeptide in the envelope of the virions (3,4,5). In order to examine the basis for the specificity and function of these hydrophobic domains, we have constructed a chimeric gene containing the signal sequence coding region of G protein fused to the bulk of the coding sequence of M protein. The fusion of the signal sequence coding region of the G protein to a protein that does not normally transverse membrane bilayers and its expression in mammalian cells would indicate whether or not the signal sequence is sufficient to initiate and maintain vectorial translocation across the membrane.

---

[1]This work was supported by Medical Research Council of Canada.
[2]Present Address: Department of Biology, M.I.T., Cambridge, MA.

The predicted amino acid sequence of M (6) reveals the presence of a sequence of the type Asn-X-Ser/Thr, which is a concensus sequence for glycosylation (7). In M protein however, this target site is not glycosylated. The attachment of a signal sequence to the M protein may insert it into the RER and thus could allow glycosylation in this site.

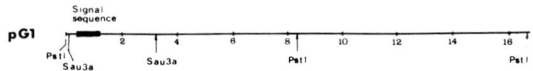

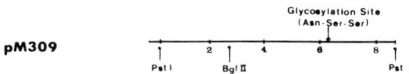

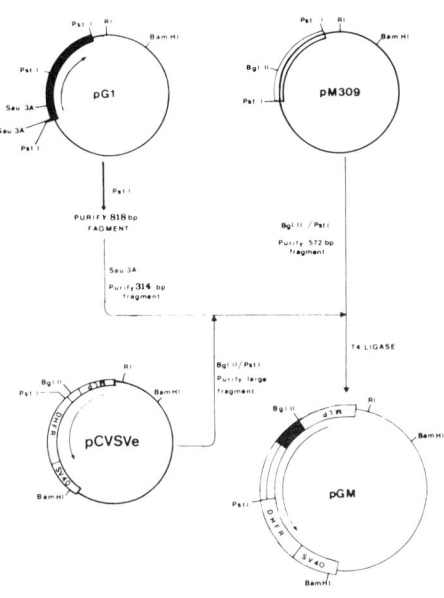

FIGURE 1. Strategy used for the construction and cloning of a G-M chimeric gene. The map coordinates for pG1 and pM-309 are from Rose and Gallione (6). All DNA fragments were purified by gel electrophoresis prior to ligation. MLP, adenovirus major late promoter, DHFR, cDNA sequence coding for the dihydrofolate reductase gene.

CONSTRUCTION OF CHIMERIC GENE

The G/M chimeric gene was constructed from the recombinant plasmids pG1 and pM309 containing cDNA copies of the complete coding sequences of the mRNAs specific for G and M protein, respectively (6) (kind gift of Dr. Jack Rose).

pG1 contains Sau3A sites at nucleotide positions 10 and 324. Thus this restriction fragment would contain the entire signal sequence coding region as well as the 5' untranslated region required for ribosome binding. This fragment would also contain sequences coding for the first 82 amino acids of the mature G protein. pM309 contains a unique Bgl II site located at nucleotide position 259. Ligation of the Sau3A pG1 fragment with Bgl II cut pM309 would produce an in phase hybrid gene containing the G specified sequences mentioned above as well as sequences coding for the 157 carboxy terminal amino acids and the entire 3' untranslated region of the M gene.

The 314 bp Sau3A fragment of pG1 was obtained by digestion with Sau3A of the 414 bp fragment generated by digesting pG1 with Pst I. The 572 bp Bgl II/Pst I fragment from pM309 was obtained by digesting pM309 with these enzymes. These fragments were ligated directly into the mammalian expression vector pCVSVe (gift of Drs. R. Kaufman and P. Sharp). pCVSVe contains sequences coding for dihydrofolate reductase and was used to express this protein in CHO cells (8). This vector also contains the major late transcription promotor of adenovirus, the SV40 72 base pair repeat, and the SV40 poly A site. pCVSVe contains unique Bgl II and Pst I cleavage sites suitable for insertion of foreign DNA.

The fusion of the Sau3A fragment with the Bgl II/Pst I cut pM309 fragment would generate a product having a Bgl II cohesive end at the 5' end and a Pst I cohesive end at the 3' end. Thus cloning the hybrid gene so constructed into Bgl II/Pst I cut pCVSVe would place the gene in the correct orientation for transcription just downstream from the regulatory sequences of the vector. This cloning strategy is summarized in Fig. 1. Several clones of the expected size (approximately 900 bp larger than Bgl II/Pst cut pCVSVe) were isolated following transformation of E. coli LE293. Several of these are shown in Fig. 2A. Digestion of recombinants containing Bam H1 show that pGM03, 04 and 07 contain the fused G/M gene while pGM12 contains just the pM insert (Fig. 2A).

Fusion of G to M in the correct orientation and ligation into pCVSVe was determined by digestion of the pGM clones with Bgl II/Pst I, seen in Fig. 2B. pGM 07, generates a fragment coincident with the fragment generated from Bgl II/Pst I cut pM309. pGM03, 04, on the other hand produced a fragment of lower mobility. Since pGM07 was shown to contain an insert of the size expected for the G-M hybrid, it must contain the G specific fragment in the incorrect orientation while pGM03, 04 contain the GM hybrid in the correct orientation.

EXPRESSION OF CHIMERIC GENE IN MAMMALIAN CELLS

pGM03 was introduced into COS-1 cells by the calcium phosphate transfection procedure (9,10). The cells were labeled with ^{35}S methionine 40-48 hours after transfection and cell extracts were prepared. The cell extracts were immunoprecipitated with anti M antibody and analyzed by SDS-polyacrylamide gel electrophoresis.

As shown in Fig. 3, lane h, a protein having an apparent molecular weight of approximately 36000 was precipitated by

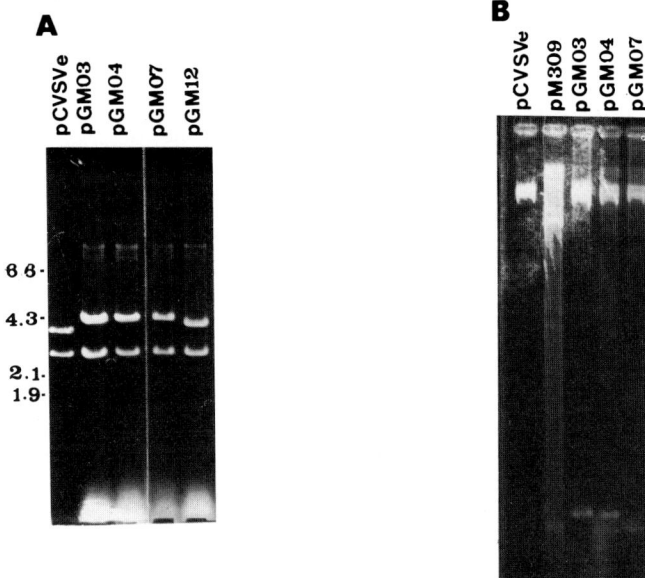

FIGURE 2. Digestion of recombinant plasmids. Ligated recombinant plasmids constructed as described in Fig. 1 were used to transform E. coli LE392 to tetracycline resistance. Positive colonies were selected at random from the plates, grown to saturation in 5 ml of L broth containing the selective antibiotic and plasmids purified and digested with Bam HI (panel A) or with Bgl II and Pst I (panel B). The size markers (in Kilobase pairs) are from a sample of Hind III digested λ DNA run in a separate well.

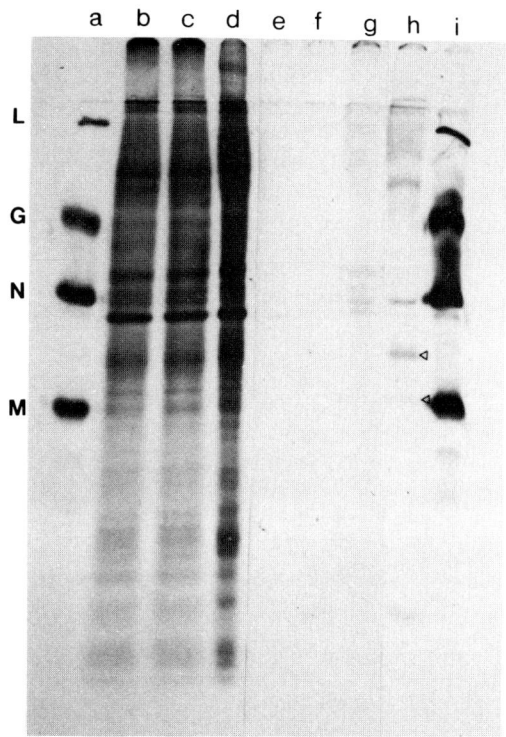

FIGURE 3. Polyacrylamide gel analysis of proteins immunoprecipitated from COS-1 cells transfected with pCVSVe or pGM-03. [^{35}S]-methionine labelled cell extracts prepared from COS-1 cells transfected with pCVSVe or the recombinant plasmid pGM03 were immunoprecipitated and analyzed on a 12% polyacrylamide gel. Lane a, [^{35}S]-methionine labelled VSV; lane b, sample of whole cell extract from COS-1 cells; lanes c and d, whole cell extract from COS-1 cells transfected with pCVSVe and pGM03, respectively; lanes e and f, pCVSVe transfected cells immunoprecipitated with normal antiserum and anti-M antiserum, respectively; lanes g and h, pGM03 transfected cells immunoprecipitated with normal antiserum and anti-M antiserum, respectively. The open arrowheads indicate two protein species, have apparent molecular weights of approximately 36 K and 32 K respectively, specifically immunoprecipitated from pGM03 transfected cells by anti-M antiserum.

anti-M from pGM03 transfected cells. This protein was not precipitated from these cells with normal rabbit serum nor from pCVSVe transfected cells immunoprecipitated with anti-M

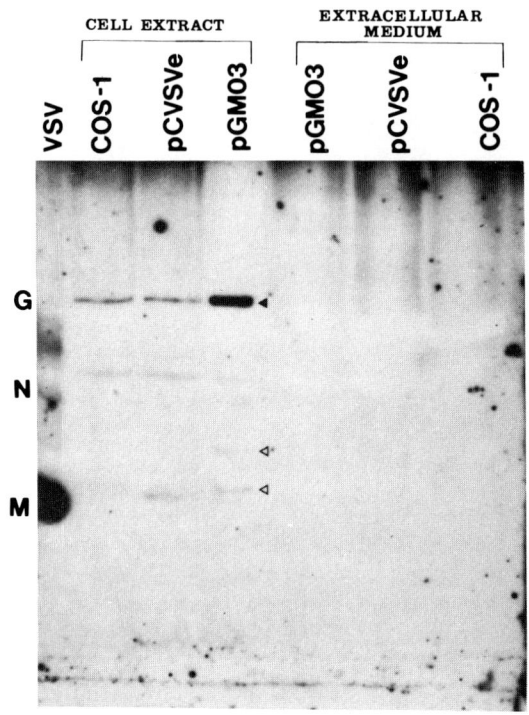

FIGURE 4. Immunoblotting of COS-1 transfected cells with anti-M antibody. COS-1 cells, grown in 60 mm culture plates were transfected with pCVSVe or pGM03. 48 hours post-transfection, the cells were collected, washed, and lysed with sample buffer. One half of each sample was run on a 10% polyacrylamide gel. For the extracellular medium, the growth medium was collected from each plate, precipitated with acetone and the precipitate was collected, washed, and solubilized in sample buffer. One half of the total precipitated material from each sample was run on the same gel. Following electrophoresis, the proteins were transferred onto nitrocellulose and reacted with anti-M antibody followed by ^{125}I-protein A (11). The open arrow heads indicate protein species reactive with anti-M antibody and only present in the pGM03 transfected cells.

or normal serum (lanes e and f). Therefore, this protein appears to be specifically precipitated by anti-M only from cells containing the chimeric gene. A minor protein having a molecular weight of approximately 31-32,000 was also immunoprecipitated by anti-M antibody.

These results were confirmed by immunoblotting as shown in Fig. 4, lanes b-d. In this case there were two protein species having molecular weights of approximately 32,000 and 36,000 immunoreactive with anti M which were present in pGM03 transfected cells but not present in cells transfected with pCMSVe or from nontransfected cells. A 80,000D protein present in pGM03 transfected cells was also present in the other cell extracts.

SECRETION OF M DERIVED PROTEINS

The culture medium was collected from cells 40-48 hrs following transfection and analyzed by immunoblotting with anti M antibody. No M specific polypeptides seemed to be secreted by cells transfected with pGM03 (Fig. 4, lanes e-g).

The 36,000D polypeptide immunoprecipitated by antibody therefore may represent a hybrid protein containing NH_2-terminal peptide of G fused onto a truncated M protein. Experiments are in progress to determine if the signal sequence has been processed and the hybrid protein is glycosylated in vivo.

ACKNOWLEDGEMENTS

We thank Drs. J. Rose, R. Kaufmann, P. Sharp, L. Prevec and K. Ghosh for kind help.

REFERENCES

1. Sabatini, D.D., Kreibich, G., Morimoto, T., Adesnik, M. (1982). J. Cell Biol. 92, 1.
2. Blobel, G. (1980). Proc. Natl. Acad. Sci. 77, 1496.
3. Ghosh, H.P. (1980). Rev. Infec. Dis. 2, 26.
4. Irving, R.A., Toneguzzo, F., Rhee, S.H., Hofmann, T. and Ghosh, H.P. (1979). Proc. Natl. Acad. Sci. 76, 570.
5. Capone, J., Toneguzzo, F. and Ghosh, H.P. (1982). J. Biol. Chem. 257, 16.
6. Rose, J.K. and Gallione, C.J. (1981). J. Virol. 39, 519.
7. Sharon, N. and Lis, H. (1982). In "The Proteins" H. Neurath, R.N. Hill, eds. Vol. V, p. 1. Academic Press, N.Y.
8. Kaufman, R.J. and Sharp, P.A. (1982). Mol. and Cell Biol. 2, 1304.

9. Graham, F.L. and Wan Der Eb, A.J. (1973). Virology 52, 456.
10. Wigler, M., Pellicer, A. Silverstein, S., Axel, R., Urlaub, G. and Chasin, L. (1979). Proc. Natl. Acad. Sci. 76, 1373.
11. Towbin, T., Staehelin, T. and Gordon, J. (1970). Proc. Natl. Acad. Sci. 75, 715.

INTRACELLULAR PROCESSING OF VESICULAR STOMATITIS VIRUS AND NEWCASTLE DISEASE VIRUS GLYCOPROTEINS[1]

Trudy G. Morrison and Lori J. Ward

Department of Molecular Genetics and Microbiology
University of Massachusetts Medical Center
Worcester, Massachusetts 01605

Intracellular pathways have evolved to insure that newly made proteins arrive at their proper destinations (1). Plasma membrane glycoproteins follow a pathway that generally involves insertion of nascent polypeptides into the membrane of the rough endoplasmic reticulum (RER) and subsequent transport through the Golgi membranes and finally insertion into the plasma membrane (1,2,3,4).

Perhaps all plasma membrane glycoproteins follow a similar path. By virtue of possessing a signal sequence (2) which insures the proteins' insertion into the membrane of the RER, plasma membrane glycoproteins may enter a membrane flow which ultimately results in insertion into the plasma membrane. However, in polarized cells such as MDCK cells, the glycoprotein (G) of vesicular stomatitis virus (VSV) is inserted on one side of the cell, the basolateral side, while the glycoproteins of paramyxoviruses and myxoviruses are inserted on the apical side of the cell (5,6,7). Such a result argues for two pathways to the cell surface in at least these cells.

We explored the idea that plasma membrane glycoproteins are processed in different ways in non-polarized chicken embryo cells. We reasoned that different pathways may be reflected in different kinetics of transport to the cell surface. We, therefore, compared the transit times of the VSV G protein and the Newcastle disease virus (NDV) hemagglutinin-neuraminidase (HN) glycoprotein to the Golgi membranes and to the cell surface of chick embryo cells. Our data show that the transit time of the HN protein to the cell surface is much longer than that of the G protein. The difference in the transit time can be accounted for by the time it takes the HN glycoprotein to reach the Golgi membranes.

[1] This work was supported by NIH Grant AI-13847.

We first asked how long it takes pulse-labelled glycoprotein to reach the cell surface. Our assay for cell surface glycoprotein utilizes antibody binding to intact cells. Monolayers of infected cells were pulse-labelled with [^{35}S]-methionine or pulse-labelled and chased with cold methionine. After various lengths of chase, washed monolayers were incubated with anti-NDV or VSV antisera. Molecules on the outside surface of cells will bind antibody while internal proteins will not. Unbound antibody is washed away, the cells are lysed and immunobeads containing goat anti-rabbit IgG used to precipitate immune complexes. Cell surface molecules will be precipitated while internal polypeptides including glycoprotein which has not yet reached the cell surface will not precipitate.

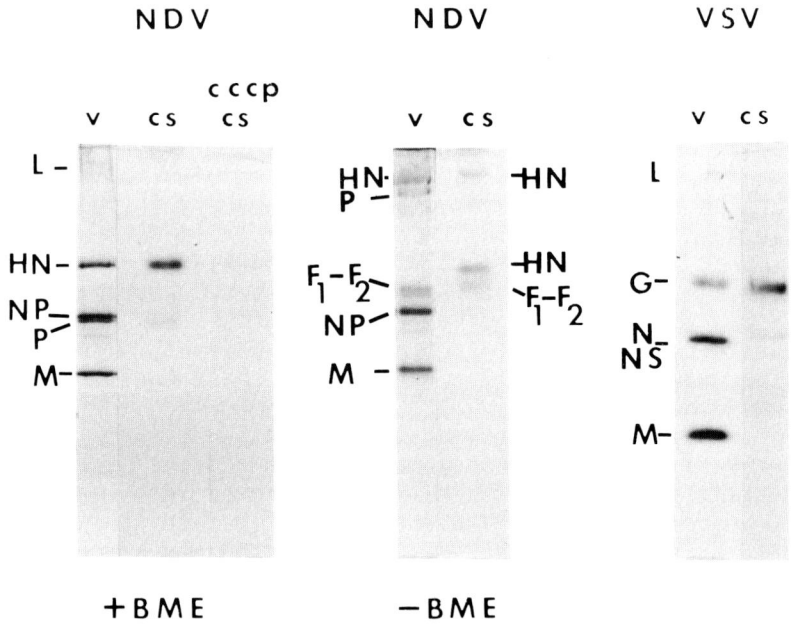

Figure 1: Antibody Binding Assay of Cell Surface Molecules.
v = virion proteins, cs = cell surface proteins, cccp = carboxyl cyanide m-chlorophenylhydrazone.

Figure 1 shows the result of such an assay. The NDV HN protein is precipitated after such a procedure, as well as a polypeptide with a molecular weight of 56,000 daltons. That this polypeptide is the fusion protein is shown by the fact that it comigrates with the F_1-F_2 complex if electrophoresed under non-reducing (-BME) conditions. The internal polypeptides, NP, P, and M proteins, are not precipitated. It should be noted that while virion-associated HN protein is resolved only as a multimer seen at the top of the gel if electrophoresed under non-reducing conditions (8), cell surface HN is seen in both monomeric and multimeric forms.

That this protocol precipitates only glycoproteins at the cell surface and not glycoproteins which have not yet reached the cell surface is shown by the use of CCCP. An inhibitor of oxidative phosphorylation, CCCP is known to block intracellular migration of glycoproteins (9,10). Addition of CCCP to cells just after the pulse-label blocks the migration of the labelled protein to the cell surface (9,10). In fact, no NDV proteins are detected by our assay after CCCP addition.

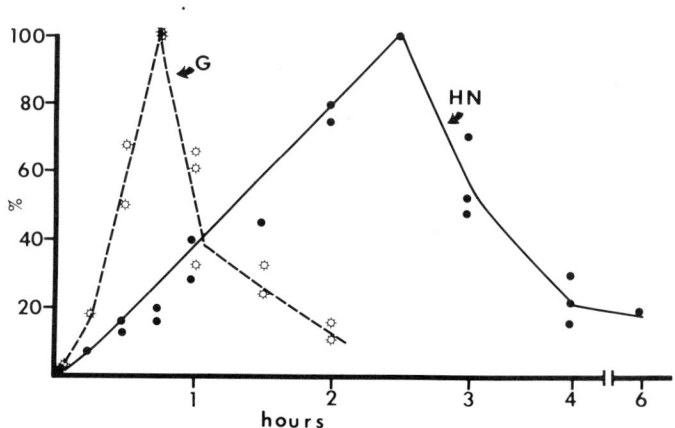

Figure 2: Appearance of Glycoproteins at the Cell Surface.
Values are expressed as the percentage of maximal amount precipitated after onset of chase.

This protocol also results in the precipitation of only the VSV G protein and not the internal proteins, N, NS, M and L.

Using this assay, we were able to follow the kinetics of HN and G protein at the cell surface. After varying lengths of chase, pulse-labelled G protein appears at the cell surface and then disappears as the protein is incorporated into virions. Note the VSV G enters the plasma membrane rapidly with a half-time of approximately 27 minutes and exits rapidly as reported previously (11,12). However, the HN protein enters the plasma membrane more slowly, with a half-time of approximately 78 minutes, and exits quite slowly. Surface HN can be detected by 6 hours after the onset of the chase.

We next asked how long it takes the two molecules to

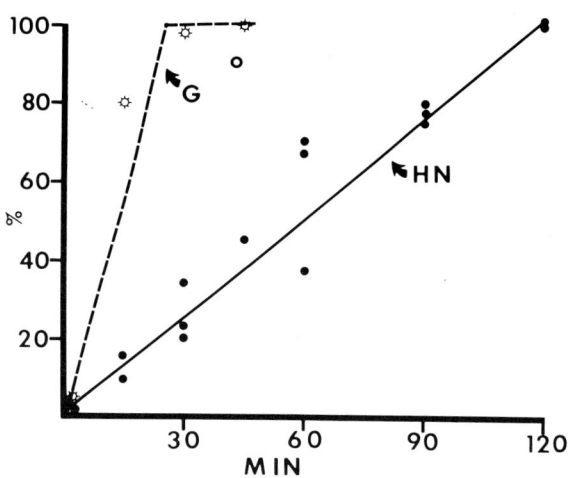

Figure 3: Appearance of Endoglycosidase H-resistant glycoprotein.
Values are plotted as percent of total glycoprotein.

reach the trans-Golgi membranes. In the trans-Golgi, the carbohydrate on glycoprotein is converted to resistance to endoglycosidase H by the addition of periferal sugars (4). We, therefore, asked how long it takes the two glycoproteins to acquire endoglycosidase H resistance. Figure 2 shows a plot of the appearance of endoglycosidase H-resistant glycoprotein with time of chase.

The G acquires endoglycosidase H resistance very quickly. The NDV HN protein becomes resistant quite slowly. As reported previously, the G protein acquires endoglycosidase H resistance with a half-time of approximately 13 minutes (11,12,13). In contrast, the HN protein acquires endoglycosidase H resistance with a half-time of approximately 60 minutes.

In summary, the G protein reaches the cell surface with a half-time of approximately 27 minutes and reaches the trans-Golgi membranes with a half-time of approximately 13 minutes. Thus, the G migrates from the Golgi membranes to the cell surface with a half-time of approximately 14 minutes. By contrast, the HN reaches the cell surface with a half-time of approximately 78 minutes and the trans-Golgi membranes with a half-time of approximately 60 minutes. Thus, the HN protein migrates from the Golgi membranes to the cell surface with a half-time of approximately 18 minutes, a time not significantly different from that of the VSV G protein. Thus, the difference in the kinetics of intracellular transport of the two glycoproteins resides primarily in the transit from the RER to the trans-Golgi membranes.

If all plasma membrane glycoproteins were transported by similar mechanism, the kinetics of transport should be identical. Results presented here, therefore, argue for different pathways to the cell surface.

References

1. Sabatini, D.D., Kreibich, G., Morimoto, T., and Adesnik, M. (1982). J. of Cell Biol. 92, 1.
2. Blobel, G., and Dobberstein, B. (1975). J. Cell Biol. 67, 835.
3. Palade, G.E. (1975). Science 189, 347.
4. Rothman, J.E. (1981). Science 213, 1212.
5. Rodriguez-Boulan, E., and Sabatini, D.D. (1978). Proc. Natl. Acad. Sci. USA 75, 5071.
6. Roth, M./G., Fitzpatrick, J.P., and Compans, R. (1979). Proc. Natl. Acad. Sci. 76, 6430.

7. Roth, M.G., Compans, R., Giusti, L., Davis, A.R., Nayak, D.P., Gething, M.J., and Sambrook, J. (1983). Cell 33, 435.
8. Schwalbe, J., and Hightower, L. (1982). J. of Virol. 41, 947.
9. Tartakoff, A., and Vassalli, P. (1979). J. Cell. Biol. 83, 284.
10. Fries, E., and Rothman, J. (1980). Proc. Natl. Acad. Sci. USA 77, 3870.
11. Madoff, D.H., and Lenard, J. (1982). Cell 28, 821.
12. Strous, J., and Lodish, H.F. (1980). Cell 22, 709.
13. Schmidt, M., and Schlesinger, M. (1980). J. Biol. Chem. 255, 3334.

RESCUE OF VSV IN PERSISTENTLY INFECTED L CELLS BY SUPERINFECTION WITH VACCINIA[1]

Patricia Whitaker-Dowling and J. S. Youngner

Department of Microbiology
School of Medicine
University of Pittsburgh
Pittsburgh, PA 15261

Mouse L cells persistently infected with vesicular stomatitis virus (VSV) are resistant to superinfection with VSV and other viruses (1). Interferon (IFN) plays a part in the modulation of this persistent infection, since treatment of the cells with anti-IFN serum converts the persistent infection to a lytic infection (2). Thacore and Youngner showed that vaccinia can rescue VSV from the inhibitory effects of IFN in L cells (3), and recently we have reported that this rescue occurs at the level of VSV protein synthesis (4). We have also identified a specific kinase inhibitory factor (SKIF) produced by vaccinia that inhibits the IFN-induced, dsRNA-dependent protein kinase (4).

In this study we identify the point at which the replication of superinfecting wild-type VSV (wt-VSV) is blocked in persistently infected L cells (L_{VSV}) and show that vaccinia is able to reverse the block to wt-VSV growth. Infection with vaccinia produces a slight but reproducible stimulation of the growth of the persistent virus (VSV-PI) in L_{VSV} cells.

VACCINIA MEDIATED RESCUE OF VSV GROWTH

Coinfection with vaccinia (MOI = 10) has a profound effect on growth of wt-VSV in L_{VSV} cells (Figure 1). By 24 hours after infection with wt-VSV alone (MOI = 10), little or no replication of the wt-VSV is observed; however, L_{VSV} cells doubly infected with wt-VSV and vaccinia produce approximately 100 times more wt-VSV than singly infected cells, a yield similar to that seen in normal L cells. The growth of VSV-PI

---

[1]This work was supported by U.S. Public Health Service Research Grant AI-06264 from the National Institute of Allergy and Infectious Diseases.

is stimulated by about 10-fold by infection with vaccinia. This stimulation occurs whether or not L_{VSV} cells are also superinfected with wt-VSV.

Since both VSV infection and high doses of IFN can affect the penetration of VSV into L cells (5,6), we compared the ability of ^{35}S-labeled wt-VSV (MOI = 10) to enter L_{VSV} cells and uninfected L cells. Uninfected L cells and L_{VSV} cells take up virus at the same rate: 3.5% of the radioactive virus enters the cells after incubation at 37°C for 1 hour.

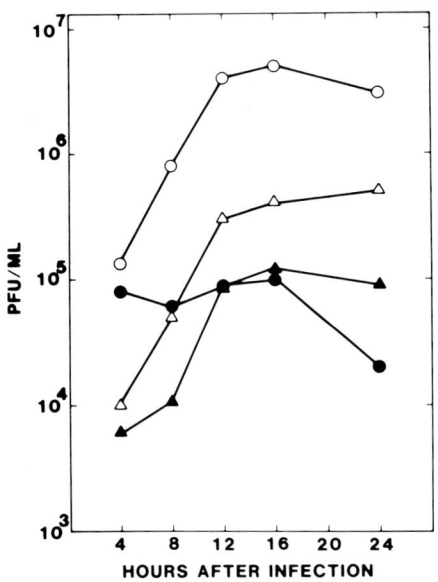

FIGURE 1. Growth of VSV in L_{VSV} cells with and without vaccinia. Monolayers of L_{VSV} cells in 24-well trays were mock- or vaccinia-infected (MOI = 10). After 2 hr at 37°C the cells were mock- or VSV-infected (MOI = 10) in the presence of 5 µg/ml of actinomycin D. The inoculum was removed after 1 hr at 37°C, the cells were washed 2 times with medium and recultured in medium containing actinomycin D. At the indicated times after VSV infection, the fluid from individual wells was harvested and infectivity assayed in primary chicken embryo cells at 37°C. VSV-PI and wt-VSV are easily distinguished by plaque size. VSV-PI from mock-infected cells (▲), VSV-PI from cells superinfected with vaccinia (△), wt-VSV from cells superinfected with wt-VSV (●), and wt-VSV from cells superinfected with wt-VSV + vaccinia (○).

THE EFFECT OF VACCINIA ON VSV GENE EXPRESSION

To determine if vaccinia has any effect on VSV RNA synthesis, primary transcription by superinfecting wt-VSV was measured in the presence or absence of coinfecting vaccinia by labeling L_{VSV} cells with ^{3}H-uridine in the presence of actinomycin D and cycloheximide. Vaccinia produces only a modest stimulation of the incorporation of ^{3}H-uridine into TCA precipitable material (Figure 2). This analysis is complicated by the fact that early RNA synthesis by vaccinia is only partially sensitive to inhibition by actinomycin D in the presence of cycloheximide (unpublished observation). Therefore, we measured primary transcription in L_{VSV} cells by isolating the RNA and testing its ability to direct the synthesis of VSV protein in an <u>in vitro</u> rabbit reticulocyte lysate (Figure 3). Vaccinia has little or no influence on the

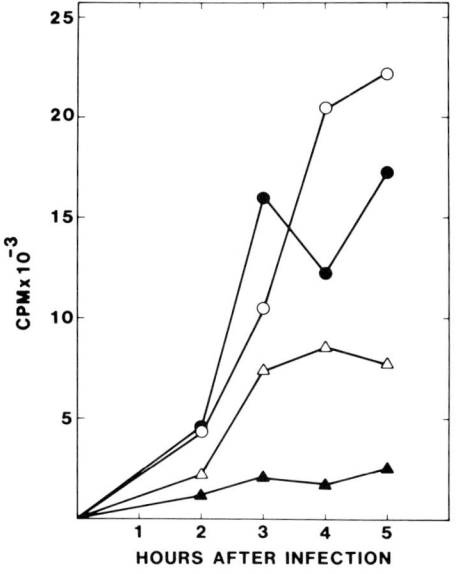

FIGURE 2. Primary transcription by VSV in L_{VSV} cells with and without vaccinia. Monolayers of L_{VSV} were infected with vaccinia as described in Figure 1. After 2 hr at 37°C, the cells were mock- or VSV-infected (MOI = 10) in the presence of actinomycin D and cycloheximide (200 µg/ml). The inoculum was removed after 1 hr at 37°C, and the cells were washed, and recultured in medium containing actinomycin D, cycloheximide, and ^{3}H-uridine (10 µCi/ml). At the indicated times after VSV infection, cultures were lysed in 1% SDS and the ^{3}H-uridine incorporated into TCA precipitable material determined. Uninfected (▲), vaccinia infected (∆), wt-VSV infected (●), and wt-VSV + vaccinia infected (O).

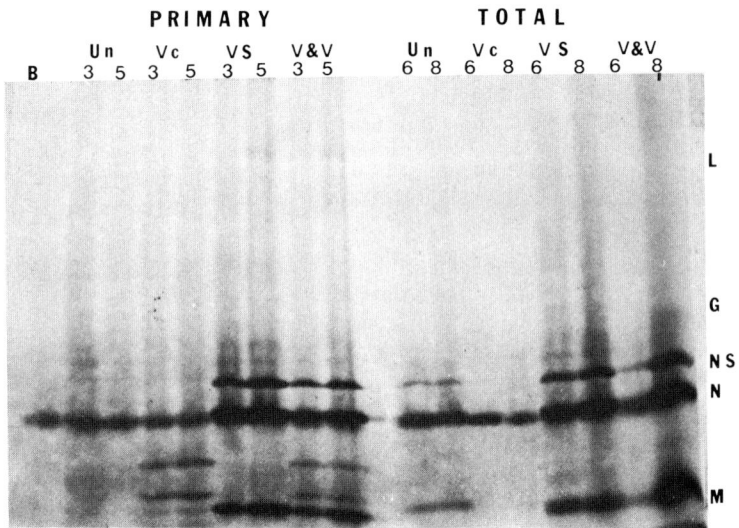

FIGURE 3. The amount of functional VSV mRNA in L_{VSV} superinfected with VSV and vaccinia. Monolayers of L_{VSV} were infected with wt-VSV and vaccinia as described in Figure 2 (to measure the level of functional primary transcripts) or as described in Figure 1 (to measure the amount of total VSV mRNA). RNA was isolated at the indicated hours after VSV infection by phenol extraction followed by ethanol precipitation. The RNA preparations were assayed for the ability to direct protein synthesis in a rabbit reticulocyte lysate translation system. An equal amount of purified RNA was added to each reaction to give a final concentration of 50 μg/ml of lysate. Products synthesized during 60 minutes of incubation at 30°C were analyzed by SDS-PAGE. B indicates the lane which received lysate to which no exogenous RNA was added. Un refers to mock-infected L_{VSV}, Vc indicates vaccinia infected L_{VSV} cells, VS refers to VSV infected cells and V&V indicates doubly infected L_{VSV} cells.

amount of functional VSV primary mRNA isolated from these cells. In addition, similar analysis of total VSV mRNA was performed using RNA isolated at 6 or 8 hours after VSV infection from singly and doubly infected L_{VSV} cells. Again, vaccinia coinfection has little effect on the amount of functional total VSV mRNA (Figure 3).

Analysis of protein synthesis in L_{VSV} cells superinfected with wt-VSV and vaccinia revealed that vaccinia produces a dramatic stimulation of the synthesis of VSV protein (Figure 4). Persistently infected cells were mock-infected, singly infected with wt-VSV or vaccinia, or doubly infected with VSV + vaccinia; the cells were labeled with ^{35}S-methionine for 30 minutes at 2, 4, and 6 hours after infection. Examination of extracts from these cells by SDS-PAGE revealed that little or no VSV protein is synthesized in cells infected with VSV alone. However, large amounts of all five VSV proteins are produced in cells doubly infected with VSV + vaccinia.

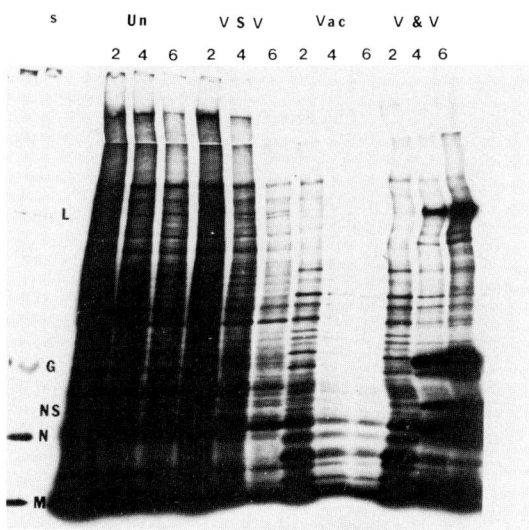

FIGURE 4. SDS-PAGE analysis of protein synthesis in L_{VSV} cells superinfected with wt-VSV and vaccinia. Monolayers of L_{VSV} cells were infected with vaccinia and VSV as described in Figure 1. At 2, 4, and 6 hr after VSV infection the cells were recultured in methionine-free medium to which ^{35}S-methionine (20 µCi/ml) had been added. After 30 minutes at 37°C, cell extracts were prepared and analyzed by SDS-PAGE as described (4). Un refers to mock-infected L_{VSV}, VSV refers to wt-VSV infected L_{VSV}, Vac to cells infected with vaccinia, and V&V to cells infected with VSV + vaccinia. The lane labeled S received a standard of ^{35}S-labeled VSV proteins. The numbers at the top of each lane indicate the hour after VSV infection that the cultures were labeled with ^{35}S-methionine.

VACCINIA-MEDIATED INHIBITION OF THE IFN-INDUCED PROTEIN KINASE

The ability of vaccinia to rescue VSV in IFN-treated cells has been correlated with the SKIF-mediated inhibition of the IFN-induced protein kinase activity (4). Mock-infected, singly infected, and VSV + vaccinia-infected L_{VSV} cells were assayed to determine if the IFN-induced protein kinase activity was present and if vaccinia infection resulted in an inhibition of the kinase activity in these cells. Kinase activity was detected by the phosphorylation of P1, a protein of M_r 67,000, whose phosphorylation has been correlated with IFN-induced, dsRNA-dependent protein kinase activity (7). The IFN-specific kinase is detectable in L_{VSV} cell extracts in the presence of dsRNA, although in lower amounts than seen in extracts from normal L cells treated with 100 units/ml of IFN

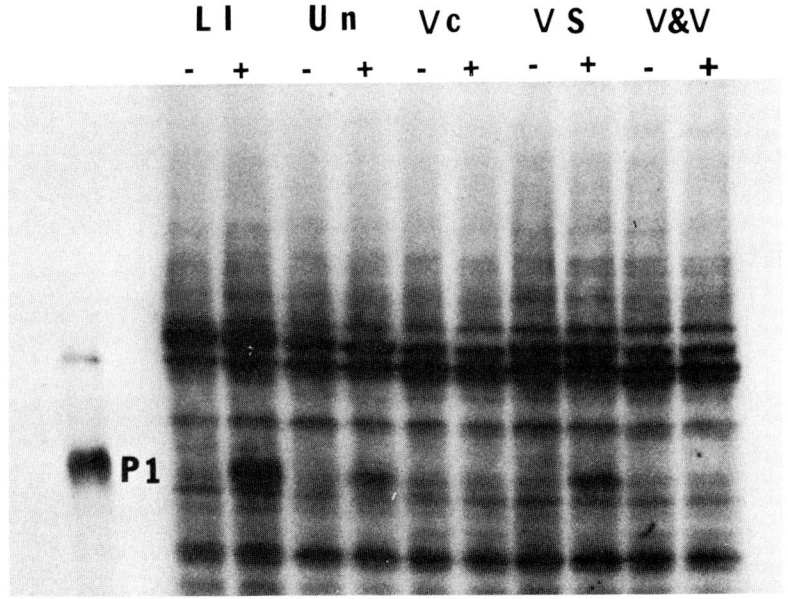

FIGURE 5. Protein kinase activity in L_{VSV} superinfected with VSV and vaccinia. Monolayers of L_{VSV} were mock-infected (Un), singly infected with VSV (VS) or vaccinia (Vc), or doubly infected with VSV + vaccinia (V&V) as described in Figure 1. At 2 hr after VSV infection, S10 lysates were prepared as described (8). An S10 lysate was also prepared from uninfected normal L cells which had been treated for 24 hr with 100 units/ml of mouse IFN (LI). The extracts were incubated at 30°C for 30 minutes in the presence of 0.4 mCi/ml γ-^{32}P-labeled ATP with (+) or without (-) the addition of 0.4 μg/ml of poly I:C. The products of this reaction were analyzed by SDS-PAGE. P1 marks the migration position of the 67,000 molecular weight protein phosphorylated by the IFN-induced protein kinase (7).

(Figure 5). As in IFN-treated cells, this kinase function is inhibited in extracts from L_{VSV} cells infected with vaccinia.

DISCUSSION

We have shown that coinfection with vaccinia reverses the inhibition of the growth of superinfecting wt-VSV in L cells persistently infected with VSV. Vaccinia has little or no direct effect on wt-VSV RNA synthesis, as measured by primary transcription or total transcription, but dramatically increases the synthesis of wt-VSV protein. The mechanism of rescue of VSV by vaccinia in L_{VSV} cells appears to be the same as the mechanism of rescue of VSV in IFN-treated L cells, namely, stimulation of VSV protein synthesis. It is interesting to note that IFN-induced protein kinase activity is present in L_{VSV} cells and that vaccinia is able to inhibit this enzyme in L_{VSV} cells as it does in IFN-treated L cells.

Although vaccinia also stimulates the growth of endogenous VSV-PI in persistently infected cells, the effect is less dramatic than the effect on superinfecting wt-VSV. This difference in extent of rescue of VSV-PI may be due to the expression of genetic differences between wt-VSV and VSV-PI, e.g., the altered RNA synthesis pattern of VSV-PI reported by Frey and Youngner (9). An alternative explanation is that there are simply fewer VSV-PI genomes available in L_{VSV} cells for rescue by vaccinia.

ACKNOWLEDGMENT

The authors are grateful to Lenna Liu for excellent technical assistance and to Bruce A. Phillips for helpful discussion.

REFERENCES

1. Youngner, J. S., Dubovi, E. J., Quagliana, D. O., Kelly, M., and Preble, O. T. (1976). J. Virol. 19, 90.
2. Youngner, J. S., Preble, O., and Jones, E. (1978). J. Virol. 28, 6.
3. Thacore, H. R., and Youngner, J. S. (1973). Virology, 56, 505.
4. Whitaker-Dowling, P. and Youngner, J. S. (1983). Virology (in press).
5. Whitaker-Dowling, P., Youngner, J. S., Widnell, C. C., and Wilcox, D. K. (1983). Virology (in press).
6. Whitaker-Dowling, P., Wilcox, D., Widnell, C., and Youngner, J. S. (1983). Proc. Natl. Acad. Sci. USA 80, 1083.

7. LeBleu, B., Sen, G., Shaila, S., Cabrer, B., and Lengyel, P. (1976). Proc. Natl. Acad. Sci. USA 73, 3107.
8. Lewis, J. A., Mangheri, L. R. and Esteban, M. (1983). Proc. Natl. Acad. Sci. USA 80, 26.
9. Frey, T. K. and Youngner, J. S. (1982). J. Virol. 44, 167.

THE EFFECT OF VESICULAR STOMATITIS VIRUS
ON ADENOVIRUS TYPE 2 REPLICATION[1]

James Remenick and John J. McGowan

Department of Microbiology
Uniformed Services University of the Health Sciences
Bethesda, Maryland   20814

VSV INHIBITION OF EUKARYOTIC MACROMOLECULAR SYNTHESIS

Vesicular Stomatitis Virus (VSV) is an enveloped rhabdovirus containing an RNA-dependent RNA polymerase (1,2). The 42 S negative stranded genome RNA is sequentially transcribed to yield as an initial product a plus-strand 47-nucleotide leader which is neither capped nor polyadenylated (2,3). Transcription of the leader RNA is required for the syntheis of the five monocistronic messenger RNAs which are capped and polyadenylated (1,2).

The Viral Function Responsible for Inhibition.
Eukaryotic RNA,DNA and protein syntheis are drastically altered by an infection with VSV. The general characteristics of VSV inhibition of host cellular macromolecular synthesis have been well characterized (4,5,6). Experiments to determine the viral function and the cellular target that interact to cause the phenomenon known as shut-off have recieved considerable attention during the last few years (5,6,7,8). Studies have used UV-irradiation, ts mutants, and DI particles of VSV to determine that transcription of at least leader RNA is required for the inhibition of cellular macromolecular synthesis (5,6,7,9). Others previously (10) used a DNA dependent in vitro transcription system (11) made with extracts from HeLa cells to directly test whether VSV leader

---

[1] This work was supported by Uniformed Services University of the Health Sciences Protocol No. C7349. The opinons or assertions contained herein are the private ones of the authors and are not to be construed as official or reflecting the views of the Department of Defense or the Uniformed Services University of the Health Sciences.

RNA is the viral function required for the inhibition of cellular RNA. Using VSV RNAs transcribed in vitro it was determined that the VSV leader RNA was at least 30 times more effective in the inhibtion of the specific initiation of RNA transcription in vitro then any other RNA tested (10). More recent data obtained by independent investigators confirms these previous results using two strains of VSV (12).

Previous studies (6,10) have tentatively identified RNA polymerase II and III as the cellular target for the inhibiton of transcription. However, the precise identification of the cellular traget(s) affected by VSV infection has not been made. Use of these in vitro systems currently offers us our best opportunity to investigate the cellular target in the shut-off phenomenon.

Studies of the VSV inhibitor of cellular protein and DNA synthesis have also implicated the WT leader sequence as a possible inhibitor (5,9,10). The use of in vitro translation systems are yielding additional information on the shut-off function required by VSV to inhibit cellular protein synthesis (15). VSV inhibition of cellular DNA synthesis has not been studied in depth (5). The lack of an in vitro replication system which would reflect the cellular process of DNA synthesis has hampered further study. However, we describe below a system which allows us to approach the effects of VSV on DNA synthesis in vitro.

Adenovirus Replication In Vitro: A Model for Eukaryotic DNA Synthesis. Our previous findings that VSV drastically inhibits cellular DNA synthesis in vivo by transcription of at least leader RNA has promted us to investigate the mechanism by which cellular DNA synthesis is inhibited. Although the capacity of VSV to shut-off cellular DNA synthesis could be secondary to its effect on RNA synthesis, it seems unlikely based on the almost identical rates of inhibition of both cellular RNA and DNA synthesis (5).

The sequence of steps required for DNA synthesis in eukaryotic cells is too complex to dissect in vivo. The interacting components can only be analyzed in vitro under specific conditions. Unfortunately, a system which supports the initiation of cellular DNA synthesis does not presently exist. A simple model for the accurate in vitro replication of adenovirus DNA has been developed (13). Since adenovirus relies primarily on cellular processes for its DNA synthesis, we have chosen this model to analyze the effects of VSV on eukaryotic DNA replication. To this end, experiments have been designed to exploit the in vitro replication system developed for adenovirus DNA.

Nuclear extracts from adenovirus-infected HeLa cells preserve the characteristics of adenovirus DNA replication (13). The nuclear extracts require exogeneous adenovirus DNA-terminal protein complex as a template, they replicate strands in a 5' to 3' direction, and simultaneously displace the complementary strand (13). The general characteristics of adenovirus replication in vivo and in vitro are well established (for review,14) which makes it possible to assess the biological relevance of the synthetic reactions in vitro.

USE OF THE AD 2 REPLICATION SYSTEM WITH VSV RNA

The Specificity of Cell Extracts. We first determined the nature of the DNA synthesized from the adenovirus type 2 (AD 2) DNA template using HeLa cell extracts. Extracts of HeLa cells were obtained as described (13) and mixed with Xba-1-cleaved AD 2 DNA. These five Xba-I-cleaved restriction fragments were replicated in vitro in the presence of ^{32}P-dCTP for 60' at 31^0C. The labeled products were resolved by electrophoresis on 0·7% agarose gels and visualized by autoradiography (data not shown). Typically 40-55% of the radioactivity was incorporated into the C and E terminal fragments of newly synthesized viral DNA. These terminal fragments represent less than 16% of the total genome yet contain greater than 50% of the total activity incorporated. Therefore, our reaction conditions were specific and suitable for the studies below.

Effect of VSV Plus-Strand RNA on AD 2 Replication: Kinetics of the Reaction. Previously it was (5) reported that VSV transcription is required to shut-off MPC-11 and L cell DNA synthesis. To test this possibility, VSV genome RNA was transcribed in vitro and the resulting syntheisized plus strand RNAs were separated form the viral nucleocapsids by ultracentrifugation (10). Phenol extracted purified in vitro transcribed VSV RNAs were then added into the AD 2 replication system and the amount of incorporation into acid precipitable counts determined (10).

Figure 1 shows the kinetics of replication of the AD 2 DNA template in the presence of increasing concentrations of VSV plus-strand RNA. The relatively low concentration (12 ng/ul) actually stimulated the incorporation of ^{32}P-dCTP

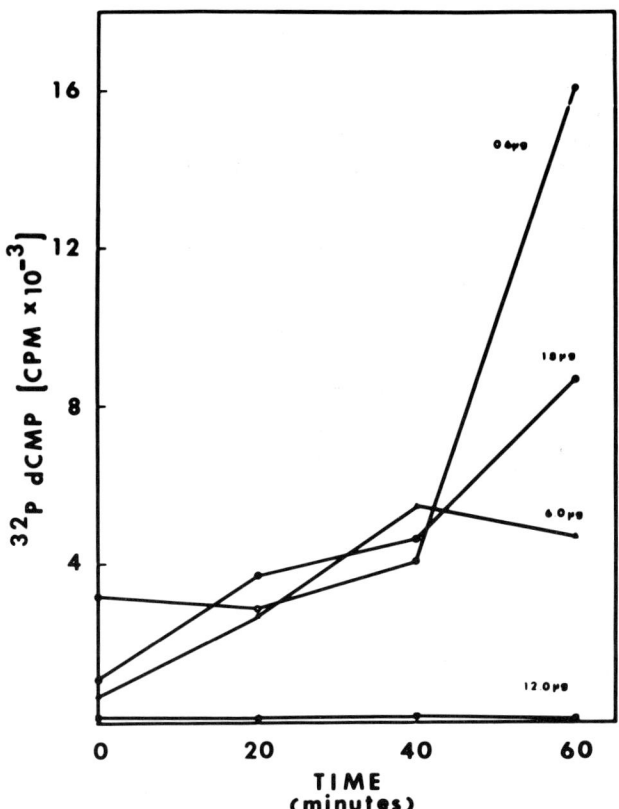

Figure 1. The effect of VSV plus-strand RNA on adenovirus type 2 replication in vitro. Purified VSV was transcribed in vitro (10) and added into the AD 2 replication extract prepared as described (13). The concentrations indicatated were added into a 50 ul reaction and at the times indicated 10 ul aliquots were removed and the amount of radioactivity incorporated determined (10).

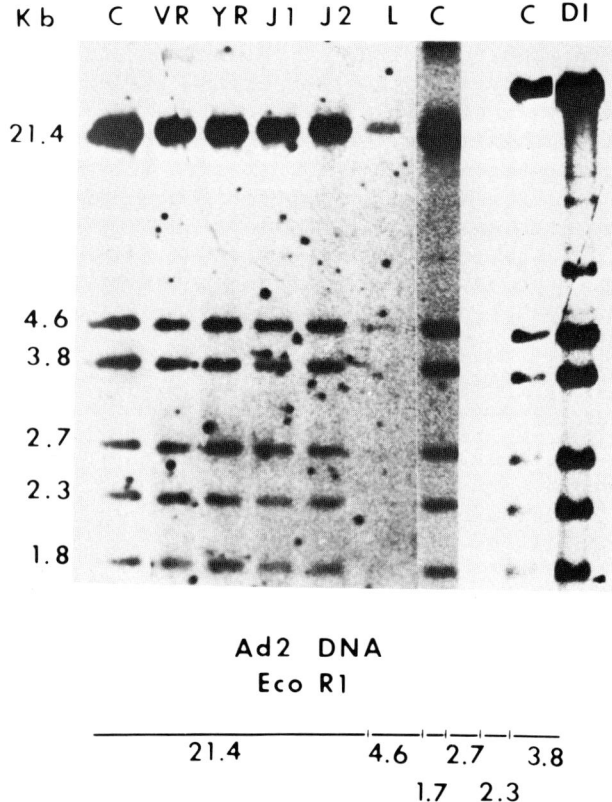

Figure 2. Effect of various viral and non-viral RNAs on AD 2 replication in vitro. Total VSV RNA (excluding leader), VR; yeast RNA, YR; WT leader RNA, L; and synthetic oligonucleotides containing portions of the WT leader RNA, J-1, and J-2; were added into the replication reaction at the same concentration (15 ng/ul). The DI leader RNA was added at a concentration of 30 ng/ul. The reactions were incubated at 31^0 C for 60'. Samples were processed as described and visualized by autoradiography (13).

into acid preciptiable counts. Higher concentrations (36-120 ng/ul) of VSV in vitro transcribed RNA did not significantly inhibit the reaction. A reduction of >95% in total DNA synthesis was observed in the presence of the highest concentration of VSV RNA tested (240 ng/ul). The kinetics of the replication reaction were quite similar for each concentration tested despite the degree of inhibition obtained. In the experiments described below we test individual species of RNA to define which VSV plus-strand RNA(s) is capable of inhibiting the specific replication of AD 2 DNA.

Comparative Effect of Various Viral and Nonviral RNAs on AD 2 DNA replication. The effects of various viral RNA transcripts and several nonviral RNAs were tested at the same concentration for their capacity to inhibit AD 2 specific DNA synthesis. In each case preparations of VSV WT plus-strand leader RNA, DI leader, VSV incomplete and complete mRNA excluding leader, yeast RNA, synthetic oligonucleotides J-1 or J-2 were added to the reaction. The reaction mixture contained: HeLa-cell extracts, AD 2 DNA template, the appropriate buffer and deoxynucleoside triphosphates required for AD 2 replication in vitro. Incorporation of ^{32}P-dCMP into AD 2 specific DNA was determined by electrophoresis on 0.8% agarose gels, which were analyzed by autoradiography. The radioactivity in the gel slices was quantitated by Cerenkov counting and scanning the gels with a densitometer.

A representative autoradiogram is shown in Figure 2. The relatively low concentration of WT leader added (15 ng/ul) markedly inhibited replication of the AD 2 specific DNA by >90%. The only other RNA tested at this concentration which resulted in >24% reduction of AD 2 specific DNA was J-1 which is a synthetic oligonucleotide made from sequence between nucleotides 18 to 25 of the WT leader. The addittion of the DI leader in concentrations which were 3.5 times higher than the molar amount of VSV WT leader resulted in no loss of AD 2 specific DNA synthesis. Indicating that a sequence unique to the WT leader is responsible for the inhibition of AD 2 DNA synthesis.

ROLE OF VSV PLUS STRAND LEADER RNA IN HOST-SHUTOFF

The relationship between the inhibition of cellular RNA synthesis and cellular DNA synthesis by VSV remains poorly defined. The availability of techniques to study the specific replication of AD 2 in vitro have allowed us to begin investigations on the inhibitory action of VSV specific RNA products in greater depth. Adenovirus replication is dependent to a large extent on the host cellular enzymes. Therefore, it has provided us with an opportunity to determine whether the inhibition of eukaryotic DNA synthesis by VSV occurs via a cascading mechanism or by a direct effect on the cellular replication mechanisms.

We tested the effect of adding preformed viral RNA products on the in vitro replication of AD 2 specific DNA. At the concentrations tested only VSV leader RNA has exhibited a very strong capacity to inhibit AD 2 DNA replication (0.5 uM). Other RNAs tested (specifically DI leader RNA) at concentrations of 3.5 - 6.0 uM higher than WT leader did not significantly effect the reaction. Current dogma suggests that AD 2 replication is not affected by the addittion of purified RNASE, does not require an RNA primer, but is dependent on the binding of an 80,000 molecular weight protein to the 5' termini of the DNA to initiate synthesis (13,14). We have found when T_1 RNASE is added into the reaction with leader that DNA synthesis is restored to some degree. The DNA synthesis observed with WT leader in the presence of T_1 RNASE is primarily repair polymerase activity and not the result of the specific initiation of replication as determined by the lack of incorporation of isotope into the terminal fragments of AD 2 DNA. We have subsequently found that our RNASE is contaminated with DNASE which makes it difficult to speculate further.

Experiments are now in progress to define the sequences unique to the WT leader which are important in inhibition of AD 2 replication in vitro. We will also determine if there is a specific interaction of the WT leader with some component of the replication extract. In addittion, studies of VSV's effect on the replication of adenovirus in tissue culture cells will ascertain whether the results obtained to date are the consequences of an in vitro system or truly representative of the events in vivo.

ACKNOWLEDGEMENTS

We express our appreciation to Mr. Theodore Osgood for his technical assistance. We also wish to thank Paul M. Williams for his help in some of the preliminary studies.

REFERENCES

1. Banerjee, A.K., G. Abraham, and R. Colonno (1977) J. Gen. Virol. 34, 1.
2. Abraham, G., and A.K. Banerjee (1976) Proc. Natl. Acad. Sci. USA 73, 1584.
3. Ball, L.A., and C.N. White (1976) Proc. Natl. Acad. Sci. USA 73, 442.
4. Marvaldi, J.M., J. Sekellick, P.I. Marcus, and J. Lucas-Lenard (1978) Virology 84, 127.
5. McGowan, J.J., and R.R. Wagner (1981) J. Virol. 38, 356
6. Weck, P.K., and R.R. Wagner (1979) J. Virol. 30, 410.
7. Wertz, G.W., and J.S. Younger (1972) J. Virol. 9, 85.
8. Yaoi, Y., H. Mitsui, and M. Amano (1970) J. Gen. Virol. 8, 165.
9. Duningan, D.D., and J.M. Lucas-Lenard (1983) J. Virol. 45, 618.
10. McGowan, J.J., S.U. Emerson, and R.R. Wagner (1982) Cell 28, 325.
11. Manley, J.L., A. Fire, A. Cano, P.A. Sharp and M.L. Gefter (1980) Proc. Natl. Acad. Sci. USA 77, 3855.
12. Grinnell, B.W., and R.R. Wagner (1983) J. Virol. (in press).
13. Challberg, M.D., and T.J. Kelly Jr. (1979) Proc. Natl. Acad. Sci. USA 76, 655.
14. Winnacker, E.L. (1978) Cell 14, 761.

INHIBITION OF ADENOVIRUS AND SV40 DNA SYNTHESIS BY VESICULAR STOMATITIS VIRUS[1]

M. E. Reichmann, Harlan B. Scott II and Delphine Krantz

Department of Microbiology
University of Illinois
Urbana, Illinois 61801

Inhibition of host cell DNA synthesis by lytic viruses is generally considered to be an indirect result of the inhibition of host cell protein synthesis (1,2). However, recent studies of VSV infected mouse cells indicated that the kinetics of the inhibitions of protein and nucleic acid synthesis are sufficiently different to suggest, that the two events are unrelated (3). Although certain hypotheses have been advanced about the putative action of VSV, the complexity of the eukaryotic cell presents difficulties in the design and interpretation of meaningful experiments. In the experiments described in this communication, the problem was approached by using DNA viruses as targets for DNA synthesis inhibition by VSV. SV40 was selected because of the stringent dependency of its DNA synthesis on the host cell cycle, histones and DNA polymerase. This dependency suggested similarities between the cellular and SV40 DNA replicative mechanism, and hopefully also a correspondence in the inhibitory mechanisms by VSV. Adenovirus, although it replicates in a much more autonomous way independent of the cell cycle, can serve as a convenient source of an _in vitro_ DNA replicating system for inhibition studies.

VSV inhibited SV40 DNA synthesis in Vero cells to approximately the same extent as that of cellular DNA. Unlike in normal SV40 infections, incompletely supercoiled SV40 DNA species accumulated in the VSV inhibited system to the extent that they were detectable by steady state labeling. A similar accumulation of incomplete SV40 DNA supercoils was also observed when protein synthesis was inhibited with cycloheximide in the absence of VSV.

Dana and Nathans (4) have shown that a short pulse (10-15 min) of SV40 DNA labeling established a linear

---

[1]This work was supported by grant AI 12070 from the National Institute of Allergy and Infectious Diseases.

gradient of radioactivity in the molecules synthesized during the pulse. The radioactive counts increased with distance from the single origin of replication. Using the slope of this gradient as a measure of DNA chain elongation rates we found that this rate remained unchanged during inhibition by VSV. The absence of any effect on the rate of DNA synthesis and the similarity in the electrophoretic profiles of cycloheximide and VSV inhibited SV40 DNA synthesis, suggested that the major effect of VSV on SV40 DNA synthesis was through inhibition of protein synthesis.

Horwitz et al. (5) have demonstrated that adenovirus DNA synthesis became independent of protein synthesis late in the infectious cycle. If DNA synthesis inhibition by VSV were indirectly associated with the inhibition of protein synthesis, late adenovirus DNA replication would be expected to be insensitive to VSV. We report here, however, that late adenovirus DNA synthesis in HeLa cells, although not very sensitive to cycloheximide, was extensively inhibited by VSV.

INHIBITION OF SV40 DNA SYNTHESIS BY VSV.

The effect of SV40 DNA synthesis was investigated in Vero cells synchronized by a double thymidine block. The cells were infected with SV40 after the first release and with VSV 4 hrs after the second release. Incorporation of ^{3}H thymidine into intracellular SV40 DNA (Hirt supernatants) was followed at various times after infection (Table 1). The extent of incorporation into cellular DNA (Hirt precipitates) was determined (data not shown). The extent of SV40 DNA synthesis inhibition was very similar to that of cellular DNA, increased as the VSV infection progressed and reached a maximum of approximately 70%. The experiment was not continued beyond a time when more than 20% of the cells became permeable to trypan blue, indicating extensive cell killing.

Characterization of intracellular SV40 DNA synthesized in the presence and absence of VSV or cycloheximide.
SV40 infected cells in the presence and absence of VSV or cyclohexmide were labeled with ^{32}P, intracellular SV40 DNA was isolated from Hirt supernatants and electrophoresed as described in the legend to Fig. 1. Lanes a and b show profiles of newly synthesized SV40 DNA in the absence of either VSV or cycloheximide. In lanes d and e were SV40 DNAs isolated from cells which were superinfected with VSV for the specified periods of time. Lane c represents SV40 DNA obtained from VSV-free cells treated with cycloheximide 2 hrs

TABLE 1
INCORPORATION OF ^{3}H THYMIDINE INTO SV40 DNA (HIRT SUP.) IN SYNCHRONIZED VERO CELLS WITH AND WITHOUT VSV INFECTION

Time (hr) after release	CPM SV40 + ts31VSV	SV40
4 VSV added	--	--
5	19,150	19,280
7	13,225	26,250
9	15,030	34,200
11	11,110	33,320
13	4,510	15,830

Vero cells in 60 mm plastic petri plates were synchronized by a double thymidine block as described by Pages et al. (10). Immediately following the first release, the cells were infected with 50 PFU/cell of SV40 strain 776 for 90 min at 39°C. Following the second release, at the indicated time the cells were superinfected with 5 PFU/cell of VSV mutant tsG31 at 39°C for 30 min. This mutant inhibited SV40 DNA synthesis as well as the wild type but cell detachment and death (as defined by exclusion of trypan blue) were delayed. In this way it was possible to follow the effect of VSV infections over a longer period of time (up to 8 hrs at least 80% of the cells were viable). One hour before the indicated time 1.5 μCi/ml of ^{3}H thymidine (60-80 Ci/mM) were added to the medium. Following a two hour incubation, the cells were washed with cold saline and intracellular SV40 DNA was isolated by the Hirt procedure (11).

before the pulse. In lanes a and b the heaviest fast moving band corresponds to the supercoiled form of SV40 DNA and the lighter band in the middle of the lanes represents the circular relaxed forms. The slower moving bands near the top of the gel probably contain the dimeric forms described by Sundin and Varshavsky (6). In addition to these bands, lanes c, d and e also exhibit ladder-like patterns characteristic of partially supercoiled DNA forms with varying numbers of twists. The similarity of the SV40 DNA profiles obtained by inhibition of protein synthesis with cycloheximide (lane c) to those obtained from VSV superinfeted cells (lanes d and e), suggested that VSV may not effect DNA synthesis directly.

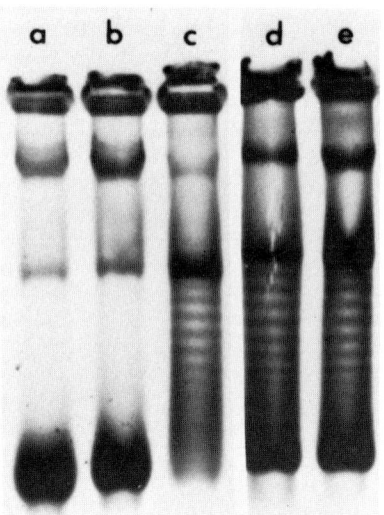

FIGURE 1. Electrophoretic profiles of ^{32}P labeled intracellular SV40 DNA in the presence and absence of either VSV or cycloheximide.

Vero cells, synchronized and infected with SV40 as described in Table 1, were either superinfected, mock infected of treated with 50 µg/ml of cycloheximide, 4 hrs after release. The cells were labeled at the intervals indicated with 0.1 mCi/ml of ^{32}P. Intracellular SV40 DNA was isolated by the Hirt procedure (11) and electrophoresed at 4 V/cm for 12 hrs in a horizontal 0.8% Agarose (Seakem) gel, using 8.9 mM tris base, 8.9 mM boric acid and 0.25 mM EDTA as buffer.
Lanes a and b: SV40 DNA labeled 10-12 and 6-8 hrs after release respectively; lane c: SV40 DNA in the presence of cycloheximide labeled 6-8 hrs after release; lanes d and e: SV40 DNA in the presence of VSV labeled 10-12 and 6-8 hrs after release respectively.

<u>Relative SV40 DNA elongation rates in the presence and in the absence of VSV.</u>
The possible existence of a direct effect on DNA synthesis was examined by comparing SV40 DNA chain elongation

rates in the presence and absence of VSV. These rates were
measured, using a method originally applied by Dana and
Nathans (4) to locate the origin of replication in SV40 DNA.
In this procedure a linear radioactive gradient, with counts
increasing as a function of distance from the origin of
replication, was established by short (10-15 min) pulse
labeling (^{3}H thymidine) of non-synchronized nascent SV40
DNA molecules. At a given duration of the pulse, the slope
of the gradient reflects the rate of DNA chain elongation,
since this rate alone determines the proportion of molecules
whose synthesis is finished during the pulse. These slopes
were determined by measuring radioactive counts in
restriction fragments obtained with a mixture of HindII and
HindIII enzymes, suitable normalization of the counts with a
randomly labeled ^{32}P control and corrections for
thymidine content based on the known SV40 DNA nucleotide
sequence (4,7). The relative slopes obtained in the presence
and absence of VSV, respectively were $4.55 \pm 0.17 \times 10^{-3}$
and $4.63 \pm 0.23 \times 10^{-3}$ per nucleotide after a 10 min
pulse and $4.20 \pm 0.17 \times 10^{-3}$ and $4.0 \pm 0.08 \times 10^{-3}$
after a 15 min pulse. These results demonstrated that VSV
inhibition of SV40 DNA synthesis did not effect the chain
elongation rates. The identical elongation rates in the
presence and absence of VSV and the similarity of the
electrophoretic profiles of intracellular SV40 DNA in VSV
infected and cycloheximide treated cells, suggested that the
VSV effect was primarily indirect, through inhibition of
protein synthesis.

INHIBITION OF ADENOVIRUS DNA SYNTHESIS BY VSV.

Horwitz et al. (5) demonstrated, that late adenovirus
DNA synthesis was independent of protein synthesis. If the
action of VSV on DNA synthesis was indirect through the
inhibition of protein synthesis, late adenovirus DNA
synthesis should be insensitive to superinfection with VSV.
Adenovirus DNA synthesis in the HeLa cells 19 hrs after
infection, was followed by incorporation of ^{3}H thymidine in
the presence or absence of either VSV or cycloheximide. High
molecular weight DNA and adenovirus DNA were separated by
high salt precipitation (Hirt procedure). Incorporation of
radioactive counts was measured in both fractions. As the
data in Table 2 indicate, incorporation of ^{3}H thymidine
into either fraction was relatively insensitive to
cycloheximide, although host protein and DNA synthesis were
strongly inhibited in control experiments under these
conditions (98% and 75% respectively). The insensitivity to
cycloheximide suggested that the radioactive counts in both

TABLE 2
INCORPORATION OF ^{3}H THYMIDINE INTO ADENOVIRUS DNA IN NON-SYNCHRONIZED HELA CELLS 19 HRS AFTER INFECTION AND IN DOUBLY INFECTED CELLS THUS AFTER VSV INFECTION.

Type of infection	CPM	
	Sup.	Prec.
Adeno	67,380	180,490
Adeno + VSV	33,430	66,040
Adeno + Cyclohex.	61,240	131,510

Subconfluent cultures of HeLa cells on 100 mm diameter plastic tissue culture plates (approximately 1 x 10^7 cells) were mock infected or infected with approximately 50 PFU/cell of adenovirus type 2. Twelve hours later, the cells were either infected with VSV tsG33 5 PFU/cell at 39°C or mock infected. When required, two hours prior to pulsing cycloheximide was added to a final concentration of 50 µg/ml to the media covering the cells. The cells were pulsed at 19 hours post adenovirus infection with media containing 1 µCi/ml ^{3}H thymidine (77.2 Ci/mmol) for 30 min. After the pulse, the cells were rinsed and assayed for ^{3}H thymidine incorporation by the Hirt procedure (11). The precipitates in solution were sonicated with three 10 second pulses with a Heat Systems-Ultrasonics Inc. Sonifier Cell Disruptor Model 185 setting 4.5. An aliquot of the supernatant and of the sonicated precipitates was precipitated with 10% trichloroacetic acid (TCA) in the presence of 200 µg bovine serum albunim (BSA) as carrier and counted in a toluene based scintillation fluid.

fractions represent adenovirus DNA synthesis. Extensive integration of adenovirus-like sequences into host cell high molecular weight DNA has been reported in the literature (8,9). At 36 hrs after infection, the adeno-genome equivalents incorporated into host cell DNA were estimated at 500-1000 copies (8). The association of the radioactive counts with adenovirus DNA sequences in the high molecular weight DNA fraction (Table 2) was further demonstrated by annealing experiments. After sonication, the DNA was annealed to cellulose nitrate filters containing varying quantities of denatured non-radioactive adenovirus DNA. With excess cold DNA, approximately 82% of the radioactive counts were retained by the filters. DNAs labeled in the presence and absence of cycloheximide followed similar hybridization kinetics (Fig. 2).

Adenovirus DNA synthesis was inhibited by VSV in doubly infected cells. Seven hours after VSV infection the extent of inhibition in the Hirt supernatant was 51% (Table 2). In the Hirt precipitate the apparent inhibition was 64%, but not all the counts incorporated into this fraction in the presence of VSV were in adeno-DNA sequences. Annealing experiments of the sonicated high molecular weight DNA fraction indicated that only 24% of the radioactive counts were retained by the filters, suggesting that 76% of the counts might be associated with cellular DNA sequences (Fig. 2). A suitable correction of the data in Table 2, based on the annealing data, resulted in 90% inhibition of adenovirus DNA sequences in the high molecular weight DNA fraction. Unlike in the absence of VSV, extensive synthesis of cellular DNA seems to have taken place. A quantitative comparision with parallel uninfected controls indicated that cellular DNA synthesis, as measured by ^{3}H thymidine incorporation, was completely restored in the mixed infection.

The data presented here suggested that the primary cause of SV40 DNA synthesis inhibition by VSV was the inhibition of protein synthesis. Because of extensive similarities between SV40 and host cell DNA synthesis it is likely that the inhibition of protein synthesis, on which both types of DNA synthesis are stringently dependent, is the primary factor in their inhibition by VSV. On the other hand, the more autonomous synthesis of adenovirus DNA might be inhibited by a different mechanism and could involve steps unique to DNA synthesis of this virus.

The observation that in adenovirus infected cells, superinfection with VSV restored cellular DNA synthesis, is difficult to explain. Since either virus alone inhibited cellular DNA synthesis, it appeared that their ability to do so was compromised in the mixed infection. At present it is

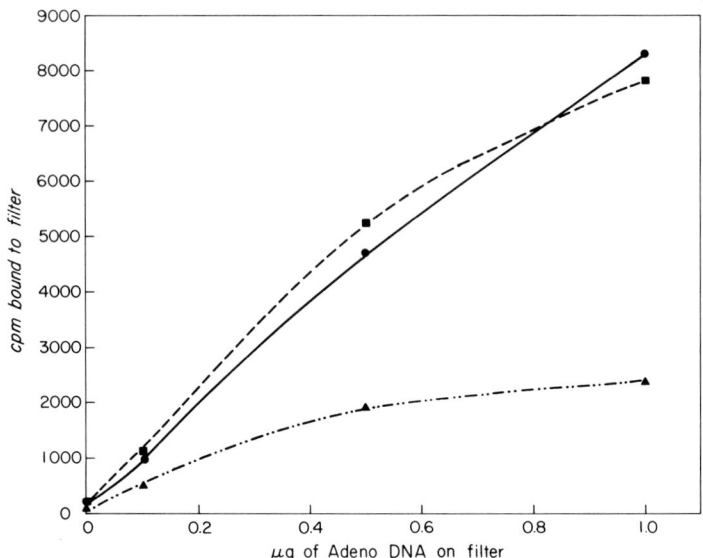

FIG. 2. HeLa cells were infected as in Table 2 except that the Hirt precipitates were not sonicated, but were dialyzed against 10 mM Tris-HCl (pH 8.0), 150 mM NaCl, and 10 mM EDTA. The DNA was extracted by the procedure of Maki et al. (12).

Adenovirus DNA purified from virions was sonicated and denatured with NaOH at a final concentration of 0.2 M for one hour. To neutralize the solution 2 M Tris-HCl was added. The DNA was passed through a filter made of Gene Screen (New England Nuclear). The filter was washed with 1 x SSC (0.15 M NaCl, 0.015 M sodium citrate, pH 7.0) and air dryed. The DNA was baked onto the filter for 2-4 hours at 80-100°C. Smaller filters were punched out of this filter and placed into siliconized conical eppendorf tubes.

The hybridization was carried out esentially as described in the Gene Screen manufacters instruction booklet. The procedure entailed prehybridizing the filters with a hybridization solution of 50% formamide (deionized), 2 x SSC, 0.05 M Na_2HPO_4/NaH_2PO_4, pH 6.5, 0.02% polyvinyl-pyrrolidone (M.W. 40,000), 0.02% BSA, 0.02% ficoll (M.W. 400,000) and denatured salmon sperm DNA (100 μg/ml) overnight at 42°C. That was followed by hybridization with sonicated and denatured DNA purified from the Hirt precipitates (10,000 cpm/filter) in the hybridization solution for 22 hrs at 42°C. The filters were then washed twice with a 0.3 M NaCl, 0.06 M Tris-HCl, pH 8.0, 0.002 M EDTA solution for 5 min at room temperature, twice with a 0.3 M NaCl, 0.06 M Tris-HCl, pH

not clear whether cellular DNA synthesis is completely restored in the sense that all nucleotide sequences are synthesized in the proper temporal order. If that were the case, it is not clear how the two viruses effect each others infectious cycle so as to negate their influence on these cellular events. Experiments are in progress to provide answers to these questions.

REFERENCES

1. Tamm, I. (1975). Am. J. Path. 81, 163.
2. Hand, R., and Tamm, I. (1974). "Cell cycle controls" Academic Press, New York.
3. McGowan, J. J., and Wagner, R. R. (1981). J. Virol. 38, 356.
4. Danna, K. J., and Nathans, D. (1972). Proc. Natl. Acad. Sci. U.S.A. 69, 3097.
5. Horwitz, M. S., Brayton, C., and Baum, S. G. (1973). J. Virol. 11, 544.
6. Sundin, O., and Varshavsky, A. (1981). Cell 25, 659.
7. Buchman, A. R., Burnett, L., and Berg P. (1981). "DNA tumor viruses" Cold Spring Harbor, New York.
8. Tyndall, C., Younghusband, H. B., and Bellett, A. J. D. (1978). J. Virol. 25, 1.
9. Newmann, R., and Doerfler, W. (1981). J. Virol. 37, 887.
10. Pages, J., Manteuil, S., Stehelin, D., Fiszman, M., Marx, M., and Girard, M. (1973). J. Virol 12, 99.
11. Hirt, B. (1967). J. Mol. Biol. 26, 365.
12. Maki, R., Roeder, W., Traunecker, A., Sidman, C., Wabl, M., Raschke, W., and Tonegawa, S. (1981). Cell 24, 353.

8.0, 0.002 M EDTA, 0.5% SDS solution for 30 min at 60°C, and twice with a 0.003 M Tris base solution for 30 min at room temperature. The filters were dried and counted in a toluene based scintillation fluid. DNA from adenovirus infected cells in the presence ■---■ and absence ●——● of cycloheximide. DNA from cells doubly infected with adenovirus and VSV ▲---▲.

EXPRESSION OF MEASLES VIRUS RNA IN BRAIN TISSUE[1]

Knut Baczko,[2] Martin Billeter[3] and Volker ter Meulen[2]

[2]Institute of Virology, University of Würzburg,
D - 8700 Würzburg, F.R.G.
[3]Institute for Molecular Biology I,
University of Zürich, Switzerland

Measles virus is a human pathogen which causes both acute and subacute infections, and has also been implicated in multiple sclerosis. Following acute measles, the individual develops a life-long immunity to reinfection. However, in some cases, the virus establishes a persistent infection in the central nervous system (CNS) which gives rise to the slowly progressing disease subacute sclerosing panencephalitis (SSPE). Current evidence suggests that virus persistence in SSPE is associated with a specific lesion in the production of virus matrix (M) protein, although all other polypeptides are produced. The lack of this major structural polypeptide is thought to disrupt the virus maturation process and lead to a strictly cell-associated phenotype, thus accounting for the slowly progressing nature of the disease. In order to assess the involvement of measles virus in SSPE, we decided to clone the virus genomic RNA, so producing specific probes with which to examine virus expression within diseased tissue.

CLONING PROCEDURE

Genomic RNA was isolated from intracellular nucleocapsids produced in lytically infected MA160 cells (1). The cloning procedure was straight forward and used standard methods. In brief, genomic RNA was polyadenylated at its 3' end (2) and then used as template for oligo-dT primed cDNA synthesis by reverse transcriptase. Single stranded cDNAs were converted to double strands, tailed and joined to plasmid pBR322 at the PstI site. E. coli cells were transfected with this DNA and recombinant clones were screened by <u>in situ</u> hybridization using a probe obtained by reverse transcription

---

[1]This work was supported by the Deutsche Forschungsgemeinschaft, Hertie-Stiftung, the Schweizerische Nationalfonds (No. 3.045.81), and the Kanton of Zürich.

of polyadenylated 50 S MV RNA. 15 clones were selected with inserts ranging from 500 to 1000 bp in size. These were further characterized by Northern blotting. Three reacted with uninfected cell RNA, the other 12 reacted only with measles virus infected cell RNA. These clones could be divided into four groups, according to their hybridization behaviour, as shown in Figure 1.

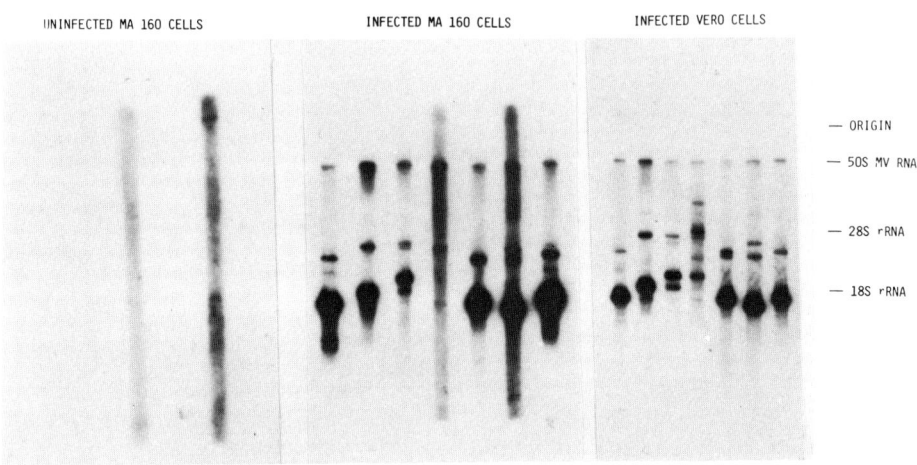

Figure 1. Identification of MV-specific RNA species from lytically infected MA 160 and Vero cells.
Cytoplasmic RNAs were electropohoresed in a vertical 1 % agarose slab gel containing 6 mM methylmercury hydroxide, and blotted directly to nitrocellulose. Filters were baked, cut into small strips and hybridized with individual nick-translated 32P labeled cDNA clones. Clones 1-4 are derived from virus genomic RNA and habe been assigned tentatively as: 1, Nucleocapsid; 2, Haemagglutinnin; 3, Fusion; 4, L. clones P, M, N are derived from virus m RNA (3).

Three size classes of virus- specific RNA were detected. The largest comprises 50S genome-size molecules, and the smallest, most abundant species constitutes the virus mRNA. In addition all clones also recognised RNA species of intermediate size (is-RNA). Clones of group 1 showed an identical hybridzation pattern to that obtained using cloned copies of sequences contained in the mRNA for nucleocapsid protein (kindly provided by Dr. S. Rozenblatt) (3). These clones also cross-hybridized

EXPRESSION OF MEASLES VIRUS RNA

to the nucleocapsid specific DNA itself. The sequence of the first several hundred nucleotides has been determined (Fig. 2) and the start of N-mRNA was located by S1 mapping.

```
                                                    ↓                                          100
ACCAAACAAAGTTGGGTAAGGATAGTTCAAATCAATGATCATCTTCTAGTGCACTTAGGATTCAAGATCCTATTATCAGGGACAAGAGCAGGATTAGGGA

                   ALU                                      BAM                                200
TATCCGAGATGGCCACACTTTTAAGGAGCTTAGCATTGTTCAAAAGAAACAAGGACAAACCACCCATTACATCAGGATCCGGTGGAGCCATCAGAGGAAT
               MET ALA THR LEU LEU ARG SER LEU ALA LEU PHE LYS ARG ASN LYS ASP LYS PRO PRO ILE THR SER GLY SER GLY GLY ALA ILE ARG GLY ILE

                                                                                               300
CAAACACATTATTATAGTACCAATCCCTGGAGATTCCTCAATTACCACTCGATCCAGACTTCTGGACCGGTTGGTCAGGTTAATTGGAAACCCGGATGTG
LYS HIS ILE ILE ILE VAL PRO ILE PRO GLY ASP SER SER ILE THR THR ARG SER ARG LEU LEU ASP ARG LEU VAL ARG LEU ILE GLY ASN PRO ASP VAL

                                                                                               400
AGCGGGCCCAAACTAACAGGGGCACTAATAGGTATATTATCCTTATTTGTGGAGTCTCCAGGTCAATTGATTCAGAGGATCACCGATGACCCTGACGTTA
SER GLY PRO LYS LEU THR GLY ALA LEU ILE GLY ILE LEU SER LEU PHE VAL GLU SER PRO GLY GLN LEU ILE GLN ARG ILE THR ASP ASP PRO ASP VAL

                                                     KPN                                       500
GCATAAGGCTGTTAGAGGTTGTCCAGAGTGACCAGTCACAATCTGGCCTTACCTTCGCATCAAGAGGTACCAACATGGAGGATGAGGCGGACCAATACTT
SER ILE SER LEU LEU GLU VAL VAL GLN SER ASP GLN SER GLN SER GLY LEU THR PHE ALA SER ARG GLY THR ASN MET GLU ASP GLU ALA ASP GLN TYR PHE

                                                                                               600
TTCACATGATGATCCAATTAGTAGTGATCAATCCAGGTTCGGATGGTTCGAGAACAAGGAAATCTCAGATATTGAAGTGCAAGACCCTGAGGGATTCAACATGAT
SER HIS ASP ASP PRO ILE SER SER ASP GLN SER ARG PHE GLY TRP PHE GLU ASN LYS GLU ILE SER ASP ILE GLU VAL GLN ASP PRO GLU GLY PHE ASN MET
```

Figure 2. Sequences of the 5' terminal region of MV positive strand RNA and of the region encoding the N-terminal of the nucleocapsid protein. The point where N mRNA starts is marked by an arrow.

The sequence of the first 9 nucleotides of the cDNA matched perfectly the 3' sequence of genomic RNA. This proved that group 1 cDNAs are derived from the exact 3' end of genomic measles virus RNA. Group 2 cDNAs hybridized to a mRNA of approximately 2100 nucleotides (Fig. 3) which was not detected using any of the clones derived from measles virus mRNAs (P,N,M) provided by Dr. Rozenblatt. Group 3 clones hybridzed to the same mRNA as group 2, but also recognized an additional mRNA of about 2650 nucleotides in size. Experiments are in progress to identify the proteins encoded by these mRNAs which we believe are the haemagglutinin (H) and fusion protein (F) for groups 2 and 3 respectively. Group 4 is most probably derived from an area towards the 5' end of the genome. This clone hybridizes to several RNAs in the A(-) fraction which are probably defective RNAs. No hybridization is detectable to the A(+) fraction. This clone could represent part of the L gene. Transcripts of this gene are known to be present only in low amounts in the A(+) fraction. The nature of the is-RNA is not yet fully understood. Some of these molecules are polyadenylated and are found in the A(+) fraction (Fig.3). These bands are of positive polarity, as intermediate-sized RNA hybridizes to a recloned M13 probe of negative sense, derived from group 1 inserted sequences. This suggests that these RNAs are polycistronic mRNAs or polymerase read-throughs, which are also known to occur in other negative strand viruses (4,5). The existence of these measles virus polycistronic RNAs permits the determination of the order in which proteins are encoded on the virus genome. This we have tentatively determined as: L-2(H?)-3(F?)-M-P-N. This includes the reassignment of an H clone as P (6).

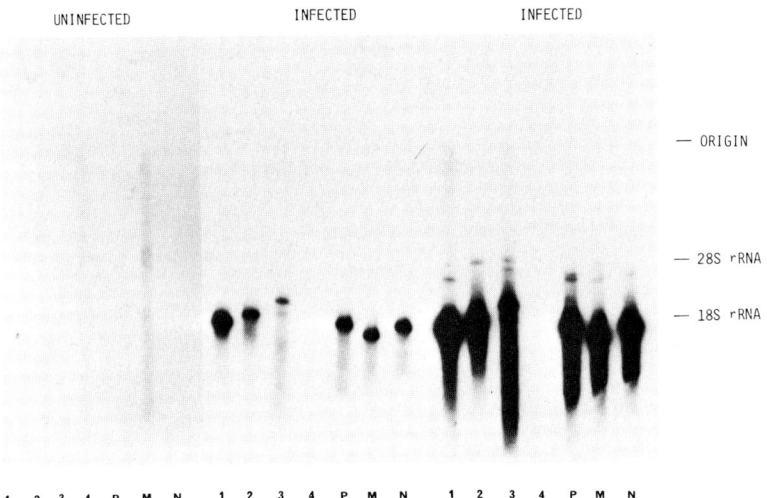

Figure 3. Northern blot of oligo-dT selected A(+) cytoplasmic RNAs from Vero cells, hybridized to nick-translated, cloned cDNAs as in Fig. 1.

EXAMINATION OF BRAIN RNA

The available cDNAs were used to characterize virus RNA in measles virus infected rat brains. In these studies suckling rats were infected with the neurotropic measles virus strain (CAM) which induced an acute encephalitis after an incubation period of 10-15 days. Infectious virus could be isolated from diseased brain. In addition, another group of suckling rats were intracerebrally inoculated with a lytic measles virus rescued from an SSPE cell line (MF). This persistently infected cell line produced M protein and released infectious virus following physiological trauma (7). These animals also developed an acute encephalitis after an incubation period of approximately two weeks.

RNA extracted by the guanidium isothiocyanate technique (8) was examined by the method of Northern blotting using the cloned measles virus DNAs. The pattern of virus-specific RNAs expressed in the CAM-infected brains was similar to that observed in lytically infected cells in vitro. All three size classes of virus-specific RNA (with the exception of L message) could be detected with each probe. However, the results were different using RNA from SSPE-MF infected rat brain (Fig. 4). All mRNAs except P and L could be detected, although the amount of matrix protein mRNA was greatly reduced and there was a high concentration of the intermediate-sized P-M transcription read-through. This indicates that expression of M-mRNA by this SSPE virus may be decreased in CNS tissue.

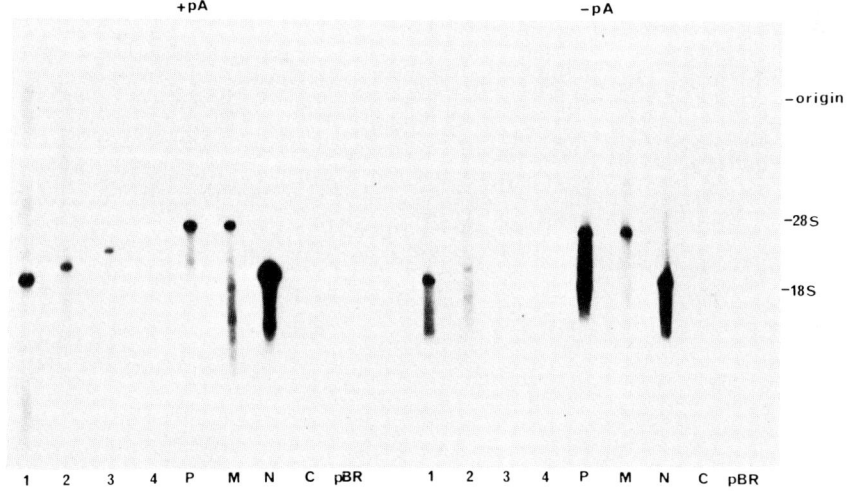

Figure 4. Northern blot of oligo-dT selected RNAs from SSPE-MF infected rat brain hybridized to nick-translated, cloned cDNAs as in Fig. 1 and 2. Control cDNAs C = cDNA of corona mRNA 7 (kindly provided by Dr. Siddell), pBR = pBR322 without insert.

The presence of M protein specific mRNA does not imply that functional matrix protein is necessarily synthesised within the infected brain. One SSPE cell line is known in which M protein mRNA is not utilized correctly in translation reactions (9). Further experiments are in progress to investigate the activity of these mRNAs in in vitro protein synthesising reactions. This study demonstrates the usefulness of these techniques in analysing the state of measles virus gene expression in diseased tissue.

Acknowledgement.
We thank Dr. S. Rozenblatt for making his cDNA clones available to use, and Dr. U.G. Liebert for performing the animal studies.

REFERENCES

1. Baczko, K., Billeter, M., and ter Meulen, V. (1983). J. gen. Virol. 64, 1409.
2. Sippel, A.E. (1973). Eur. J. Biochem. 37, 31.
3. Rozenblatt, S., Gesang, C., Lavie, V., and Neumann, F.S. (1982). J. Virol. 42, 790.
4. Herman, R.C., Adler, S., Lazzarini, R.A., Colono, R.C., Banerjee, A.K., and Westphal, H. (1978). Cell 15, 735.
5. Collins, P.L., and Wertz, G.W. (1983). PNAS 80, 3208.
6. Bellini, W.J., Richardson, C., Meyers, C., and Lazzarini, R.A., This volume.
7. Carter, M.J., Willcocks, M.M., and ter Meulen, V., This volume.
8. Chirgwin, J.M., Przybyla, A.E., MacDonald, R.J., and Rutter, W.J. (1979). Biochemistry 18, 5294.
9. Carter, M.J., Willcocks, M.M., and ter Meulen, V. (1983). Nature 305, 153.

TEMPORAL CHANGES OF MEASLES VIRAL PROTEINS IN HELA CELLS ACUTELY AND PERSISTENTLY INFECTED WITH MEASLES VIRUS[1]

Karen K. Y. Young and Steven L. Wechsler

Department of Molecular Virology
The Christ Hospital Institute of Medical Research
2141 Auburn Avenue
Cincinnati, Ohio 45219

Measles virus, a paramyxovirus, normally causes an acute infection with accompanying classic symptoms of rash and fever. Occasionally, however, the virus can generate a persistent infection. The neurological disorder subacute sclerosing panencephalitis (or SSPE) is one clinical manifestation of persistent measles virus infection. SSPE is the best studied example in man of a persistent viral disease caused by a virus that normally causes an acute infection (1). As such, measles virus persistence offers an excellent system for the study of viral persistence in man.

The study of persistent infections has been greatly aided by the establishment of persistently infected tissue culture cell lines in the laboratory. By comparing persistently and acutely infected cell lines, one may glean some information about the possible cause and/or consequence of persistent infections. This report presents some preliminary results of one such study in progress in our laboratory.

The purpose of this study was to examine the fate, with time, of newly synthesized viral proteins during persistent and acute infections. Acutely and persistently infected HeLa cells were subjected to a one-hour pulse of radioactive labeling, then chased for varying lengths of time with medium containing no radioactive label. The resultant labeled intracellular viral proteins were isolated by immune precipitation and examined by SDS-PAGE.

Acute infection was accomplished by infecting HeLa cells with wild type measles virus at a multiplicity of infection of one. Two HeLa cell lines persistently infected with measles virus (2, 3, 4) were studied. The first

---

[1] This work was supported by NIH Grant AI-18647

persistently infected cell line, K11, produces extracellular virus at a low level (a productive persistent infection). The second line, K11A, was derived from K11 after extended culture in the presence of measles specific antisera. Unlike K11, K11A produces no detectable extracellular virus (a non-productive persistent infection). However, virus can occassionally be recovered from K11A cells following extensive co-cultivation with suitable permissive cells (5). In this regard K11A closely resembles the brain cells of patients with SSPE, which also produce no directly detectable infectious virus yet from which SSPE virus has occassionally been isolated by co-cultivation techniques (1). We have previously reported differences, detected by pulse labeling followed by SDS-PAGE, between the M (matrix), H (hemagglutinin), and NP (nucleocapsid) proteins of these persistently infected cells compared to acutely infected cells (6). Differences were also seen between the two persistently infected cell lines.

PULSE-CHASE EXPERIMENTS

Pulse-chase experiments have previously been done with cells acutely infected with wild type measles virus (7, 8). In the study reported here, the results for the acutely infected cells were identical to those of the productive persistently infected cell line K11. Therefore, due to space limitations, gels of the acutely infected cells are not shown.

Stability of viral proteins with time. The major viral proteins from acutely infected HeLa cells and K11 cells appeared to be stable throughout the chase period (Figure 1). On the other hand, the viral M (matrix) protein band from K11A cells diminished in intensity and eventually disappeared during the chase (Figure 2). At the end of the labeling period (lane 0), the M protein formed a prominent band on the gel. Two hours into the chase, the intensity of the M protein band was greatly diminished. The band was barely visible after a chase of four hours and by the end of the six-hour chase, it had all but disappeared. Therefore, it appears that the M protein in K11A is less stable than that in both acutely infected HeLa cells and the other persistently infected line.

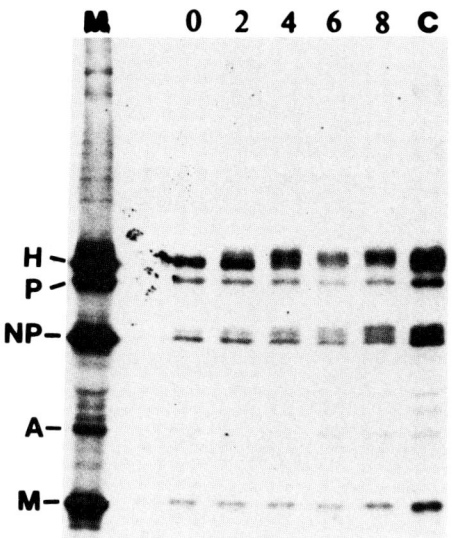

Figure 1. Fluorogram of a pulse-chase experiment in K11 cells. Cell monolayers were labeled with [^{35}S]methionine for one hour as previously described (7, 9). Cells subjected to chase periods were washed three times with 1 ml of medium supplemented with a large excess of cold methionine (150 mg/L) and 10% fetal calf serum and then reincubated in the same medium. After either a pulse labeling period alone, or a pulse followed by a chase, the monolayers were harvested, immune precipitated with anti-measles antiserum (9), run on SDS-PAGE (10) and processed for fluorography. Viral proteins: H, hemagglutination protein [80,000 daltons (80K)]; P, nucleocapsid associated phosphoprotein (70K); NP, nucleocapsid protein (60K); M, matrix protein (37K); A, cellular actin (43K). L, involved in transcription (200K) and F_1 (41K) and F_2 (20K), the 2 components of the fusion protein are not consistently detected and were not analyzed. Gel lanes: M, measles proteins as marker; 0, no chase; 2, 2 hr chase; 4, 4 hr chase; 6, 6 hr chase; 8, 8 hr chase; C, continuous labeling for 8 hours.

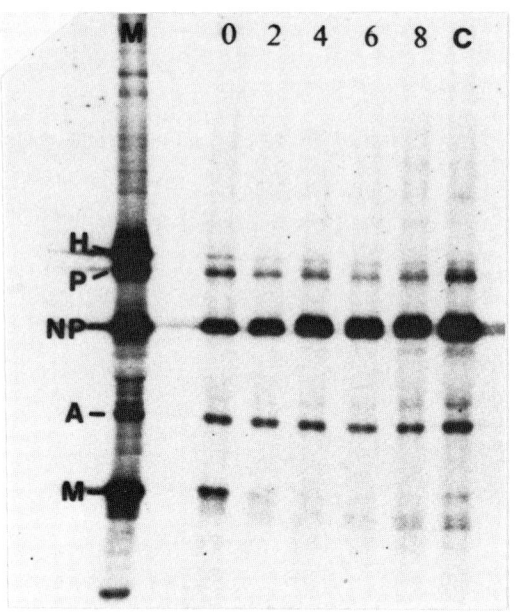

Figure 2. Fluorogram of pulse-chase labeled K11A cells. Conditions were identical to those of Figure 1, but with K11A cells.

Processing of viral proteins. Two viral proteins, the major nucleocapsid protein (NP) and viral hemaglutinin (H), in acutely infected cells and K11 were processed over a period of time (Figure 1). In the course of the chase, the initial band of each protein gradually disappeared with the concommitant appearance and subsequent increase in intensity of a new band of slightly higher molecular weight just above the original band. The NP and H proteins are phosphorylated (11) and glycosylated (7, 12), respectively. The new protein bands are probably the more highly phosphorylated and glycosylated forms of their respective precursor proteins (6, 13).

The situation in K11A was somewhat different. The processing of NP appeared to proceed normally (Figure 2). (This is seen more clearly in a shorter exposure of this gel, shown in Figure 3). On the other hand, there was no detectable processing of the H protein during the chase period. As late as eight hours into the chase, no new protein band appeared above the original H protein band.

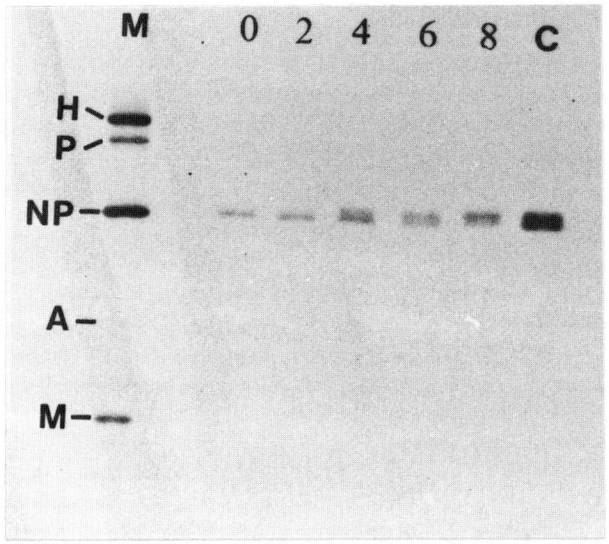

Figure 3. A shorter exposure of the gel in Figure 2.

Therefore, the H protein is either not processed or is processed very slowly. In addition, there also appears to be less H protein present in the K11A cells. It is not known whether the failure of the H protein to be processed (or its reduced quantity) in K11A cells is related to the instability of the M protein in these cells.

DISCUSSION

We have previously reported alterations in the electrophoretic mobilities or quantities of the M, H, and NP proteins in the persistently infected cell line K11A compared to cells acutely infected with wild type measles virus (6, 13). Using pulse-chase experiments, we have reported here two additional significant differences between the viral proteins of these non-productive, persistently infected K11A cells compared to the viral proteins of cells acutely infected with measles virus. The viral M protein in K11A cells was unstable. In addition, the H protein in K11A cells did not undergo the processing observed during acute measles virus infection.

A second persistently infected cell line, K11, that does produce some infectious extracellular virus, had a stable M protein and normal processing of H protein. Since the K11A cell line was derived from K11 cells, it is possible that the alterations in M and H detected here may be involved in the biological difference between K11 and K11A, i.e., the change from a productive persistent infection to a non-productive persistent infection.

The cause of the instability of the K11A M protein is not known. The protein may be inherently unstable, it may be unstable due to temperature sensitivity (the pulse-chase experiments were performed at $37^{\circ}C$), or it may be more susceptible to proteolytic degradation. It is also possible that the K11A cell M protein is normal, but that the K11A cell itself has changed in such a way as to degrade that particular protein faster than normal.

The lack of processing of the H protein and the reduced quantity of H protein in K11A cells appeared to be reflected in the biological properties of these cells. K11A cells have reduced surface viral immunofluorescence and reduced hemadsorption ability compared to K11 cells and HeLa cells acutely infected with measles virus (2, 3, 6). These differences are consistent with either reduced amounts of the viral hemagglutination (H) protein in the cell membrane or abnormal processing of this protein.

The lack of processing of the H protein and the instability of the M protein in K11A cells may be coincidental or they may both be linked to the mechanism of non-productive persistence. The instability of the M protein could also be involved in the lack of processing and reduced quantity of H. Since the M protein may be involved in anchoring H protein in the cell membrane, an unstable M protein could result in H protein being lost from the membrane. This would not only reduce the total amount of H present, but would eliminate processing events that occur while H is in the cell membrane.

Reduced H protein and lack of its processing could be involved in the mechanism of non-productive viral persistence. K11A was derived from K11 through extensive culture in the presence of antisera directed against measles virus antigens. This process may have selected in favor of persistently infected cells which produce little or no cell surface viral antigens (as suggested by the antigenic modulation model for viral persistence (14)). Should presence of viral proteins on the cell surface be necessary for alignment of viral nucleocapsids and subsequent viral budding, selection in favor of cells which lack viral proteins on the cell surface would also be selection in favor of cells which cannot produce extracellular viruses.

K11A cells are of special interest because of the close similarity of the non-productive viral persistent state to that in the brain cells of patients with SSPE. In both systems, no detectable infectious virus is produced, and virus can be recovered only following extensive co-cultivation with cells permissive for measles virus. The instability of the K11A cell M protein also appears similar to that of the M protein in SSPE infected brain cells. A great deal of evidence supports the theory that in SSPE the viral M protein is present in greatly reduced amounts (1), and that this may be due to synthesis of an unstable M protein (1, 15, 16). Thus, we have both an *in vivo* and an *in vitro* persistent measles infection with similar biology and an apparently similar molecular defect, instability of the M protein.

It is tempting to speculate that mutations in the M protein are involved in the mechanism of establishment of non-productive persistent infections. However, it is also possible that persistent measles virus infection leads to alterations in the measles virus M protein. If this is the case, then instability of the M protein may be involved in the maintanence of non-productive measles virus persistent infections.

More than one mechanism probably exists for the generation of non-productive persistently infected cells. We have found that late passages of K11 cells produce fewer extracellular virus particles than earlier passages (as much as a 40-fold reduction in 150 passages). Yet these cells contain M proteins of normal stability and their H proteins are processed in an apparently normal manner. As this diminution of virus yield occurs as a result of extensive serial passage in the absence of antisera, the possibility of an alternate pathway to the non-productive state cannot be ruled out. Experiments are in progress to test out the various hypotheses.

ACKNOWLEDGMENTS

We wish to thank Dr. R. Rustigian for providing us with various passages of K11 and K11A cells.

REFERENCES

1. Wechsler, S. L., and Meissner, H. C. (1982). In "Progress in Medical Virology" (J. L. Melnick, ed.), Vol. 28, p. 65. S. Krager, Basel.
2. Rustigian, R. (1962). Virology 16, 101.
3. Rustigian, R. (1966). J. Bacteriol. 92, 1792.
4. Rustigian, R. (1966). J. Bacteriol. 92, 1805.
5. Rustigian, R., Winston, S. H., and Darlington, R. W. (1979). Infec. Immun. 23, 775.
6. Wechsler, S. L., Rustigian, R., Stallcup, K. C., Byers, K. B., Winston, S. H., and Fields, B. N. (1979). J. Virol. 31, 677.
7. Wechsler, S. L., and Fields, B. N. (1978). J. Virol. 25, 285
8. Graves, M. C. (1981). J. Virol. 38, 224.
9. Wechsler, S. L., Weiner, H. L., and Fields, B. N. (1979). J. Immunol. 123, 884.
10. Laemmli, U. K. (1970). Nature. 227, 680.
11. Bussell, R. H., Waters, D. J., and Seals, M. K. (1974). Med. Microbiol. Immunol. 160, 105.
12. Mountcastle, W. E., and Chopin, P. W. (1977). Virology 78, 463.
13. Wechsler, S. L., Meissner, H. C., Ray, U. R., Weiner, H. L., Rustigian, R., and Fields, B. N. (1981). In "The Replication of Negative Strand Viruses" (H. L. Bishop and R. W. Compans, ed.), p. 615. Elsevier North Holland Inc.
14. Joseph, B. S., and Oldstone, M. B. A. (1975). J. of Exp. Med. 142, 864.
15. Lin, F. H. and Thormar, H. (1980). Nature. 285, 490.
16. Machamer, C. E., Hayes, E. C., and Zweernik, H. J. (1981). Virology. 108, 515.

MODULATION OF MEASLES VIRUS-SPECIFIC PROTEIN SYNTHESIS BY CYCLIC NUCLEOTIDES AND AN INDUCER OF ADENYLATE CYCLASE[1]

Steven J. Robbins, Jessica Randle, and Jon Eagle

Virology and Immunology Section
Queensland Institute of Medical Research
Herston, Brisbane, Queensland 4006

Measles virus (MV) replication in mammalian cells is inhibited by a variety of biological and chemical agents including specific antibody (1,2), interferon (3-5), cytochalasin B (6,7), anaesthetics (8,9), prostaglandins (10), and cyclic nucleotides (10-12). The mechanisms by which these substances inhibit MV replication is unknown or poorly understood. However, certain virus replication processes are affected similarly by many of these inhibitors. These processes include (1) assembly of nucleocapsids and virions, (2) intracellular nucleocapsid distribution, (3) fusion of cells, and (4) virus-specific protein synthesis.

The results of other studies indicate that the establishment and/or maintenance of persistent MV infections involves tissue- or cell-specific restriction of virus replication (13-16). Interestingly, the virus replication processes affected by the inhibitors mentioned above also appear to be affected in persistent infections. It has been suggested that a common bio-molecular mechanism may be involved in the inhibition or restriction of MV replication observed in these studies (11,12).

Previous workers have shown that the synthesis of certain MV structural proteins is restricted in tissues from the central nervous system (CNS) of patients with subacute sclerosing panencephalitis or SSPE (17,18). Similar restriction of MV-specific protein synthesis has also been described in virus-carrying cell lines derived from biopsied SSPE tissue (19). While the synthesis of a number of MV-specific proteins was shown to be decreased in these studies, the virus M protein was most significantly affected.

In a separate study, the synthesis of the M protein was

---

[1] Supported by a grant from the National Health and Medical Research Council of Australia and by funds from the Queensland State Government.

shown to be inhibited by treatment of acutely infected human amnion cells (AV3) with adenosine 3',5' monophosphate (cyclic AMP)(11). A subsequent study showed that treatment of acutely infected human neural cells with an inhibitor of phosphodiesterase (papaverine) also inhibited the synthesis of the virus M protein (12). These findings suggested that elevated intracellular concentrations of cyclic AMP in CNS tissues may be involved in restricting the synthesis of the M protein in SSPE.

We have been interested in the mechanism by which cyclic AMP inhibits the production of MV structural proteins and what role this may play in the evolution of persistent MV infections. In this paper, we describe the effects of cyclic AMP and an inducer of adenylate cyclase (dopamine) on MV-specific protein synthesis in persistently infected AV3 cells (16).

In our initial experiments, we examined the synthesis of uninfected AV3 cell proteins following exposure to various concentrations of cyclic AMP or dopamine. Concurrently, we examined the synthesis of MV-specific proteins in acutely and persistently infected cells which were also treated with the chemicals. Figure 1 shows the relationship between the synthesis of cellular and virus-specified proteins and the cyclic nucleotide concentration used to treat the cells. Similar results were obtained for dopamine although the concentrations used were approximately 10 fold lower (data not shown). While significant inhibition of cellular protein synthesis was not seen when concentrations of up to 10 mM cyclic AMP were used to treat uninfected cells, the synthesis of viral proteins was clearly inhibited in both acutely and persistently infected cells. The inhibition was more pronounced in acutely infected cells and paralleled the decreased production of infectious particles reported earlier (11).

While the synthesis of the virus M protein was markedly affected in cyclic AMP-treated acutely infected cells (data not shown), there was only a generalized inhibition of all virus structural proteins in persistently infected cells (see Figure 2). At cyclic AMP concentrations greater than or equal to 500 μM, the HA, P, and F_1 viral proteins were not readily detected by immunoprecipitation and SDS-PAGE. However, since these proteins were not produced at high levels in untreated cells, it was difficult to evaluate the actual extent of their inhibition.

The parameters used for the treatment of the persistently infected cells with cyclic AMP and dopamine in these experiments were based on those used in earlier studies for the treatment of acutely infected cells, i.e. 24 hours of exposure followed by radio-labelling and immunoprecipitation (11). The production of the M protein in the persistently infected cells

MEASLES VIRUS-SPECIFIC PROTEIN SYNTHESIS

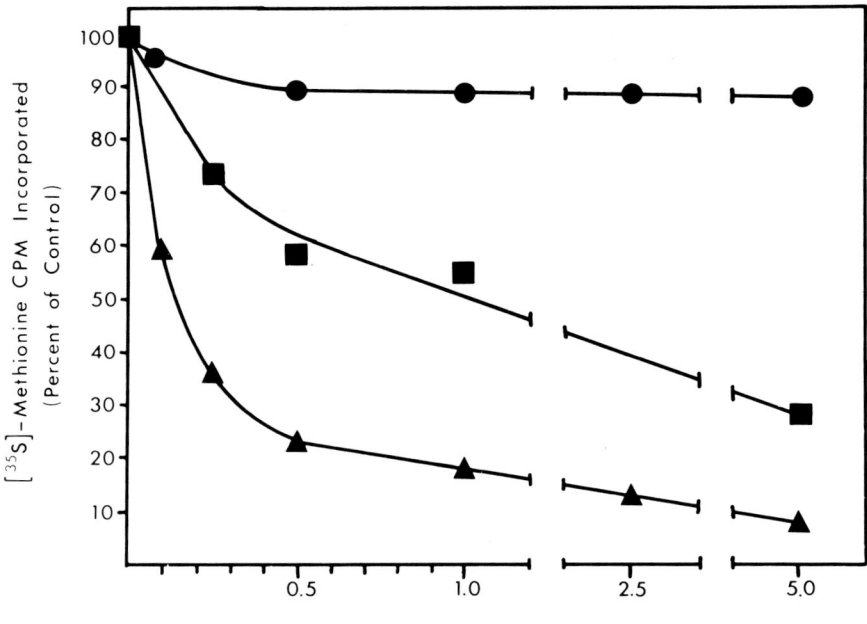

Figure 1. Measles virus- and cell-specific protein synthesis in uninfected (●), acutely infected (▲), and persistently infected (■) human amnion cells (AV3) following exposure of cells to different concentrations of cyclic AMP for 24 hours. All cells were labelled with [^{35}S]methionine at 40 μCi/ml for 4 hours in the presence of cyclic AMP. Cell-specific incorporation of radio-isotope was determined by counting TCA-precipitable materials in a beta-scintillation counter; virus-specific isotope incorporation was determined by similarly counting immunoprecipitable materials from acutely and persistently infected cells.

suggested that the treatment protocol was not sufficient to inhibit the protein's synthesis or that the biosynthetic step involved was no longer susceptible to the inhibitory effects of the chemical. Although treatment of the cells with cyclic AMP and/or dopamine for up to one week further inhibited the overall synthesis of the viral proteins, it did not appear to specifically block the synthesis of any particular protein (data not shown). This seemed to indicate that the biosynthetic step involved in the inhibition of virus M protein production in acutely infected cells was not susceptible to

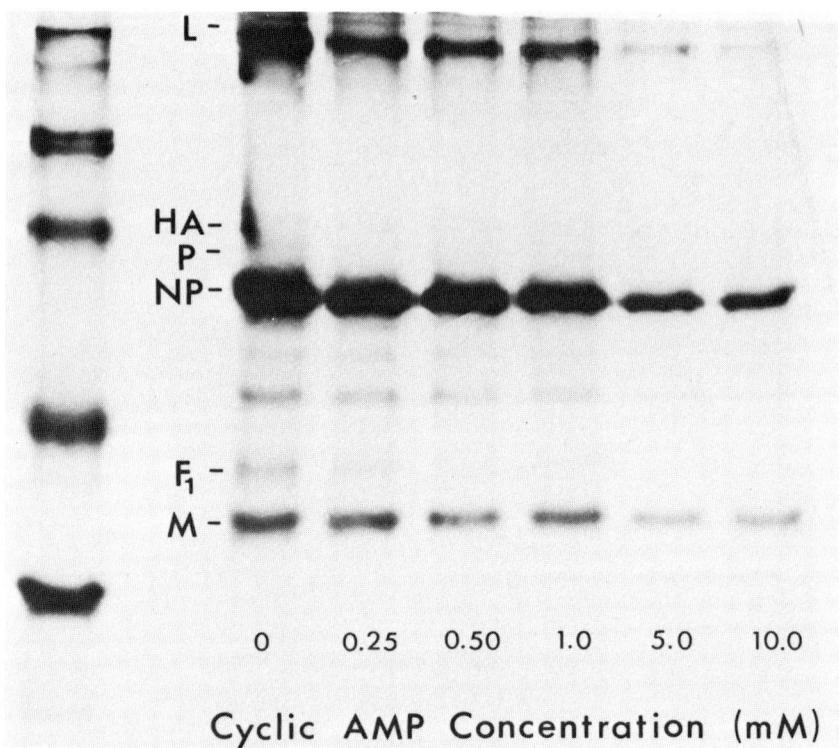

Figure 2. Immunoprecipitates of [^{35}S]methionine-labelled cyclic AMP-treated persistently infected AV3 cells as analyzed by SDS-PAGE and fluorography. Radio-labelled molecular weight marker proteins are shown in the left of the figure. Immunoprecipitation was carried out as previously described (11) using uniform conditions for each sample.

inhibition in persistently infected cells. Although this result seems paradoxical, it may indicate that initiation of M protein production is the step affected by cyclic AMP and that once initiated (transcribed? translated?) it is no longer as susceptible to the inhibitory effects of the chemical. Nevertheless, there does seem to be a generalized inhibitory effect of cyclic AMP on MV-specific protein production.

Interestingly, chronic treatment of the AV3/MV cells with cyclic AMP and/or dopamine also resulted in a pronounced redistribution of intracellular virus antigens. As shown in Figure 3, substantial numbers of MV inclusion bodies in cell nuclei were observed in such chemical-treated cells. With prolonged treatment, dissapation of cytoplasmic inclusion

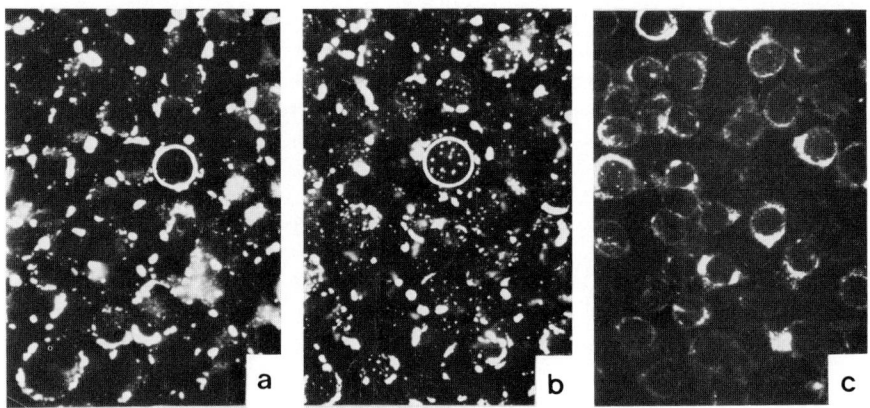

Figure 3. Expression of MV antigens in persistently infected AV3/MV cells as detected by immunofluorescence (a) and after treatment of 1 mM cyclic AMP for 4 days (b) or 250 µM cyclic AMP for 6 weeks (c). The encircled areas in (a) and (b) encompass single cell nuclei showing the relative levels of MV antigens.

bodies also occurred. The significance of these observations is unclear. It is perhaps worth noting that intranuclear MV inclusion bodies occur frequently in SSPE CNS tissues and have previously been proposed as diagnostic markers of the disease (20).

Although the exact mechanism by which cyclic AMP inhibits MV replication was not determined in the present study, the modulatory effect of the chemical on MV-specific protein synthesis and intracellular antigen distribution in persistently infected cells seems to support the hypothesis that cyclic nucleotide levels may play regulatory roles in such infections.

REFERENCES

1. Rustigian, R. (1962). Virology 16, 101.
2. Rustigian, R. (1966). J. Bact. 92, 1805.
3. Oddo, F. G., Sinatra, A., Tomasino, R. M., and Chiarini, A. (1965). Arch. Ges. Virusforsch. 16, 148.
4. Mirchamsy, H., and Rapp, F. (1969). J. Gen. Virol. 4, 513.
5. Jacobson, S., and McFarland, H. F. (1982). J. Gen. Virol. 63, 351.
6. Follett, E. A. C., Pringle, C. R., Pennington, T. H., and Shirodaria, P. (1976). J. Gen. Virol. 32, 163.

7. Stallcup, K. C., Raine, C. S., and Fields, B. N. (1983). Virology 124, 59.
8. Knight, P. R., Nahrwold, M. L., and Bedows, E. (1980). Antimicrob. Agents Chemother. 17, 890.
9. Knight, P. R., Nahrwold, M. L., and Bedows, E. (1981). Antimicrob. Agents Chemother. 20, 298.
10. Dore-Duffy, P. (1982). Prostaglandins Leukotrienes and Medicine 8, 73.
11. Robbins, S. J., and Rapp, F. (1980). Virology 106, 317.
12. Miller, C. A., and Carrigan, D. R. (1982). Proc. Natl. Acad. Sci. (USA) 79, 1629.
13. Menna, J. H., Collins, A. R., and Flanagan, T. D. (1975). Infect. Immun. 11, 152.
14. Cremer, N. E., Oshiro, L. S., and Hagens, S. J. (1979). J. Gen. Virol. 42, 637.
15. Wild, T. F., and Dugre, R. (1978). J. Gen. Virol. 39, 113.
16. Rapp, F., and Robbins, S. J. (1981). Intervirol. 16, 149.
17. Hall, W. W., and Choppin, P. W. (1979). Virology 99, 443.
18. Hall, W. W., and Choppin, P. W. (1981). N. Engl. J. Med. 304, 1152.
19. Lin, F. H., and Thormar, H. (1980). Nature (London) 285, 490.
20. Martinez, A. J., Ohya, T., Jabbour, J. T., and Duenas, D. (1974). Acta Neuropath. 28, 1.

Viral Proteins: Antigenic and Functional Analyses

FINE STRUCTURAL ANALYSIS AND PHOSPHORYLATION SITE DETERMINATION IN VSV NS PROTEIN[1]

Lorraine L. Marnell and Donald F. Summers

Department of Cellular and
Molecular Biology
University of Utah
Salt Lake City, Utah 84132

NS protein is a phosphoprotein with a molecular weight of 25,000 (1) and it has been suggested that phosphorylation of this small enzyme subunit may be involved in regulation of the VSV RNA polymerase enzyme function by mediating the switch from transcription to replication (2,3,4,5). We decided that the only way to precisely relate the changes in enzyme function to phosphorylation changes was to determine the sites of phosphorylation in NS protein for active and inactive enzyme with the ultimate goal of determining if changes in sites of phosphorylation correlated with changes in enzyme function.

ANALYSIS BY POLYACRYLAMIDE GEL ELECTROPHORESIS

First, we have studied how the highly acidic nature of NS protein, predicted from the DNA sequence (1), affects its behavior in polyacrylamide gels. NS protein migrates anomalously on sodium dodecyl sulfate (SDS) polyacrylamide gels (6,7). NS protein has also been shown to exhibit at least two molecular weight forms in different gel systems which contain urea (2,3,5). Under no conditions has NS protein been shown to migrate as its true molecular weight of 25K, as determined by DNA sequence data (1). The reason for this aberrant migration is unclear but it has been suggested that phosphorylation might cause the aberrant migration of NS protein either by changing the charge of the protein (2, 5); through the formation of phosphodiester bonds which leads to dimers of NS protein (8); or because of it bound SDS poorly

---

[1]This work was supported by NIH grant 08-RIAI-12316B and NSF grant PCM-8110986-A02.

due to its highly acidic nature as indicated by DNA sequence data (1). A schematic representation of the amino acid sequence of NS protein deduced from the cDNA sequence is shown in Fig. 1.

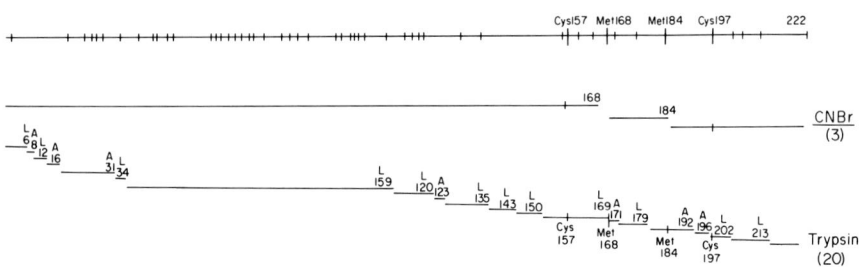

FIGURE 1. NS protein amino acid sequence (1). On the first line the short vertical lines represent aspartate and glutamate residues. Second line shows the 3 CNBr peptides. Third line shows tryptic peptides, ending with arginine (A) or lysine (L).

In order to establish what the basis for the aberrant electrophoretic behavior of NS protein was, we analyzed VSV proteins on polyacrylamide gels using the detergent cetyltrimethylammonium bromide (CTAB) instead of SDS (9). CTAB is a cationic detergent and might be expected to bind to the acidic residues of NS protein and perhaps result in a migration which more closely corresponded to its true molecular weight. Figure 2 shows polyacrylamide gel analysis of VSV proteins in three different buffer systems. Figure 2A,B shows that NS protein migrates as a 48K protein in our standard Tris-glycine buffered gel system. When the buffer was changed to an acidic buffer (Fig.2C,D) NS protein migrated with an apparent molecular weight of 35K. When the gels were made with the acidic buffer and the detergent in the samples and the gels was changed to CTAB, NS migrated with an apparent molecular weight of 32K (Fig.2E,F,G). Positive identification of NS protein was made by using VSV intracellular RNPs, labeled with [^{32}P], in which NS is the only radiolabeled protein (Fig.2B,D,G). In the CTAB system M protein migrates slightly slower than its true molecular weight of 26K so in Figure 2E M and NS proteins were not resolved. There is also no evidence of more than one form of NS protein on either acid-SDS gels or CTAB gels (data not shown). Since in the CTAB gels NS protein migrated very close to its true molecular weight the hypothesis that it migrates anomalously on SDS gels because it binds SDS poorly appears to be substantiated by these data, and the suggestion that it might be a dimer appears unlikely.

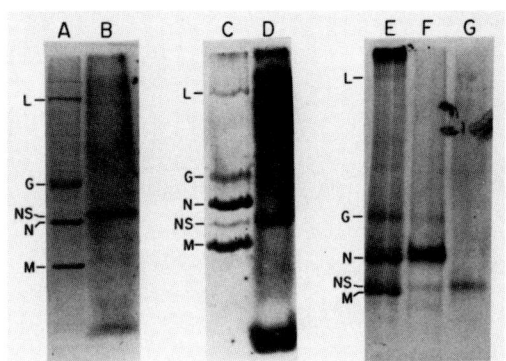

FIGURE 2. PAGE analysis of NS protein. Intracellular RNPs radiolabelled with [^{35}S]methionine (lanes F) or [^{32}P] orthophosphate (lanes B,D,G); or virions radiolabeled with [^{35}S]methionine (lanes A,C,E) were analyzed by SDS Tris-glycine PAGE (lanes A,B)(10), SDS β alanine PAGE (lanes C,D)(9) or CTAB β alanine PAGE (lanes E,F,G)(9).

Analysis of CNBr Cleavage Products. The deduced amino acid sequence of NS protein showed that most of the acidic groups were located on the amino-terminal half of the protein (1). Therefore we analyzed this domain of NS protein further by analyzing the CNBr cleavage products for the amount of phosphorylation of these CNBr peptides and for their migration on SDS polyacrylamide gels. The sequence predicts cleavage at methionines 168 and 184 creating a large peptide of a molecular weight of 18,028, containing the N-terminus of the protein, and 2 small peptides (MW, 4943 and 2139).

The sequence of NS protein predicts that it will contain a total of 75 aspartate and glutamate residues, 90% of which are in the large CNBr peptide. This large N-terminal fragment should also contain [^{35}S]cysteine. Therefore, in order to identify the large acidic terminal peptide, NS protein was isolated from intracellular RNPs (which contain only one form of NS protein, (2)), radiolabeled with [^{3}H]aspartate and [^{3}H]glutamate, [^{35}S]cysteine, or [^{32}P] orthophosphate, cleaved with CNBr and analyzed by PAGE (12). Figure 3, lane 1, shows that residual uncleaved NS protein migrated with an apparent molecular weight of 45Kd. The large major CNBr peptide migrated with a molecular weight of 32Kd. This large peptide contains the majority of the [^{35}S]cysteine (Fig. 3, lane 1); the acidic residues (Fig. 2, lane 2); and the phosphoamino groups (Fig 3., lane 3), as predicted by the sequence.

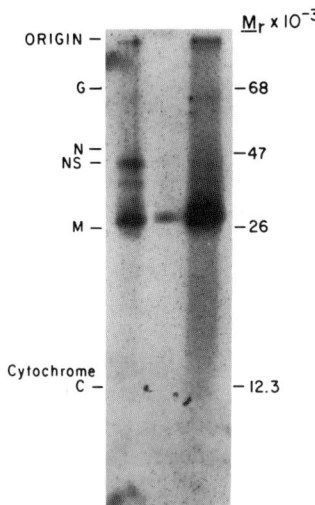

FIGURE 3. PAGE analysis of NS CNBr peptides. NS protein from intracellular RNPs (11) was isolated from SDS Tris -glycine polyacrylamide slab gels, cleaved 18hr with 5mg CNBr, dried and analyzed by SDS-urea phosphate PAGE (12). NS was radiolabeled with either [^{35}S]cysteine (lane 1), [^{3}H]aspartate and [^{3}H]glutamate (lane 2), or [^{32}P] orthophosphate (lane 3).

The sequence also predicts the existence of a smaller [^{35}S]cysteine-containing peptide. This peptide is shown in Fig.4A, migrating right above cytochrome C. This peptide contains no detectible [^{32}P], [^{3}H]aspartate or [^{3}H]glutamate (Fig. 3). Therefore our analysis showed that NS protein is highly acidic and contains the majority of its acidic residues and phosphoamino groups on its large N-terminal CNBr peptide. Furthermore, this large CNBr peptide migrates with an apparent molecular weight of 32Kd rather than 18Kd as the sequence would predict.

Analysis of Peptides on CTAB Gels. In order to obtain a more accurate estimate of the molecular weight of the CNBr peptides we analyzed the peptides using the detergent CTAB instead of SDS for PAGE. NS protein, radiolabeled with [^{35}S]cysteine, was cleaved with CNBr and analyzed by SDS-urea PAGE (12)(Fig. 4A). One can see that there were two major peptides migrating with apparent molecular weights of 33Kd and 15Kd. The 33Kd peptide peptide was then isolated from the SDS gel and analyzed by CTAB PAGE (Fig. 4B). Under these conditions it migrated with an apparent molecular weight of 17Kd, very close to the molecular weight (18Kd) predicted by the DNA sequence (1). This provided further evidence that this is the acidic N-terminal peptide predicted by the sequence. Furthermore it shares with NS protein the property of aberrant migration on SDS PAGE, probably because of its highly acidic nature.

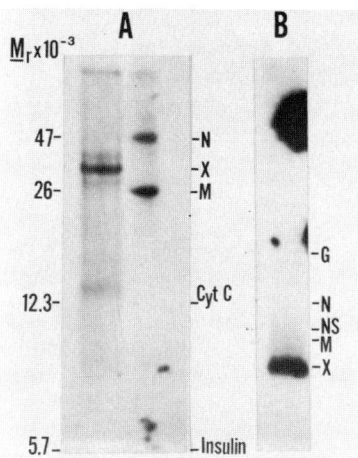

FIGURE 4. CTAB PAGE analysis of NS CNBr peptides. NS protein from intracellular RNPs, radiolabeled with [^{35}S]cysteine were analyzed by SDS-urea PAGE (A,lane 1), the large band (X) was cut out of lane 1, electroeluted, concentrated and analyzed by CTAB PAGE (B). Lane A1, [^{35}S]cysteine RNPs. Lane A2, [^{35}S]methionine virions. B, peptide X from A.

ANALYSIS OF NS PROTEIN TRYPTIC PEPTIDES

After we had localized the sites of phosphorylation of NS protein to the amino terminal CNBr peptide we did a more detailed analysis in order to more precisely localize the sites. For this purpose we chose tryptic peptide analysis by high performance liquid chromatography (HPLC)(13). By using the amino acid sequence of NS protein deduced from the DNA sequence (1), the nature of the tryptic peptides could be predicted, and one could determine which specific amino acids these peptides contained. Once the peptides were identified their specific sites of phosphorylation could be determined. From the sequence, 20 tryptic peptides were predicted; three of the tryptic peptides could be positively identified by labeling with [^{35}S]methionine, [^{35}S]cysteine and [^{3}H]lysine. The sequence predicts that peptide 13 should have all the above labels, peptide 16 should have only [^{35}S]methionine label and peptide 18 should have [^{35}S]cysteine and [^{3}H]lysine.

Purified NS protein, radiolabeled with either [^{35}S]methionine or [^{3}H]lysine, was digested with trypsin and the peptide samples were combined and analyzed by HPLC (13), as shown in Figure 5. The data show that NS protein radiolabeled with [^{35}S]methionine contained two tryptic peptides (fractions 22,30). The first peak (faction 6) contained a highly positive charged residue, and was shown not to be a peptide by further analysis (data not shown). Two methionine-containing tryptic peptides were consistent with the prediction made from the DNA sequence. However, Figure 5 also shows that both peptides radiolabeled with [^{35}S]methionine also contained [^{3}H]lysine instead of only one as predicted by the sequence.

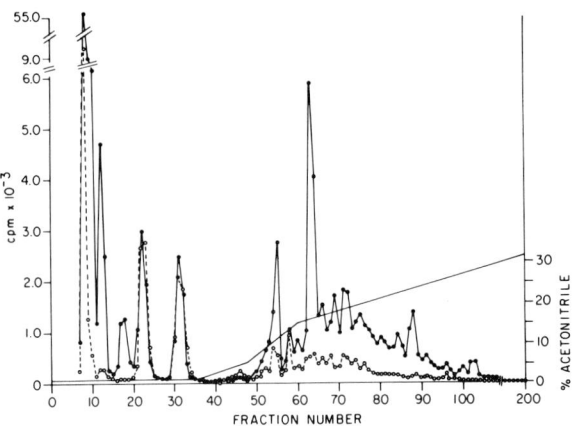

FIGURE 5. HPLC analysis of NS protein. NS protein was purified from intracellular RNPs by SDS PAGE, performic acid treated and digested with trypsin (20mg/ml,4hr,20C).Peptide samples, radiolabeled with either [^{35}S]methionine (O---O) or [^{3}H]lysine (O---O), were combined and analyzed by reversed phase HPLC. Peptides were eluted from a Whatman C18 column in a gradient of acetonitrile in 0.1% phosphoric acid.

To further analyze the tryptic peptides, NS protein was labeled with [^{35}S]cysteine and Figure 6A shows the HPLC profile of the tryptic peptides. One can see that instead of two peptides labeled with [^{35}S]cysteine as predicted, that there are at least three major and two minor peptides, In addition, peptide 13 was predicted to contain both [^{35}S]cysteine and [^{35}S]methionine, but there were no peptides observed that contained both amino acids (Fig. 5 and 6).

It was possible that the three or more peaks of [^{35}S]cysteine seen by HPLC analysis resulted from the additon of one or more phosphate residues to one or both of the cysteine peptides predicted by the cDNA sequence. The addition of a phosphate residue(s) to a tryptic peptide might be expected to change the behavior of this peptide on HPLC, leading to two or more new peaks eluting from the column. To test this possibility, we treated [^{35}S]cysteine labeled NS tryptic peptides with bacterial alkaline phosphatase before HPLC analysis. This treatment had been shown to remove about 50% of the label from [^{32}P]-radiolabeled NS protein (data not shown). The HPLC profile of the phosphatase-treated NS peptides (Fig.6B) shows no change from the untreated sample (Fig.6A). Therefore, the presence of phosphate groups on the cysteine-containing peptides did not appear to affect the elution of these peptides from HPLC suggesting that there were more that two cysteine-containing tryptic peptides in NS protein.

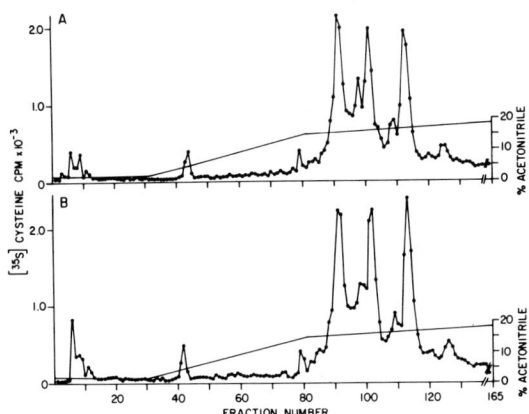

FIGURE 6. HPLC analysis of NS protein. A: NS protein, purified and analyzed as in Fig. 5 except radiolabeled with [^{35}S]cysteine. B: same as A except that after trypsin treatment the peptides were treated with bacterial alkaline phosphatase.

We have also attempted to identify which tryptic peptides of NS protein were phosphorylated. Tryptic peptides, radiolabeled with either [^{32}P] orthophosphate or [^{3}H]aspartate and [^{3}H]glutamate, were analyzed together by HPLC (Fig. 7). One can see that there were at least 11 peptides containing apartate and glutamate, confirming the highly acidic nature of NS protein, predicted by the sequence

(1). We attempted to identify the large 9.9Kb acidic tryptic peptide using several different types of size exclusion chromatography in buffers of pH 1.5, 3.0, 7.4 and in 67% acidic acid, but in each case the peptides aggregated and were eluted in the void volume. Fig. 7 also shows that there were 4 major phosphopeptides (peaks at fractions 110-111, 117, 123 and the shoulder at fraction 120), two of which contained acidic residues (peak 110-111, shoulder 120). Additional analysis (data not shown) has shown that neither of the [^{35}S]methionine-containing peptides coelute with the phosphopeptides. However the first phosphopeptide (Fig 7, peak 110-111) coelutes with the third major [^{35}S]cysteine-containing peptide (data not shown). More specific identification of these peptides and the localization of the phosphate residues in the sequence will have to wait until the descrepencies between the published cDNA sequence (1) and the results of our analyses of the tryptic peptides are resolved.

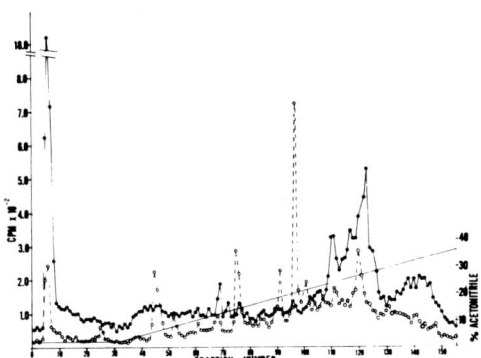

FIGURE 7. HPLC analysis of NS protein. NS protein was purified and analyzed as in Fig. 5 except radiolabeling was done with [^{32}P] orthophosphate (O---O) or [^{3}H]aspartate and [^{3}H]glutamate (O---O).

Although the sequence predicts that only one of the two methionine peptides should contain a lysine, our analysis showed that both contained lysines. This could have been caused by incomplete digestion, however other studies have shown that digestion was complete by 2hr (data not shown) whereas in all the data shown digestion was for at least twice that long. The possibility does exist that the phosphate groups interfered with the specificity of the trypsin, preventing cleavage at lys179.

Our data also showed that neither methionine peptide contained a cysteine residue. Incomplete digestion would have increased the probability that they were in the same peptide so that cannot be the cause of the discrepency. The hetergeneity of the cysteine peptides was probably not caused by phosphates disturbing the elution patterns since treatment of the peptides with alkaline phosphatase did not change the elution patterns. Incomplete cleavage could account for the extra cysteine peptides but not for the fact that methionine and cysteine do not coelute.

Strain differences between the strain sequenced and the one analyzed for tryptic peptides was not the reason for the discrepencies, since analysis of the tryptic peptides of the sequenced strain also showed the same diffrences (data not shown).

In summary, we have shown that specific tryptic peptides, predicted from the published sequence (1), cannot be identified by specific amino acid radiolabeling and analysis by HPLC. Work is in progress to clone the NS gene in order to sequence this VSV gene and resolve the descrepencies shown above. We have also confirmed the highly acidic nature of NS protein and shown that this is very likely the cause of its aberrant migration on SDS PAGE. In addition, we have shown the phosphates residues are located on the large CNBr N-terminal peptide. Further work is in progress to isolate the large CNBr peptide and cleave it with trypsin in order to further pinpoint the positions of the phosphoamino acids. In addition, differences between virion, cytoplasmic soluble and cytoplasmic-RNP NS protein phosphorylation sites will be studied.

REFERENCES

1. Gallione, C.J., Greene, J.R., Iverson, L.E., and Rose, J.K. (1981). J. Virol. 39, 529-535.
2. Clinton, G.M., Burge, B.W., and Huang, A.S. (1978). J. Virol. 27, 340-346.
3. Clinton, G.M., Burge, B.W., and Huang, A.S. (1979). Virology 99, 84-94.
4. Hsu, C., Morgan, E.M., and Kingsbury, D.W. (1982). J. Virol. 43, 104-112.
5. Kingsford, L., and Emerson, S.U. (1980). J. Virol. 33, 1097-1105.
6. Hunt, L.S., and Summers, D.F. (1976). J. Virol. 20, 637-645.
7. Naeve, C.W., Kolakofsky, C.M., and Summers, D.F. (1980). J. Virol. 33, 856-865.
8. Evans, D., Pringle, C.R., and Szilagyi, J.F. (1979).J.

Virol. 31, 325-333.
9. Maizel, J.V., Jr. (1971). Methods in Virology 5, 180-246.
10. Toneguzzo, F., and H.P. Ghosh. (1978). Proc. Natl. Acad. Sci. U.S.A. 75, 715-719.
11. Rubio, C., Kolakofsky, C., Hill, V.M., and Summers, D.F. (1980). Virology 105, 123-135.
12. Swank, R.T., and Munkres, K.D. (1971). Anal. Biochem. 39, 462-477.
13. Fullmer, C.S., and Wasserman, R.H. (1979). J.Biol.Chem. 254, 7208-7212.

MAPPING PHOSPHATE RESIDUES REQUIRED FOR VSV TRANSCRIPTION ON THE NS PROTEIN MOLECULE[1]

C.-H. Hsu and D.W. Kingsbury

Division of Virology and Molecular Biology
St. Jude Children's Research Hospital
P.O. Box 318, Memphis, Tennessee 38101

NS PHOSPHORYLATION AND VSV TRANSCRIPTION

Vesicular stomatitis virus (VSV) transcription seems to be activated by phosphorylation of a minor component of the virus nucleocapsid, the NS protein (1,2). Nucleocapsids that transcribe rapidly contain a highly phosphorylated form of NS, designated NS2, which migrates ahead of the more abundant, less heavily phosphorylated form, NS1, in some electrophoretic conditions (1-3). An extra cluster of chymotryptic phosphopeptides, denoting specific sites of phosphorylation, also distinguishes NS2 from NS1 molecules (2). Partial enzymatic dephosphorylation of NS2 converts it to NS1, removes these extra phosphates, and reduces transcriptase activity about 80% (2). In this work, we have located phosphorylation sites common to NS1 and NS2 in the primary structure of the protein and we have obtained some indications of where NS2-specific phosphates reside.

CLEAVAGE AT METHIONINE

The NS protein has a molecular weight of about 22K (K = 1,000), according to the base sequence of its gene (4). Based on the predicted amino acid sequence (4), we expected three cyanogen bromide (CNBr) fragments (Fig. 2). When $^{32}PO_4$-labeled NS was treated with CNBr, a large peptide with a molecular weight of about 17K, and a smaller, 4K peptide, were seen, corresponding to the expected N-terminal and C-terminal fragments, respectively (Figs. 1A and 2). The third, internal fragment that should have been generated may

---

[1]Supported by research grant AI 05343 from the National Institute of Allergy and Infectious Diseases, by Cancer Center Support Grant CA 21765 from the National Cancer Institute, and by ALSAC.

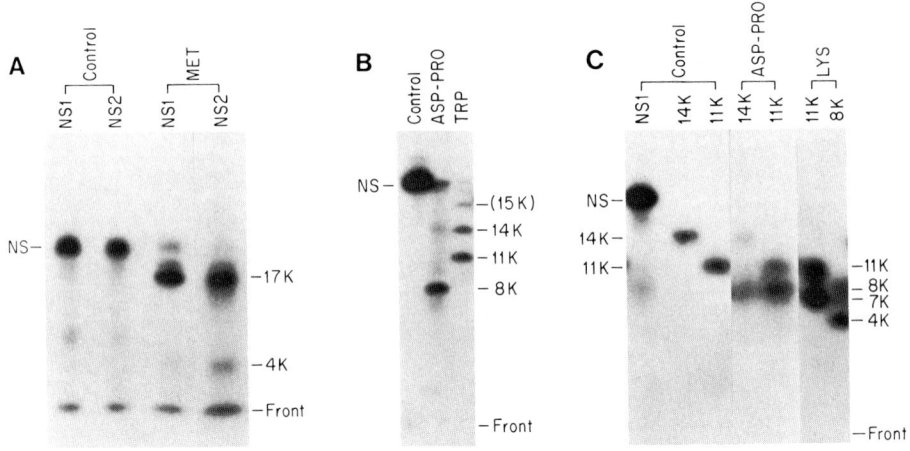

FIGURE 1. Polyacrylamide gel electrophoresis of [^{32}P]phosphate-labeled NS (2) and its cleavage products. In all cases, wet gels were exposed to X-ray film without fixing or staining, to obviate losses of low molecular weight peptides. Ovalbumin (45K), trypsin (24K), lysozyme (14K), Aprotinin (6.5K) and insulin B chain (3.5K) were used as unlabeled molecular weight markers. A. NS1 and NS2 from virions were isolated separately by gel electrophoresis and cleaved at MET with CNBr in 70% formic acid for 2 hr at 24°C in the dark, before electrophoresis in a 10% polyacrylamide gel. B. NS1 from virions was cleaved at ASP-PRO in 70% formic acid at 50°C for 16 hr (6), or at TRP with 3-bromo-3-methyl-2-[(2-nitrophenyl)thio]-^{3}H-indole (BNPS-skatole) in 70% acetic acid, for 16 hr at 24°C in the dark (7) and fragments were separated by electrophoresis in a 10% to 18% polyacrylamide gradient gel. C. The 14K and 11K products of BNPS-skatole treatment were extracted from gels and cleaved at ASP-PRO as in panel B; the same 11K peptide and the 8K product of ASP-PRO cleavage were cleaved at LYS$_{35}$ with trypsin (20 μg/ml in 0.01 M Tris-HCl, pH 8.0, 37°C, 16 hr.). Products were separated as in panel B.

have migrated in the radioactive band at the electrophoretic buffer front, since it is expected to have a low molecular weight (about 1.6K). The 4K fragment from NS2 molecules contained more radioactive phosphate than its counterpart from NS1 molecules, suggesting that this fragment contains some sites that are specific to NS2. It also appears that the 17K fragment is the main repository of phosphorylated sites common to both NS1 and NS2 molecules (2).

CLEAVAGE AT TRYPTOPHAN

To further localize phosphorylated residues in the 17K CNBr fragment, ^{32}P-labeled NS molecules were cleaved at TRP. Two major peptides with apparent molecular weights of 14K and 11K were obtained in addition to a larger peptide, tentatively assigned a molecular weight of 15K (Fig. 1B). The 14K and 15K peptides appear to be products of incomplete cleavage at the three TRPs in NS (Fig. 2). (The expected 7K, 3K and 1K fragments located at the C-terminal side of NS were seen after longer autoradiographic exposures; evidently, they contain considerably less radiophosphate than the 11K peptide; data not shown.)

CLEAVAGE AT ASPARTATE-PROLINE

Formic acid treatment of either the 11K or 14K fragment, under conditions relatively specific for the single ASP-PRO linkage in NS, yielded one main ^{32}P-labeled product, an 8K peptide (Fig. 1C). The same result was obtained after ASP-PRO cleavage of intact NS (Fig. 1B), indicating that most of the phosphate residues in NS and in the 11K and 14K fragments reside in this 8K region. The 8K fragment appears to be located at the N terminus of the NS molecule, because it was also obtained when the 17K CNBr fragment was cleaved at the ASP-PRO site (data not shown).

CLEAVAGE WITH TRYPSIN

To locate the phosphorylated sites in the 11K peptide and its 8K subfragment more precisely, each was digested separately with trypsin. The only trypsin-sensitive bond in either peptide is at LYS_{35}, since LYS_{50} is followed by a PRO, rendering it resistant to tryptic digestion (5). The 11K peptide yielded a ^{32}P-labeled fragment about 7K in molecular weight, whereas the 8K piece yielded a 4K subfragment (Fig. 1C). Based on the site specificities of the cleavage reactions that yielded these 7K and 4K subfragments, they appear to reside between residues 35 and 78 (Fig. 2). Comparing these findings with our previous chymotryptic peptide maps of NS1 and NS2 and with the predicted phosphoaminoacid compositions of these chymotryptic peptides, we note that there are five peptide bonds between residues 35 and 78 which are potentially chymotrypsin-sensitive (at LEU_{39}, PHE_{40}, TYR_{53}, PHE_{54} and TYR_{75}) and that the two predicted products (peptides 41-53 and 55-75) contain both phosphoserine and phosphothreonine, characteristic of the two groups of consistently phosphorylated chymotryptic peptides found in

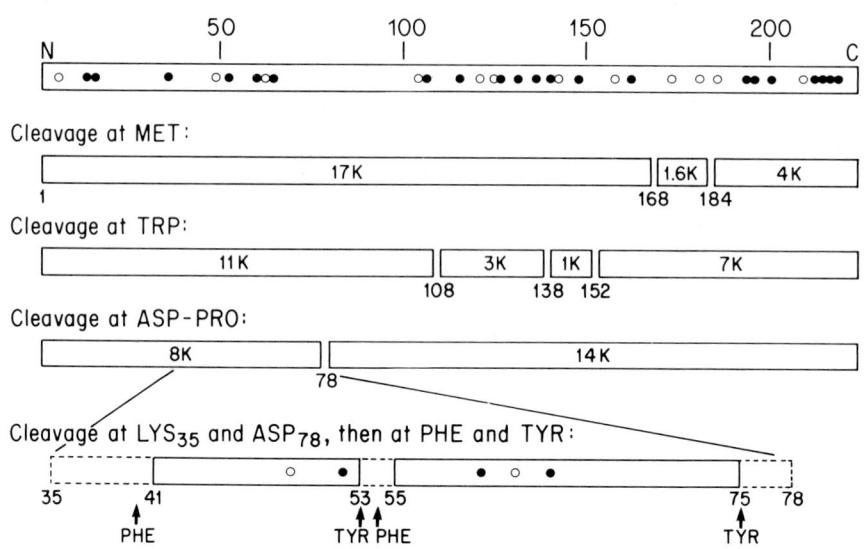

FIGURE 2. Diagram of the primary structure of the NS molecule and its expected cleavage products, based on the data of Gallione et al. (4). In the top band, N is the amino terminus and C is the carboxyl terminus. Open circles are the positions of THR residues and filled circles designate SER residues. The peptides expected from specific cleavages are shown below. Numbers below gaps represent amino acid residue positions at sites of cleavage.

both forms of NS (2). In addition to having an additional SER, peptide 55-75 is twice as rich in acidic amino acids as peptide 41-53, so we hypothesize that the former represents the previously designated electronegative chymotryptic peptide 1 and the latter represents the less electronegative chymotryptic peptide 2. Each of these peptides had been further resolved into subspecies that probably reflect variable phosphorylation or chymotryptic cleavage (2).

CONSTITUTIVE AND VARIABLE PHOSPHORYLATION SITES

Fig. 2 summarizes the deductions we have made from these cleavage studies. We believe that most of the sites in the NS molecule that are constitutively phosphorylated reside between residues 41 and 75, buried in the tertiary structure of the molecule and thereby protected from enzymatic dephosphorylation. Similar findings have been made by J.C. Bell

(Ph.D. Thesis, McMaster University) and L. Prevec (personal communication). The secondary sites of phosphorylation that are specific for NS2 molecules and which fully activate the viral transcriptase evidently reside elsewhere. The 4K C-terminal CNBr peptide may contain such sites. This peptide was more heavily phosphorylated in NS2 molecules (Fig. 1A), and it exhibits a high local concentration of SER residues (4), which may be exposed to the environment and the protein kinases that reside there.

ACKNOWLEDGMENTS

H.A. McDaniel, Jr. provided skillful technical assistance.

REFERENCES

1. Kingsford, L., and Emerson, S.U. (1980). J. Virol. 33, 1097.
2. Hsu, C.-H., Morgan, E.M., and Kingsbury, D.W. (1982). J. Virol. 43, 104.
3. Clinton, G.M., Burge, B.W., and Huang, A.S. (1979). Virology 99, 84.
4. Gallione, C.J., Greene, J.R., Iverson, L.E., and Rose, J.K. (1981). J. Virol. 39, 529.
5. Keil, B. (1971). In "The Enzymes" (P.D. Boyer, ed.), vol 3, p. 249. Academic Press, New York.
6. Shih, T.Y., Stokes, P.E., Smythers, G.W., Dhar, R., and Oroszlan, S. (1982). J. Biol. Chem. 257, 11767.
7. Hunziker, P.E., Hughes, G.J., and Wilson, K.J. (1980). Biochem. J. 187, 515.

MATRIX (M) PROTEIN REQUIREMENT FOR THE BINDING OF VESICULAR STOMATITIS VIRUS RIBONUCLEOCAPSID TO SONICATED PHOSPHOLIPID VESICLES[1]

John R. Ogden and Robert R. Wagner

Department of Microbiology
University of Virginia Medical School
Charlottesville, Virginia 22908

The vesicular stomatitis virus (VSV) matrix (M) protein is intimately associated with both the ribonucleocapsid core and the envelope of the virus. It is, however, not exposed at the exterior surface of the intact virion, since it is refractory to protease digestion and ^{125}I-labeled lactoperoxidase iodination (1, 2). The use of lipophilic monofunctional and bifunctional cross-linking reagents in intact virions demonstrates that the M protein is in close proximity to the inner surface of the envelope bilayer but penetrates little, if at all, into the envelope (3). Moreover, *in vitro* binding studies with lipid vesicles have shown the M protein to be a peripheral membrane protein, which reconstitutes only with vesicles containing acidic phospholipid components (4, 5).

Earlier studies have shown that the major structural proteins (N, G, and M) of VSV take independent pathways to the site of virus assembly at the plasma membrane of the infected cell (6, 7, 8). Appearance of M protein at the cell membrane of an infected cell causes decreased mobility of the already inserted glycoprotein (G) (9). The M protein is believed to function in virus maturation as a bridge between cytoplasmic viral ribonucleocapsids and regions of the cellular membrane into which the G protein has been inserted (10). This process may occur through the interaction of the highly positively charged M protein (11) with the phosphatidylserine (PS) polar headgroups located in the inner leaflet of the plasma membrane. Analysis of the phospholipid composition of the VSV envelope revealed that PS comprises 18% of the total

---

[1]This work was supported by PHS grant AI-1112 from the NIAID, grant MV-9E from the ACS and grant PCM-88-00494 from the NSF. JRO is a postdoctoral trainee supported by PHS Training Grant CA-9109 from the NCI.

phospholipid and is asymmetrically distributed in the bilayer with 85% located in the inner leaflet (12).

We have developed an *in vitro* model system to study the role of M protein in the binding of VSV nucleocapsid core structures to membrane bilayers. Nucleocapsid cores, with or without M protein, are allowed to bind to sonicated unilamellar vesicles (SUV) composed of acidic and/or neutral phospholipids. Core structures are also allowed to react with negatively charged SUV previously saturated with M protein. Our results indicate that M protein is required for the optimal binding of lipid vesicles to ribonucleocapsid cores. We have also noted that the best binding occurs when M protein is reconstituted with the negatively charged vesicle first, but this may be only a steric effect.

PREPARATION OF VSV RIBONUCLEOCAPSID CORE STRUCTURES

Solubilization of VS virions in solutions containing nonionic detergents and varying ionic strengths can selectively remove the viral envelope and specific viral proteins from the ribonucleocapsid core (13, 14). Ribonucleocapsid cores with M protein attached (RNP/M) were prepared by suspending whole VS virions (Indiana serotype) in 10 mM Tricine (pH 7.5) containing 30 mM octylglucoside. The insoluble cores were removed from the envelope components and soluble G protein by ultracentrifugation. Analysis by SDS-polyacrylamide gel electrophoresis revealed that only the N, NS, L, and M proteins were associated with the RNP/M complex. Electron microscopic studies showed the RNP/M complex to be a condensed structure, slightly thinner and longer than an intact virion. It has a density of about 1.30 g/cm^3.

Ribonucleocapsid core structures without M protein (RNP) were prepared by solubilizing whole VS virion in 10 mM Tricine (pH 8.0) containing 1% Triton X-100, 0.25 M NaCl and dithiothreitol (0.2 mg/ml). Ultracentrifugation was able to separate the core structures from the soluble envelope, G protein and M protein. Analysis by SDS-PAGE revealed that only the N, NS, and L proteins remain associated with the RNP. Electron microscopic examination revealed the RNP to be a long, strung out structure several times longer and thinner than an intact virion. The density of RNP is about 1.26 g/cm^3. Core structures were stored in calcium-free and magnesium-free phosphate buffered saline (PBS) at 4°C until used.

PREPARATION OF SONICATED UNILAMELLAR PHOSPHOLIPID VESICLES (SUV)

Small sonicated unilamellar phospholipid vesicles (SUV) were prepared and purified to homogeneity by the method of Barenholz et al. (15). Lipid mixtures of either 100 mol% 1,2-dimyristoyl-3-sn-phosphatidylcholine (DMPC) or 50 mol% 1,2-dipalmitoyl-3-sn-phosphatidylcholine (DPPC) plus 50 mol% bovine brain phosphatidylserine (BPS), both containing trace amounts of [^{14}C]DPPC as a marker, were dried and lyophilized overnight. The lipid preparations were resuspended in PBS at 35°C; each sample was sonicated for 3 min and then centrifuged for 185 min at 165,000 x g and 31°C. The upper two-thirds of the suspension, containing the SUV, was collected and used as the source of phospholipid vesicles. Vesicles range in size from 300-500Å.

RECONSTITUTION OF RIBONUCLEOCAPSIDS WITH NEUTRAL OR ACIDIC PHOSPHOLIPID VESICLES

Purified ribonucleocapsid cores without (RNP) or with (RNP/M) matrix protein were mixed with SUV of 100 mol% DMPC. The reaction mixtures were incubated for 6h at 37°C, then layered on top of a 0-66% sucrose gradient in PBS and ultracentrifuged to equilibrium. Figures 1A and 1B illustrate the density gradient analysis of the reaction mixtures. In both cases only a low level of binding, accompanied by a moderate density shift, occurred. The presence of M protein on the ribonucleocapsid core had no effect on the level of binding of neutral phospholipid vesicles to ribonucleocapsid core structures (Table 1).

Either RNP or RNP/M structures were incubated for 6h at 37°C with mixed SUV containing 50 mol% DPPC and 50 mol% BPS and then centrifuged to equilibrium in a 0-66% sucrose gradient. Figures 1C and 1D represent the analysis of these gradients. RNP structures bound the same low level of negatively charged vesicles as they had neutral vesicles (Table 1). The RNP/M structures, on the other hand, were able to bind much greater amounts of negatively charged vesicles (Fig. 1D and Table 1) than neutral vesicles (Fig. 1B). The presence of the M protein on the ribonucleocapsid also greatly increased the capability of VSV ribonucleocapsids to bind negatively charged vesicles (Fig. 1D); a large density change accompanied the increased binding. The concentration of salt had no effect on the binding by RNP/M of DPPC/BPS-SUV, indicating that the reaction was more specific than that which took place without M protein (data not shown).

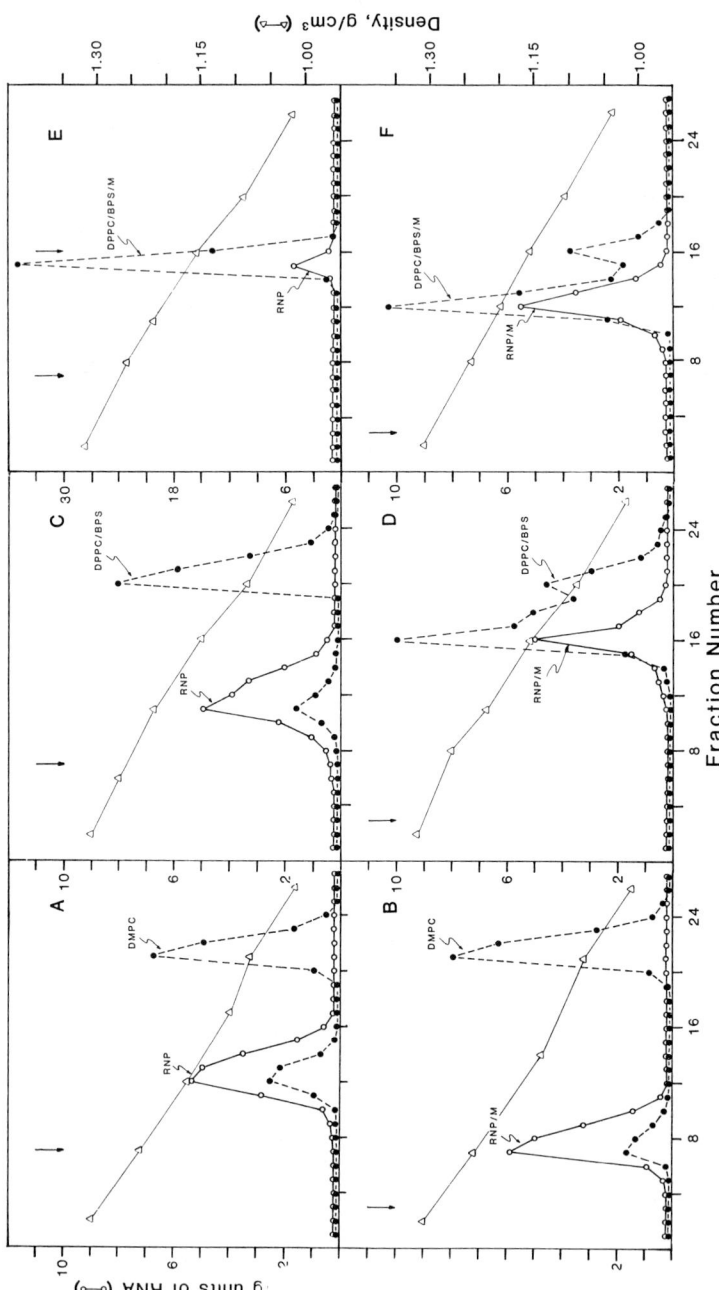

FIGURE 1. Sucrose density gradient analysis of the reconstitution mixtures of (A) RNP and DMPC-SUV, (B) RNP/M and DMPC-SUV, (C) RNP and DPPC/BPS(1:1)-SUV, (D) RNP/M and DPPC/BPS (1:1)-SUV, (E) RNP and DPPC/BPS(1:1)-SUV previously saturated with M protein, and (F) RNP/M and DPPC/BPS(1:1)-SUV previously saturated with M protein. The vertical arrows indicate where lipid-free RNP or RNP/M core structures would band in the gradient.

TABLE 1
DENSITIES AND BINDING CAPACITIES OF VSV RIBONUCLEOCAPSIDS AND PHOSPHOLIPID VESICLES

Sample	Density (g/cm³)	Picomole of lipid bound/ μg unit of RNA[a]
VSV	1.17	
RNP/M[b]	1.30	
RNP[c]	1.26	
DMPC-SUV[d]	1.08	
DPPC/BPS-SUV[e]	1.09	
DPPC/BPS/M-SUV[f]	1.15	
RNP + DMPC	1.17	417
RNP/M + DMPC	1.23	307
RNP + DPPC/BPS	1.22	313
RNP/M + DPPC/BPS	1.15	1,988
RNP + DPPC/BPS/M	1.16	6,203
RNP/M + DPPC/BPS/M	1.19	1,892

[a] Length of RNA found in one μg of whole virion.
[b] Nucleocapsid containing the N, NS, L, and M proteins.
[c] Nucleocapsid containing the N, NS, and L proteins.
[d] Sonicated unilamellar vesicles of DMPC.
[e] Sonicated unilamellar vesicles of 50 mol% DPPC and 50 mol% BPS.
[f] DPPC/BPS-SUV reconstituted with M protein (mole ratio 100:1).

RECONSTITUTION OF RIBONUCLEOCAPSIDS WITH NEGATIVELY CHARGED PHOSPHOLIPID VESICLES PREVIOUSLY SATURATED WITH M PROTEIN

Matrix protein, purified by the method of Zakowski et al. (5), was reconstituted with DPPC/BPS-SUV at a mole ratio of 1:100; this is a ratio which will ensure saturation of the vesicles with M protein (4). The reconstituted vesicles were then purified by sucrose density ultracentrifugation. After dialysis against PBS, the vesicles were stored at 31°C under N_2 and in the dark.

RNP or RNP/M structures were mixed with negatively charged phospholipid vesicles previously saturated with M protein (DPPC/BPS/M). The mixtures were incubated for 6h at

$37^{\circ}C$ and then centrifuged for 18h in a 0-66% sucrose gradient in PBS. Figures 1E and 1F illustrate the density gradient analysis of the two binding reactions. The presence of the M protein on the vesicles greatly enhanced the capability for RNP structures to bind negatively charged phospholipid vesicles (compare Fig. 1C and 1E). The RNP structures were also able to bind three times as much phospholipid (Fig. 1E) as the RNP/M structures (Fig. 1F). This may be due to the increased area for binding afforded by the strung out structure of the RNP, rather than to any inherent tendency for more efficient binding.

Figures 1D and 1F illustrate that the RNP/M condensed structure has a maximum binding capacity and that the addition of M protein to the vesicle before binding does not increase this capacity. Whether M-M interactions are also involved in Figure 1F could not be determined. As before, a large density change accompanied the binding of phospholipid vesicles by RNP and RNP/M.

Table 1 summarizes the results of the various binding experiments. The mixture of RNP structures with neutral (DMPC) or negatively (DPPC/BPS) charged vesicles resulted in only a low level of binding, which in the case of the DPPC/BPS vesicles was easily reversed by the removal of salt from the reaction mixture (data not shown). The reaction of RNP/M structures with the neutrally charged DMPC vesicles also did not lead to a high level of binding. Only when the combination of both M protein and negatively charged vesicles occurred was a high level of binding achieved. The combination of RNP/M and DPPC/BPS vesicles achieved six-fold greater binding than did the reaction of DPPC/BPS vesicles with RNP; the absence of the M protein reduced the binding efficiency considerably.

When M protein is on both the ribonucleocapsid (RNP/M) and the vesicle (DPPC/BPS/M), there is no significant increase in binding when compared to RNP/M and DPPC/BPS vesicles without M protein. This suggests that the condensed structure of the RNP/M complex allows for only a limited number of vesicles to bind and that the addition of M protein to the vesicles does not alter this limit. Possibly, the M-protein binding sites of RNPs are saturated.

Whereas RNP structures were inefficient at binding negatively charged (DPPC/BPS) vesicles, the previous addition of M protein to the vesicles greatly enhanced the binding capabilities (more than eighteen-fold). It would appear that the RNP can bind DPPC/BPS/M vesicles better than the RNP/M structures, but the binding efficiency may be more a consequence of the increased binding area offered by the strung out RNP than to any inherent tendency for more efficient binding.

Our results indicate that M protein is essential to achieve optimal binding between VSV ribonucleocapsids and sonicated phospholipid vesicles. This suggests that the M protein may interact with the negatively charged phosphatidylserine headgroups of the inner leaflet of the VS viral envelope. This interaction may be in addition to any association with the transmembrane G protein and may be a stabilizing force to help the virion maintain its shape.

REFERENCES

1. Schloemer, R.H., and Wagner, R.R. (1975). J. Virol. 16, 237.
2. McSharry, J.J. (1977). Virology 83, 482.
3. Zakowski, J.J., and Wagner, R.R. (1980). J. Virol. 36, 93.
4. Wiener, J.R., Pal, R., Barenholz, Y., and Wagner, R.R. (1983). Biochemistry 22, 2162.
5. Zakowski, J.J., Petri, W.A., Jr., and Wagner, R.R. (1981). Biochemistry 20, 3902.
6. Knipe, D.M., Baltimore, D., and Lodish, H.F. (1977). J. Virol. 21, 1128.
7. Knipe, D., Lodish, H.F., and Baltimore, D. (1977). J. Virol. 21, 1140.
8. Knipe, D.M., Baltimore, D., and Lodish, H.F. (1977). J. Virol. 21, 1149.
9. Johnson, D.C., Schlesinger, M.J., and Elson, E.L. (1981). Cell 23, 423.
10. Dubovi, E.J., and Wagner, R.R. (1977). J. Virol. 22, 500.
11. Carroll, A.R., and Wagner, R.R. (1979). J. Virol. 29, 134.
12. Patzer, E.J., Wagner, R.R., and Dubovi, E.J. (1979). CRC Crit. Rev. Biochemistry 6, 165.
13. Emerson, S.U., and Wagner, R.R. (1972). J. Virol. 10, 297.
14. Newcomb, W.W. and Brown, J.C. (1981). J. Virol. 39, 295.
15. Barenholz, Y., Gibbes, D., Litman, B.J., Goll, J., Thompson, T.E., and Carlson, F.D. (1977). Biochemistry 16, 2806.

COMPARATIVE NUCLEOTIDE SEQUENCE ANALYSIS OF THE
GLYCOPROTEIN GENE OF ANTIGENICALLY ALTERED RABIES VIRUSES[1]

W. H. Wunner, C. L. Smith, M. Lafon,[2] J. Ideler,[3]
and T. J. Wiktor

The Wistar Institute of Anatomy and Biology
36th Street at Spruce
Philadelphia, Pennsylvania 19104

The glycoprotein (G) of rabies virus is the major viral antigen responsible for the induction and binding of virus-neutralizing (VN) antibodies. Antigenic differences in G among laboratory strains of rabies virus and street virus strains detectable by the use of monoclonal antibodies (1, 2) suggest that rabies virus can escape anti-viral immunity by gradually changing the antigenicity of G. A detailed analysis of the antigenic structure of rabies virus G is therefore necessary to understand the relationship of one virus strain to another in terms of antigenic cross-reactivity and cross-protective immunity.

Recently, an operational map of rabies virus G of the CVS-11 strain was described which delineated three functionally independent antigenic sites based on the grouping of G variant rabies viruses which were resistant to neutralization by one or more of a panel of anti-G monoclonal antibodies (3). Our present goal is to physically map the functionally independent antigenic sites on rabies virus G. In this report, we present an expanded operational map of G from the ERA strain of rabies virus which delineates five functionally distinct antigenic sites, and identify nucleotide changes in the G gene which code for amino acid substitutions corresponding to site III epitopes in the operational map.

---

[1] This work was supported by Research Grants AI-18883 and AI-09706 from the National Institutes of Allergy and Infectious Diseases.
[2] Present address: Institute Pasteur, Service Rage 25, rue du Dr. Roux, 5724 Paris Cêdex 15.
[3] Present address: Department of Virology, Agriculture University, Wageningen, The Netherlands.

CONSTRUCTION OF THE OPERATIONAL ANTIGENIC MAP OF ERA STRAIN VIRUS GLYCOPROTEIN

Twenty-one monoclonal antibodies specific for rabies virus G and capable of neutralizing ERA strain virus were obtained from hybridomas as previously described (3, 4). These were used to isolate 52 antigenically unique mutant viruses from parental stock ERA virus as previously described (3, 5) except that a plaquing system was used in which BHK-S13 cells were suspended in agarose (6). The 52 neutralization-resistant variant viruses were arranged into groups as shown in Figure 1 by their pattern of reactivity in cross-neutralization tests with a panel of 25 monoclonal antibodies. Variants that lost reactivity to a number of

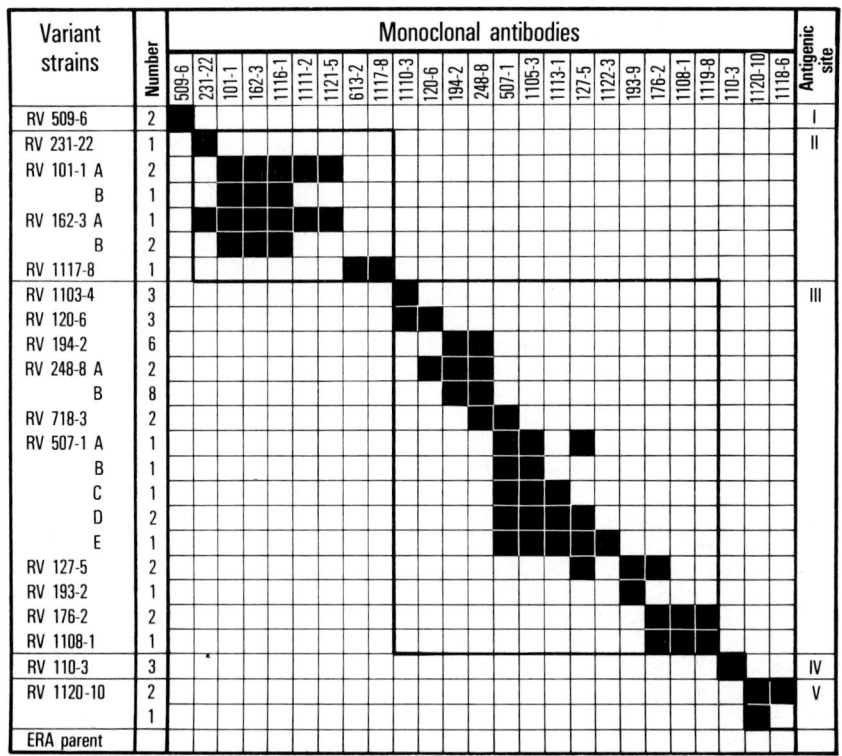

FIGURE 1. Operational antigenic map of ERA G. Neutralization-resistant variants (RV) were selected using monoclonal antibodies, and then tested for susceptibility (□) or resistance (■) to neutralization by antibodies in the panel. RV viruses are identified by the number of the monoclonal antibody used for the selection.

antibodies simultaneously were considered operationally linked by the epitopes recognized by those antibodies. Conversely, variants which were resistant to only one antibody revealed an epitope which was considered operationally independent of any other epitopes recognized by the available monoclonal antibodies. The antigenic map of ERA virus G confirms antigenic sites I, II and III previously described for CVS virus (3). Two additional antigenic sites (designated IV and V) were defined by monoclonal antibodies 110-3 and 1120-10 which neutralized ERA virus, but not CVS virus.

NUCLEOTIDE SEQUENCE ANALYSIS OF VARIANT VIRUS GLYCOPROTEIN GENES

Virion RNAs were isolated from mutant viruses representing the five operationally defined antigenic sites of ERA strain virus G and transcribed in dideoxy-chain termination sequencing reactions (7). Transcripts were primed with synthetic cDNA oligomers corresponding to the cloned nucleotide sequence of mRNA specific for ERA strain G (8) or CVS sequences which we have determined by direct cDNA sequencing using strain common primers (unpublished data). The cDNA products of the four base-specific reaction mixtures (A, C, G, and T) for each cDNA primer-virion RNA hybrid were screened in groups of A's, C's, G's, and T's in polyacrylamide gels. Our analysis has shown that mutations resulting in amino acid residue changes were infrequent and that a single residue change may be responsible for an altered structure on the antigen directly complementary to the combining site for a given VN antibody.

LOCATION OF ANTIGENIC SITE III IN THE RABIES VIRUS GLYCOPROTEIN

A single base change in the nucleotide sequence of RV 194-2, RV 248-8, RV 507-1, and RV 120-6 at nucleotide positions 1062 (G→T), 1062 (G→T), 1072 (T→G), and 1133 (G→A), respectively, leads to an altered amino acid in the variant G. Three of these mutations are shown in Figure 2. The base change at position 1133 for RV 120-6 is not shown. A single base change (G→A) at position 1062 has also been detected for the RV 194-2 variant of CVS virus (data not shown). This change in the second base of the codon C G G for arginine-333 of the parental CVS G sequence to C A $\overline{G}$ in CVS RV 194-2 confirmed the glutamine substitution in $\overline{the}$ tryptic peptide replacing arginine-333 as previously reported (9).

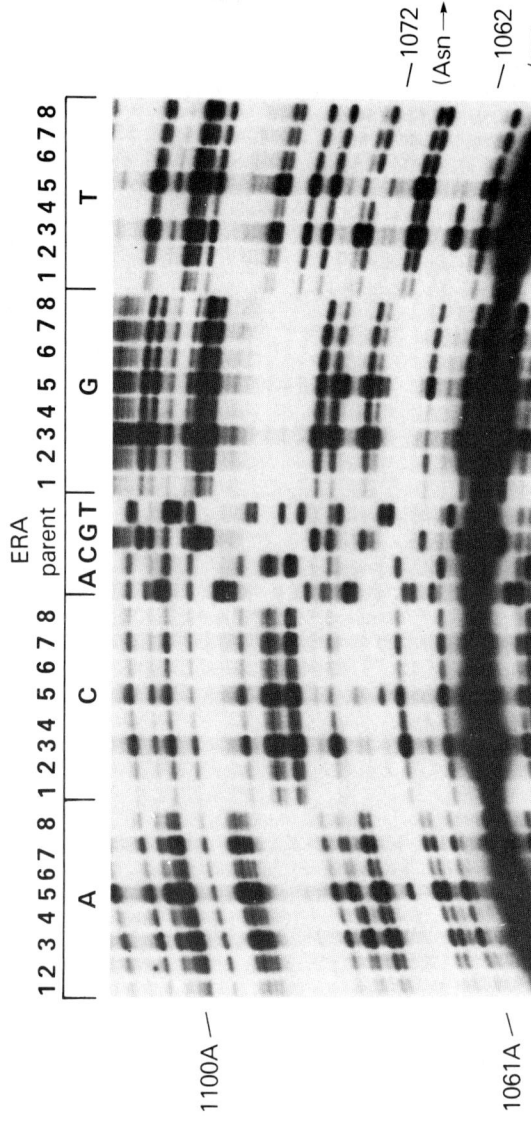

FIGURE 2. Comparative nucleotide sequence analysis of the G gene of antigenic variant viruses. Virion RNAs from RV 509-6 (1), RV 101-1 (2), RV 194-2 (3), RV 248-8 (4), RV 507-1 (5), RV 120-6 (6), RV 110-3 (7), RV 1120-10 (8), and ERA strain parent virus were incubated in sequencing reactions with cDNA primer 1001-1018. Products of the A, C, G, and T sequence-specific reactions were separated in 8% polyacrylamide gels. The oligonucleotide primer and transcripts are numbered to correspond with the nucleotide sequences of the cloned G sequences (ref. 9).

The location of unique amino acid changes in the primary sequence of G is schematically shown in Figure 3. Variant viruses that have been assigned to site III of the operational epitope map (Figure 1 and ref. 3), show clustering of mutations in the nucleotide sequence suggesting that residues involved in the binding of VN antibodies which delineate antigenic site III occupy positions close to each other in the G molecule.

An additional amino acid change which has been described (10) is indicated in Figure 3 and represents the substitution of glutamic acid in CVS variant RV 231-22 replacing lysine-198 in the parental G.

The mapping of mutations in G of rabies virus will permit us to locate functional epitopes within antigenic sites. This study is a preparation for interpreting relationships between rabies and rabies-related viruses and between rabies viruses that exhibit antigenic changes as a result of passage in different animal species.

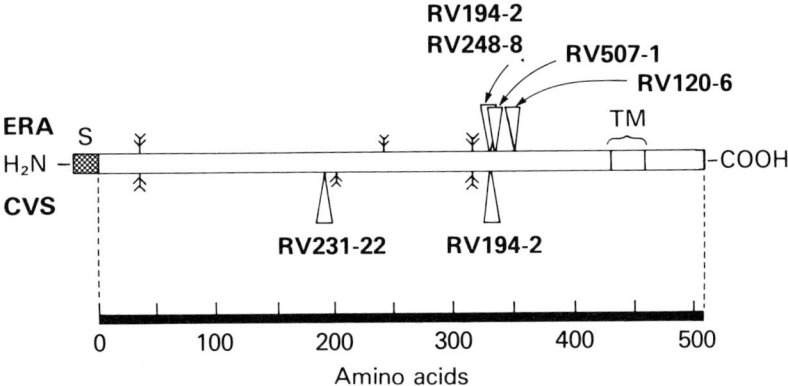

FIGURE 3. Map locations of altered amino acids in the glycoprotein sequence of rabies variant (RV) viruses. The diagram shows the location of mutations (darts) of RV viruses and carbohydrate attachment sites (✯) in ERA and CVS strains above and below the open bar, respectively. The signal peptide (S) and transmembrane (TM) segments are also shown.

ACKNOWLEDGMENTS

We are grateful to Maureen Devlin and Kimmerly Otte for excellent technical assistance, to Dr. Kim Arndt of the Chemistry Department, University of Pennsylvania, for his guidance through the manual synthesis of oligonucleotide primers, and to Dr. Ponzy Lu for providing facilities to do the DNA synthesis manually.

REFERENCES

1. Flamand, A., Wiktor, T. J., and Koprowski, H. (1980). J. gen. Virol. 48, 105.
2. Wiktor, T. J., Lafon, M., Dietzschold, B., and Wunner, W. H. (1983). In "Advances in Gene Technology: Molecular Genetics of Plants and Animals" (W. J. Whelan and J. Schultz, eds.), in press. Academic Press, New York.
3. Lafon, M., Wiktor, T. J., and Macfarlan, R. I. (1983). J. gen. Virol. 64, 843.
4. Wiktor, T. J., and Koprowski, H. (1978). Proc. Natl. Acad. Sci. U.S.A. 75, 3938.
5. Wiktor, T. J., and Koprowski, H. (1980). J. exp. Med. 152, 99.
6. Sedwick, W. D., and Wiktor, T. J. (1967). J. Virol. 1, 122.
7. Sanger, F., Nicklen, S., and Coulsen, A. R. (1977). Proc. Natl. Acad. Sci. U.S.A. 74, 5463.
8. Anilionis, A., Wunner, W. H., and Curtis, P. J. (1981). Nature 294, 275.
9. Dietzschold, B., Wunner, W. H., Wiktor, T. J., Lopes, A. D., Lafon, M., Smith, C. L., and Koprowski, H. (1983). Proc. Natl. Acad. Sci. U.S.A., 80 70.
10. Dietzschold, B., Wunner, W. H., Smith, C. L., and Verrichio, A. (1983). In "Negative Strand Viruses" (R. Compans and D. H. L. Bishop, eds.) (this volume), Academic Press, San Francisco.

VARIATION IN GLYCOSYLATION PATTERN OF G PROTEINS AMONG
ANTIGENIC VARIANTS OF THE CVS STRAIN OF RABIES VIRUS[1]

B. Dietzschold, W. H. Wunner, M. Lafon, C. L. Smith, and
A. Varrichio

The Wistar Institute of Anatomy and Biology
36th Street at Spruce
Philadelphia, Pennsylvania 19104

Glycoproteins from different fixed rabies virus strains vary greatly in their electrophoretic mobilities (1). Sodium dodecyl sulfate-polyacrylamide gel electrophoresis (SDS-PAGE) of the structural proteins from virions of ERA and Flury HEP strains has revealed single bands of glycoproteins which differ by approximately 5,000 daltons. Analysis of virion proteins of CVS and PM strains has shown that two forms of the glycoprotein exist which are designated GI and GII. The slower migrating GI of CVS corresponded in electrophoretic mobility to the glycoprotein of ERA and the faster migrating GII of CVS corresponded to the glycoprotein of Flury HEP virus. Comparative tryptic peptide analysis of CVS GI and GII has demonstrated that all peptides derived by trypsin digestion of GI comigrated with those of GII (1). Thus, GI and GII appear to differ in their extent of glycosylation.

We have recently determined the structural relationship between GI and GII. In this paper, we report on the investigation of two CVS variant viruses with antigenically altered glycoproteins affecting glycosylation.

CHARACTERIZATION OF CVS VARIANT VIRUS GLYCOPROTEINS

Two antigenic variants of the CVS-11 parent virus which resisted neutralization by monoclonal antibody 231-22 and 194-2 as previously described (2) contained glycoproteins with uniquely altered electrophoretic mobilities in SDS-PAGE compared with parent virus glycoprotein (Fig. 1). Two glycoprotein species, GI and GII with molecular weights of 67,000 and 62,000, respectively, can be observed in CVS

[1] This work was supported by Research Grants AI-09706 and AI-18883 from the National Institutes of Allergy and Infectious Diseases.

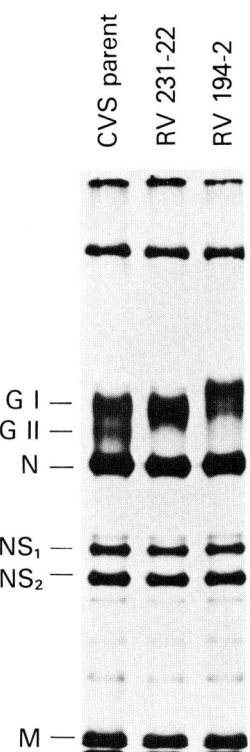

FIGURE 1. Electrophoretic analysis of structural proteins of CVS-11 parent and variant viruses. CVS-11 parent virus and the antigenic variants RV231-22 and RV194-2 were labeled with [^{35}S]methionine and the purified virions were subjected to SDS-PAGE as described (1).

parent. Glycoprotein GII is absent in RV231-22 virus. Both species are present in RV194-2 virus, but they exhibit slower electrophoretic mobilities compared with GI and GII of parent virus.

In order to confirm previous results comparing tryptic peptides of CVS GI and GII (1), the virus-specific glycoproteins were isolated from tunicamycin-treated cells individually infected with CVS-11, RV231-22, and RV194-2 virus (data not shown). In every case, immunoprecipitated virus-specific glycoproteins migrated in SDS-polyacrylamide gels faster than virion-associated glycoprotein. Furthermore, only one glycoprotein species with the same molecular weight for all three viruses could be isolated from these cells.

This suggests that the differences in the electrophoretic pattern of virion-associated G proteins represent variation in sugar content of these proteins.

ANALYSIS OF TRYPTIC GLYCOPEPTIDES

Glycoproteins from CVS-11, RV231-22, and RV194-2 were labeled with [^{3}H]glucosamine or [^{14}C]glucosamine, purified by SDS-PAGE, carboxymethylated, and exhaustively digested with trypsin to determine the chemical relationship between GI and GII in the various viruses. The tryptic glycopeptides were resolved by reverse phase high-pressure liquid chromatography (HPLC). Figure 2a shows that two major peaks of glucosamine-labeled tryptic glycopeptides are present in GI of CVS-11 (#1, #4) and one minor peak (#2), but only one major peak could be resolved from GII of CVS-11. Two additional minor peaks (#3, #5) regularly observed as shoulders to peak #4 probably represent a heterogeneity of glycopeptides related to #4 (or were not resolved). However, the retention time of the single GII peak was identical to peak #4 of GI. The pattern of tryptic glycopeptides obtained from the single glycoprotein of RV231-22 virus (Fig. 2b) is similar to that of GI of CVS parent. In Figure 2c, an additional glycopeptide peak (#6) was obtained from the glycoprotein (GI plus GII) of RV194-2 virus.

To localize the various glycopeptides within the deduced amino acid sequence of the CVS-11 glycoprotein (3), the isolated peptides were subjected to automated Edman degradation. The amino acid sequence for glycopeptide #1 (Table 1) was the same for all three viruses. Only residues 1, 2 and 4 of glycopeptide #1, which matched the amino acid sequence of CVS glycoprotein at positions 200, 201, and 203, were obtained. However, this partial sequence was sufficient to map the glycopeptide to the predicted glycosylation site II at asparagine 204 of the CVS glycoprotein. The first two residues of glycopeptide #2 were shown to match the start of the sequence of glycopeptide #1. Therefore, it is believed that peptide #2 is an incompletely cleaved peptide of glycopeptide #1. With the exception of residue 6, the entire amino acid sequence of peptide #4 was obtained. This sequence matches residues 314 to 320 of the CVS-11 glycoprotein and contains the predicted glycosylation site III at asparagine 319. Regarding glycopeptide #6 of RV194-2, the first five residues and residues 8 and 9 were obtained. These amino acids match residues 153, 154, 155, 156, 157, 160, and 161 of the CVS-11 glycoprotein. Since this amino acid sequence in the parent glycoprotein does not contain a glycosylation signal, it was suspected that an undetectable

(2a)

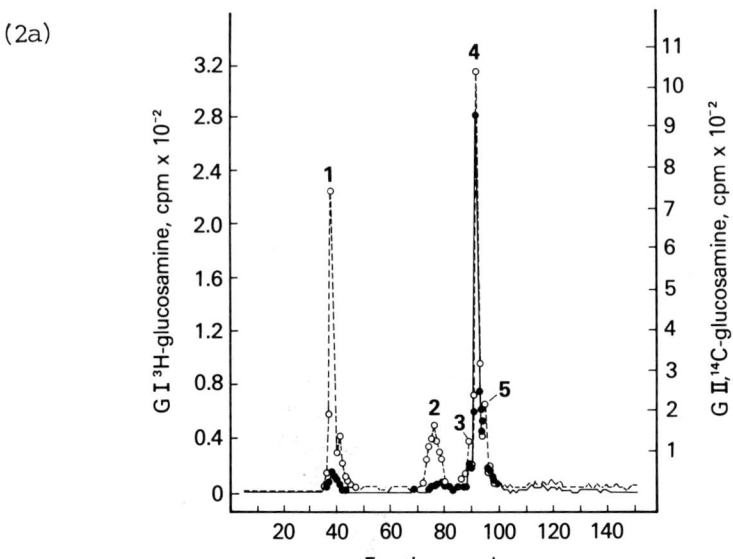

(2b)

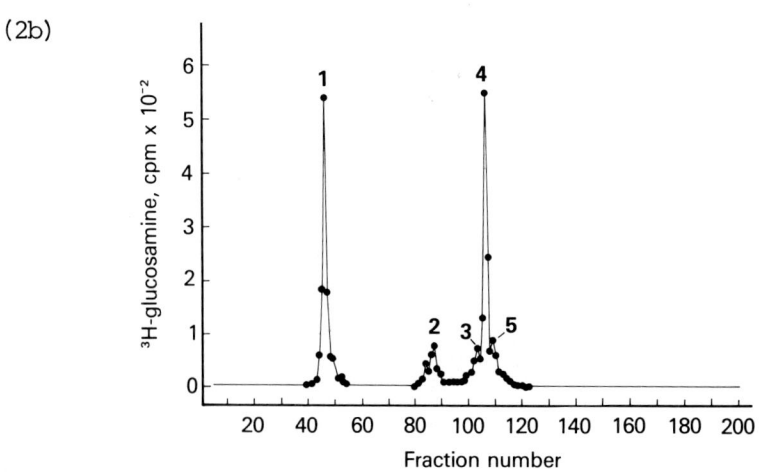

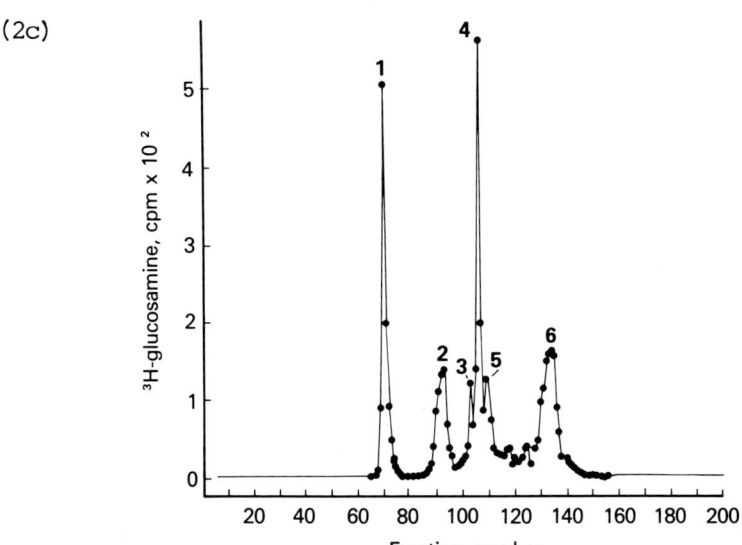

FIGURE 2. HPLC analysis of glucosamine containing tryptic peptides of glycoproteins of CVS-11 (2a), RV231-22 (2b) and RV194-2 (2c). Glycoprotein species, GI and GII, were separated by SDS-PAGE. GI labeled with ^{3}H (●——●) glucosamine and GII labeled with ^{14}C (o---o) glucosamine were mixed, carboxymethylated and exhaustively digested with trypsin. Tryptic peptides were separated by HPLC using a Beckman Ultrasphere ODS column. HPLC conditions were as follows: Solvent A was .1M $NaClO_4$, Solvent B was 70% acetonitrile, 30% H_2O, .1% H_3PO_4; a gradient of 0 to 90% B was run for 50 min at a flow rate of 1.5 ml/min at ambient temperature. Glycopeptides from RV231-22 and RV194-2 were prepared in a similar way as described for CVS, except that GI and GII of RV194-2 were not separated.

amino acid substitution occurred at residue 158 of the RV194-2 virus glycoprotein.

NUCLEOTIDE SEQUENCES ENCODING PREDICTED GLYCOSYLATION SITES

Overall, this direct amino acid sequencing data did not precisely identify the glycosylation sites in the different glycopeptides because asparagine, the initial amino acid which signals such a site was never obtained. Therefore, we

TABLE 1
AMINO ACID SEQUENCES OF GLYCOPEPTIDES OBTAINED BY EDMAN DEGRADATION

Glycopeptide no.	Virus	Sequence
1	CVS, RV231-22, RV194-2	200 AS-G
2	CVS, RV231-22, RV194-2	AS
4	CVS, RV231-22, RV194-2	314 AYTIF-K
6	RV194-2	153 VFPGG--SG

determined the nucleotide sequences of regions in the parental and variant glycoprotein genes which encode the glycosylation signals asn-x-ser or asn-x-thr. We employed the dideoxy sequencing method (4) and used synthetic DNA primers as described elsewhere (5) matching nucleotides 8-24, 553-569, and 1001-1018 of the cloned glycoprotein gene to determine nucleotide sequences containing the predicted glycosylation sites I, II, and III, respectively. Furthermore, to confirm the putative extra glycosylation site of RV194-2 glycoprotein, a synthetic DNA primer corresponding to nucleotides 377-393 was also used.

The nucleotide sequence of the CVS-11 glycoprotein gene in the region of the predicted glycosylation site I (data not shown) revealed the presence of a glycosylation signal (asn-leu-ser) at asparagine 37 in each of CVS-11 parent, RV231-22, and RV194-2 virion RNAs. However, since none of the isolated glycopeptides from the glycoproteins of CVS-11 parent, RV231-22, and RV194-2 viruses possess the amino acid sequence of site I, we believe the first glycosylation signal is not utilized for carbohydrate attachment.

Regarding glycosylation site II of CVS-11 glycoprotein, nucleotide sequencing confirmed the location of the predicted glycosylation signal at asparagine 204 (3). Moreover, since the glycosylation signal sequence was unambiguous it is unlikely that the two glycoprotein species in CVS parent virus are derived from two separate gene transcripts. Rather it is likely that the two different glycoprotein forms reflect different conformations as previously suggested (6). Direct nucleotide sequence analysis of this region in the glycoprotein gene of CVS-11 parent, RV194-2 and RV231-22 viruses revealed a A→G base change in the codon for residue 198 of RV231-22 as shown in Figure 3. The amino acid change from lysine to glutamic acid is located five residues from asparagine 204 in CVS glycoprotein. As a result of this mutation, it appears that carbohydrate attachment occurs at asparagine 204 in the GII species of RV231-22 glycoprotein. This site is apparently not utilized in the GII form of parental CVS glycoprotein. One reason for this may be that the conformation of GII conceals residue 204 making it inaccessible for carbohydrate attachment. Rules established by Chou and Fasman (7) predict that this single amino acid change in the deduced primary sequence alters the secondary structure of the glycoprotein in this region and presumably allows glycosylation site II to be utilized in the GII species.

The predicted glycosylation signal at asparagine 319 for the glycoproteins of CVS, RV231-22 and RV194-2 viruses was confirmed by nucleotide sequencing of the glycosylation site 3 region of the glycoprotein gene (data not shown). This glycosylation site is highly conserved and utilized in all rabies virus strains examined and is also present at the same location in vesicular stomatitis virus (VSV) G (8).

Nucleotide sequencing of the region coding for extra putative glycosylation site in RV194-2 virus glycoprotein revealed the expected amino acid change at residue 158 resulting in a substitution of lysine with asparagine (AAG→AAT). This change in the primary sequence creates a new glycosylation site and confirms the presence of such a site predicted by the analysis of peptide #6 from RV194-2 virus glycoprotein.

The data presented here have shown that 1) all glycosylation signals may not be utilized and 2) glycosylation may be dependent on the secondary structure of the molecule. The glycosylation site which is always glycosylated in rabies virus strains and in VSV is highly conserved. At this point, we do not know whether the variation in glycosylation among antigenic variants will impair biological functions of the protein.

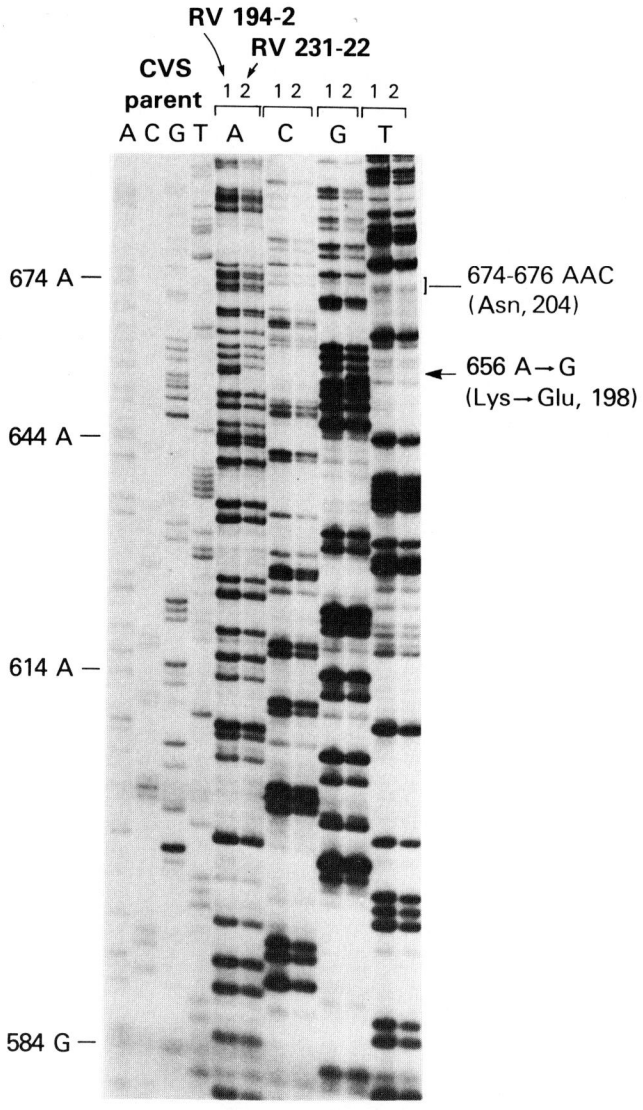

FIGURE 3. Comparative nucleotide sequence of the region encoding glycosylation site II. Virion RNAs from CVS-11 parent virus and antigenic variants RV194-2 and RV231-22 were incubated with DNA primer complementary to bases 553 to 569 in dideoxy-chain terminating sequencing reactions. Products of the A, C, G, and T sequence-specific reactions were separated in 8% polyacrylamide sequencing gel. The nucleotides of primer and transcripts are numbered to correspond with the nucleotide sequence of the cloned glycoprotein gene.

ACKNOWLEDGMENTS

We are grateful to Claire Kirk, Maureen Devlin and Kimmerly Otte for excellent technical assistance.

REFERENCES

1. Dietzschold, B., Cox, J. H., and Schneider, L. G. (1979). Virology 98, 63.
2. Lafon, M., Wiktor, T. J., and Macfarlan, R. I. (1983). J. Gen. Virol. 64, 843.
3. Yelverton, E., Norton, S., Obijeski, J. F., and Goeddel, D. V. (1983). Science 219, 614.
4. Sanger, F., Nicklen, S., and Coulsen, A. R. (1977). Proc. Natl. Acad. Sci. USA 74, 5463.
5. Wunner, W. H., Smith, C. L., Lafon, M., Ideler, J., and Wiktor, T. J. (1983). In "Negative Strand Viruses" (R. Compans and D. H. L. Bishop, eds.), Academic Press, San Francisco (in press).
6. Dietzschold, B., Wiktor, T. J., MacFarlan, R., and Varrichio, A. (1982). J. Virol. 44, 595.
7. Chou, P. Y., and Fasman, G. D. (1974). Biochemistry 13, 222.
8. Rose, J. K., Doolittle, R. F., Anilionis, A., Curtis, P. J., and Wunner, W. H. (1982). J. Virol. 43, 361.

CHANGE IN PATHOGENICITY AND AMINO ACID SUBSTITUTION
IN THE GLYCOPROTEIN OF SEVERAL SPONTANEOUS
AND INDUCED MUTANTS OF THE CVS STRAIN OF RABIES VIRUS[1]

I. Seiff[1], M. Pepin[2], J. Blancou[2], P. Coulon[1]
and A. Flamand[1]

1. Bât 400, Université Paris-Sud, 91405 Orsay Cedex, France
2. Centre National d'Etudes sur la Rage, BP n°9,
   54220 Malzeville, France

We have previously demonstrated that 90 % of spontaneous or 5 Fu induced mutants of the CVS strain of rabies virus selected for the resistance to neutralization by monoclonal antibodies n° 194-2 and 248-8 (from the Wistar collection) were avirulent or attenuated in adult mice (1, 2). Those two monoclonal antibodies recognize site IIIa of the CVS glycoprotein (3). Using the same technique, Dietzschold et al (4) isolated 2 spontaneous non pathogenic mutants of the ERA and CVS strains of rabies virus. They found that the mutant glycoproteins differed from the wild-type by one amino acid substitution at position 333, an arginine being replaced by a glutamine (CVS) or by an isoleucine (ERA). Because we possess a collection of mutants which differ in their attenuation, we decided to investigate if they were mutated in the same position.

The nucleotide sequence of the 3' terminal region of the glycoprotein gene was determined for 4 avirulent (AvO_1, AvO_4, AvO_{10} and AvO_{11}) and two attenuated strains (AtO_1 and AtO_2). AvO_1, AvO_4, AtO_1 and AtO_2 are 5 Fu induced mutants (2) while AvO_{10} and AvO_{11} are spontaneous. The CVS parental strain and VO_{67} (V = Virulent, O = Orsay), a 5 Fu induced mutant resistant to neutralization by monoclonal antibodies n° 194-2 and 248-8 which retains its virulence, were also included in this study.

In order to precise the difference between attenuated and avirulent phenotypes, the pathogenic power of the mutants was also investigated, by intramuscular or intracerebral inoculation of 3 weeks-old mice with increasing doses of the virus.

---

[1]This work was supported by the Centre National de la Recherche Scientifique through the L.A. 040086 and through the ATP "Etats-Unis 1983", and by the Ministère de la Recherche et de l'Industrie through the ATP n° 83V0085.

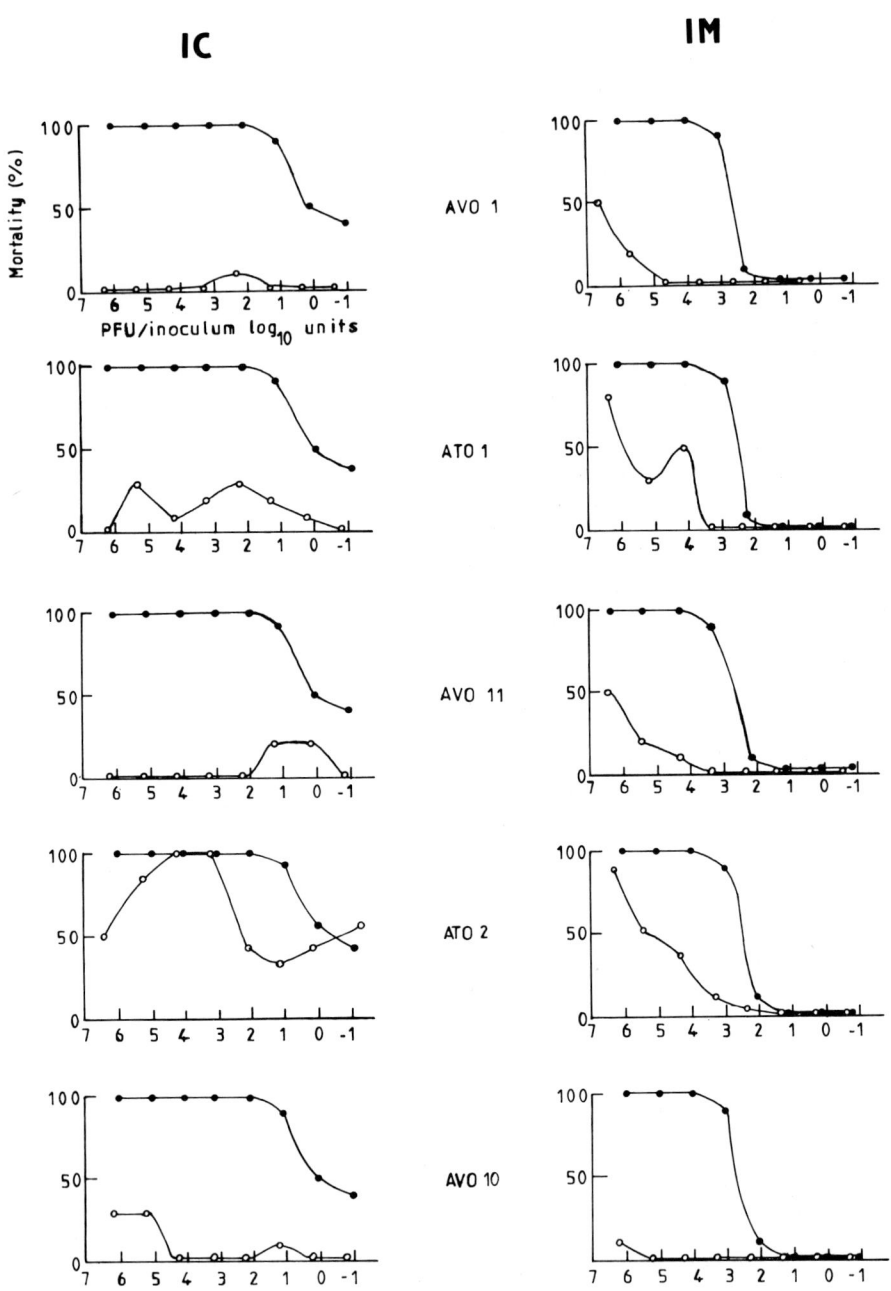

FIGURE 1. Residual pathogenicity of AvO and AtO mutants for young mice.

Series of ten 3 weeks-old mice were inoculated intracerebrally (IC) or intramuscularly (IM) with 30 µl (IC) or 50 µl (IM) of increasing dilutions of CVS (●—●) or mutant strain (O—O).

```
              KFPIYTIPDKLGPWSPIDIHHLRCPNNLVVEDEGCTNLSGFSYMELKVGY      50

              ISAIKVNGFTCTGVVTEAETYTNFVGYVTTTFKRKHFRPTPDACRAAYNW

              KMAGDPRYEESLQNPYPDYHWVRTVRTTKESLIIISPSVTDLDPYDKSLH     150

              SRVFPSGKCSGITVSSTYCSTNHDYTIWMPENPRPGTPCDIFTNSRGKRA

              SNGNKTCGFVDERGLYKSLKGACRLKLCGVLGLRLMDGTWVAMQTSDETK     250

              WCSPDQLVNLHDFRSDEIEHLVVEELVKKREECLDTLESIMTTKSVSFRR
    Gln                                       Q
    Arg       LSHLRKLVPGFGKAYTIFNKTLMEADAHYKSVRTWNEIIPSKGCLKVGGR     350
    Gly                                       G
              CHPHVNGVFFNGIILGPDDRVLIPEMQSSLLRQHMELLESSVIPLMHPLA

              DPSTVFKEGDEAEDFVEVHLPDVYKQISGVDLGLPNWGKYVLMTAGAMIG     450

              LVLIFSLMTWCRRANRPESKQRSFGGTGGNVSVTSQSGKVIPSWESYKSG

              GEIRL                                                  505
```

FIGURE 2. Aminoacid sequence of the glycoprotein of mutant and parental CVS strains.
Complete amino acid sequence is drawn according to Yelverton et al (6). The location of the primer is indicated by a thick line. The nucleotide sequence of AvO_1 and AtO_1 was determined throughout the underlined region. The brackets enclose the membrane-anchoring region. The arrows refer to the potential glycosylation sites. The amino acid exchange in mutants AvO_1, AvO_4, AvO_{11}, AtO_1 and AtO_2 is presented above the wild-type arginine 333 (R = arginine ⟶ Q = glutamine) ; the amino acid exchange in mutant AvO_{10} is presented under the wild-type (R = arginine ⟶ G = glycine).

Fig. 1 shows that none of the mutants was completely avirulent for young mice, although the difference with the fully pathogenic CVS strain was always evident. This was not surprising in view of our previous finding that non pathogenic mutants AvO_1 and AvO_2 retain their virulence for suckling mice (1). When older mice were inoculated, the difference between attenuated and avirulent mutants became more evident : animals injected with increasing doses of avirulent mutants showed no symptoms and always survived the inoculation while mice inoculated with attenuated viruses had comportmental and paralytic symptoms leading eventually to the death of the animals.

A synthetic oligonucleotide complementary to the region of the gene corresponding to amino acids 308-312 was used to prime the transcription of cDNA from a mutant or wild type RNA template, by the dideoxy chain termination procedure (5). Around 25 % of the glycoprotein gene could be sequenced by this method (Fig. 2). The amino acid sequence of the Orsay CVS strain, deduced from the base sequence, was identical to that published by Yelverton et al (6). The six avirulent and attenuated mutant differ from the wild type in the triplet coding for amino acid 333 : in mutants AvO_1, AvO_4, AvO_{11}, AtO_1 and AtO_2, a G   A transition at base 1062 was found, the arginine being replaced by a glutamine ; in mutant AvO_{10}, a C   G tranversion at base 1061 was observed, the arginine being replaced by a glycine (Fig. 3). No other difference between wild-type and mutants was evidenced.

No difference, whatsoever, could be detected between VO_{67} (also mutated in site III but still virulent) and CVS. The difference is therefore probably located before amino acid 319, an observation which suggests that site III would correspond to a folding of the molecule.

Our results confirm the crucial role of arginine 333 in the pathogenicity of rabies virus. They do not explain why our mutants differ in their attenuation. Our hypothesis is that attenuated viruses, which are less frequent than avirulent mutants among the resistants to neutralization by antibodies n° 194-2 and 248-8, would have a second mutation, partially correcting the loss of the arginine. Another explanation is that some mutant stocks would be a mixture of non pathogenic viruses and fully pathogenic revertants, giving an intermediate phenotype when injected to animals. The fact that no contaminating sequence could be detected in AtO_1 or AtO_2 in the CAG triplet (Fig. 3) indicate that either the amount of revertant virus was too low to be visible or that the reverse mutation was in another region of the gene. Experiments are in progress to clarify this point.

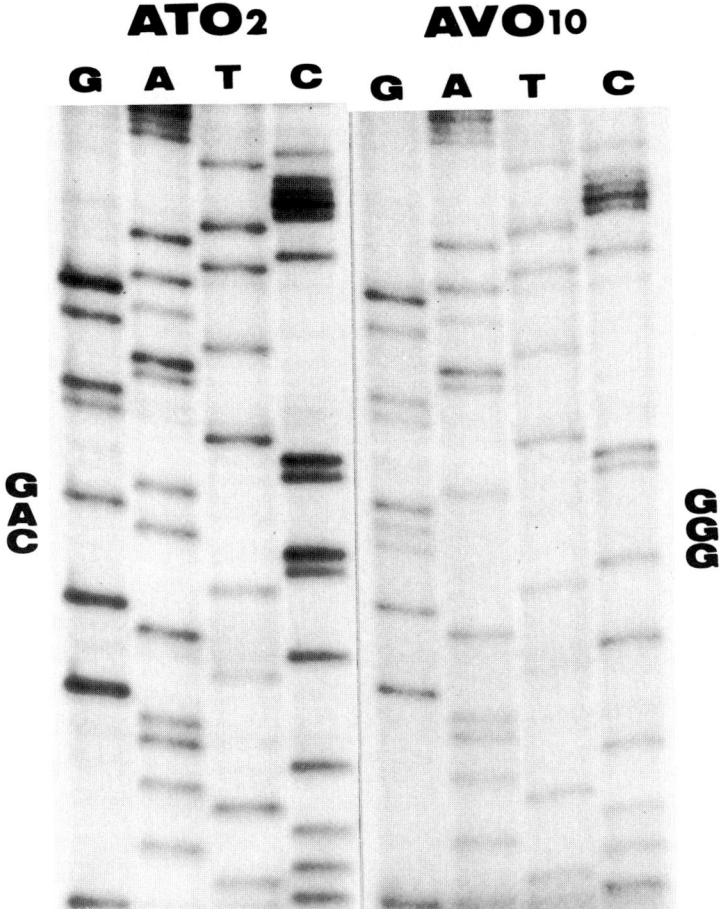

FIGURE 3. Identification of the nucleotide changes leading to loss of pathogenicity.

The synthetic oligonucleotide 5'-d(GTCCCAGGGTTTGGAAA)3' was used to prime the transcription of RNA extracted from AvO_{10} or AtO_2 purified virions, in the presence of dideoxynucleotides. The CGG sequence found in the CVS parental strain is replaced by CAG (AtO_2) or GGG (AvO_{10}).

ACKNOWLEDGMENTS

The authors are greatly indebted to D.H.L. Bishop for the synthesis of the oligonucleotide primer. They also wish to thank W.H. Wunner for the communication of part of the nuclotide sequence of the CVS glycoprotein gene, and Ph. Vigier for helpful discussions. The excellent technical assistance of J. Benejean, Ch. Thiers, J. Gagnat and J. Alexandre is gratefully acknowlegded.

REFERENCES

1. Coulon, P., Rollin, P., Aubert, M., and Flamand, A. (1982). J. Gen. Virol. 61, 97-100.
2. Coulon, P., Rollin, P.E., and Flamand, A. (1983). J. Gen. Virol. 64, 693-696.
3. Lafon, M., Wiktor, T.J., and MacFarlan, R.I. (1983). J. Gen. Virol. 64, 843-851.
4. Dietzschold, B., Wunner, W.H., Wiktor, T.J., Lopès, A.D., Lafon, M., Smith, C.L., and Koprowski, H. (1983). Proc. Natl. Acad. Sci. USA. 80, 70-74.
5. Sanger, F., Nicklen, S., and Coulson, A.R. (1977). Proc. Natl. Acad. Sci. USA. 74, 5463-5467.
6. Yelverton, E., Norton, S., Obijeski, J.F., and Goedel, D.V. (1983). Science. 219, 614-620.

HOW MANY FORMS OF THE NEWCASTLE DISEASE VIRUS P PROTEIN ARE THERE?[1]

L. E. Hightower, G. W. Smith, and P. L. Collins[2]

Microbiology Section, The Biological Sciences Group, The University of Connecticut Storrs, Connecticut 06268

PROPERTIES OF P

The P proteins of Newcastle disease virus (NDV) are the second most abundant proteins synthesized during infection and the P gene is second in the transcriptional order (1). There is biochemical (2,3) and genetic (4,5) evidence that P is involved in viral RNA synthesis, but the molecular details of its involvement are unknown. P is generally described as a nucleocapsid-associated phosphoprotein. However, this descriptive veneer glosses over a rich variety of molecular forms. Virions contain four variants of P with different isoelectric points, only the two most acidic forms of which are phosphorylated (6). P proteins form disulfide-linked multimers large enough to be trimers and several small nonstructural proteins which appear to be related to P have been identified (7,8).

FORMS OF P IN INFECTED CELLS

Secondary cultures of chicken embryo (CE) cells were either mock-infected (Figure 1A,D) or infected with strain AV at 5 PFU/cell (Figure 1B,C)

---

[1]This work was supported by Public Health Service grant HL23588 and National Science Foundation grants PCM78-08088 and PCM81-18285. We benefited from the use of a cell culture facility supported by Public Health grant CA14733.
[2]Present address: Department of Microbiology and Immunology, University of North Carolina, Chapel Hill, North Carolina 27514.

and exposed to [^{32}P]orthophosphate (Figure 1C,D) from 3 to 6 hours post-infection (p.i.) or to [^{35}S]methionine (Figure 1A,B) from 5 to 6 hours p.i. Radioactive proteins were extracted and analyzed on two dimensional polyacrylamide gels using isoelectric focusing in the horizontal dimension followed by SDS-PAGE in the vertical dimension after the methods of O'Farrell (9). Only the gel regions from pH 5.8 to 6.2 and 53 kilodaltons to 56 kilodaltons are shown in Figure 1 and in subsequent figures. Five forms of P were identified among the [^{35}S]methionine-labeled proteins from infected cells (Figure 1B). Spots 1-4 occupied the same positions on gels as the four P variants from virions. The fifth form of P, identified by partial digest peptide mapping, sometimes streaked and sometimes focused into a well-formed spot intermediate in mobility between spots 3 and 4, and more acidic than either. Form 5 was phosphorylated (Figure 1C) as were forms 3 and 4; however, unlike the latter two variants, form 5 was detected in extracts of infected cells but not virions. Much larger amounts of form 2 relative to the other forms of P were obtained from virions than from infected cells (6). See Figure 3A for patterns of P isolated from virions. The dark horizontal streak across Figure 1B was the nucleocapsid protein, which did not focus under these conditions.

At present, we can only speculate on the biologic roles of the various forms of P in virions and infected cells. Based on its relatively high abundance, we suggest that form 1 may have primarily a structural role in nucleocapsids. Form 2,

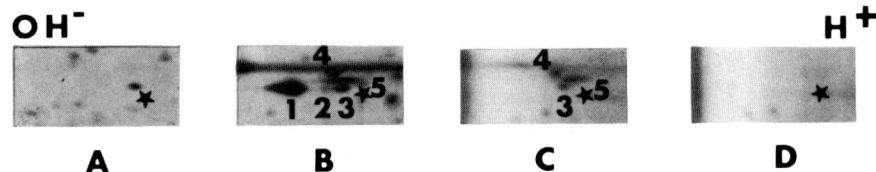

Figure 1. Forms of [^{35}S]methionine-labeled and ^{32}P-labeled P extracted from infected CE cells and analyzed on two dimensional gels. The stars mark the same position in each fluorogram (^{35}S) or autoradiogram (^{32}P).

which is relatively enriched in virions, may be involved in folding of the nucleocapsid during virion assembly. Form 5, found in infected cells but not virions, may be involved in replication whereas forms 3 and 4 may be required for both transcription and replication.

FORMS OF P MADE IN CELL-FREE PROTEIN SYNTHESIZING SYSTEMS

mRNA was extracted from mock-infected (Figure 2A), virulent strain AV-infected (Figure 2B), and avirulent strain B1-infected CE cells at 9 hours p.i. and purified on oligo-dT cellulose columns as described previously (10). The mRNA was used to program nuclease-treated reticulocyte lysates. Two dimensional gel analyses of the [^{35}S]methionine labeled P protein synthesized in these lysates are shown in Figure 2. The overall pattern of P proteins of strain AV made in vitro was remarkably similar to the pattern from infected cells (Figure 1B), indicating that the mechanisms that generated the P variants operated in the reticulocyte lysate as well as in infected cells. In particular, the P proteins made in vitro included a fifth form similar in pI and mobility to form 5 from infected cells, and the in vitro products were relatively deficient in form 2. The similarities between in vivo and in vitro P patterns of strain AV suggest that mechanisms at the level of translation or perhaps mRNA heterogeneity as well as post-translational modifications may be responsible for P variants.

Since strain-specific differences in the patterns of P from infected cells have been reported

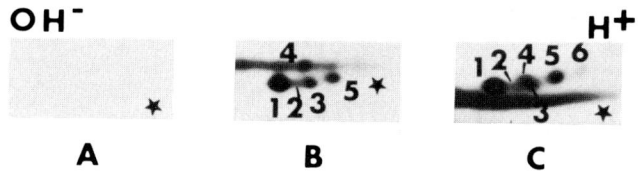

Figure 2. Forms of the [^{35}S]methionine-labeled P proteins synthesized in vitro and analyzed on two dimensional gels.

(11), we compared the P proteins synthesized in reticulocyte lysates programmed with mRNA from strain B1 (Figure 2C) to those of strain AV. Although strain B1 has a slightly smaller nucleocapsid protein than strain AV and slightly larger P proteins than strain AV (12), there are strong similarities in the overall patterns of P variants made in vitro. For both strains, form 1 was the most abundant and a fifth form was produced. The triangular cluster of forms 2, 3, and 4 of strain AV also appeared to be present among the products of strain B1 mRNA; however, the cluster was much tighter and these three forms were only partially resolved. Strain B1 produced a sixth spot which may be an additional form of P, but it has not been mapped. There was a hint of a sixth spot in the in vitro pattern from strain AV (Figure 2B) as well, but it was partially hidden in the streak of nucleocapsid protein.

COMPOSITION OF P MULTIMERS FROM VIRIONS

We used a modification of the standard O'Farrell protocol for two dimensional gel electrophoresis to determine the composition of the disulfide-linked trimers of P from purified strain AV virions. The strategy was to omit mercaptoethanol from

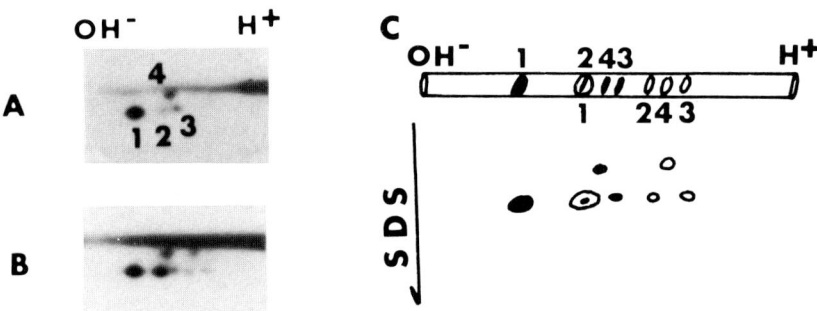

Figure 3. Analysis of P multimers from [^{35}S] methionine-labeled AV virions using isoelectric focusing under either reducing (A) or nonreducing (B) conditions in the horizontal dimension followed by reducing SDS-PAGE in the vertical dimension. Panel C shows a diagrammatic sketch of Panel B.

the isoelectric focusing dimension to allow P proteins to focus as disulfide-linked molecules. Under these conditions, the net charge of the P multimers might be different enough from P monomers to separate them. As shown in Figure 3, this expectation was fulfilled. The pattern of four P variants from [^{35}S]methionine-labeled purified virions analyzed under normal reducing conditions is shown in Figure 3A. The same preparation separated under nonreducing conditions during isoelectric focusing is shown in Figure 3B. Under these conditions, both P monomers and disulfide-linked multimers are expected since only a portion of virion P can be extracted as multimers (6). The characteristic pattern of P monomers and four additional spots which appeared to be a duplicate of the monomer pattern but shifted to the acidic side were obtained. Our interpretation of this pattern is that the shifted P variants represent disulfide-linked homomultimers of P. Other data suggest that these are at least homotrimers (6). Since each spot experienced the same magnitude of acidic shift, we suggest that the subunits in all four of the homotrimers associate in a similar fashion. These data are consistent with the findings of Chinchar and Portner (2) that the disulfide-linked portions of P are in a highly basic, protease-resistant part of the molecule. Neutralization or shielding of some of the basic amino acid residues in the disulfide-linked multimers could have produced the acidic shift observed by us. Further, our results imply that modified forms of P such as phosphorylated forms 3 and 4 either associate only with the same modified forms or that all subunits in a P multimer are modified in the same way after multimer formation. The strategy shown in Figure 3 should be generally applicable to the analysis of complex disulfide-linked multimers.

A GLIMPSE OF THE MOLECULAR ANATOMY OF P

Virus-specific proteins with apparent molecular weights of 36,000 and 33,000 which appear to be P-related have been detected in NDV-infected cells and in cell free systems. Similar proteins have been detected in mumps virus-infected cells (13). The two forms made in vitro can be labeled with fmet-tRNA, and they share several tryptic peptides

Protease-Resistant Domain   **Protease-Sensitive Domain**

Basic
S-S
RNA Binding Site

Acidic
Met Poor
Modified

```
SP1   HOOC···─────────────┤                    30K
SP2   HOOC···───────────┤                      24K
                     ├──────────────NH2        33K
                   ├────────────────NH2        36K
P     HOOC─────────────────────────NH2         53K
```

Figure 4. Molecular anatomy of the P protein. The relative sizes based on electrophoretic mobilities of the 53 kilodalton P proteins and smaller forms are drawn to scale. The placement of SP1 and SP2 at the exact C-terminus is speculative. The 33 and 36 kilodalton in vitro products are shown.

which are also common to P protein (7). Therefore, they are thought to be overlapping fragments from the N-terminus of P. These fragments are more acidic than P as determined by isoelectric focusing and are relatively poor in methionine, containing only 2 of the 10 methionyl tryptic peptides of full-size P. Additional comparisons of these fragments and P using radioactive amino acids other than methionine for peptide mapping would be useful. The 36 kilodalton proteins made in vivo and in vitro have very similar behavior on two dimensional gels; however, we are less certain about the identity of the 33 kilodalton in vivo and in vitro products. The 33 kilodalton protein made in infected cells has been reported to be highly basic (8) whereas the 33 kilodalton in vitro product is more acidic than full-size P. The possibility that the similarity in electrophoretic mobilities of these proteins is coincidental needs further evaluation. The 36 kilodalton fragment is phosphorylated in vivo (11) and thus contains the site(s) for at least one type of P modification.

Our data on P-related fragments are summarized in Figure 4 along with data obtained by Chinchar and Portner (2). Taken together, these results begin to reveal the molecular anatomy of P. Chinchar and Portner treated nucleocapsid-associated P with S. aureus protease V8 and obtained two highly basic fragments SP1 and SP2 which contained the cysteine residues involved in disulfide bonding of P multimers. These fragments remained associated with cores and presumably also contained a RNA-binding site of P (14). SP1 and SP2 have different partial digest peptide maps than the N-terminal fragments made *in vitro* and thus represent the part of P close to and possibly including the carboxyl terminus.

The P protein in its many forms is likely to hold keys to the assembly and biological activities of the nucleocapsid of paramyxoviruses. We are just beginning to appreciate the complexity of this protein.

ACKNOWLEDGMENTS

We thank Peter Guidon Jr. for help with gel analyses, Mary Jane Spring for photography, and Gayle Hightower for aid in preparing the manuscript.

REFERENCES

1. Collins, P. L., Hightower, L.E., and Ball, L. A. (1980), J. Virol. 35, 682.
2. Chinchar, V. G., and Portner, A. (1981). Virology 115, 192.
3. Hamaguchi, M., Yoshida, T., Nishikawa, K., Naruse, H., and Nagai, Y. (1983). Virology 128, 105.
4. Madansky, C. H., and Bratt, M. A. (1982). J. Virol. 37, 317.
5. Peeples, M. E., and Bratt, M. A. (1982). J. Virol. 41, 965.
6. Smith, G. W., and Hightower, L. E. (1981) J. Virol. 37, 256.
7. Collins, P. L., Wertz, G. W., Ball, L. A., and Hightower, L. E. (1982). J. Virol 43, 1024.
8. Chambers, P., and Samson A. C. R. (1982). J. Gen. Virol. 58, 1.
9. O'Farrell, P. H. (1975). J. Biol. Chem. 250, 4007.

10. Collins, P. L., Hightower, L. E., and Ball, L. A. (1978). J. Virol. 28, 324.
11. Chambers, P., and Samson, A. C. R. (1980). J. Gen. Virol. 50, 155.
12. Collins, P. L., and Hightower, L. E. (1982). J. Virol. 44, 703.
13. Herrler, G., and Compans, P. W. (1982). Virology 119, 430.
14. Raghow, R., and Kingsbury, D. W. (1979). Virology 98, 267.

FOUR FUNCTIONAL DOMAINS ON THE HN GLYCOPROTEIN OF NEWCASTLE DISEASE VIRUS[1]

Ronald M. Iorio and Michael A. Bratt

Department of Molecular Genetics and Microbiology
University of Massachusetts Medical School
Worcester, Massachusetts 01605

COMPETITION ANTIBODY BINDING RADIOIMMUNOASSAYS

A panel of nine monoclonal antibodies specific for the hemagglutinin-neuraminidase (HN) glycoprotein of Newcastle disease virus have been prepared by fusion of SP2 myeloma cells with spleen lymphocytes from a BALB/c mouse immunized with intact, UV-inactivated virus (1). These antibodies have been used in competition antibody binding radioimmunoassays in which each monoclonal antibody was assayed for its ability to compete for the binding of ^{125}I-labeled antibodies to microtiter wells coated with intact virions. Antibodies that compete for the binding of the labeled antibody with an affinity similar to that of the homologous antibody are assumed to be recognizing overlapping antigenic determinants. As shown in Table 1, these studies delineate four epitopes on the surface of the HN molecule with epitopes 2 and 3 being partially overlapping in a topological sense. This is remarkably similar to studies with monoclonal antibodies directed against the HN molecules of Sendai (2) and parainfluenza viruses (3) which have also identified four antigenic determinants (including two which partially overlap) on the surface of the molecule.

FUNCTIONAL INHIBITION STUDIES

The functional relevance of each of the four epitopes on the surface of the HN molecule has been determined by assaying each antibody's neutralization of infectivity, inhi-

---

[1]This investigation was supported in part by a grant from the National Institute of Allergy and Infectious Disease (AI 12467), an instititutional BRSG grant (RR-05712) and a postdoctoral fellowship from the National Multiple Sclerosis Society to Ronald M. Iorio.

TABLE 1
EPITOPE ASSIGNMENTS OF ANTI-HN MONOCLONAL ANTIBODIES

Competing Antibody

^{125}I-Antibody	$HN1_a$	$HN1_b$	$HN1_c$	$HN2_a$	$HN2_b$	$HN3_a$	$HN4_a$	$HN4_b$	$HN4_c$	Epitope
$HN1_a$	+	+	+	−	−	−	−	−	−	1
$HN1_b$	+	+	+	−	−	−	−	−	−	1
$HN1_c$	+	+	+	−	−	−	−	−	−	1
$HN2_a$	−	−	−	+	+	±	−	−	−	2
$HN2_b$	−	−	−	+	+	±	−	−	−	2
$HN3_a$	−	−	−	±	±	+	−	−	−	3
$HN4_a$	−	−	−	−	−	−	+	+	+	4
$HN4_b$	−	−	−	−	−	−	+	+	+	4
$HN4_c$	−	−	−	−	−	−	+	+	+	4

+ > 60% competition;   ± 50-60% competition;   − < 50% competition

hition of neuraminidase, with both fetuin and neuraminlactose as substrate, and inhibition of hemagglutination. These data are summarized in Table 2.

The neutralizing activity of each of the nine anti-HN monoclonal antibodies was assayed by the commonly-used end-point dilution method in which 100 pfu were treated with serial two-fold dilutions of antibody from an initial concentration of 100 ug/ml. Comparison of the specific neutralizing activity of these antibodies can be very misleading because of the limitation imposed by the use of a 50% neutralization point. For example, antibody $HN1_c$, despite its relatively low specific neutralizing activity, is the most potent neutralizer of infectivity as evidenced by its 0.9% persistent fraction of nonneutralized virus. Conversely, antibody $HN4_c$, although having the highest specific neutralizing activity, is the weakest neutralizer on the basis of its extremely high persistent fraction. Moreover, although antibodies to all four epitopes are capable of neutralizing viral infectivity, they do so only to an apparently epitope-specific extent, with antibodies to epitope 1 the strongest, those to epitopes 2 and 3 intermediate and those to epitope 4 the weakest neutralizers.

With the goal of determining the role of each of the four epitopes in the neuraminidase (NA) activity of the HN molecule, two-fold serial dilutions of each antibody were assayed in NA inhibition assays with the large molecular weight molecule, fetuin, as substrate. As in the case of neutralization of infectivity, the specific inhibiting activity is misleading; the percent activity remaining following treatment with a saturating amount of antibody is much more indicative of the importance of that epitope to NA activity. The data also suggests a correlation between inhibition of NA and neutralization of infectivity.

In an effort to make a finer antigenic analysis of the role of the HN epitopes in the NA activity of the molecule, NA inhibition assays were also performed with the much smaller substrate neuraminlactose. These efforts were complicated by the apparent nonspecific enhancement of NA activity by ascites fluids, an effect which has also been seen (with both substrates) with a parainfluenza virus (3). Thus, Table 2 shows the percent activity remaining following treatment of virus with a saturating amount of an affinity-purified preparation of each antibody. Only antibodies to epitope 2 are capable of inhibiting NA with the smaller substrate. As a control, the affinity-purified preparations were shown to inhibit NA with fetuin to an extent similar to that obtained with the ascites preparations (data not shown).

TABLE 2
FUNCTIONAL INHIBITION STUDIES

Antibody	spec. neut.[a] activity	% persistent fraction	spec. NA[b] inh. titer (Fetuin)	% NA activity remaining (Fetuin)	% NA activity remaining (Neuraminlactose)	specific[c] HI-activity
HN1a	1280	1.6+0.9	65	3.7	130.2	8200
HN1b	640	2.7+0.8	16	15.1	105.8	3000
HN1c	640	0.9+0.6	48	1.8	127.4	6100
HN2a	640	5.3+2.0	39	14.8	41.5	2000
HN2b	640	3.4+1.2	38	11.3	29.7	900
HN3a	2560	6.4+2.1	48	31.9	92.0	12
HN4a	2560	13.8+4.7	2	42.6	97.5	8
HN4b	2560	7.6+2.9	<2	67.8	109.3	17
HN4c	5120	16.9+2.7	<2	60.2	116.2	4

[a]The reciprocal of the highest dilution of antibody giving a 50% reduction in plaque number relative to an anti-NP control.
[b]The reciprocal of the highest dilution of ascites fluid, on a per mg of antibody per ml basis inhibiting 50% of the neuraminidase activity.
[c]The reciprocal of the highest dilution of ascites fluid, on a per mg of antibody per ml basis inhibiting four HA units of virus.

An ascites fluid preparation of each of the antibodies was next assayed for its ability to inhibit the agglutination of chicken erythrocytes by 4 HA units of virus. Antibodies to epitope 1 had the highest specific HI activities while antibodies to epitope 2, although also inhibitory of HA, have at least a three-fold lower activity. Antibodies to epitopes 3 and 4 have specific HI activities of less than 20, comparable to those obtained with control antibodies specific for the nucleocapsid protein (NP).

Table 3 presents a summary of the functional properties of the four epitopes on the surface of the HN glycoprotein of NDV. On the basis of this, we have constructed a tentative linear map of their relative positions with respect to the HA and NA sites (Figure 1). Epitope 1 is placed closest to the HA site because antibodies to it have the highest HI titers. Epitope 2 is closest to the NA site because antibodies to it are the only ones that inhibit NA with the smaller substrate, but is placed between the two sites because antibodies to it can inhibit HA as well. Antibodies to epitopes 3 and 4 are only able to inhibit NA with the larger substrate and do not inhibit HA. Thus, they are placed the most distal to the HA site but with epitope 3 closest to the NA site as it topologically overlaps with the second epitope. We are using this map as a working model of the antigenic structure of the HN glycoprotein of NDV.

EVIDENCE FOR SEPARATE SITES FOR HA AND NA

Recent evidence from mutant and neuraminic acid analog analyses suggests that the NA and HA sites are located on different active sites on the HN molecule of paramyxoviruses (4-7). This study provides further antigenic evidence in support of this view. Although antibodies to each of the epitopes are capable of inhibiting NA with the larger substrate, antibodies to epitopes 3 and 4 are not able to inhibit the HA activity of the virus. Thus, this is evidence of an antigenic distinction between the HA and NA activities on the HN molecule. Even more convincing, however, is the failure of antibodies to epitope 1 to inhibit NA with neuraminlactose as substrate despite their extremely high HI titers. This is evidence for an antigenic separation between the two sites on the surface of the HN glycoprotein of NDV and is similar to that found with Sendai virus (2).

TABLE 3
SUMMARY OF FUNCTIONAL INHIBITION PROPERTIES OF HN EPITOPES

HN Epitope	Infectivity	Neuraminidase Fetuin	Neuraminlactose	Hemagglutination
1	+++	+++	-	+++
2	++	++	++	++
3	++	++	-	-
4	+	+	-	-

FIGURE 1
LINEAR MAP OF HN EPITOPES

ACKNOWLEDGEMENTS

We acknowledge helpful discussions with L. Edward Cannon, Robert Woodland, Allen Portner, Mark Peeples, Trudy Morrison and Raymond Welsh and the technical assistance of Judith Brackett, and the help of Anne Chojnicki in the preparation of the manuscript.

REFERENCES

1. Iorio, R.M. and Bratt, M.A. (1983). J. Virol. 48, in press.
2. Orvell, C., and Gradien, M. (1982). J. Immunol. 129, 2779.
3. Yewdell, J., and Gerhard, W. (1982). J. Immunol. 128, 2670.
4. Portner, A. (1981). Virology 115, 375.
5. Smith, G.W., and Hightower, L.E. (1980). In "Animal Virus Genetics" (B. Fields, R. Jaenisch, and C.F. Fox, eds.), p. 623, Academic Press, New York.
6. Smith, G.W., and Hightower, L.E. (1982). J. Virol. 42, 657.
7. Smith, G.W., and Hightower, L.E. (1983). J. Virol. 47, 385.

MAPPING MUTANT AND WILD-TYPE M PROTEINS OF NEWCASTLE DISEASE VIRUS (NDV) BY REPEATED PARTIAL PROTEOLYSIS[1]

Mark E. Peeples,[2] and Michael A. Bratt
Department of Molecular Genetics and Microbiology
Worcester, Massachusetts 01605

Partial proteolysis in SDS-polyacrylamide gels by the procedure of Cleveland et al (1) is a valuable tool in comparing proteins. However, information about the nature of the partial proteolysis fragments and the section of the molecule represented by them is not provided by this procedure. We have expanded upon this procedure by subjecting the membrane protein (M) of NDV to partial digestion with S. aureus V8 protease, and then isolating the products, redigesting them, and analyzing the resulting fragments. This has allowed us to construct a map of wild-type (AV-WT) M, and by comparison of this map with that of a ts mutant M, we have begun to locate the alteration in the mutant protein.

MAPPING AV-WT M PROTEIN

Figure 1 shows partial proteolysis of AV-WT M using increasing concentrations of protease. In addition to the undigested 41 K (kilodalton) M, 11 major fragments were generated. Each fragment was then cut from the gel and redigested with a higher protease concentration. Figure 2 shows that redigestion of band 1 (uncut M) on the left, again yields all 11 fragments; fragments 8 and 9 are difficult to separate upon redigestion. Redigestion of fragment 2 yields fragment 4 (not 3), 8/9, and 12. Fragment 3 yields fragments 4, 8/9 and 11. Fragment 4 yields fragments 8/9 and 11. At the protease concentrations used here, fragment 9 yields no new fragments on redigestion, but digestion at higher protease concentrations yields fragment 11 which is a final digestion product (data not shown).
Comparing fragments generated by redigesting each initial fragment, we generated the map of M in Figure 3;

---

[1] Supported by Public Health Service grant AI 12467 (National Institute of Allergy and Infectious Diseases).
[2] Present address: Department of Immunology/Microbiology, Rush Medical College, Chicago, Illinois 60612.

four S. aureus V8 sites are designated A,B,C and D. An example of the way the map was constructed follows. 2 is generated from M (fragment 1) by the loss of 2K; it has arbitrarily been placed at the right end. Further digestion of 2 generates 4, but not 3; thus 4, but not 3, is contained within 2. Therefore, 3 must come from the other end, the left end of the molecule. A fragment the size of 4 is generated from both 2 and 3 and can be explained as two fragments of similar molecular weight resulting from approximately equivalent fragments lost from 2 and 3: the left end to protease site A (2K), and the fragment between sites C and D (2K).

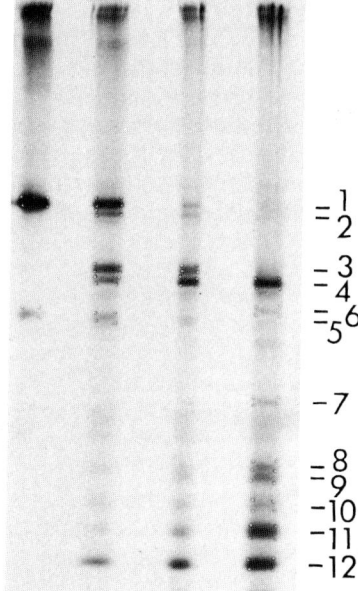

FIGURE 1. Partial proteolysis pattern of the M of AV-WT. A ^{35}S-methionine labeled cell lysate was displayed on a 10% SDS-polyacrylamide gel (2) and autoradiographed to locate M. Gel slices containing M were rehydrated, placed on top of a 15% SDS-polyacrylamide gel with different concentrations of S. aureus V8 protease, and digested (1). Major fragments are identified, including the uncut M, #1.

Fragments on the map in figure 3 are shown in their most likely configuration, although a few ambiguities (due in part to our inability to separate redigested fragments 8/9) exist at the internal sites. Fragment 7 and several minor fragments have not been placed on the map.

LOCATING THE CHANGE IN THE M PROTEIN OF TS MUTANT D1

Virions of group D ts mutants are less infectious and less hemolytic than AV-WT (3) and cells infected by them package less fusion protein (F_{1+2}) into virions (4). In addition, mutant D1's M migrates in SDS-polyacrylamide gels with an apparent molecular weight of 40K compared to 41K for the AV-WT M, regardless of temperature of synthesis (Figure 4 and reference 5). Revertant analysis has revealed that altered M migration and decreased packaging of F_{1+2} are clearly related to D1's originally selected ts phenotype (6).

Since D1's complete M migrates more rapidly, it seemed

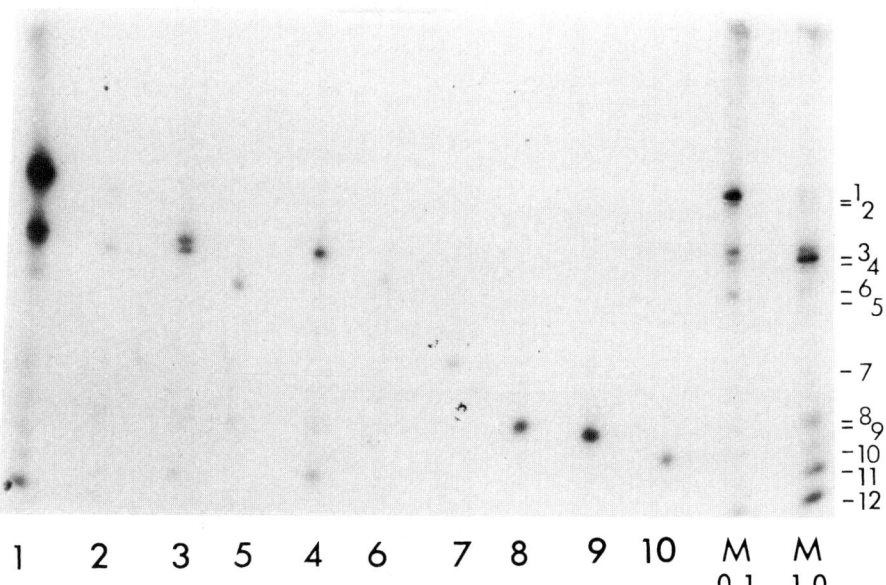

FIGURE 2. Redigestion of partial proteolysis products of AV-WT's M. Major fragments from the digests shown in Fig. 1 were isolated from the track where they were present in greatest abundance. They were redigested by the procedures described for Fig. 1, but using the next higher protease concentration. Products of the digestion of M with 0.1 and 1.0 ug of protease provide markers (on the right).

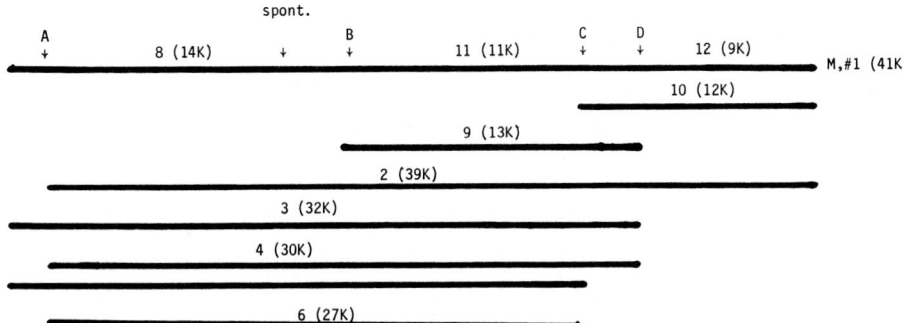

FIGURE 3. S. aureus V8 protease map of the M protein of AV-WT. Protease sites were located by determining which large fragments contained smaller fragments, and then aligning overlapping fragments. The smallest fragments which could not be digested further were placed on the full map, and the four protease sites they define identified as A,B,C and D. A spontaneous cleavage site whose products are often found in the absence of protease is also shown. Two approximately 2K fragments between the left end and A, and between C and D, have not been isolated but are presumed to exist because of size differences in fragments 1 and 2, and also 10 and 12.

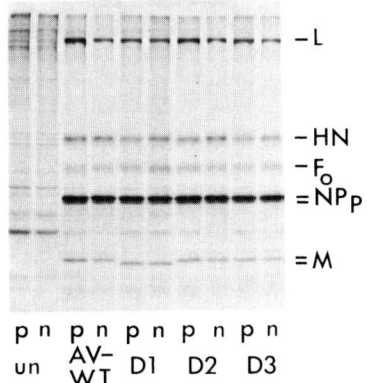

FIGURE 4. AV-WT and mutant proteins in chicken embryo cells. Cells incubated at permissive (p) or nonpermissive (n) temperature for 6 hours were labeled for 1 hour with 20 uCi/ml ^{35}S-methionine in medium lacking methionine, lysed in sample buffer, and displayed on a 10% SDS-polyacrylamide gel (2). Uninfected (un) cells, and cells infected with AV-WT, or ts mutants D1, D2, or D3 are presented.

likely that some of its partial fragments might do so, as well. Figure 5 shows that, in addition to the uncut M (fragment 1), four D1 fragments, 2, 5, 10 and 12, migrate faster. In the map of AV-WT's M in Figure 3, fragments 2, 10 and 12 are a nested set of fragments; each contains the 9K fragment 12 located at an end of the molecule. This conclusion has recently been confirmed by digesting M from AV-WT and D1 with S. aureus V8 in solution; under these conditions D1's fragment 12 is smaller than that of AV-WT, while fragment 3 from both sources is identical (data not shown). Thus the lesion in the M protein of D1 clearly seems to lie at an end of the molecule.

Fragment 5 also reflects the increased migration rate of

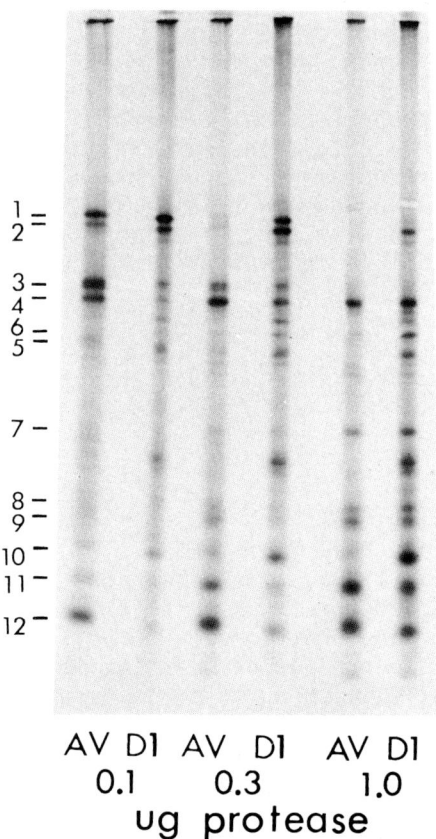

FIGURE 5. Partial proteolysis of the M protein of AV-WT and mutant D1. M was isolated and digested as in Fig. 1. The 12 major fragments are indicated along the side.

D1's M. It is generated spontaneously in the absence of protease as a result of fixing the gel in acetic acid and drying with heat (data not shown). Fragment 5 must include the altered region of D1's M and must therefore include the right end of the molecule. The spontaneous breakage site must be approximately in the position shown in Figure 3.

By repeated partial proteolysis of the M of AV-WT with S. aureus V8, we have deduced a map of this protein. By comparing the M of AV-WT and mutant D1 by partial proteolysis, we have located the altered portion of D1's M to a terminal region of the molecule. This region of M may either directly or indirectly interact with the F glycoprotein which is required for hemolysis and infectivity (7). Such an interaction is indicated by studies with the group D mutants and their revertants altered in the packaging of F and its functioning in virions (4, 6).

We do not yet know which end of M is involved in this mutation and whether the alteration is located at the extreme or within the terminal fragment. However, it is tempting to speculate that the mutation is in the carboxy terminus of M and represents a nonsense mutation causing premature termination during translation. Analysis of the carboxy terminus of AV-WT, D1, and its three revertants whose M's migrate like that of AV-WT (6) will be interesting.

ACKNOWLEDGEMENTS

We thank Judy Brackett and Rhona Glickman for technical help and Anne Chojnicki for preparing this manuscript.

REFERENCES

1. Cleveland, D.W., Fischer, S.G., Kirschner, and Laemmli, U.K. (1977). J. Biol. Chem. 252, 1102.
2. Laemmli, U.K. (1970). Nature (London). 227, 680.
3. Peeples, M.E. and Bratt, M.A. (1982. J. Virol. 42, 440.
4. Peeples, M.E. and Bratt, M.A. Newcastle disease virus infectivity depends on the quantity of fusion glycoprotein, submitted for publication.
5. Peeples, M.E., Gallagher, J.P., Bratt, M.A. (1981). In "The Replication of Negative Strand Viruses" (D.H.L. Bishop and R.W. Compans, eds.), p. 567. Elsevier-North Holland, Inc., New York.
6. Peeples, M.E. and Bratt, M.A. Revertants of the group D temperature sensitive mutants of Newcastle disease virus; coreversion of M protein size and incorporation of F_{1+2} into virions, submitted for publication.
7. Nagai, Y. and Klenk, H.-D. (1977). Virology 77, 125.

STRUCTURAL CHARACTERIZATION OF HUMAN PARAINFLUENZA VIRUS 3

Douglas G. Storey and C. Yong Kang

Department of Microbiology and Immunology
University of Ottawa, School of Medicine
Ottawa, Ontario, Canada K1H 8M5

Human parainfluenza virus type 3 (HPIV3) infections have great clinical significance because they occur early in life, are readily transmissible, and are likely to recur [1,2,3,]. Despite its medical importance, little is known about the molecular biology of this virus.

The most thoroughly studied parainfluenza virus is Sendai virus, a murine parainfluenza type 1 virus. Sendai virus has been shown to have 9 major structural proteins with molecular weights ranging from 145 K to 34.7 K when grown in cultured cells [4,5,6]. The biological functions and structural locations have been assigned to the major polypetides. Sendai virus contains either two or three glycoproteins depending on the host cell of origin. The largest glycoprotein, molecular weight of 69K, has both haemagglutinin and neuraminidase activities and is designated HN [5,7]. The small glycoprotein, molecular weight of 53K is involved in virus induced haemolysis, cell fusion and initiation of infection [7]. This glycoprotein is designated F. Sendai virions grown in cultured cells, have an intermediate size glycoprotein of 65K molecular weight, which has been identified as F_0, the inactive precursor of the fusion protein [7]. Three nonglycosylated proteins have been associated with the nucleocapsid and are involved in RNA polymerase activity [8]. These proteins are L protein a high molecular weight protein, P, the polymerase protein, and the NP the nucleocapsid subunit protein. It is of interest to note that the three nucleocapsid proteins are all phosphorylated [6,9]. The smallest structural protein with a molecular weight of 34.7 K is the matrix protein M. This protein is responsible for maintaining the structural integrity of the virus [5].

Preliminary works on bovine, ovine, and human parainfluenza virus type 3 [10,11,12,13] have indicated that there are similarities in the structural proteins to those of Sendai virus and the other paramyxoviruses. The characteristic which distinguishes the parainfluenza type 3 viruses from Sendai virus is that parainfluenza type 3 viruses do not need trypsin in the medium in order to form plaques [10]. This observation indicates that only the cleaved fusion protein is incorporated into the virion. Indeed, for all strains of parainfluenza type 3 viruses tested, only 2 glycoproteins are associated with the virion [10,11,12,13].

Growth and Purification of HPIV3. One difficulty when working with HPIV3 is the low virus yields when grown in cultured cells. In order to overcome this problem HPIV3 strain C243 (provided by the Laboratory Center for Disease Control, Ottawa, Canada) has been grown in monolayer cultures of $LLCMK_2$ cells. The growth curve of the virus is shown in Figure 1. A titre of 9 X 10^7 plaque forming units/ml of culture medium was achieved within 24 hours and maintained for at least 50 hours post infection. The formation of polykaryocytes appeared at 12 hours post infection. Cell destruction became prominent 50-60 hours post infection.

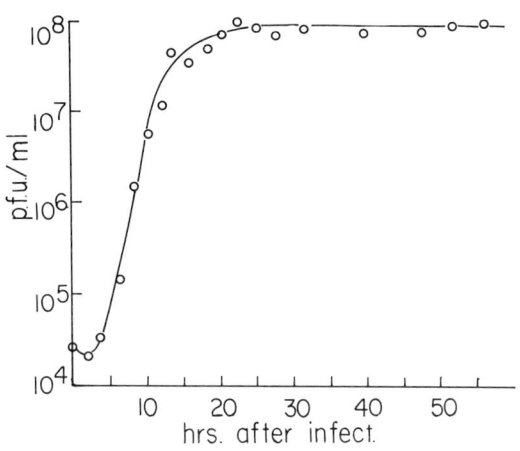

FIGURE 1. Growth curve of HPIV3 in LLC MK_2 cells. Cell cultures (7 X 10^6 cells) were infected at a multiplicity of infection of 3. Samples were assayed for infectivity by plaque assay using LLC MK_2 cells.

PROTEINS OF HPIV3

The structural proteins of the virion were labeled with radioactive precursors starting immediately after the adsorption period of one hour. The virus was grown in the presence of ^{14}C-glucosamine to label the glycosylated proteins, ^{32}P-orthophosphate to label the phosphorylated proteins and ^{35}S-methionine or ^{14}C-amino acids mixture to label all the structural proteins. The culture fluid was harvested 48 hours after infection. To remove the cellular debris the culture fluid was clarified by low speed centrifugation. The labeled virus was then pelleted by high speed centrifugation, resuspended in sample buffer and then analyzed by SDS polyacrylamide gel electrophoresis.

Identification and Molecular Weight Determination of the structural Proteins of HPIV3.
Bovine, ovine and human parainfluenza type 3 viruses have been reported to contain 5, 6, and 8 structural proteins respectively (10,12,13). These proteins range in molecular weight from 17K to 125K. The variation amongst the parainfluenza viral proteins made it necessary to identify the structural proteins of HPIV3 and determine the molecular weights of the proteins. HPIV3 was labeled with ^{14}C-amino acids mixture or ^{35}S-methionine for 48 hours. The radioactively labeled HPIV3 was purified and the structural proteins were analyzed in a 15% polyacrylamide gel with a 3% stacking gel containing 0.1% SDS. Proteins of known molecular weights have been used as references. These include myosin (H-chain) 200K, phosphorylase B 92.5K, bovine serum albumin 68K, ovalbumin 43K, chymotrypsinogen 25.7K, and B-lactoglobulin 18.4K.

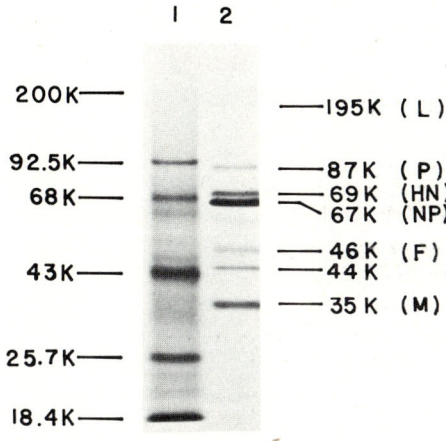

FIGURE 2. Polyacrylamide gel analysis of marker proteins and of HPIV3 proteins. ^{14}C-amino acids labeled myosin (200K), phosphorylase B (92.5K), bovine serum albumin (68K), ovalbumin (43K), chymotrypsinogen (25.7K) and B-lactoglobulin (18.4K) (1) and ^{35}S-methionine labeled HPIV3 (2) were analyzed by electrophoresis in 15% polyacrylamide gel containing 0.1% SDS.

HPIV3 contains 7 major and 2 minor structural proteins. The major proteins have molecular weights of 195K, 87K, 69K, 67K, 46K, 44K, and 35K, for L, P, HN, NP, F, an unknown protein, and M respectively. The minor proteins have molecular weights of 60K and 20K (Figure 2). The major structural proteins were assigned on the basis of similarities to Sendai virus structural proteins and on this study (see below).

To determine whether ^{35}S-methionine labeling represents the total proteins in correct proportions, a comparison was made between the viral proteins labeled with ^{35}S-methionine, or with a ^{14}C-amino acids mixture, and when the viral proteins where stained with Coomassie Brilliant Blue. The three methods represent similar molar quantities of the proteins as were found for the proteins of Sendai virus [6,14]. Thus, most experiments were carried out with ^{35}S-methionine.

The Phosphorylated Proteins of HPIV3. Phosphorylated proteins are often found in the enveloped negative strand RNA viruses. These include orthomyxo, paramyxo and rhabdoviruses. The P, NP and M proteins of Sendai and Newcastle disease virus are the predominantly phosphorylated proteins [9,15,16]. The HN and F_0 of Sendai virus are also phosphorylated [9,16].

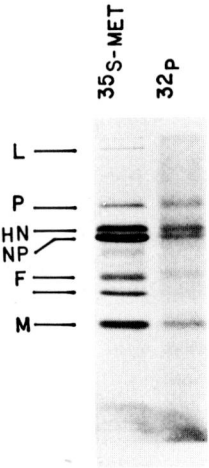

FIGURE 3. Analysis of the phosphorylated proteins of HPIV3. ^{32}P-orthophosphate labeled HPIV3 proteins were analyzed in 15% polyacrylamide gel containing 0.1% SDS. ^{35}S-methionine labeled HPIV3 proteins were coelectrophoresed for identification.

When HPIV3 was grown in the presence of ^{32}P-orthophosphate, five of the 7 major virion proteins P, HN, NP, F and M were phosphorylated (Figure 3). Of these proteins the P and M proteins were the prominently labeled. Phosphorylated proteins have been implicated as playing a role in the replication and assembly of the paramyxoviruses [9,15,16]. It therefore is not surprising to find that the nucleocapsid and envelope proteins of HPIV3 are phosphorylated. Futher study of these proteins will be important in the understanding of viral replication strategy.

Glycoproteins of HPIV3. Paramyxoviruses characteristically have at least two glycoproteins associated with their virions, the HN and F proteins. The parainfluenza type 3 viruses are no exception. Ovine, bovine and human parainfluenza type 3 viruses all have 2 glycoproteins associated with their virions (10,12,13). The larger glycoprotein has a molecular weight of 69K-73K. The smaller glycoprotein has a molecular weight of 51K-55K depending on the virus.

The glycoproteins of HPIV3 were identified by ^{14}C-glucosamine labeling. Two proteins of molecular weights 69K and 46K were predominantly labeled with the radioactive precursor (Figure 4). These correspond to the HN and F proteins respectively. Another small protein of approximately 20K molecular weight also had some incorporation of the ^{14}C-glucosamine (unpublished observation).

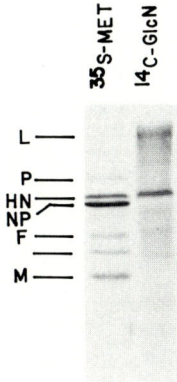

FIGURE 4. Identification of HPIV3 glycoproteins. ^{14}C-glucosamine labeled HPIV3 proteins were analyzed in 15% polyacrylamide gel containing 0.1% SDS. ^{35}S-methionine labeled HPIV3 proteins were coelectrophoresed for the reference of each protein.

The envelopes of the paramyxoviruses are associated with several enzyme activities including neuraminidase, haemagglutinin, haemolysin and cell fusion activities. These activities are attributed to the glycoproteins present on the virion surface (5,6,7). The larger glycoprotein of Sendai, SV5 and Newcastle disease viruses is associated with the hemagglutinin and neuraminidase activities, and the smaller glycoprotein is responsible for the fusion and haemolysing activities. The major glycoprotein of HPIV3 corresponds to the HN protein of Sendai and SV5 in size and in the molar quantity incorporated into the virion. Thus, the 69K molecular weight protein has been designated HN. Paramyxovirus F_0 protein has been shown to be cleaved post translationally into two products, designated F_1 and F_2, which are held together by disulfide bonds [18]. Under reducing conditions, cleavage of the disulfide bonds is expected, and 2 proteins are seen when the cleaved product is analyzed by SDS-PAGE. The samples in figure 4 were reduced with mercaptoethanol prior to electrophoresis. Therefore, the 46K molecular weight glycoprotein corresponds to the F_1 of other paramyxoviruses. A minor protein with an estimated molecular weight of 20K was constantly found in samples of HPIV3. This protein was labeled when the virion was grown in the presence of ^{14}C-glucosamine (unpublished observation). This may indicate that the 20K molecular weight protein is the F_2 protein These findings suggest that the putative F_0 protein of HPIV3 may have a molecular weight of about 66K.

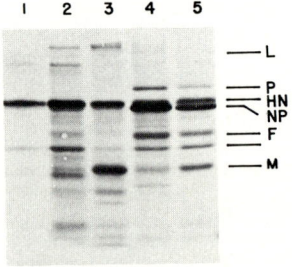

FIGURE 5. Location of HPIV3 proteins in the virion. ^{35}S-methionine labeled HPIV3 proteins were analyzed in 15% polyacrylamide gel containing 0.1% SDS after various treatments: (lane 5) total virion proteins; (lane 4) pelleted nucleocapsid protiens after triton X-100 and 1 M NaCl treatment; (lane 3) soluble envelope components after 2% triton X-100 and 1M NaCl treatment; (lane 2) envelope components after removal of M protein by dialysis to remove salt and centrifugation; (lane 1) HN protein recovered from the supernatant of high speed centrifugation to remove remaining envelope components.

Location of the Virion Proteins. The envelope proteins of SV5 and Sendai virus have been separated from the nucleocapsids by disruption with Triton X-100 in the presence of 1M KCl [6,17]. Dialysis was done to remove the KCL, resulting in the aggregation of the M protein and the aggregated protein was separated from other envelope components by centrifugation. In order to determine the location of each protein in the virion of HPIV3, a similar procedure was used. Figure 5, lane 5 represents the standard proteins of HPIV3 labeled with ^{35}S-methionine. The virions were disrupted by 2% Trition X-100 and 1M NaCl, which solubilized the HN, F and M proteins and leaves the nucleocapsids in the pellet after ultracentrifugation. Figure 5, lane 4 represents the proteins associated with the pelletable nucleocapsids. These proteins closely resemble the proteins found in the standard virus except there is a reduction in the amount of HN and M proteins. The envelope proteins are represented in lane 3. It should be noted that the L, P, and NP proteins are not present in lane 3. Thus, the partial disruption of the virions indicates that the L, P and NP proteins are associated with the nucleocapsid, and the HN and M proteins are associated with the envelope of the virion. Track 2 represents the proteins remaining in the supernatant after dialysis to remove the salt and centrifugation of the envelope components to pellet the M protein. The M protein is virtually eliminated from this sample. The HN protein can be further separated from the envelope proteins by sucrose velocity sedimentation (lane 1). The high molecular weight proteins in lanes 1, 2, and 3 may represent polymers of the HN proteins. Similar polymers have been reported for mumps virus [19,20]. It is of interest to note that the 20K molecular weight protein remains associated with the envelope proteins, and is not present in the lane with the pelleted nucleocapsid proteins. Thus, it is possible that 20K protein may represent a subunit of F_o. The F protein and the 44K protein are present in all samples shown in figure 5 with the exception of the purified HN sample. In this sample only a small amount of the 44K protein is present. The 44K protein is also associated with Sendai virus and a previous report suggested that it is a cellular polypeptide incorporated into the virion [8]. The Sendai virus polypeptide is solubilized with 2% Triton X-100 and so remains associated with the envelope proteins. The origin and function of the 44K HPIV3 protein remains to be determined. The F protein of HPIV3 has not yet been isolated in pure form, because it is very tightly associated with other envelope components of the virion.

Summary. The structural proteins of human parainfluenza type 3 were studied. These proteins were designated on the basis of the work reported here, and by comparison with the properties of other paramyxovirus structural proteins. The L protein was designated on the basis of its high molecular weight and its association with the nucleocapsid. The P and the NP proteins have been assigned on the basis of their phosphorylation and by the partial disruption of the virion, which showed that they were associated with the nucleocapsids. The HN and F proteins were identified as glycoproteins on the basis of their incorporation of ^{14}C-glucosamine and the HN was shown to be an envelope protein. The M protein was identified by its phosphorylation, association with the virion envelope and aggregation under low salt conditions. We suggest that the 20K molecular weight protein is a subunit of fusion protein. Further work is necessary to assign the identity and function of the 44K molecular weight protein.

ACKNOWLEDGMENTS

This study was supported by grant MA-7696 from the Medical Research Council of Canada.

REFERENCES

1. Parrot, R.H., Vargosko, A.J., Kim, H.W., Bell, J.A., and Chanock, R.M. (1962). Am. J. Public Health. 52, 907.
2. Hope-Simpson, R.E. (1981). J. Hygiene. 87, 393.
3. Glezen, W.P., Loda, F.A., Clyde, W.A., Senoir, R.J., Sheaffer, C.I., Conley, W.G., and Denny, F.W. (1971). J. Ped. 78, 397.
4. Mountcastle, W.E., Compans, R.W., Caliguri, L.A., and Choppin, P.W. (1970). J. Virol. 7, 47.
5. Scheid, A., and Choppin, P.W. (1975). In "Negative Strand Viruses" (B.W.J. Mahy and R.D. Barry, eds.), p. 177. Academic Press, London.
6. Lamb, R.A., and Mahy, B.W.J. (1975). In "Negative Strand Viruses" (B.W.J. Mahy and R.D. Barry, eds.), p. 65. Academic Press, London.
7. Scheid, A., and Choppin, P.W. (1974). Virol. 57, 475.
8. Lamb, R.A. Mahy, B.W.J., and Choppin, P.W. (1976). Virol. 69, 116.
9. Lamb, R.A. (1975). J. Gen. Virol. 26, 249.
10. Shibuta, H., Kanda, T., Adachi, A., and Yogo, Y. (1979). Microbiol. Immunol. 23, 617.
11. Shibuta, H., Kanda, T., Hazama A., Adachi, A., and Matamoto, M. (1981). Infect. Immun. 34, 262.
12. Morein, B., Sharp, M., Sundquist, B., and Simons, K. (1983). J. Gen. Virol. 64, 1557.
13. Guskey, L.E., and Bergtrom, G. (1981). J. Gen. Virol. 54, 115.
14. Marx, P.A., Portner, A., and Kingsbury, D.W. (1974). J. Virol. 13, 107.
15. Smith, G.W., and Hightower, L.E. (1981). J. Virol. 37, 256.
16. Lamb, R.A., and Choppin, P.W. (1977). Virol. 81, 382.
17. Scheid, A., Caliguri, L.A., Compans, R.W., and Choppin P.W. (1972) Virol. 50, 640.
18. Matsumoto, T. (1982). Microbiol. Immunol. 26, 285.
19. Merz, D.C., Server, A.C., Waxham, M.N., and Wolinsky, J.S. (1983). J. Gen. Virol. 64, 1457.
20. Herrler, G., and Compans, R.W. (1983). J. Virol. 47, 354.

CHARACTERIZATION OF STRUCTURAL PROTEINS OF PARAINFLUENZA VIRUS 3 AND mRNAs FROM INFECTED CELLS

George B. Thornton, Jayasri Roy, and Amiya K. Banerjee

Roche Institute of Molecular Biology
Roche Research Center
Nutley, New Jersey 07110

The human parainfluenza viruses, members of the paramyxovirus family, are second only to respiratory syncytial virus as causes of lower respiratory tract disease in young children. The most characteristic condition these agents produce in infants and young children is croup, i.e. swelling of the larynx leading to obstructed breathing and a hoarse cough. The epidemiological patterns and clinical manifestations of parainfluenza infections, in adults as well as children, emphasize the protective effect of neutralizing antibodies as well as lack of complete or long-lasting immunity. In order to gain insight into the molecular basis of pathogenicity of these viruses, we have initiated a project to study in detail the structure and function of the genes of parainfluenza virus type 3 (PI-3). Some recent progress has been made on the characterization of PI-3 in several laboratories (1-3).

VIRUS STRUCTURAL PROTEINS

PI-3, HA-1, strain 47885 (kindly provided by Dr. Robert Chanock, National Institutes of Health, Bethesda, MD) was plaque purified in BSC-1 cells (kindly provided by Dr. Robert Simpson, Waksman Institute, Rutgers University, New Brunswick, NJ). The virus was grown in BSC-1 cells (m.o.i. 2 Pfu/cell) in Dulbecco's modified Eagles medium from GIBCO (H21) containing 10% fetal calf serum. After 48 hrs, the virus from the supernatant was purified (4). For labeling with ^{35}S-methionine or ^{3}H-glucosamine, the medium contained one-fourth the usual level of unlabeled methionine and lacked glucose, respectively.

Fig. 1 shows analyses of PI-3 structural proteins by polyacrylamide gel electrophoresis. As shown in lane B, at least six distinct polypeptides (1 to 6) with some minor bands (shown by arrows) were visualized by autoradiography. Virus labeled with ^{3}H-glucosamine (lane C) specifically labeled polypeptides Nos. 3 and 5, indicating that these represent the viral glycoproteins. The nature of the band migrating near the origin has not been explored. By comparison with published data (1,3), these polypeptides probably represent the HN and F_1, respectively. Two additional bands (shown by arrows) may represent unreduced glycoproteins HN (top) or F_0 (bottom). The polypeptide No. 4 appeared to be the viral nucleocapsid protein (NP), since, as shown in lane E, purified nucleocapsid prepared from Triton-disrupted virions migrated at the same rate as polypeptide No. 4. The estimated molecular weights and possible identities are shown in Table 1. From these results, it seems that the molecular composition of PI-3 bears close resemblance to other members of the paramyxovirus family (5).

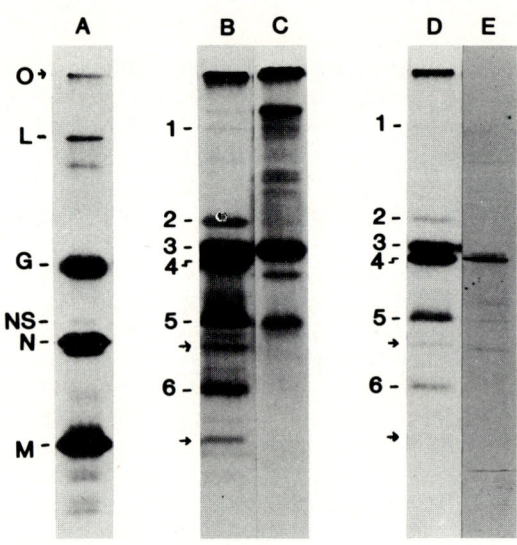

Fig. 1. Polyacrylamide gel electrophoresis of PI-3 structural proteins. (A) marker ^{3}H-methionine VSV; (B) ^{35}S- methionine PI-3; (C) ^{3}H-glucosamine PI-3; (D) ^{35}S-methionine PI-3; (E) purified RNP prepared from ^{35}S-methionine PI-3. Fluorography was carried out following electrophoresis. Nos. 1 to 6 represent PI-3 structural proteins. L,G,NS,N,M represent VSV structural proteins. Arrows indicate unidentified proteins.

TABLE 1

CHARACTERIZATION OF PI-3 STRUCTURAL PROTEINS

Protein No.	1	2	3	4	5	6
Estimated Mol.Wt. (x 10^{-3})	220	86	72	70	55	36
Glycoprotein	no	no	yes	no	yes	no
Protein Assignment	L	P	HN	NP	F_1*	M

*F_0, 68,000 (see lane C, bottom arrow, Fig. 1)

VIRUS-SPECIFIC mRNAs

To isolate virus-specific mRNAs, cells were infected with PI-3; 15 hrs and 20 hrs post-infection actinomycin D (5 µg/ml) and ^{3}H-uridine (20 µc/ml) were added, respectively. Cells were harvested at 24 hrs post-infection and total mRNA from infected cells was isolated according to the method of Glisin et al. (6). Poly(A)$^+$ and poly(A)$^-$ fractions were separated by oligo(dT)-cellulose chromatography. RNA fractions were then analyzed by electrophoresis in 1.2% agarose containing methyl-mercury as the denaturing agent (7).

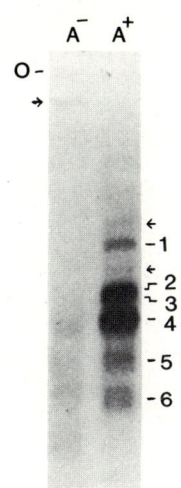

Fig. 2. Analysis of mRNAs isolated from PI-3 infected cells. ^{3}H-uridine-labeled total RNA was isolated from PI-3-infected cells and separated into poly(A)$^+$ and poly(A)$^-$ fraction by oligo(dT)-cellulose chromatography. Electrophoresis of each fraction was carried out in 1.2% agarose gel containing methyl mercury. Fluorography was carried out following electrophoresis. Nos. 1 to 6 represent PI-3 mRNAs. Arrows indicate unidentified RNA species.

As shown in Fig. 2, RNA lacking poly(A) (A$^-$ lane) contained a large RNA species (shown by arrow) with an estimated molecular weight of 5.4×10^6. This RNA species probably represents the full-length genome RNA of PI-3. The poly(A)-containing mRNA (A$^+$ lane) contains at least six mRNA species. The second prominent band contains two distinct mRNA species (Nos. 2 & 3) when seen at lower exposure. Two additional RNA species (shown by arrows) can also be seen. Using HeLa cell 28S and 18S ribosomal RNA markers in the same gel, the estimated molecular weights of each of the RNA species were calculated and are shown in Table 2.

TABLE 2

CHARACTERIZATION OF PI-3 mRNAs FROM INFECTED CELLS

mRNA No.	1	2	3	4	5	6
Estimated Mol.Wt.	1.65×10^6	1.0×10^6	0.92×10^6	0.76×10^6	0.5×10^6	0.3×10^6

TABLE 3

HYBRIDIZATION OF PI-3 GENOME RNA WITH ^{3}H-mRNA

Tube No.	^{3}H-mRNA µl	Genome RNA (0.01 µg/µl)	Treatment	Acid Precipitable (cpm)	RNase Resistance %
1	5	-	None	13,620	-
2	5	-	70°C, 2 hr + RNase	1,863	13.6
3	5	5	"	13,840	100

Hybridization was carried out in 1 x SSC, as described earlier (8).

Thus, it seems that possibly six monocistronic mRNAs were present in the infected cells. In vitro translation of each of the mRNA species should reveal the identity of the individual mRNAs. Finally, to demonstrate that the mRNA species were indeed virus-specific, the poly(A)$^+$-mRNAs were hybridized with excess of PI-3 genome RNA. As shown in Table 3, the entire ^{3}H-labeled mRNA entered into duplex as determined by resistance to digestion with RNase A & T$_1$. However, a small proportion (14%) of radioactivity self-annealed indicating the presence of possible high degree of secondary structure in the mRNAs.

ACKNOWLEDGMENT

We thank L. Guskey, University of Wisconsin, Milwaukee for generous help and advice in the purification of PI-3.

REFERENCES

1. Guskey, L.E., and Bergstrom, G. (1981). J. Gen. Virol. 54, 115.
2. Shibuta, H., Kanda, T., Adachi, A., and Yogo, Y. (1979). Microbiol. Immunol. 23, 617.
3. Morein, B., Sharp, M., Sundquist, B., and Simons, K. (1983). J. Gen. Virol. 64, 1557.
4. Banerjee, A.K., and Rhodes, D.P. (1973). Proc. Nat. Acad. Sci. (U.S.A.) 70, 3566.
5. Choppin, P.W., and Compans, R.W. (1975). In "Reproduction of Paramyxoviruses" (H. Fraenkel-Conrat and R.R. Wagner, eds.). p. 99, Vol. 4, Plenum Publishing Corp., New York.
6. Glisin, V., Crkvenjakov, R., and Byus, C. (1974). Biochemistry 13, 2633.
7. Bailey, J.M., and Davidson, N. (1976). Anal. Biochem. 70, 75.
8. Rhodes, D.P., Abraham, G., Colonno, R.J., Jelinek, W., and Banerjee, A.K. (1977). J. Virol. 21, 1105.

CHARACTERISATION OF MUMPS VIRUS
PROTEINS AND RNA[1]

E.J.B. Simpson, J.A. Curran, S.J. Martin
E.M. Hoey and B.K. Rima

Department of Biochemistry
The Queen's University of Belfast
Belfast BT9 7BL, N. Ireland

PROTEINS IN MUMPS VIRUS INFECTED CELLS

Mumps is a human paramyxovirus. We have studied the polypeptides induced after virus infection by labelling with ^{35}S-methionine, ^{35}S-cysteine, ^{14}C-amino acids, ^{32}P-orthophosphate and ^{3}H-glucosamine/mannose in order to elucidate the number, types and approximate molecular weights of the virus induced proteins. We have found in the present study (Fig. 1) and earlier (1) that the virus induces in Vero cells a large 'L' protein (mol. wt. 160K); the nucleocapsid protein N (70K), which is phosphorylated and appears susceptible to proteolysis so that it is often found as an 57-60K phosphoprotein in purified nucleocapsids; the polymerase 'P' protein (45-47K), which frequently migrates as a double band in SDS-PAGE; the matrix protein M (40K) which at least in one strain migrates as a double band (we have not yet conclusively demonstrated that the P and M proteins are phosphorylated in our virus/host system); the glycoproteins HN (80K) and FO (69K) which appears to be processed by a trypsin like enzyme to F1 (61K) and F2 (10-14K), both of which are glycosylated. These are the structural proteins of mumps virus (2,3). The unique nature of the L, HN, N, M and P proteins has been demonstrated by limited proteolysis (4,5) carried out as described by Cleveland (6), We have carried out the classical method of limited proteolysis with V8 protease as well as the strip method described in Fig. 2 which has the advantage that all the induced proteins may be analysed in one gel. Furthermore, the uniqueness of these proteins has been demonstrated by two dimensional tryptic digest peptide mapping (data not shown). These analyses have

---

[1]This work was supported by the Medical Research Council, (U.K.)

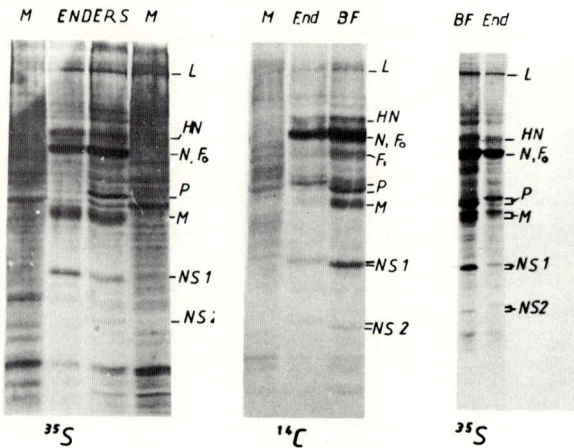

Figure 1. SDS-PAGE (8-15%) analysis of infected and mock infected (M) cell lysates radiolabelled as indicated

not been applied to the FO protein which comigrates with N, or to the F1 protein as this appears to be very poorly labelled with ^{35}S-methionine or cysteine even in pulse-chase experiments. Limited proteolysis of ^{3}H-glucosamine/mannose labelled proteins has demonstrated the distinct nature of the HN and FO and the relationship between FO and F1 (data not shown).

NON-STRUCTURAL PROTEINS

In addition to the structural proteins we have also detected two non structural proteins (1) confirmed by others (4,5) which we now propose to call NS1 and NS2. These are easily detected (Fig. 1), especially when ^{35}S-cysteine or ^{14}C-amino acids are used for NS1 labelling and ^{35}S-methionine for NS2 labelling. Limited proteolysis analysis of the whole of a first dimension gel strip (Fig. 2) indicated, besides the fact that the N protein contaminates positions above as well as below its own major band, that there is an apparent relationship between the P, and the NS1 and NS2 proteins. This has also been reported by Herrler and Compans (5). We have confirmed these data with conventional limited proteolysis for both the Enders and Belfast (BF) strains of mumps virus and it is of interest to note that the P as well as the NS1 and NS2 proteins show migration differences between these two strains and a comparison of the limited proteolysis

MUMPS VIRUS PROTEINS AND RNA 335

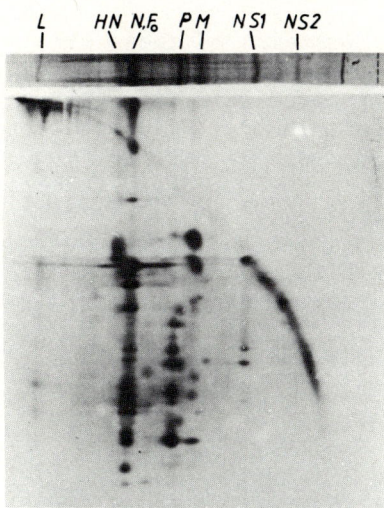

Figure 2. Limited proteolysis of whole first dimension gel strip digested with V8 protease analysed in 15% gel.

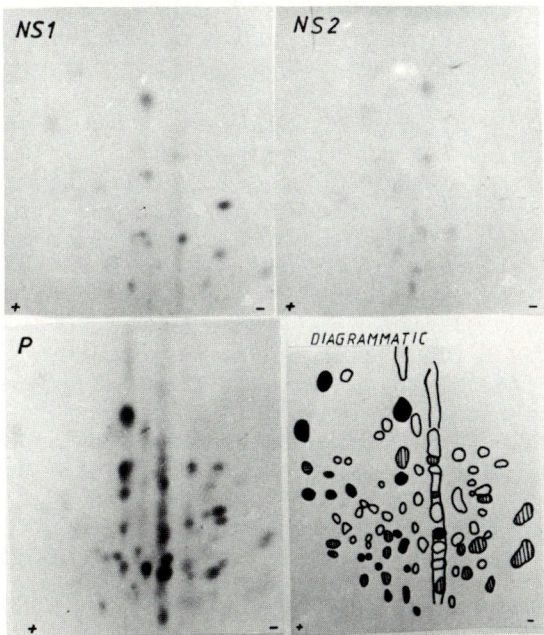

Figure 3. 2 D tryptic peptide maps of P, NS1 and NS2 plus diagram of shared spots in P, NS1 and NS2 (●); in P and NS1 (◉), in P only (O) and in P and NS2 (⊖)

patterns of the P proteins of these strains shows a difference in one band obtained after digestion with V8 protease (data not shown). It appeared more difficult to corroborate the relationship between the P, NS1 and NS2 proteins after labelling with ^{35}S-cysteine because P and NS2 label poorly with this precursor. We employed, therefore, 2D tryptic digest peptide mapping (7) to obtain a more direct proof of the relationship between the P protein and the non-structural proteins. ^{35}S-methionine labelled P, NS1 and NS2 proteins were cut out of gels and subjected to digestion, electrophoresis and chromatography and the 2D peptide maps were prepared for each of the three proteins (Fig. 3). Combinations of the P, NS1 and NS2 digests were also analysed in order to confirm the identity of spots with similar migrations. The following conclusions were drawn from the peptide maps: all the spots in the NS2 digest were present in the P protein; all except one of these were also present in the NS1 protein. All the spots in the NS1 protein were also found in the P protein digests. It thus appears that the P, NS1 and NS2 proteins contain similar tryptic peptides with NS2 peptides being found in both the larger proteins. The question is how the NS proteins are derived from P. Pulse-chase experiments with ^{35}S-methionine (data not shown) did not indicate a precursor-product relationship between P, NS and NS2. The P as well as the non-structural proteins appeared to be detectable after labelling with short pulses and were stable after chase periods. Furthermore, pretreatment of infected cells with proteolysis inhibitors such as PMSF, TPCK, TLCK, $ZnCl_2$ and iodoacetamide did not lead to the disappearance of the NS1 and NS2 bands. It seems therefore that these proteins are not derived from P by proteolytic processing and that their relationship to P may be similar to that between the P protein of Newcastle disease virus and two non-structural proteins which appear to be translated from the P mRNA and to represent the N termini of the P protein (8). Recently, it has also been shown that the non-structural C protein found in Sendai virus infected cells, which has been described as a unique gene product (7), is translated from an mRNA which comigrates with the RNA coding for the P protein of Sendai virus (9). Similarly, we have identified a non-structural 37K protein related to the P protein in measles virus infected cells (10).

MESSENGER RNA AND <u>IN VITRO</u> TRANSLATION

PolyA containing RNA labelled with 5-^{3}H-uridine in the presence of actinomycin D has been extracted from infected Vero cells by the method of Glisin (11) modified as

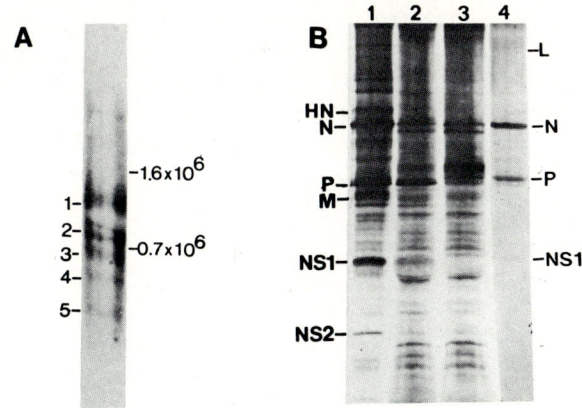

Figure 4A. Methylmercury agarose gel of ^{3}H-urd labelled polyA$^+$ RNA from mumps virus infected cells.
Figure 4B. SDS-PAGE (8-15%) of <u>in vivo</u> (1,4) and <u>in vitro</u> (2,3) synthesised proteins labelled with ^{35}S-met. Lanes 1, 2: BF-inf. cells; Lanes 3,4: Enders inf. cells.

described earlier (12). The length estimates of such RNA preparations have been made after methylmercuryhydroxide agarose gel electrophoresis by comparison to ribosomal RNA samples. Fig. 4 shows the banding pattern obtained from mumps virus infected cells. The length of the molecular species in bands 1 to 5 are approximately 3000, 2100, 1750, 1400 and 1000 nucleotides. The band of 2100 nucleotides length may actually consist of two species as it appears as a doublet in some gels. This material is the starting material for cloning experiments underway and it has also been used in <u>in vitro</u> translation experiments (Fig. 4B). The products directed by mRNA from infected cells in rabbit reticulocyte lysates are identifiable as the mumps N, P M and NS1 proteins. An extra band migrating ahead of the N protein may represent the unglycosylated form of the HN or FO protein. The identity of the P and N proteins has been demonstrated by limited proteolysis of the proteins synthesized <u>in vivo</u> and <u>in vitro</u>. The identity of the P and NS1 bands is further indicated by the mobility difference of these proteins in the BF and Enders strains which has also been observed <u>in vitro</u>.

STOECHIOMETRIC ANALYSIS

In conclusion, it appears that the NS1 and NS2 proteins are related to P but not derived from it by limited proteolysis. Provided that all proteins are stable, as they were shown to be in pulse-chase experiments, a scan of the ^{14}C labelled proteins analysed by densitometry of fluorograms on preflashed films will give an indication of the molar ratios in which various proteins are synthesised in the infected cell. When we use the rate of synthesis of the N protein as reference (1.0) the rates of synthesis of the P, NS1 and NS2 proteins in Enders infected Vero cells were 0.38, 0.38 and 0.27, respectively. In BF infected cells the figures for P, NS1 and NS2 were 0.36, 0.65 and 0.19, respectively. It is clear from these data that the NS1 and NS2 proteins are major products in the infected cell and are very prevalent. In the BF strain NS1 is made in greater copy numbers than P and the observation that the NS proteins are not derived from P by proteolysis and the fact that they appear to by synthesized in vitro allows for interesting speculation about the mechanisms that control the synthesis of different proteins from possibly only one cytoplasmic mRNA.

ACKNOWLEDGEMENTS

We thank Ms C. Lyons for assistance with the cell culture work. J.A.C. was supported by a Dept. of Education (N.I.) post-graduate scholarship.

REFERENCES

1. Rima, B.K., Roberts, M.W., McAdam, W.D., and Martin, S.J. (1980). J. gen Virol. 46, 501.
2. Örvell, C. (1978). J. gen. Virol. 41, 527.
3. McCarthy, M., and Johnson, R.T. (1980) J. gen. Virol. 46, 15.
4. Naruse, H., Nagai, Y., Yoshida, T., Hamaguchi, M., Matsumoto, T., Isomura, S., and Suzuki, S. (1981). Virology 112, 119.
5. Herrler, G., and Compans, R.W. (1982). Virology, 119, 430.
6. Cleveland, D.W. Fisher, S.G., Kirschner, M.W., and Laemmli, U.K. (1977). J. biol. Chem. 252, 1102.
7. Lamb, R.A. and Choppin, P.W. (1977) Virology 84, 469.
8. Collins, P.H., Wertz, G.W., Ball, L.A., and Hightower, L.E. (1982). J. Virol. 43, 1024.
9. Dethlefesen, L., and Kolakofsky, D. (1983). J. Virol. 46, 321.
10. Rima, B.K., Lappin, S.A., Roberts, M. W., and Martin, S.J. (1981). J. gen. virol. 56. 447.
11. Glisin, V., Crkvenjakov, R., and Byus, C. (1974) Biochemistry 13, 2633.
12. Martin, S.J., and ter Meulen, V. (1976) J. gen. Virol. 32, 321.

ANALYSIS OF THE ANTIGENIC STRUCTURE AND FUNCTION OF
SENDAI VIRUS PROTEIN NP[1]

K.L. Deshpande and A. Portner

Division of Virology and Molecular Biology
St. Jude Children's Research Hospital
Memphis, Tennessee 38101

TOPOGRAPHICAL ANALYSIS OF SENDAI VIRUS PROTEIN NP

Sendai virus contains a single-stranded RNA genome of negative polarity, which is coated by viral proteins NP, P and L to form the nucleocapsid structure (1). The functions of these proteins have yet to be assigned; however, by analogy to vesicular stomatitis virus (VSV) (2) and Newcastle disease virus (3,4), NP, the major protein species, appears to be a structural protein, while accessory proteins P and L appear to function together as the viral RNA polymerase.

The antigenic structure of NP was probed by using twelve anti-NP monoclonal antibodies in competitive-binding ELISA's (Table 1). Three antigenic sites were delineated. Sites II and III were nonoverlapping and competition was reciprocal with respect to antibodies delineating each site. Site III contains only one antibody. Antibodies to site I did not compete with antibodies to sites II and III, however, in a few instances site II and III antibodies competed nonreciprocally with antibodies to site I. These results suggest that either site I physically overlaps with sites II and III, or more likely, that the binding of antibodies to sites II or III allosterically inhibit the binding of antibodies to site I. The nonreciprocal competitions within site I may have resulted from either epitope modification after binding of the competing antibody, or from

---

[1] This work was partially supported by Public Health Service research grant AI-11949 from the National Institute of Allergy and Infectious Diseases, by CORE grant CA-21765 from the National Cancer Institute, and by ALSAC. K.L.D. is supported by a National Research Service Award (CA-09346) from the National Cancer Institute.

TABLE 1

ANTIGENIC SITE DETERMINATIONS OF SENDAI VIRUS PROTEIN NP USING PROTEIN A-PURIFIED MONOCLONAL ANTIBODIES

Antigenic Site	Antibody Isotype	Competing Antibody	Competition Reaction with Peroxidase-labeled Antibody											
			M4	M6	M10	M13	M17	M18	M19	WS16	M5	M7	S11	M8
I	IgG[a]	M4	+[b]	+	+	−[c]	+	+	+	−	−	−	−	−
	IgG	M6	+	+	+	+	+	+	−	−	−	−	−	−
	IgG	M10	+	+	+	−	+	+	+	+	−	−	−	−
	IgG	M13	+	−	+	+	+	+	+	+	−	−	−	−
	IgM	M17	+	+	+	+	+	+	−	+	−	−	−	−
	IgM	M18	+	+	+	−	+	+	+	+	−	−	−	−
	IgG	M19	+	+	+	+	+	+	+	+	−	−	−	−
	IgG	WS16	+	+	+	+	+	+	+	+	−	−	−	−
II	IgG	M5	−	−	−	−	−	−	−	−	+	+	+	−
	IgG	M7	+	−	−	−	+	−	−	−	+	+	+	−
	IgG	S11	−	−	−	−	+	+	−	−	+	+	+	−
III	IgG	M8	+	−	−	−	+	+	−	−	−	−	−	+

Note: Labeled antibodies, used at concentrations which were 50-75% saturating, were incubated with duplicate serial dilutions of competitor in the competitive-binding ELISA (5). Hybridoma anti-NP antibodies were produced by fusing SP2/0-Ag14 plasmacytoma cells with splenocytes from BALB/c mice previously immunized with the Enders strain of Sendai virus disrupted in 1.5% n-Octyl-β-d-glycopyranoside.

[a] All the monoclonal antibodies were of the K light chain class.
[b] +, indicates ≥50% competition at a concentration which caused 100% competition for its homologous labeled antibody.
[c] −, indicates <50% competition.

the location of an epitope within a groove on NP. Since site I contains the largest number of epitopes, it is probable that this part of NP is most accessible to antibody binding, and it also may be the most antigenic.

EFFECTS OF ANTI-NP ANTIBODIES ON SENDAI VIRUS TRANSCRIPTION

It is not understood how NP participates in Sendai virus transcription, therefore, we tested the effect of the anti-NP antibodies on in vitro transcription (Table 2). The following antibodies inhibited transcription by at least 50%: M4, M19, and WS16 (site I), S11 (site II) and M8 (site III).

In surveying the transcription inhibition caused by each antibody, the extent of transcription inhibition did not always correlate with antibody concentration and titer. In general, low antibody concentrations and titers [M10, M13, M17, 18 (site I) and M7 (site II)] caused negligible (10-30%) transcription inhibition. However, with antibody concentrations of approximately 1 mg/ml [M6, M19 (site I), and M5, S11 (site II)] transcription inhibition varied from 16-60%. This shows that a high concentration of antibody did not always inhibit transcription. The antibodies appear to have similar affinities for NP (from comparing plots of antibody dilution versus ELISA activity, data not shown), therefore it is likely that only specific epitopes on NP are important in inhibiting transcription. This concept is supported by studies with monoclonal antibodies to influenza virus protein NP (5) and VSV protein N (6).

KINETICS OF TRANSCRIPTION INHIBITION BY ANTI-NP ANTIBODIES

To learn more about how the anti-NP antibodies affect Sendai virus transcription, M4, M8 and S11 were examined for time dependent effects on in vitro transcription (Fig. 3). When the antibodies were added to transcribing nucleocapsids (Fig. 3B and D), complete transcription inhibition occurred in all cases. This effect was similar to that observed with antibodies to influenza virus protein NP (5) and VSV protein N (7,8), and indicates that elongation of nascent RNA was inhibited. Ribonuclease contamination of the antibodies was excluded as a possible cause for loss of transcriptase activity since labeled product RNA was stable when incubated with each of the three antibody preparations (data not shown).

In contrast, when nucleocapsids were incubated with the antibodies before substrates were added (Fig. 3A and C), M4 and M8 caused almost complete transcription inhibition,

TABLE 2

SENDAI VIRUS TRANSCRIPTION BY PURIFIED NUCLEOCAPSIDS WITH ASCITIES FLUIDS CONTAINING ANTI-NP MONOCLONAL ANTIBODIES[a]

Antigenic Site	Monoclonal Antibody	Antibody Titer[b]	Antibody Concentration in Ascites Fluid[c] (mg/ml)	Transcription Inhibition (%)[a]
I	M4	437,400	3.6	78
	M6	145,800	1.0	22
	M10	48,600	0.2	10
	M13	48,600	0.2	10
	M17	145,800	0.6	32
	M18	16,200	0.1	6
	M19	145,800	0.8	60
	WS16	437,400	5.2	58
II	M5	145,800	1.2	16
	M7	48,600	0.2	22
	S11	145,800	0.9	55
III	M8	1,312,200	7.5	75

[a]5 µl ascities fluid was reacted with 50 µg nucleocapsid in 50 µl at 30° for 5 min before addition of 50 µl substrates (9). Reaction mixtures were incubated for 2.5 hrs at 30°. Greater concentrations of ascites fluids were avoided because they caused nonspecific transcription inhibition. Similar results were obtained using protein A-purified antibodies (data not shown), indicating that any decrease in transcription was not due to nucleases in the ascites fluids.

[b]Antibody titers were determined by indirect ELISA (5).

[c]Monoclonal antibody concentrations were measured by ELISA (10).

while S11 inhibited transcription by only 50%. This result was repeated in three separate experiments and contrasts with that of Fig. 3D. Therefore, we suggest that the nucleocapsid structure undergoes a conformational change after exposure to substrates which increases susceptibility to inhibition by S11. Such a conformational change would indicate a role for changes in NP configuration in the control of Sendai virus gene expression.

Since M4 and M8 inhibited transcription to a similar extent, it is possible that the binding of each antibody affects the same site on NP. Since M8 competed nonreciprocally with M4 in the competitive-binding studies, it is possible that epitopes M4 and M8 are proximal to each other, however, due to the nonreciprocal nature of the competition we propose that M8 binding leads to transcription inhibition after causing a conformational change in the M4 epitope. Since epitope S11 (site II) was distinct from

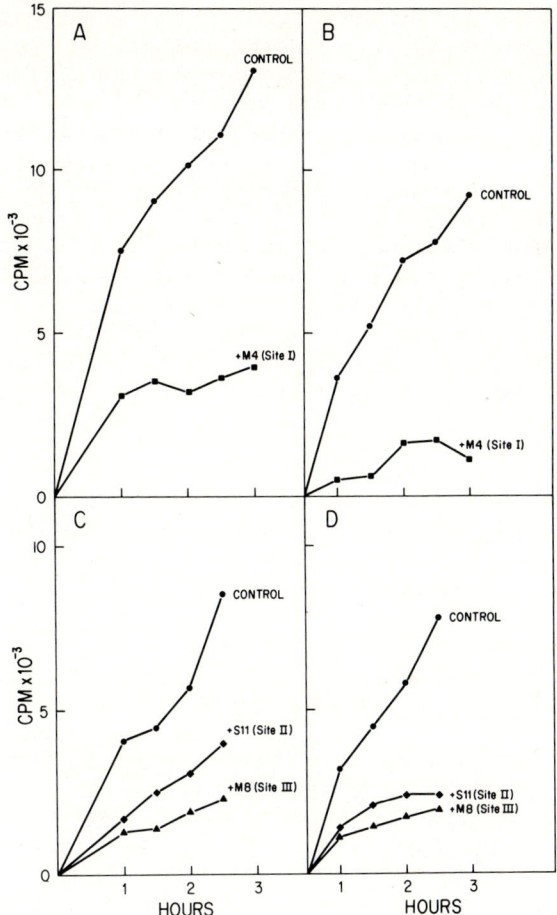

FIGURE 1. Kinetics of transcription inhibition by anti-NP antibodies in ascites fluid. (A) 18.0 µg M4 (■) were incubated with 50 µg nucleocapsids at 30° 5 min prior to the addition of substrates. (•) designates untreated control for all panels. (B) 18.0 µg M4 were added to reactions containing 50 µg nucleocapsids 25 min after transcription was initiated. ^{3}H-UTP (at 1/100 the concentration of unlabeled UTP) was added 5 min after antibody addition to measure RNA made only after antibody addition. (C) 21.0 µg M8 (▲) or 4.7 µg S11 (♦) were incubated with 30 µg nucleocapsids under the conditions of panel A. (D) 21.0 µg M8 or 4.7 µg S11 were incubated with 30 µg nucleocapsids under the conditions of Panel B. Incorporation of ^{3}H-UTP into TCA precipitable counts was determined from duplicate samples at transcription initiation and at the times indicated. Ascites fluid containing antibodies to influenza virus was used in each experiment as a negative control, and found to have no effect on transcription (data not shown).

epitopes M4 (site I) and M8 (site III) and correspondingly inhibited transcription to a lesser extent, it is possible that S11 affects transcription by a mechanism different from that caused by the binding of M4 and M8.

Possible mechanisms to explain how the anti-NP monoclonal antibodies inhibit transcription include: steric hinderance of transcriptase phassage along the nucleocapsid structure; distortion of the nucleocapsid structure by antibody interaction with the NP molecule; and hindrance of possible NP movement required for exposing the RNA template to the transcriptase. Since all monoclonal antibodies specific for NP did not affect transcription in the same manner, we may infer that epitopes differ with respect to their roles in nucleocapsid structure and/or function.

ACKNOWLEDGMENTS

We thank Ruth Ann Scroggs for skillful technical assistance and Dennis W. Metzger for production of the hybridomas and for determination of antibody isotype and subclass.

REFERENCES

1. Choppin, P.W. and Compans, R.W. (1975). In "Comprehensive Virology" (H. Fraenkel-Conrat and R.R. Wagner, eds.), Vol. 4, p. 95. Plenum Press, New York.
2. Emerson, S.U. and Yu, Y. (1975). J. Virol. 14, 1348.
3. Hamaguchi, M., Yoshida, T., Nishikawa, K., Naruse, H., and Nagai, Y. (1983). Virology 128, 105.
4. Chinchar, V.G. and Portner, A. (1981). Virology 115, 192.
5. Van Wyke, K.L., Bean, W.J., Jr., and Webster, R.G. (1981). J. Virol. 39, 313.
6. De, B.P., Tahara, S.M., and Banerjee, A.K. (1982). Virology 122, 510.
7. Carrol, A.R. and Wagner, R.R. (1978). Virology 25, 657.
8. Harmon, S.A. and Summers, D.F. (1982). Virology 120, 194.
9. Chinchar, V.G. and Portner, A. (1981). Virology 109, 59.
10. LeFrancois, L. and Lyles, D.S. (1982). Virology 121, 157.

Note added to proof: We recently have found, through Western blot analysis, that antibodies M5, M7, and S11 are specific for nucleocapsid protein P, and not NP.

MONOCLONAL ANTIBODIES AS PROBES OF THE ANTIGENIC STRUCTURE AND FUNCTIONS OF SENDAI VIRUS GLYCOPROTEINS.[1]

A. Portner
Division of Virology and Molecular Biology
St. Jude Children's Research Hospital
Memphis, TN 38101 USA

Paramyxoviruses initiate infection by the action of two envelope glycoproteins which cover the surface of the virion. The hemagglutinin-neuraminidase (HN), the larger of the two, has cell-binding, hemagglutinating and neuraminidase activities (1), while the other virus glycoprotein F is involved in virus penetration and has hemolysis and cell fusion activities (2). Antibodies to HN and F inhibit the biological activities associated with these molecules and appear to play an important role in preventing and limiting virus infection (3,4). Recent studies have defined the action of antibodies on HN (5-8) and F (8) more precisely by showing that hybridoma monoclonal antibodies differ in their ability to suppress viral functions. In the present study, we have used the interaction of monoclonal antibodies specific for HN and F of Sendai virus to investigate the role of individual epitopes in the expression of biological activity.

FUNCTIONAL ANALYSIS OF HN WITH MONOCLONAL ANTIBODIES; BIOLOGICAL ACTIVITIES OF HN MAP AT DIFFERENT EPITOPES.

We analyzed the antigenic structure of HN by competitive binding assays (ELISA) in which purified mouse hybridoma monoclonal antibodies, labeled with peroxidase, were competed with unlabeled antibodies for virus immobilized on polystyrene wells. Using ten anti-HN monoclonal antibodies, the competition studies delineated four non-overlapping antigenic sites on the HN molecule (data not shown). Antibodies used in our analysis bound to their respective epitopes in enzyme immunoassays (EIA) with similar titers

---

[1]This work was supported by Public Health Service Research GRant AI-11949 from the National Institute of Allergy and Infectious Diseases, and by CORE CA-21765 from the National Cancer Institute.

($\approx 10^{-5}$) and binding curves, suggesting that antibody concentrations and functional affinities of different monoclonal antibody preparations were about the same.

To relate the biological activities of HN to the antigenic structure, and thus, physically distinct regions on the HN molecule, anti-HN antibodies were analyzed for their reactivity in neuraminidase (NI), hemagglutination (HI) and hemolysis inhibition (HLI) tests (Table 1). In these tests, antibodies binding to topologically distinct sites exhibited unique patterns of reactivity. Thus, antibodies binding to HN antigenic site I inhibited hemagglutination and neuraminidase activities, whereas antibodies to site II inhibited neither activity. Antigenic sites III and IV had different roles in hemagglutinating and neuraminidase activities: antibodies recognizing site III inhibited hemagglutination but were devoid of NI activity, while antibodies reacting with antigenic site IV displayed the reverse pattern, inhibiting neuraminidase activity but not hemagglutination. The observed differences between the effects of the different antibodies clearly depends on the region of the HN molecule to which the antibody binds.

Although hemolysis activity (a measure of fusion) is an F glycoprotein function, anti-HN antibodies may indirectly inhibit hemolysis by inhibiting virus adsorption which is a prerequisite for hemolysis. Recent reports (7,8), however, indicate that anti-HN antibodies may inhibit hemolysis by other mechanisms. Therefore, we tested anti-HN antibodies to antigenic sites II and IV which lack HI activity to determine whether hemolysis inhibition occurs with antibody-treated virions not blocked in absorption to erythrocytes. Hemolysis inhibition was observed by anti-HN antibodies reacting with antigenic site II but not with site IV. This indicates that antibodies binding specifically to HN site II can interfere with F-mediated fusion, and thus prevent virus penetration. As expected, antibodies to antigenic sites I and III inhibit adsorption and therefore inhibit hemolysis.

To determine whether the antigenic structure of HN was different on the surfaces of infected cells and virions, we tested the binding of anti-HN antibody by EIA and immunofluorescence (IF). In both cases, anti-HN binding was identical, indicating that the antigenic structure of HN was unaltered by the budding process (Table 1). However, as we will show, this is not the case with the F protein.

ANALYSIS OF F GLYCOPROTEIN FUNCTION AND STRUCTURE WITH MONOCLONAL ANTIBODIES.

Competitive binding assays delineated three antigenic sites on the F molecule. Sites I and II were non-over-

TABLE 1.
REACTIVITY OF ANTI-HN MONOCLONAL ANTIBODIES WITH SENDAI VIRIONS AND INFECTED CELLS[a]

Antigenic Site	Virions				Infected Cells		
	EIA[e]	HI[b]	NI[c]	HLI[d]	EIA[e]	Fluorescence Surface	Cyto
I	+	+	+	+	+	+	+
II	+	-	-	+	+	+	+
III	+	+	-	+	+	+	+
IV	+	-	+	-	+	+	+

[a] Assays were performed using a 1/100 dilution of hybridoma ascites fluid for the HI and NI and 1/200 for HLI.

[b] +, antibody completely blocked hemagglutination; -, antibody had no effect on hemagglutination.

[c] +, antibody inhibited viral neuraminidase activity by >75% (O.D. 549 nm); -, neuraminidase activity was the same as samples lacking antibody.

[d] +, complete inhibition of hemolysis activity; -, hemolysis activity was the same as virus samples without antibody.

[e] +, virions EIA titer >10^5; +, infected cells O.D. 3-fold above cells without antibody; -, the same as cells not treated with antibody.

TABLE 2.
ACTIVITY OF F MONOCLONAL ANTIBODIES WITH VIRIONS AND INFECTED CELLS

Antigenic Site	Antibody Clone	Virions				Infected Cells		
		EIA	HI	NI	HLI	EIA	Fluorescence Surface	Cyto
I	16	+	-	-	+	+	+	+
	28	+	-	-	-	-	-	+
	31	+	-	-	-	-	-	+
II	33	+	-	-	-	-	-	+
III	38	+	-	-	-	-	-	+

Antibody dilutions and +, - designations the same as Table 3.

lapping whereas site III was blocked non-reciprocally by site I antibodies (data not shown). As expected, hemagglutination and neuraminidase activities were not blocked by anti-F antibodies, since these functions are associated with HN (Table 2). To determine whether specific epitopes on the F molecule are involved in fusion activity, anti-F antibodies were incubated with Sendai virions and tested for inhibition of hemolysis. Only a single antibody, M16, binding to antigenic site I, inhibited hemolysis. This is a surprising result, since two other anti-F antibodies mapping at the same antigenic site failed to inhibit hemolysis. The results indicate that the M16 epitope may be directly involved in hemolysis activity or the binding of antibody M16 induces a conformational change in F that alters the hemolysis active site. The results also suggest that the hemolysis active site, and thus, the fusion site, may be restricted to a distinct region on an F molecule.

To compare the antigenic structure of F in infected cells and virions, we measured the binding of anti-F antibodies to virions by EIA and to infected cells by immunofluorescence and EIA (Table 2). A striking result was that only M16 binding was detected on the surface of infected cells, whereas all of our anti-F antibodies bound to whole or disrupted virions. The results suggest that some F epitopes were sequestered on the surface of the infected cell and became uncovered by the budding process.

THE RELATIONSHIP BETWEEN ANTIBODY INHIBITORY ACTIVITY AND NEUTRALIZATION OF INFECTIVITY.

We have found that monoclonal antibodies of defined epitope specificity selectively inhibit HN and F functions. To determine the relationship between these properties and the ability of anti-HN and F antibodies to inhibit virus growth, we measured virus yields (Table 3). Virus was mixed with antibodies before adsorption or added to culture fluids after virus penetration and HA titers measured at 48 h after infection. Anti-HN antibodies to antigenic sites I, II and III inhibited virus production whether added before or after virus penetration. Since antibodies to sites I and III inhibited hemagglutination, it is likely that they block virus adsorption. Antibodies to HN site II failed to inhibit hemagglutination and neuraminidase, but inhibited hemolysis, and thus, may prevent virus penetration. An interesting finding was that anti-HN antibodies to site IV did not inhibit virus adsorption but did inhibit virus production after the penetration step. The fact that neuraminidase was the only other activity inhibited by these antibodies suggests that the viral enzyme is involved in virus release.

Of five anti-F antibodies to three different antigenic sites, the only antibody that inhibited hemolysis, M16, was also the only antibody that inhibited virus growth. The results are consistent with the known function of F in virus penetration (4) and supports the concept that fusion

TABLE 3
THE EFFECT OF ANTI-HN AND F MONOCLONAL ANTIBODIES ON THE GROWTH OF SENDAI VIRUS

Antibody designation	Group	Virus Yield (HA/ml) When Antibody Added[a]		Protein specified by antibody
		Before Adsorption[b]	After Penetration[c]	
S2	I	<3	<3	HN
16		<3	<3	HN
M2	II	<3	<3	HN
11		<3	2-3	HN
M9	III	<3	<3	HN
M12	IV	41	<3	HN
14		41	<3	HN
15		41	<3	HN
20		41	<3	HN
21		41	<3	HN
M16	I	<3	≈5	F
28		54	41	F
31		41	41	F
M33	II	54	41	F
M38	III	54	27	F
M1		41	41	M
MAF		41	41	

[a]HA titers at 48 h after infection.
[b]Before adsorption to CEL cell virus inoculum (0.5 HA in 0.1 ml) was incubated for 30 min with 10 µl of undiluted antibody in ascitic fluid.
[c]CEL cells were infected for 4 h to allow time for virus adsorption and penetration. Cells were then washed 3 x with PBS, and 3.0 ml of GM added containing 30 µl of undiluted ascitic fluid.

activity may be limited to a distinct region on the F molecule (9).

These studies indicate that anti-HN and F monoclonal antibodies bind to distinct regions on the HN and F proteins, and this binding specificity results in selective inhibition of different HN and F functions. It is noteworthy that anti-HN antibodies, besides blocking virus absorption, may interfere with infectivity by blocking virus fusion or inhibit virus release by blocking neuraminidase activity. The inhibition of fusion may be a steric phenomenon in which antibodies to HN site II interfere with the function F. Another possibility is that HN undergoes a conformational change before F can be expressed and antibodies to HN site II prevent this change. The biological importance of HN neuraminidase activity is unknown. These studies indicate that it is required for virus release. With regard to the F protein, hemolysis and infectivity are inhibited by anti-F antibodies that bind to a distinct epitope. This may be the fusion active site. An interesting finding was that the antigenic structure of F is apparently different on the surface of infected cells and virions. This result implies that the F protein undergoes a conformational change during or after virus release.

ACKNOWLEDGEMENTS

Ruth Ann Scroggs provided skillful technical assistance. This work was supported by ALSAC.

REFERENCES

1. Schied, A., Caliguiri, L.A., Compans, R.W., and Choppin, P.W. (1972). Virology 50, 640.
2. Homma, M. and Ohuchi, M. (1973). J. Virol. 12, 1457.
3. Schied, A. and Choppin, P.W. (1974). Virology 57, 475.
4. Merz, D.C., Schied, A., and Choppin, P.W. (1981). Virology 109, 94.
5. Portner, A. (1981). Virology 115, 375.
6. Yewdell, J., and Gerhard, W. (1982). J. Immunol. 128, 2670.
7. Miura, N., Uchida, T., and Okada, Y. (1982). Exp. Cell. Res. 141, 409.
8. Örvell, C. and Grandien, M. (1982). J. Immunol. 129, 2779.
9. Richardson, C.D., Scheid, A., and Choppin, P.W. (1980). Virology 105, 205.

VARIATIONS IN ANTIGENIC DETERMINANTS OF DIFFERENT STRAINS OF MEASLES VIRUS

Hooshmand Sheshberadaran, Erling Norrby, and Shou-Ni Chen

Department of Virology
Karolinska Institute, School of Medicine
Stockholm, Sweden

Measles is a serologically stable monotypic virus. Polyvalent sera do not distinguish surface antigens nor internal structural proteins of different measles virus strains in acute nor in persistent infections. We have recently prepared monoclonal antibodies against five of the six virus-coded structural proteins of measles, the exception being the L protein (1, 2) and used them to characterize nine strains of measles virus giving lytic infections in vitro (3). In order to obtain a broader view of the degree of antigenic variation of the individual viral components, these monoclonal antibodies were used to epitopically characterize five persistent measles virus infections and additional SSPE-derived lytic strains not included in the previous study (3).

All the hybridomas with the exception of seven were raised against the LEC strain (referred to as LEC-KI) which has received numerous passages in our laboratory for over 10 years. The remaining seven (3 anti-hemagglutinin and 4 anti-P protein) although raised against the Edmonston (EDM) strain cross-reacted strongly with the LEC-KI strain. The persistently infected cell lines studied were: HEpPi (4) and Lu106 carrier (5), established in HEp-2 and Lu 106 cells respectively with the EDM strain; MaSSPE (6) and MaPi, established in different laboratories with the Mantooth (MAN) strain; and HNT (7), established in Vero cells with a hampster neurotropic variant of Philadelphia 26 strain. HNT is nonproductive; the remaining four being semi-productive i.e. produce moderate to low levels of plaquable virus. The lytic measles virus strains: Woodfolk (WDF), MVO, MVP, Hu2, Schwartz (SCH), Moraten (MOR), EDM-dilute passaged (EDM/DP) and MAN; have been described previously (3). The five addi-

---

[1]Present adress: Institute of Sera and Vaccines, Chengtu, Szechwan, Peoples Republic of China.

tional SSPE-derived lytic strains used in studies on the matrix protein were: Dean (DN), Zister (ZIS), Halle (HAL), McClellan (MAC) and passage 5 of the original LEC isolate (LEC-WIS) (8).

The SSPE lytic strains were all examined in their 5-10th passage after isolation. Conditions of radioimmune precipitation (RIPA), immunofluorescence (IF) and other tests were as described previously (1-3).

VARIATION IN EPITOPES ON THE SURFACE PROTEINS

The nine lytic strains studied previously (3) had shown only limited variations in their hemagglutinin (H) and no variations in their fusion (F) proteins. Limitation of variation within these proteins probably arises from functional

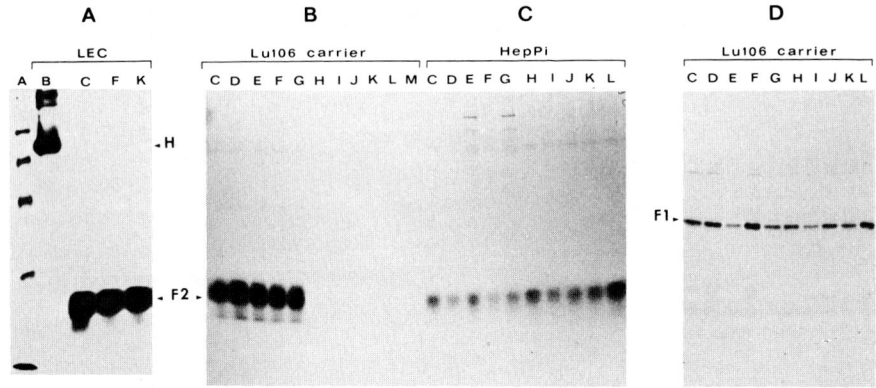

FIGURE 1. Analysis of F protein from LEC virus infected cells (A), Lu106 carrier (B and D) and HEpPi (C) persistent infections by RIPA. Infected cells were labelled either with ^{35}S-methionine (D) or with a mixture of 1,6-^{3}H-glucosamine hydrochloride and 2-^{3}H-mannose (A, B and C). Immunoprecipitates were analyzed on 12.5% polyacrylamide gels. Anti-F monoclonal antibodies used were: lane C, 9-DB10; 16-AE7; E, 16-AG5; F, 16-DC9; G, 16-EE8; H, 19-BG4; I, 19-FF4; J, 19-FF10; K, 19-GD6; L, 19-HHB4; and M, 19-HC4. An anti-H monoclonal antibody, I-44 (lane B) and mol.wt. marker proteins (lane A) are also included. The concomitantly immunoprecipitated F1 is not seen in A, B and C as it is non-glycosylated; similarly F2 is to weakly labelled by ^{35}S-methionine to be seen in D.

TABLE 1
REACTIVITY OF TEN ANTI-H MONOCLONAL ANTIBODIES WITH DIFFERENT MEASLES VIRUS STRAINS IN RIPA AND IF[a]

Hybridoma Clone No.	Lytic				Persistent	
	LEC-KI	EDM/DP	WDF, MAN	SCH, MOR, MVO, MVP, Hu2	MaSSPE, MaPi, HEpPi, Lu106 carrier	HNT
7-AG11	+	−	−	−	−	−
16-CD11	+	+[w]	+	+	+	+
16-DG3	+	+	−[b]	+	−[b]	−
Others[c]	+	+	+	+	+	d

[a] Grading of reactivity: +, distinct; +[w], weak; and −, none
[b] + in IF
[c] Hybridoma no.: I-12, I-29, I-41, I-44, 80III B2-D, 79V V17-D and 79XI C2-D
[d] + with all except 80III B2-D and 79V V17-D in both RIPA and IF (9)

constraints. Accumulation of mutations in the surface antigens in persistent infections however, could occur due to reduction of selective pressure to keep the protein functional in such systems. This appears to hold true in some instances such as: the H protein of HNT (Table 1; pattern of reactivity of the lytic strains (3)are included for comparison), the most defective of the five persistent infections studied; and the F protein of Lu106 carrier. In the latter case, two forms of the F protein were detected: a weakly glycosylated form, which reacted with all 11 anti-F monoclonal antibodies; and a strongly glycosylated form, which reacted with 5 of the 11 anti-F monoclonal antibodies (Figure 1). The differential reactivity of the two forms of the Lu106 carrier F protein with the anti-F monoclonal antibodies was confirmed by sequential RIPA wherein an antigen batch was repeatedly immunoprecipitated with one anti-F antibody until no further reactivity was observed and the resulting depleted antigen examined with another of the anti-F antibodies (data not shown). Such distinctly different forms of F were not detected in MaSSPE, MaPi and HEpPi (radioactive sugar labelled HNT F protein could not be studied).

VARIATION IN EPITOPES ON THE INTERNAL PROTEINS

Nucleocapsid (NP) and P Proteins. By analogy with other negative strand viruses, the L, NP and P proteins of measles

TABLE 2
REACTIVITY OF MEASLES VIRUS PERSISTENTLY INFECTED CELL LINES WITH ANTI-NP MONOCLONAL ANTIBODIES IN RIPA AND IF[a]

Hybridoma	Persistent Infection		
	MaPi	HNT	MaSSPE, HEpPi, Lu106 carrier
16-AC5	+	+	+
16-BB8	−[b]	−	+
16-BF2	+	+	+
16-CF7	+	−	+
16-DC3	−[b]	+	+
16-EE9	+	+	+

[a] Grading of reactivity: +, distinct; $+^w$, weak; and −, none
[b] + in IF

virus are thought to compose the transcriptase/replicase complex. Consequently alterations within them would be expected to be limited in lytic systems. However, mutations within these proteins may be of importance in evolution of persistence. Although the biological significance of the variations detected are unknown, it is of interest that in contrast to the nine lytic strains previously examined wherein no major variations were detected by the six anti-NP and six anti-P monoclonal antibodies, epitopic variations were found in two of the persistent infections. By RIPA and IF it was shown that the anti-P monoclonal antibodies detected two major variations in the HNT P protein; while the anti-NP monoclonal antibodies detected variations in the NP of HNT and of MaPi (Table 2).

Matrix (M) Protein. It has previously been shown that in contrast to the surface and the internal NP and P proteins, extensive epitopic variations are found in the M protein of measles strains giving lytic infections in vitro (3) (RIPA data included for comparison in Table 3). The persistent infections also showed variations in M epitopes (Table 3). Also shown in Table 3 are the M epitopic patterns of five additional SSPE-derived strains (the previous study examined only the MAN SSPE strain). Features of interest which emerged from this study include : 1) There was a lack of consensus of M epitopic patterns between the persistent infections. 2) A high consensus existed in the lytic SSPE M epitopic patterns as 5 out of 6 strains had the same pattern. However the fresh isolates MVO and MVP also had a similar pattern of reactivity. 3) Lytic SSPE LEC-WIS differed markedly from its derivative LEC-KI. The two strains also differed in their H epitopes (data not shown). As the passage history of LEC-KI is unclear it is not possible to assess whether the variations arose due to extensive lab-adaptation or to contamination with another strain. 4) MaSSPE and MaPi persistent infections showed the same M epitopic patterns as lytic SSPE MAN strain, the strain from which they were derived; indicating stability in M epitopic patterns. Indications of stability were also found in another study where the M epitopic patterns of LEC-KI, EDM/DP and Hu2 were seen to remain unchanged after 30 consecutive dilute passages in vitro. Furthermore lytic SSPE MAN strains from three different laboratories (two in US and one in Germany) have all been found to have the same M epitopic pattern.

The results of this study indicate that in contrast to lytic systems, the H, F, NP and P proteins of measles virus accumulate mutations in persistent infections. Furthermore the apparent conservation of M epitopic patterns in different

TABLE 3
REACTIVITY OF DIFFERENT MEASLES VIRUS STRAINS WITH ANTI-M MONOCLONAL ANTIBODIES IN RIPA[a]

Hybridoma			Vaccine	Lab	Fresh Isolates			SSPE		Persistent	
Group No.	Clone No.	LEC-KI	SCH, MOR	EDM/DP	WDF	Hu2	MVO, MVP	LEC-WIS	DN, ZIS, HAL, MAN, MAC	MaSSPE, MaPi	Lu106 carrier, HEpPi	HNT[b]
1	10-EF10	+	+	+	+	+	+	+	+	+	+	−
2	16-BB2	+	+	+	+	+	+	+	+	+	+	−
3	19-AG10	+	+	+	+	+	+	−	+	+	+	+
4	19-CC6 19-GF6	+	+	−	+	+	+[w]	+	+[w]	+[w]	+	+[w]
5	19DC5	+	+	+	+	+	−	+	−	−	+	+
6	19-DF10 19-HC5 19-HF6	+	+	+	+	−	+	−	+	+	+	+

[a] Grading of reactivity: +, distinct; +[w], and −, none
[b] Data are preliminary due to difficulties in reproducibly radiolabelling HNT M protein

systems indicates that with the aid of a larger bank of monoclonal antibodies the M epitopic patterns could be used as a marker for particular measles virus strains. Whether a correlation with particular biological properties of groups of strains will also emerge remains unknown.

ACKNOWLEDGEMENTS

We wish to thank Drs. D. McFarlin and K.W. Rammohan, NIH, Bethesda, Md. for seven of the monoclonal antibodies and three of the persistent infections; and Drs. J. Sever, NIH, Bethesda, Md. and H. Koprowski, The Wistar Institute, Philadelphia for the lytic SSPE virus strains.

REFERENCES

1. Togashi, T., Örvell, C., Vartdal, F., and Norrby, E. (1981). Arch. Virol. 67, 149.
2. Norrby, E., Chen, S.-N., Togashi, T., Sheshberadaran, H., and Johnson, K.P. (1982). Arch. Virol. 71, 1.
3. Sheshberadaran, H., Chen, S.-N., and Norrby, E. (1983). Virology. 128, 341.
4. Hayes, E.C. and Zweerink, H.J. (1978). In "ICN-UCLA Symposium of Molecular and Cellular Biology II". (J. Stevens, G. Todaro, and C.F. Fox, eds.)., p. 759. Academic Press, New York.
5. Norrby, E. (1967). Arch. ges. Virusforsch. 20, 215.
6. Dubois-Dalcq, M., Reese, T.S., Murphy, M., and Fucillo, D. (1976). 19, 579.
7. Rammohan, K.W., McFarland, H.F., Bellini, W.J., Ghenens, J., and McFarlin, D.E. (1983). J. Infect. Dis. 147, 546.
8. Barbanti-Brodano, G., Oyanagi, S., Katz, M., and Koprowski, H. (1970). Proc. Soc. Exp. Biol. Med. 134, 230.

POSITIVE IDENTIFICATION AND MOLECULAR CLONING OF THE PHOSPHOPROTEIN (P) OF MEASLES VIRUS

William J. Bellini, George Englund, Chris D. Richardson,
R. Nick Hogan, Shmuel Rozenblatt, Chester A. Meyers[1]
and Robert A. Lazzarini

Laboratory of Molecular Genetics, IRP
National Institute of Neurological and Communicative
Disorders and Stroke, NIH
Bethesda, Maryland 20205

IDENTIFICATION OF A P GENE cDNA CLONE

Measles virus, a clinically relevant infectious agent of man, contains approximately 16 Kb of genetic information within a single strand, negative polarity, RNA genome. The virus readily gives rise to persistent infections and is believed to be the causative agent for the progressive neurological disease, subacute sclerosing panencephalitis (1). Because of a longstanding interest in the molecular mechanisms of persistence, we have undertaken the molecular cloning of measles virus.

Rozenblatt et al. (2) have identified several cDNA clones of measles virus derived from oligo dT primed mRNA. One clone, Cl-G, appeared by Northern blot analysis to hybridize to a single species of mRNA, but upon _in vitro_ translation, this selected message resulted in the synthesis of two polypeptides. One of these proteins was immunoprecipitated by an anti-HA (hemagglutinin) monoclonal antibody. Thus, the clone was assigned on this basis as an HA clone.

Clone Cl-G was obtained and was sequenced by the method of Maxam and Gilbert (3). The complete sequence of Cl-G indicated that this clone contained 400 nucleotides of measles specific information and terminated in a polyadenylated region followed by GC tails. Thus, it appeared that Cl-G represented the precise 3' end of the mRNA that was reverse transcribed during the cloning event. The deduced protein sequence in all three phases indicated only a

---

[1]Present address: Laboratory of Chemoprevention, NCI, NIH, Bethesda, Maryland 20205.

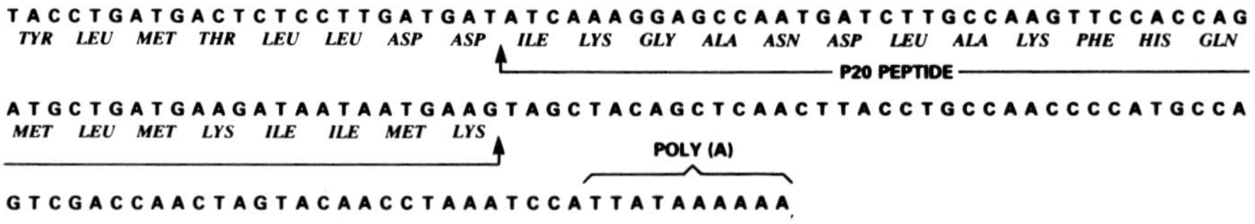

```
TACCTGATGACTCTCCTTGATGATATCAAAGGAGCCAATGATCTTGCCAAGTTCCACCAG
 TYR  LEU  MET  THR  LEU  LEU  ASP  ASP  ILE  LYS  GLY  ALA  ASN  ASP  LEU  ALA  LYS  PHE  HIS  GLN
                                        ▲─────────────────── P20 PEPTIDE ───────────────────
ATGCTGATGAAGATAATAATGAAGTAGCTACAGCTCAACTTACCTGCCAACCCCATGCCA
 MET  LEU  MET  LYS  ILE  ILE  MET  LYS  ▲           POLY (A)
GTCGACCAACTAGTACAACCTAAATCCATTATAAAAAA
```

FIGURE 1. Nucleotide and peptide sequence of C1-G used for synthetic peptide synthesis.

single open reading frame. Upon analysis, this protein sequence did not appear to contain a hydrophobic region that was expected to be present near the carboxy terminus of transmembrane glycoproteins. To positively identify the gene product of C1-G, a 20 amino acid peptide was synthesized from the sequence deduced for the carboxy terminal region (Figure 1). The peptide (P-20) was purified by HPLC, linked to keyhole limpet hemocyanin and used as an immunogen in rabbits. Rabbit antisera were purified by affinity chromatography using the P-20 peptide coupled to a solid support resin. The affinity purified anti-P-20 was compared with two anti-measles sera and several monoclonal antibodies of determined specificity in immunoprecipitation assays. Cell lysates of measles infected Vero cells were prepared as antigen following radioactive labeling with ^{35}S-methionine (4). Analysis of the immune precipitates by SDS-polyacrylamide gel electrophoresis indicated that the affinity purified P-20 antibody precipitated a single polypeptide that comigrated with the protein precipitated by the monoclonal anti-P of measles virus (Figure 2). The P protein was also precipitated by both polyvalent anti-measles sera and was clearly distinguishable from the two forms of the hemagglutinin (HA) precipitated by these sera as well as the monoclonal anti-HA.

To determine whether the protein recognized by the P-20 antibody was a phosphoprotein, immunoprecipitation of phosphorous-32 labeled measles infected cell lysates was performed. Polyacrylamide gel analysis of the immunoprecipitates of anti-P-20 serum, affinity purified P-20 antibody and monoclonal anti-P resulted in the precipitation of the identical ^{32}P-labeled protein (Figure 2). In addition, immunofluorescent studies with anti-P-20 indicated that the antibody reacted with an internal antigen of measles infected cells. The antibody stained nucleocapsid inclusions and colocalized with the antigen recognized by anti-P monoclonal antibody.

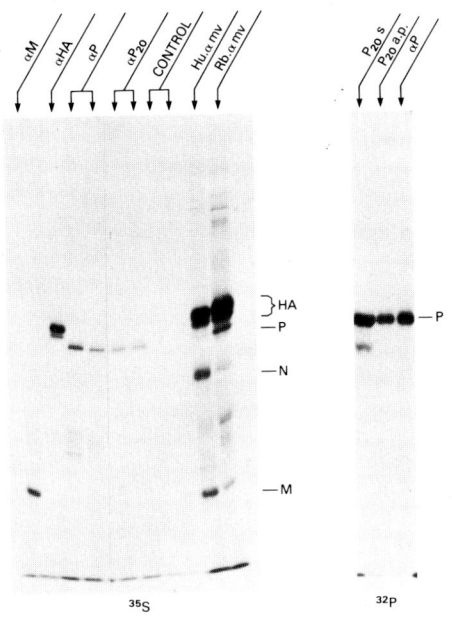

FIGURE 2. Autoradiograph of SDS-PAGE analysis of immunoprecipitated proteins from either ^{35}S-methionine labeled measles infected cell lysates (left panel) or ^{32}P-labeled measles infected cell lysates (right panel). Antisera include α M (anti-matrix protein), α HA (anti-hemagglutinin), α P (anti-phosphoprotein), α P$_{20}$ (anti-synthetic peptide, affinity purified), control (antibody not bound to affinity column from P$_{20}$ antiserum), Hu α mv (human antibody to measles virus following atypical measles), Rb α mv (rabbit anti-measles virus), P$_{20}$ s (anti-synthetic peptide serum) and P$_{20}$ a.p. (anti-synthetic peptide, affinity purified).

Clone C1-G and a related clone, 3A-8, identified by probing the measles cDNA library derived from 50s genomic RNA with C1-G, were used in hybrid arrest studies (Figure 3). Poly-A selected mRNA from measles infected Vero cells was translated in vitro using rabbit reticulocyte lysates. The assignment of the P and N (nucleocapsid) proteins was based on immunoprecipitation studies (data not shown). Increasing DNA concentrations of either C1-G or 3A-8 resulted in essentially total arrest of the translation of the P protein compared with the mock arrested mRNA preparations. The immunoprecipitation data coupled with that of hybrid arrest and immunofluorescent localization firmly establish that the nucleotide sequence contained within C1-G is measles specific and represents the coding region for the carboxy terminus of the P protein.

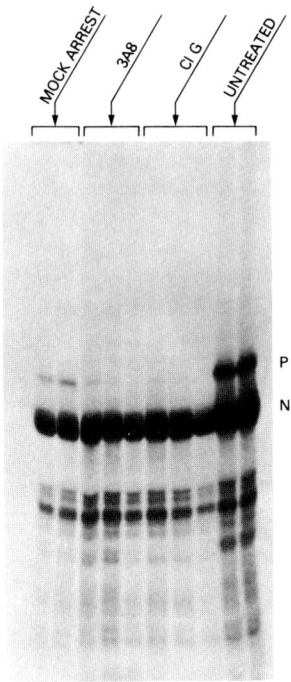

FIGURE 3. Autoradiograph of SDS-PAGE analysis of hybrid arrest and translation. Untreated, poly A selected mRNA (3 µg) was translated in vitro and the resultant peptides immunoprecipitated with rabbit anti-measles serum, mock arrest, same as untreated except taken through arrest conditions (5) with no competing DNA, 3A-8, poly A selected mRNA arrested with 1, 2 and 10 µg of 3A-8 DNA, C1-G poly A selected mRNA arrested with 5, 10, and 20 µg of C1-G DNA.

CLONING AND CHARACTERIZATION OF THE P GENE

Clones C1-G and 3A-8 were employed as probes to select a number of related clones from the 50s genomic cDNA library. Subsequent probing studies of the library have resulted in a number of cDNA clones which have been sequenced, ordered and encompass the entire P gene (Figure 4). The entire sequence of the P gene will be published elsewhere. The P cistron is comprised of 1653 nucleotides with a coding region of 506 amino acids, if the first methionine is used. Suggestive of a phosphoprotein the protein contains 15% serine plus threonine residues. This value is in good agreement with that of the P protein of Sendai virus (Kolakofsky, personal communication).

Clone 9D-8 was found to contain the 3' end of the P gene as well as to hybridize to the nucleocapsid clone, C1-15 (6). Clone 3A-8 contains the 5' end of the P cistron and extends approximately 350 nucleotides into the adjacent gene. Preliminary evidence from immunoprecipitation studies with anti-peptide serum to a region of this adjacent gene indicates that it is the M cistron (data not shown). Nucleotide sequence determination indicated an exact homology for 17 nucleotides in these boundary regions with remarkable similarity with that known for Sendai virus (7, 8, Kolakofsky, personal communication) (Table 1). By comparison with Sendai, the polyadenylation region for measles appears to be ...AAUAUUUUUUU... and the tri-nucleotide ...GAA..., the intercistronic region. Thus, it would appear that these boundary regions are highly conserved for these paramyxoviruses.

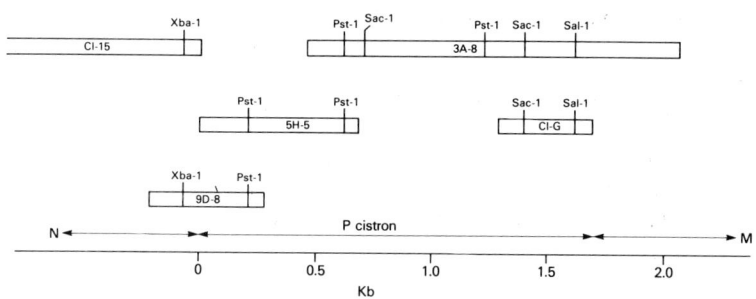

FIGURE 4. Schematic representation of cDNA clones covering the entirety of the P cistron.

TABLE 1

POLYADENYLATION AND INTERCISTRONIC BOUNDARIES
OF TWO PARAMYXOVIRUSES

Virus	Sequence
Measles	3' AAUAUUUUUU GAA UCCU..... 5'
Sendai	3' AUUCUUUUU GAA UCCC..... 5'

ACKNOWLEDGMENTS

The authors gratefully acknowledge Drs. K. W. Rammohan and D. E. McFarlin for the monoclonal antibodies to the P protein and Dr. W. Bohn for monoclonal antibody to the matrix protein. We thank Dr. Kolakofsky for sharing unpublished information concerning Sendai virus.

REFERENCES

1. Horta-Barbosa, L., Fuccillo, D. A., Sever, J. L., and Zeman, W. (1969). Nature 221, 974.
2. Rozenblatt, S., Gesang, C., Lavie, V., and Neumann, F. S. (1982). J. Virol. 42, 790.
3. Maxam, A. M., and Gilbert, W. (1980). Methods in Enzymol. 65, 499.
4. Lamb, R. A., Etkind, P. R., and Choppin, P. W. (1978). Virol. 91, 60.
5. Paterson, B. M., Roberts, B. E., and Kuff, E. L. (1977). P.N.A.S. (USA) 74, 4370.
6. Gorecki, M., and Rozenblatt, S. (1980). P.N.A.S. (USA) 77, 3686.
7. Gupta, K. C., and Kingsbury, D. W. (1982). Virol. 120, 518.
8. Dowling, P. C., Giorgi, C., Roux, L., Dethlefson, L. A., Galantowicz, M. E., Blumberg, B. M., and Kolakofsky, D. (1983). P.N.A.S. (USA) 80, 5213.

IDENTIFICATION OF A NEW ENVELOPE-ASSOCIATED PROTEIN
OF HUMAN RESPIRATORY SYNCYTIAL VIRUS[1]

Y. T. Huang, P. L. Collins, and G. W. Wertz

Department of Microbiology and Immunology
School of Medicine
University of North Carolina
Chapel Hill, N. C. 27514

Human respiratory syncytial (RS) virus is classified in the genus pneumovirus of the paramyxovirus family. The genome of RS virus is a single negative strand RNA of at least 5.6×10^6 daltons. Ten RS viral mRNAs and their encoded proteins were identified recently (1,2,3, Collins et al, this volume). Here we demonstrate that four unique viral proteins are associated with the envelope of purified RS virions as compared to three for other paramyxoviruses.

The 24 K protein is a virion structural protein. RS virus grown in the presence of (^{3}H) leucine, was purified from the medium of virus-infected HEp-2 cells by two cycles of sedimentation through sucrose gradients. The virion proteins were analyzed by SDS-PAGE. Eight polypeptides of 200 (L), 84 (G), 68 (F_0), 47 (F_1), 42 (N), 34 (P), 26 (M) and 24 kilodaltons (K) were resolved (Fig. 1A). The 42, 34 and 26 K proteins have been identified previously as the major nucleocapsid protein, nucleocapsid phosphoprotein and matrix protein, respectively (1,4,5). The 200 K protein is a component of purified nucleocapsids (4,8 this paper) and is designated the L protein by analogy to other paramyxoviruses. The 84 K species is an envelope glycoprotein (4,5,6,7,8) and is designated G. The 68 K species has been identified as the F_0 glycoprotein (9, 10 and our unpublished data), which by analogy with other paramyxoviruses is cleaved proteolytically to generate the disulfide-linked F_1 (47 K) and F_2 (20 K) subunits of the F glycoprotein. The F_2 subunit is not visible in Fig. 1A and B, but was readily detected by coelectrophoresis of (^{3}H) glucosamine-labeled virions (data not shown). Purified virions also contained a 24 K protein (Fig. 1A); a virion protein of similar electrophoretic mobility was reported previously (4,6, 7,11) but its significance was unknown.

---

[1]This work was supported by NIH Grant AI12464 and AI15134.

Comparison of (^{3}H) leucine-labeled and (^{3}H) glucosamine-labeled virions by SDS-PAGE under reducing and non-reducing conditions (data not shown) showed that the 24 K protein is unglycosylated and is distinct from the 20 K (F_2) subunit of the F glycoprotein. In addition, the 24 K protein was detected in extracts of RS virus-infected cells (Fig. 1B, lane C) and among the <u>in vitro</u> translation products synthesized in a wheat germ extract translation system programmed with poly (A)$^+$ mRNA from RS virus-infected cells (Fig. 1B, lane b). Peptide mapping confirmed that the 24 K proteins from virions, infected cell extracts, and the <u>in vitro</u> translation system were the same (Fig. 1 C & D).

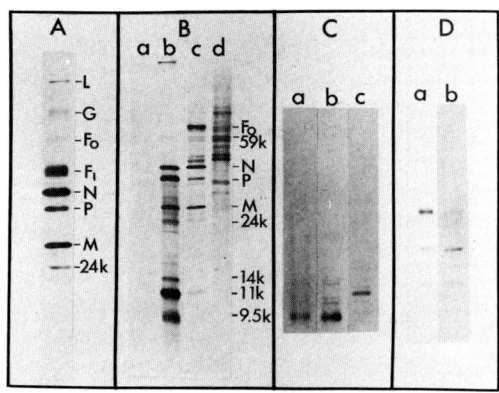

Fig. 1. Identification of the 24 K proteins. A. Proteins of purified RS virions labeled with (^{3}H) leucine were analyzed by 15% SDS-PAGE. B. Wheat germ in vitro translation system programmed with poly (A)-selected mRNA from (a) RS virus-uninfected or (b) infected, actinomycin D treated HEp-2 cells. (^{3}H) leucine-labeled products were analyzed by 15% SDS-PAGE in parallel with (^{3}H) labeled proteins extracted from (c) virus-infected and (d) uninfected act$_{35}$D-treated cells. C. Limited digest peptide mapping of the (^{35}S) cysteine-labeled 24 K protein. The 24 K protein (a) synthesized in vitro and (b) from infected cells, and (c) a comigrating cellular polypeptide were exposed to 5 ug/ml of V8 protease and analyzed by 16% SDS-PAGE. D. Limited digest peptide mapping of the (^{3}H) leucine-labeled 24 K protein. The 24 K proteins (a) extracted from infected cells and (b) from purified virions were digested with 5 ug/ml of V8 protease and analyzed by 15% SDS-PAGE.

<u>The 24 K protein is unrelated to the other viral proteins.</u> The 24 K, N, P, and M proteins were compared by tryptic peptide mapping. The profile of (^{3}H) leucine-labeled tryptic peptides separated by HPLC was different for each protein,

showing that these proteins were unrelated and that the 24 K protein was not a fragment of one of these larger proteins, (data not shown). In support of this conclusion, in vitro translation of individual viral mRNAs purified by hybrid-selection with cDNA clones demonstrated that the 24 K protein was encoded by a unique mRNA (Collins et al, this volume).

Location of the 24 K protein in the virion. A Triton detergent and salt dissociation method was used to distinguish between proteins associated with the nucleocapsid and the envelope. (^{3}H) leucine-labeled, partially purified virus was prepared in 4 aliquots and pelleted. The samples were treated, processed and analyzed as described in the figure legend. As shown in Fig. 2A, 2% Triton X-100 in 0.15M NaCl completely solubilized the G & F proteins. Triton X-100 in 0.4 M NaCl dissolved both glycoproteins as well as the majority of the M and 24 K proteins. The N, P and L proteins remained associated with the insoluble (nucleocapsid) fraction.

As a second approach to determine the virion location of the 24 K protein, nucleocapsids were purified from infected cell cytoplasmic extracts by gel filtration chromatography. The extracts were adjusted to 1% Triton and immediately loaded onto a Biogel A 15 M column (Biorad). The nucleocapsids were found to elute with the void volume and were pelleted, dissolved directly in polyacrylamide gel sample buffer and analyzed by SDS-PAGE (Fig. 2B). These results showed that three viral proteins, the N, P, and L proteins, were detected in purified nucleocapsids.

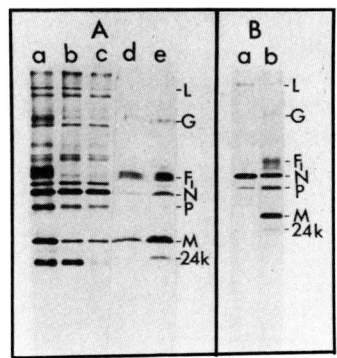

Fig. 2. Location of the 24 K$_3$ protein in the virion. A. Aliquots of partially purified (^{3}H) leucine-labeled virions were exposed to (b,d) 0.15 M NaCl containing 2% Triton and (c,e) 0.4 M NaCl containing 2% Triton. The (b,c) insoluble and (d, e) soluble fraction were separated by differential centrifugation and analyzed in parallel with (a) untreated virions

by 15% SDS-PAGE. B. Comparison by SDS-PAGE of (a) (^{3}H) leucine-labeled intracellular RS viral nucleocapsids isolated from infected cell lysates by Biorad-Biogel A 15M column chromatography and (b) (^{3}H) leucine-labeled purified RS virions.

These experiments defined three classes of virion proteins: (i) the N, P, and L proteins which were stably associated with nucleocapsids, (ii) the F and G glycoproteins which were envelope-associated and quantitatively solubilized with Triton, and (iii) the M and 24 K proteins which were almost completely solubilized with Triton and 0.4 M NaCl and, like the two glycoproteins, were not detected in column-purified nucleocapsids. Previously, a virion protein that probably corresponds to the 24 K protein was suggested to be nucleocapsid associated (4). However, the results shown here indicate that, like the other viral envelope proteins, the 24 K protein was solubilized with 2% Triton and 0.4 M NaCl.

SUMMARY

A 24 K protein, unrelated to the other viral proteins, was identified in purified virions, extracts of infected cells, and among viral polypeptides synthesized <u>in vitro</u>. Two methods determined that the 24 K protein was envelope-associated.

REFERENCES

1. Huang, Y.T. and Wertz, G.W. (1982). J. Virol. 43, 150.
2. Huang, Y.T. and Wertz, G.W. (1983). J. Virol. 46, 667.
3. Collins, P.L. and Wertz, G.W. (1983). Proc. Nat. Acad. Sci. USA, 80, 3208.
4. Peeples, M. and Levine, S. (1979). Virology 95, 137,
5. Cash, P., Pringle, C.R., and Preston, C.M. (1979). Virology 92, 375.
6. Dubovi, E.J. (1982). J. Virol. 42, 372.
7. Bernstein, J.M. and Hruska, J.F. (1981). J. Virol. 38, 278.
8. Pringle, C.R., Shirodaria, P.V., Gimenez, H.B. and Levine, S. (1981). J. Gen. Virol. 54, 173.
9. Walsh, E.E. and Hruska, J. (1983). J. Virol. 47, 171.
10. Fernie, B.F. and Gerin, J.L. (1982). Infect. Immun. 37, 243.
11. Ueba, O. (1980). Microbiol. Immunol. 24, 361.

CHARACTERIZATION OF THE GLYCOPROTEINS OF RESPIRATORY SYNCYTIAL VIRUS

Dennis M. Lambert and Marcel W. Pons

Department of Molecular Virology
The Christ Hospital Institute of Medical Research
Cincinnati, Ohio 45219

Human respiratory syncytial (RS) virus is the most important etiologic agent of lower respiratory illness in infants (1). RS virus has been tentatively placed in the genus Pneumovirus of the paramyxovirus family (2). However, RS virus differs from the paramyxoviruses in several important respects as it has no hemagglutinating or neuraminidase activity (3) and its nucleocapsid is smaller in diameter than that of paramyxoviruses (4). Surface glycoproteins of enveloped viruses are intimately involved in the adsorption and penetration steps of infection (5). Relatively little is known about the surface proteins of RS virus and there has been disagreement about the number and molecular weights of the glycoproteins present in this virus (6-10). However, the majority of reports in the literature have identified three glycopolypeptides in RS virus. No functional roles have yet been conclusively demonstrated for RS virus surface proteins although it is presumed that they are involved in adsorption and cell fusion activities. Thus it is very important to characterize these proteins in order to understand their role(s) in the infectious process and to develop means for control of this clinically important virus. In the experiments described here we have identified and partially characterized three viral-specific glycopolypeptides in purified virions and infected cells having molecular weights of approximately 90,000, 50,000 and 20,000. The effects of the drug tunicamycin as well as the effects of endoglycosidases on the synthesis and electrophoretic mobilities of these proteins were studied.

RS Virion Polypeptides. When ^{3}H-amino acid (^{3}H-AA) labeled purified virions grown in either CV-1 or HEp2 cells

---

[1] This work was supported in part by Public Health Service Grant AI14133 from NIAID and by a grant from The Thrasher Research Fund.

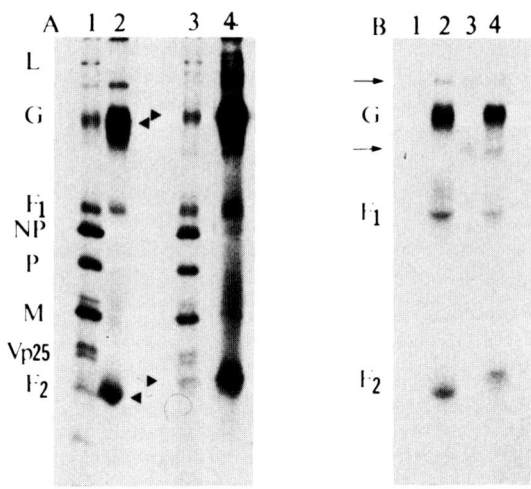

FIGURE 1 (A). Comparison of [^{3}H] amino acids ([^{3}H]AA) and [^{3}H] glucosamine-labeled proteins of purified RS virus (Long) grown in CV-1 and HEp2 cells. Virus labeled with [^{3}H]AA or [^{3}H] glucosamine (20 µCi/ml) was harvested at 48 hr p.i. and purified from the media of infected cell cultures (11). Virus pellets were resuspended in Laemmli PAGE sample buffer (12). Samples were electrophoresed in 10% polyacrylamide slab gels at 40 mA. Lane 1, [^{3}H]AA-labeled RS virions from CV-1 cells; lane 2, [^{3}H]-glucosamine-labeled virions from CV-1 cells; lane 3, [^{3}H]AA-labeled RS virions from HEp2 cells; lane 4, [^{3}H] glucosamine-labeled virions from HEp2 cells. (B) Immunoprecipitation of [^{3}H] glucosamine-labeled polypeptides from uninfected (lanes 1 and 3) and infected (lanes 2 and 4) CV-1 and HEp2 cell lysates. The procedure of Wechsler et al. (13) was used except that immune complexes were precipitated using IgGsorb (The Enzyme Center, Boston, MA).

were analyzed by SDS-polyacrylamide gel electrophoresis (SDS-PAGE), eight viral proteins were consistently observed (Fig. 1A, lanes 1 and 3). Three glycoproteins labeled with [^{3}H] glucosamine (G, F1 and F2) were identified in purified virions (Fig. 1A, lanes 2 and 4) and could be purified on lentil lectin sepharose columns (14). The three glycoproteins could be labeled equally well in HEp2 cells with [^{3}H] mannose (data not shown). Additional viral proteins labeled with [^{3}H] glucosamine did not bind to lentil lectin sepharose columns suggesting that they were not glycosylated (14). The eight polypeptides identified were assigned letter designations and had approximate molecular weights of 180,000 (L), 89,000 (G), 48,000 (F1), 42,000 (NP), 34,000 (P), 28,000 (M), 25,000 (Vp-25), and 21,000 (F2). The F2 protein was not always seen in virions labeled with ^{3}H-AA but was always heavily labeled with [^{3}H] glucosamine. The two smaller glycoproteins (F1 and F2) have been shown to be disulfide bonded subunits of an approximately 70,000 dalton protein that is the most likely candidate for the fusion protein of RS virus (10, 14).

Because of observed differences in the apparent quantities and labeling characteristics of the large glycoprotein (G) in virions grown in HEp2 cells as compared to virus grown in BSC-1 (monkey kidney) cells, Pringle et al. (9) proposed that this protein might be a host cell glycoprotein induced by RS virus replication. In order to investigate this possibility we examined immunoprecipitates of ^{3}H-glucosamine labeled uninfected and RS-infected HEp2 and CV-1 (monkey kidney) cell extracts by SDS-PAGE (Fig. 1B). All three glycoprotein species were immunoprecipitated from infected cell lysates of both cell types but not from uninfected cell lysates, suggesting that all three glycoproteins are viral encoded proteins. An interesting observation was that the G and F2 proteins of HEp2 and CV-1 grown virions and cell lysates migrated differently (Fig. 1A and B). This phenomenon was seen consistently and is most likely the result of differences in glycosylation and processing of these polypeptides in the two cell lines.

<u>Effects of Tunicamycin on Viral Infectivity and Viral Proteins.</u> The drug tunicamycin (TM) is a specific inhibitor of N-acetyl-glucosamine-lipid intermediates and inhibits glycosylation of nascent proteins (16, 17). To determine the effects of this drug on the infectivity of released RS virus we infected HEp2 and CV-1 cells at a multiplicity of 1 and treated with either 0, 0.1, 0.25, 0.5, 0.75, 1.0, 2.0, or 4.0 μg/ml TM. At 48 hr p.i. no CPE was seen in TM-treated cultures whereas the untreated monolayers exhibited extensive syncytial formation. Culture fluids were harvested and virus was titrated as previously described (11). The results (Table 1) demonstrated that release of infectious RS virus from CV-1 cells was more sensitive to TM treatment than was the release of infectious RS virus from HEp2 cells.

Similar results were obtained when purified [^{3}H]AA- or [^{35}S] methionine-labeled virus from untreated HEp2 and CV-1 cells was compared to that from TM-treated cells (0.5 - 2.0 μg/ml). The recovery of labeled virus particles from HEp2 cells was reduced by 40-60% whereas recovery of labeled virus particles from the media of treated CV-1 cultures was almost completely inhibited (approximately 5% recovered). The polypeptides of [^{35}S] methionine-labeled virions grown in untreated HEp2 or CV-1 cells and cells treated with 1.0 μg/ml TM were examined by non-reducing SDS-PAGE (Fig. 2A). No polypeptides in virions from TM-treated HEp2 cells co-migrated with either G or F1, 2 (Fig. 2A, lane 2). Two faint bands migrated more slowly than the G protein at approximately 100,000 and 94,000 daltons and might represent aggregates. However, two additional, more heavily labeled, protein bands which probably represent unglycosylated forms of the G and F proteins were observed (Fig. 2A, arrows lane 2). One band migrated at approximately 80,000 daltons and could conceivably be the unglycosylated form of the G protein. The other band migrated at approximately 55,000 daltons and could represent the unglycosylated fusion protein. Poor recovery of RS virus from TM-treated CV-1 cells did not allow observation of similar proteins (Fig. 2A, lane 4).

Incorporation of [^{3}H] glucosamine into virion glycoproteins was almost completely inhibited by 1 μg/ml tunicamycin. In order to observe the effects of TM on incorporation of [^{3}H] glucosamine in infected cells we labeled uninfected and infected CV-1 cells treated with TM at 0, 0.5, 1.0, and 2.0 μg/ml. Labeled polypeptides were immunoprecipitated from cell extracts and analyzed by SDS-PAGE under reducing conditions (Fig. 2B). Incorporation of [^{3}H] glucosamine into host cell and virion polypeptides was inhibited with increasing concentration of TM. Although TM significantly reduced the incorporation of [^{3}H] glucosamine into viral glycoproteins it was not completely inhibited. Incorporation of label into host

TABLE 1
EFFECTS OF TUNICAMYCIN TREATMENT ON RELEASE OF INFECTIOUS RS VIRUS FROM INFECTED CELLS

TM (μg/ml)	CV-1		HEp2	
	PFU/ml[a]	PFU/cell[b]	PFU/ml[a]	PFU/cell[b]
0	5.75×10^5	1.15	1.5×10^8	150.0
0.1	6.5×10^2	0.0013	9.1×10^7	94.0
0.25	1.0×10^3	0.002	6.0×10^4	0.06
0.5	7.0×10^2	0.0014	2.2×10^4	0.022
0.75	8.0×10^2	0.0016	0.8×10^3	0.0018
1.0	8.0×10^2	0.0016	2.0×10^4	0.020
2.0	5.5×10^2	0.0011	1.6×10^3	0.0016
4.0	2.5×10^2	0.0005	3.6×10^4	0.036

[a] PFU/ml values are averages of two assays per sample on HEp2 cells.

[b] PFU/cell determination made on basis of 5×10^5 CV-1 and 1×10^6 HEp2 cells/60 mm dish.

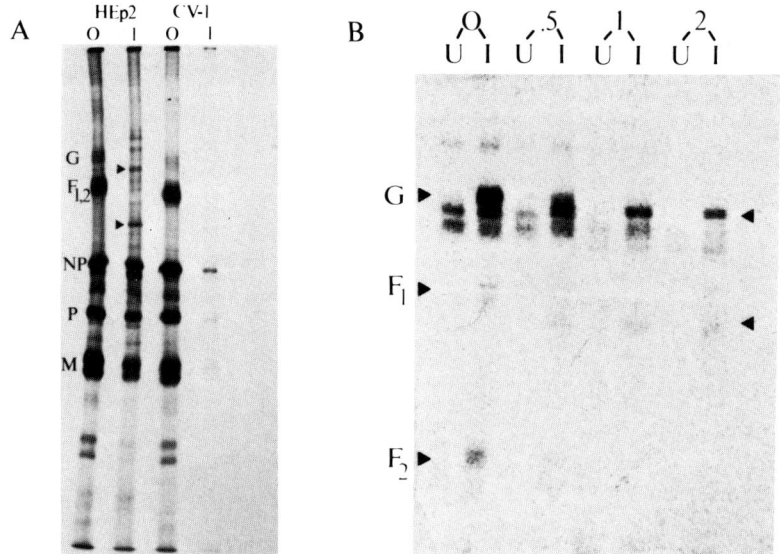

FIGURE 2A. Non-reducing SDS-PAGE analysis of ^{35}S-methionine labeled polypeptides of virions from untreated and TM-treated HEp2 and CV-1 cells. Lane 0, RS virus polypeptides from virions grown in untreated cells; lane 1, viral polypeptides from cells treated with 1 μg/ml TM.

FIGURE 2B. Effects of TM treatment on incorporation of [^{3}H] glucosamine into proteins of uninfected and infected CV-1 cells. Cells were labeled for 24 hr and harvested. U, uninfected; I, infected. Numbers (0, .5, 1, and 2) refer to the concentration of TM in μg/ml.

cell proteins was almost completely inhibited. Labeling of the G protein was less sensitive to the effects of TM than labeling of F1 and F2. In TM treated cells the G protein migrated at about 80,000 daltons and the F1 at about 41,000 daltons. The F2 protein was not detected in TM treated cells. The G and F1 proteins migrated faster in cells treated with TM (arrows), suggesting that although these proteins contained reduced amounts of carbohydrate, they were still labeled. Thus, it is possible that some TM-resistant glycosylation of RS glycoproteins occured even at relatively high concentrations of the drug. However, two observable biological functions of the virus (infectivity and cell fusion) were inhibited by relatively low doses of TM (Table 1).

Endoglycosidase Treatment of [^{3}H]AA-Labeled Virion Poly-

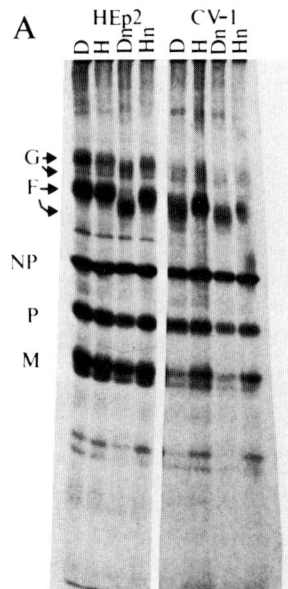 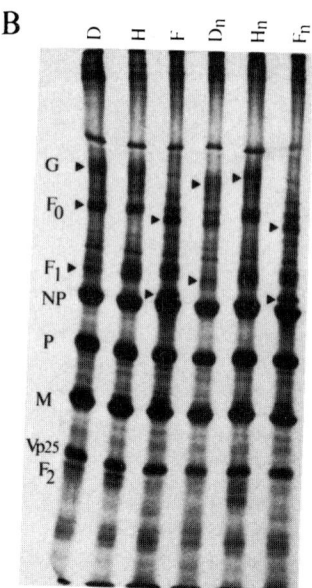

FIGURE 3. Endoglycosidase treatment of [^{3}H]AA-labeled virions. Virions were incubated with enzyme at 37°C for 40 hr. An equal volume of 2X PAGE sample buffer was added and samples were electrophoresed in 10% polyacrylamide gels under either non-reducing (A) or reducing conditions (B). Fig. 3A: HEp2, virions grown in HEp2 cells; CV-1, virions grown in CV-1 cells. Lanes D, virions treated with endo-D; lanes H, virions treated with endo-H; lanes 3 and Dn, virions treated with both endo-D and neuraminidase; lanes Hn, virions treated with endo-H and neuraminidase. Fig. 3B: [^{3}H]AA virions grown in HEp2 cells were analyzed. D, endo-D treated; H, endo-H treated; F, endo-F treated; Dn, Hn, Fn, samples were treated with endoglycosidase and neuraminidase.

peptides. Additional information concerning the nature of RS glycoproteins was obtained by treating [^{3}H]AA labeled purified virions with either endoglycosidase D (endo-D), an enzyme which cleaves complex carbohydrates (18), endoglycosidase H (endo-H), an enzyme which cleaves high mannose carbohydrates (19), and endoglycosidase F (endo-F), an enzyme which cleaves both complex and high-mannose carbohydrates(20). Treatment of labeled virions with either endo-D or endo-H had no effect on the migration of either the G or F1, 2 proteins of virions grown in HEp2 cells when electrophoresis was carried out under non-reducing conditions (Fig. 3A, lanes D and H) but there appeared to be a change in the migration of G and F1, 2 proteins of virions grown in CV-1 cells treated with endo-D (lane D). Glycoproteins of virions grown in HEp2 cells were made susceptible to digestion by endo-D by co-treatment with neuraminidase (lane Dn) suggesting that the glycoproteins of virions grown in HEp2 cells contain sialic acid residues which blocked the action of endo-D. Similarly, neuraminidase treatment along with endo-D treatment of the glycoproteins of virions grown in CV-1 cells seemed to enhance the action of the enzyme (lane Dn). Endo-H treatment alone or with neuraminidase did not alter the migration of the G or F1, 2 proteins of virions grown in either cell line (Fig. 3A, lanes H, Hn). Therefore, the carbohydrate moiety of both the G and F proteins of virus grown in either HEp2 or CV-1 cells appears to be of the complex type. Analysis of endo-F treated virions, grown in HEp2 cells, by SDS-PAGE under reducing conditions revealed altered migration of the apparent F0 and F1 proteins (the F0 migrated faster and the F1 migrated slightly slower) even without neuraminidase treatment (Fig. 3B, lane F, arrows). The G protein essentially disappeared and a new band of approximately 43,000 daltons (which migrated just above the NP protein but below the F1 protein) was observed (lanes F and Fn, arrows). Whether the 43,000 dalton protein is a subunit or a cleavage product of G (or even F1) is at present unknown. Results of endo-D and endo-H digested viral glycoproteins analyzed in reducing gels (Fig. 3B) were consistent with the results obtained in non-reducing gels (Fig. 3A). We are attempting to further clarify the relationships of these polypeptides to each other by pulse chase analysis of viral proteins in infected cells and by comparison of peptide maps of virion glycoproteins to viral proteins in infected cells. The data presented here suggests that the carbohydrate moieties of the G and F proteins are of the complex type and their presence on viral glycoproteins is necessary for the presumed biological activities of these proteins (i.e. infectivity and cell fusion)

ACKNOWLEDGEMENTS

The authors wish to express their appreciation for the excellent technical assistance of Mr. J. Hambor and Mrs. A. Lambert.

REFERENCES

1. Chanock, R.M. (1970) Science 169, 248.
2. Kingsbury, D.W., Bratt, M.A., Choppin, P.W., Hanson, R.P., Hosaka, V., ter Meulen, V., Norrby, E., Plowright, W., Rott, R., and Wunner, W.H. (1978). Intervirology 10, 137.

3. Richman, A.V., Pedreira, F.A., and Tauraso, N.M. (1971). Appl. Microbiol. 21, 1099.
4. Berthiaume, L., Joncas, J., and Pavilanis, V. (1974). Arch. Gesamte Virusforsch. 45, 39.
5. Choppin, P.W., and Scheid, A. (1980). Rev. Inf. Dis. 2, 40.
6. Pringle, C.R., and Cross, A. (1978). Nature (London) 276, 501.
7. Bernstein, J.M., and Hruska, J.F. (1981). J. Virol. 38, 278.
8. Dubovi, E. (1982). J. Virol. 42, 372.
9. Pringle, C.R., Shirodaria, P.V., Gimenez, H.B., and Levine, S. (1981). J. Gen. Virol. 54, 173.
10. Fernie, B., and Gerin, J.L. (1982). Inf. Immun. 37, 243.
11. Lambert, D.M., Pons, M.W., Mbuy, G.N., and Dorsch-Hasler, K. (1980). J. Virol. 36, 837.
12. Laemmli, U.K. (1970). Nature (London) 227, 680.
13. Wechsler, S.L., Weiner, H.L., and Fields, B.N. (1979). J. Immunol. 123, 884.
14. Lambert, D.M., and Pons, M.W. (1983). Virol. 130, (in press).
15. Gruber, C., and Levine, S. (1983). J. Gen. Virol. 64, 825.
16. Takatsuki, A., Arima, K., and Tamura, G. (1971). J. Antibiot. 24, 224.
17. Takatsuki, A., Kono, K., and Tamura, G. (1975). Agr. Biol. Chem. 39, 2089.
18. Koide, N., and Muramatsu, T. (1974). J. Biol. Chem. 249 4897.
19. Tarentino, A.L., and Maley, F. (1974). J. Biol. Chem. 249, 811.
20. Elder, J.H., and Alexander, S. (1982). Proc. Natl. Acad. Sci. USA 79, 4540.

Biology

CHARACTERIZATION OF RABIES VIRUS RECEPTOR-RICH REGIONS AT PERIPHERAL AND CENTRAL SYNAPSES[1]

Abigail L. Smith,[2,3] Thomas G. Burrage,[2] and Gregory H. Tignor[2]

School of Medicine
Yale University
New Haven, Connecticut 06510

MORPHOLOGIC STUDIES OF THE ACCUMULATION AND TRANSFER OF RABIES VIRUS AT PERIPHERAL AND CENTRAL SYNAPSES.

Rabies virus antigen and intact particles can be found in the region of neuromuscular junctions between one and six hours after infection. Co-localization of virus antigen with the motor end plate can be shown morphologically by immunoelectron microscopy and histochemically by co-localization with regions possessing high densities of the acetylcholine receptor (AChR) (1).

<u>Peripheral synapses.</u> Injection of rabies virus into either hindlimb muscles or facial muscles of the mouse is followed by accumulation of virus at synaptic junctions. Within one-two hours, the virus antigen is near motor end plates as revealed by dual fluorescent staining with antiviral and anti-AChR reagents. The co-localization becomes more precise at six hours after injection at which time viral antigen is closely associated with the nerve-muscle junction. Thin-section electron microscopy has shown that, at this time, isolated intact particles are present on both the muscle and nerve sides of the neuromuscular junction. On the muscle side of the junction, particles were sometimes enclosed in vacuoles and in other cases were free. Within the nerve terminal itself, few intact particles were found. However, by immunoelectron microscopy, accumulations of viral antigen were found to be more widespread at the junctions as revealed by labeling with a protein A-gold conjugate. Gold label was found near the junctional folds and in the nerve terminal clustered near particle-like configurations (1) and Figure 1.

---

[1] This work was supported by NIH grant AI 12541.
[2] Department of Epidemiology and Public Health.
[3] Section of Comparative Medicine.

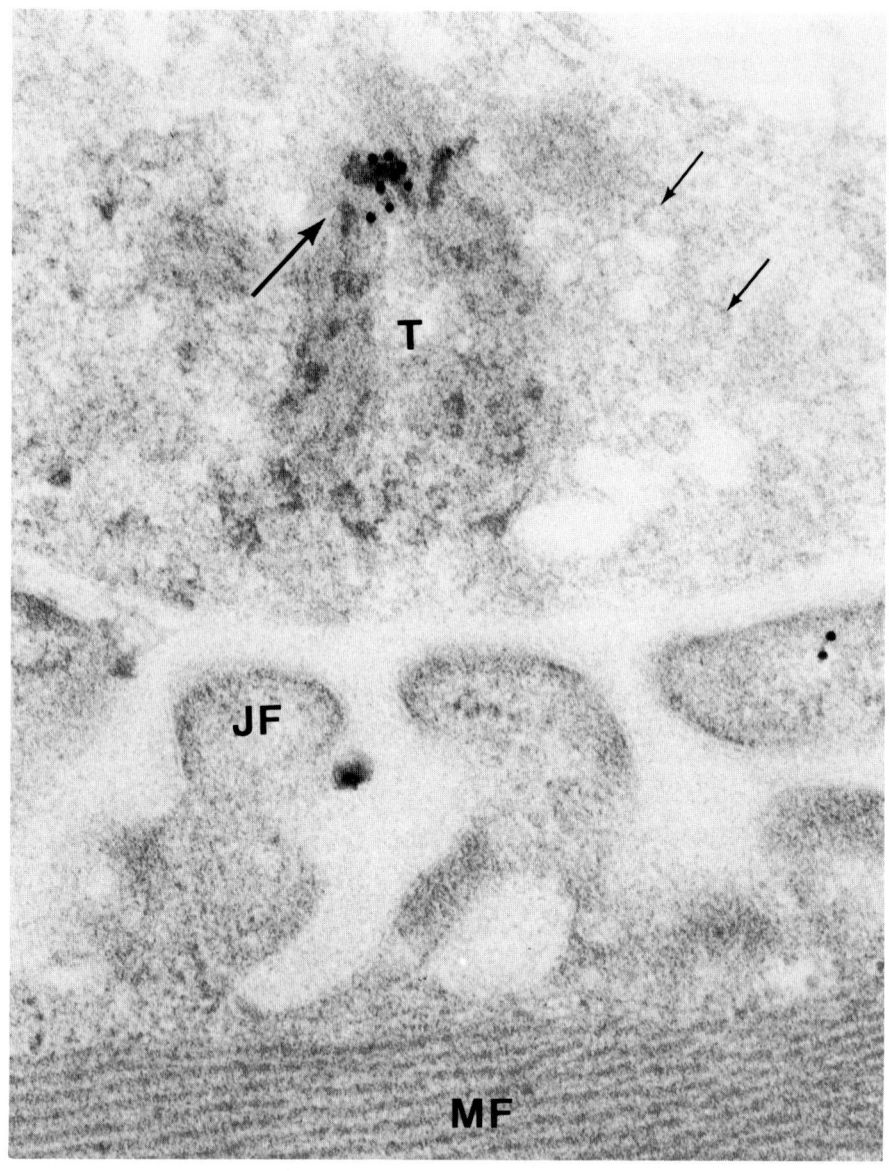

FIGURE 1. Protein A-gold labeled antigen at a neuromuscular junction. A rod-shaped profile (large arrow) with protein A-gold is in the nerve terminal (T) which has synaptic vesicles (small arrows). The muscle cell has junctional folds (JF) and myofilaments (MF). A few gold particles are associated with a junctional fold. x94,000.

Gold labeled antigen was seen being actively engulfed by neutrophils and lying within large vacuoles in macrophages (2). Gold label was also found closely associated with microtubules in large nerves. Virus particles apparently reached the neuromuscular junction by diffusion through the extracellular space near this site (1).

Central synapses. Rabies virions were frequently found at synaptic regions within the central nervous system either free in the extracellular space or enclosed within vacuoles at nerve terminals [Figure 2A]. Viral uptake was morphologically similar to that at peripheral synaptic junctions where particles were also enclosed in vacuoles [Figure 2B].

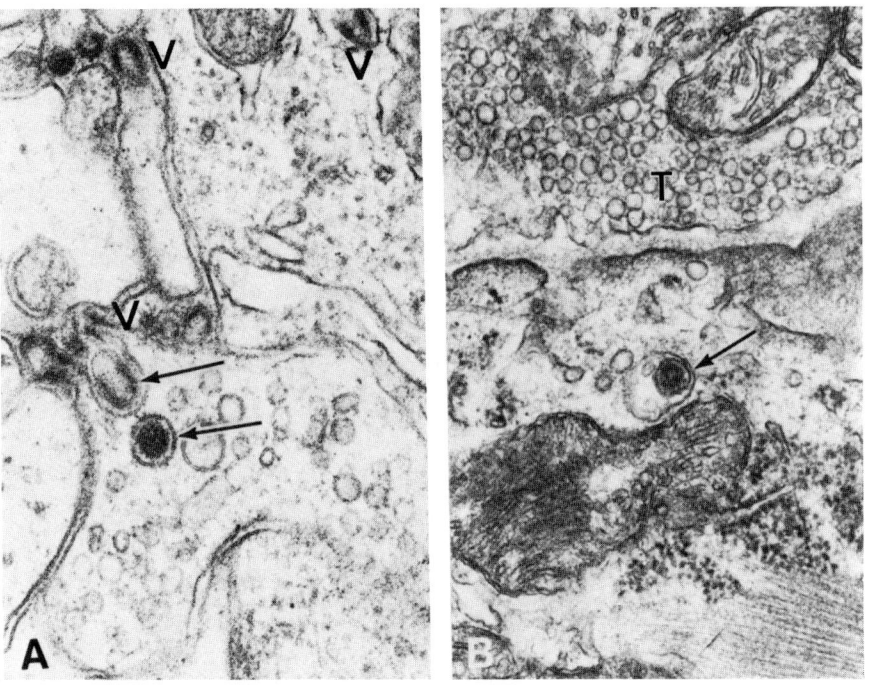

FIGURE 2. Rabies virus particles at central and peripheral synapses. A. Synaptic neuropil from the cerebellum. Rabies virus particles within vacuoles (arrows) in a terminal. Virions (V) also in the extracellular space and budding from internal membranes. x68,900. B. A particle (arrow) in the sarcoplasm of muscle. The junctional area is defined by a synaptic vesicle-laden terminal (T). x55,500.

Transfer of particles at brain synapses was frequently observed. Particles budded into coated pits consistent with receptor-mediated endocytosis [Figure 3A]. Intact particles and protein A-gold label were seen near postsynaptic densities indicating accumulation of viral proteins. [Figure 3B]. To characterize the viral receptor at these sites, we initially isolated synaptosomal fractions from infected brain tissue (3). Virions were attached to synaptosomal membranes, but particles were not attached to contaminating myelin figures or to mitochondria [Figure 3C].

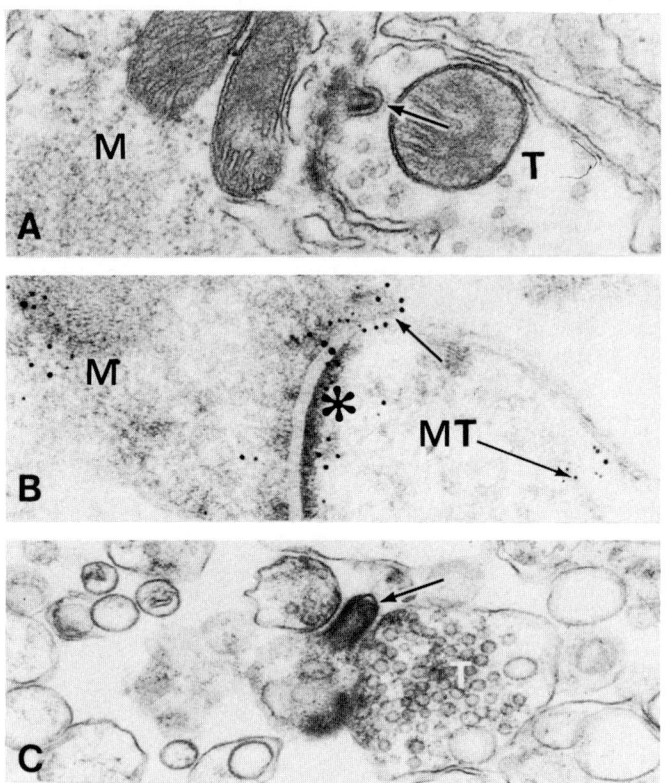

FIGURE 3. A. Particle in a coated pit of a terminal (T). Dense material is the postsynaptic surface. Viral matrix (M). x52,000. **B.** A particle (arrow) with gold label in the junctional space. Labeled matrix (M) and labeled postsynaptic density (*). x60,000 **C.** Virion-synaptosome association (arrow). Terminal (T). x74,500.

TRANSPORT AND PROCESSING OF VIRIONS AND VIRION-ASSOCIATED PROTEINS.

Rabies virions and virion-specific proteins have been localized in both the peripheral and central nervous systems using thin section electron microscopy and immunoelectron microscopy. A protein A-gold conjugate has been used to detect rabies virus antibody on thin sections of rabies virus-infected tissue embedded in Lowicryl at low temperature (2). Fixation was by paraformaldehyde and glutaraldehyde; post-fixation with osmium tetroxide was omitted. Sections were stained with uranyl acetate and lead citrate.

<u>Transport.</u> Within hours after infection of muscle tissue, gold labeled antigen was found in association with microtubules in large peripheral axons in the inoculated hind limb (1). Since a polyclonal anti-viral antibody was used in these studies, protein A-gold label was found over the glycoproteins, membrane and nucleocapsid proteins of intact particles. Since intact particles were not found in peripheral nerves at this early time, it was not possible to determine which viral proteins were associated with the microtubules. It is possible, however, that the gold label was associated with intact virions which were not in the plane of section and thus could not be morphologically identified (2). Inhibitors of microtubular function have been used in the past early after infection to show that rabies virus or subunits thereof are transported by axoplasmic flow (4,5). Viral matrices were not found within axons but were found within terminals even at 6 hours.

Within the central nervous system, gold labeled antigen was found in association with microtubules and rough endoplasmic reticulum, but not with neurofilaments or the Golgi complex (2). The latter are not known to be directly involved in axoplasmic transport. As mentioned earlier, budding into coated pits at terminals was a prominent feature of the transport process. This could result from assembly of viral proteins at the post-synaptic surface or from receptor-mediated endocytosis.

In addition to membrane-associated transport, intact virions were found either free or in vacuoles within axonal and dendritic processes. Viral matrices were not found in axons but were frequently seen in dendritic processes. Thus, transport in axons was limited to axoplasmic flow whereas transport in dendrites could occur either by axoplasmic flow or by specific assembly at synaptic sites.

<u>Processing.</u> After transfer of virions at synaptic

terminals, particles were often found with multivesicular bodies in the dendritic and cell body cytoplasm [Figure 4A]. Multivesicular bodies are thought to be an immature form of a lysosome (6). The presence of virions within these bodies is consistent with current hypotheses regarding the uptake and uncoating of enveloped viruses from this and other genera (7). Significant accumulations of gold particles were found associated with the cisternae of the endoplasmic reticulum near viral matrices. This observation was consistent with the current models of protein assembly.

IMMUNOCYTOCHEMICAL STUDIES ON RABIES VIRUS RECEPTOR-RICH REGIONS.

In earlier reports, we have suggested that sites containing high levels of the acetylcholine receptor may also contain a molecule which serves as a specific host cell receptor for rabies virus. This hypothesis is based on findings that ligands for the acetylcholine receptor and monoclonal antibody to a subunit of the receptor reduce viral attachment to muscle cells (1). As we have shown here, virus attachment also occurs at terminals in the central nervous system. Therefore, we have looked for acetylcholine receptor in the brain. Acetylcholine receptor has been previously shown to be present on the post-synaptic density of isolated synaptosomes (3). The purpose here was to determine whether the same molecule to which virus binds at peripheral synaptic regions is present at terminals in the central nervous system. The terminals of dendritic processes in the cerebellum of infected mice are sites where there are large concentrations of both intact virions and virion proteins as shown by protein A-gold label [Figure 4B]. Using an acetylcholine receptor monoclonal antibody which co-localized with virus on muscle cells (1), we found binding to proteins distributed both on the synaptic plasma membrane and within the terminal itself [Figure 4C]. Finding receptor in the terminal was an unexpected result which could suggest that the molecular address of the receptor may govern viral replicative events beyond attachment. Alternatively, the presence of receptor in the terminal may result from metabolic changes induced by the virus. Morphologic evidence of selective metabolic changes in infected neurons were found several days after infection at peripheral synapses (1). At degenerating neuromuscular junctions, nerve terminals had retracted from muscle tissue and only dense core vesicles remained intact. Co-localization studies using immunoelectron microscopy may resolve questions raised here.

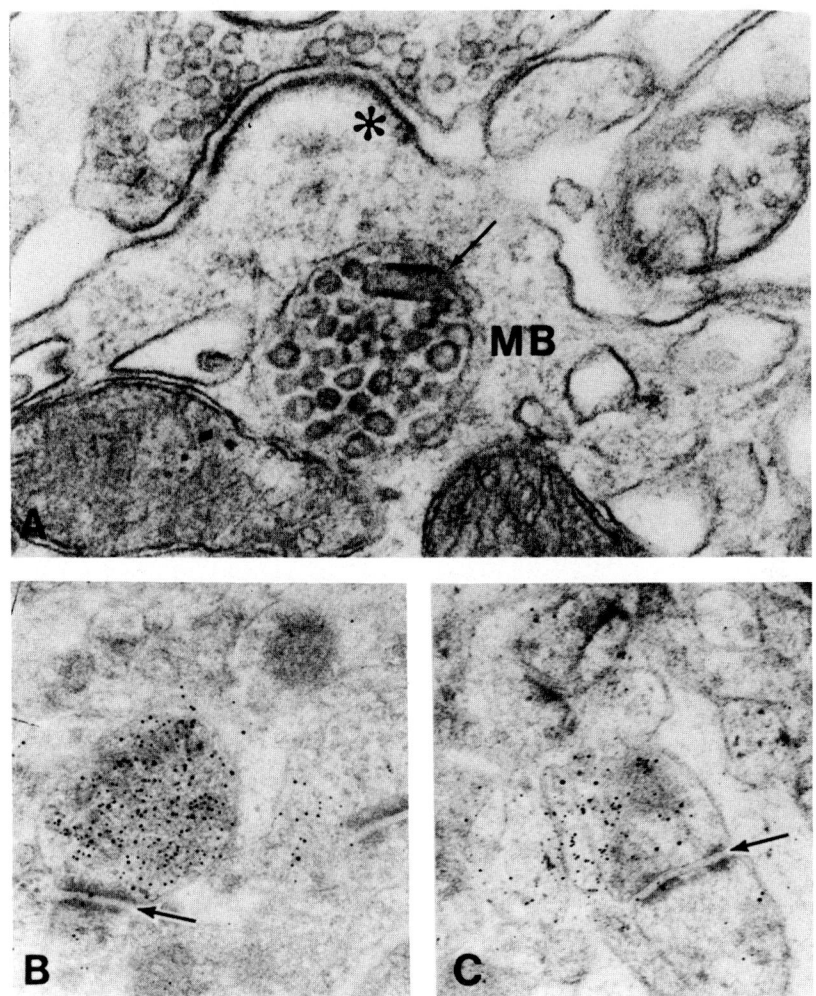

FIGURE 4. A. Rabies virion (arrow) in a multivesicular body (MB) in a dendritic process adjacent to a synaptic junction. Note the postsynaptic density (*) and synaptic vesicles in the terminal. x66,700. B. Protein A-gold label on viral particles and matrix in a terminal. Label is also present on synaptic membranes. Label also occurs on microtubules in the adjacent dendrite. x35,600. C. Terminal from infected cerebellum incubated with monoclonal antibody to the acetylcholine receptor and stained with protein A-gold. Label is on the synaptic membrane, but is largely in the terminal. Junctional space (arrow). x32,000.

ACKNOWLEDGMENTS

The technical assistance of Ms Katherine Moreno and Mr. Ruben Cedeno is appreciated. Monoclonal antibodies to the acetylcholine receptor were provided by Drs. L.L.Y. Chun and E. Hawrot. Dr. J. Meegan provided the control monoclonal antibody. K.P. Tu assisted in preparation of the manuscript.

REFERENCES

1. Burrage, T.G., Smith, A.L., Tignor, G.H., and Lentz, T.L. (1983). Submitted to Cell.
2. Burrage, T.G., Tignor, G. H., and Smith, A.L. (1983). Submitted to J. Virol. Methods.
3. Lentz, T.L., and Chester, J. (1977). J. Cell Biol. 75, 258.
4. Bijlenga, G., and Heaney, T. (1978). J. gen. Virol. 39, 381.
5. Tsiang, H. (1979). J. Neuropath. Exp. Neurol. 38, 286.
6. Rosenbluth, J., and Wissig, S. L. (1964). J. Cell Biol. 23, 307.
7. Simons, K., Garoff, H. and Helenius, A. (1982). Sci. Am. 246, 58.
8. Lentz, T. L., Burrage, T. G., Smith, A.L., Crick, J., and Tignor, G.H. (1982). Science, 215, 182.

EARLY INTERACTIONS OF RABIES VIRUS WITH CELL SURFACE RECEPTORS[1]

Kevin J. Reagan and William H. Wunner

The Wistar Institute of Anatomy and Biology
36th Street at Spruce
Philadelphia, Pennsylvania 19104

The rabies virus provides an excellent model of neurotropism whereby virus infects the central nervous system (CNS) exclusively via peripheral nerves. Infection of tissues outside the CNS, such as salivary glands, occurs by a centrifugal spread of rabies virus from the CNS, again through the peripheral nerves (1). This neurotropism imposes a severe host range limitation in vivo although in vitro the virus infects nearly all mammalian and avian cell types (2).

Since it is known that the presence of cell surface receptor sites is often a major controlling factor of tissue tropism, we have investigated early interactions of rabies virus with cells of neural and non-neural origin. We report that rabies virus utilizes a rhabdovirus-common receptor in its cellular interactions and pH-dependent fusion in vitro.

VIRUSES AND CELLS

Cultures of BHK-21 clone 13 cells and mouse neuroblastoma C1300 (NA) cells were used throughout these studies (3). The ERA strain of fixed rabies virus, its growth in BHK-21 cells, radiolabeling with [^{3}H]leucine or [^{35}S]methionine, and purification has also been described (5). The origin and purification of rabies nonpathogenic variant RV194-2, vesicular stomatitis virus (VSV), West Nile virus and reovirus type 3 are described elsewhere (4).

---

[1]This work was supported by Research Grant AI-18562 from the National Institutes of Health and the Neurovirology/Neuroimmunology Training Program (NS-07180) to the University of Pennsylvania School of Medicine from the National Institutes of Health.

STANDARDIZATION OF BINDING ASSAY

BHK-21 or NA cells at 5×10^6 cells/ml were suspended in minimal essential medium (MEM) containing 0.2% bovine serum albumin (BSA), 0.05 M HEPES (pH 7.4), and DEAE-DEXTRAN (DEAE-D; 50 µg/ml). Radiolabeled virus was added directly to cells and incubated at 4°C until an apparent equilibrium was reached (3 hours). Unattached virus was separated from cell-associated virus by centrifugation and the distribution of radiolabel between cell pellet and supernatant fractions determined by liquid scintillation spectrometry. DEAE-D was found to enhance attachment rate and was equally effective as a cell pretreatment, suggesting an action of DEAE-D on the plasma membrane rather than on the virus. These observations confirm earlier studies by Kaplan et al. (5).

The virus-cell interaction did not require divalent calcium or magnesium ions, nor was the intact cellular receptor unit (CRU) sensitive to digestion by trypsin, chymotrypsin, pronase or sialic acid-specific enzymes. Attachment was pH- and temperature-dependent. A greater level of attachment was observed at low pH but did not correlate well with an increase in infectivity.

Attachment of fixed rabies virus to BHK-21 cells obeyed the laws of mass action under the standard binding conditions used. The number of virus particles that attached per cell increased linearly as the ratio of the concentration of virus particles and cells was increased with respect to each other.

SATURATION AND COMPETITION FOR CELL SURFACE RECEPTORS

Saturation was demonstrated by binding ^{3}H-labeled virus probe to the cell surface in the presence of increasing concentrations of unlabeled homologous virus particles (Figure 1). The point at which increasing input multiplicities failed to cause a continued linear increase in specifically attached virus defines saturation of cell surface receptors. From five individual experiments it was estimated that saturation was complete at approximately $3-15 \times 10^3$ attached virions per cell.

Competition was demonstrated using excess unlabeled ERA strain rabies virus, ERA strain nonpathogenic variant RV194-2 and VSV, all of which competed with ^{3}H-labeled ERA virus for receptor sites on NA and BHK-21 cells, causing 70-82% inhibition of probe with 100 µg of competing virus. Non-competitive ligands included purified rabies virus glycoprotein (G), rabies soluble antigen (G_S) and reovirus type 3 at 100 µg of virus. West Nile Virus caused only a 27% inhibition, further suggesting the existence of separate receptor sites for the rhabdoviruses.

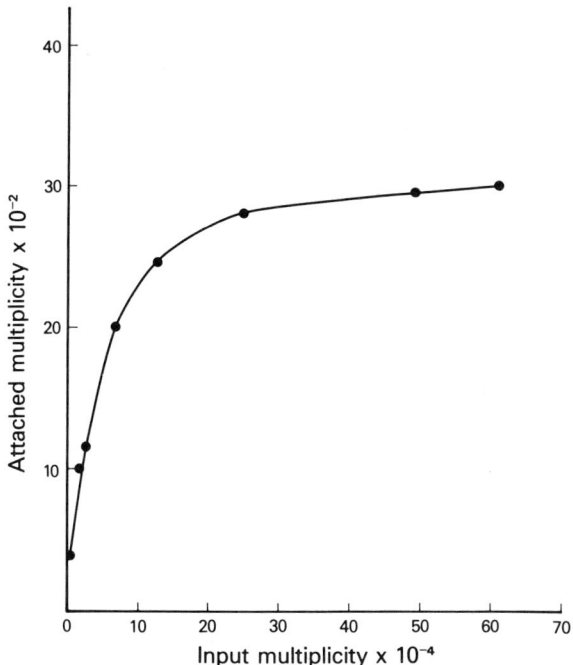

FIGURE 1. Saturation of BHK-21 cells with ERA virus. Bound virus was determined after a 3-hour incubation at 4°C.

SOLUBILIZED RECEPTOR ACTIVITY

A soluble extract was prepared by treating BHK-21 cells in suspension with 50 mM octyl-β-D-glycopyranoside (OG extract) (6) which blocked the attachment of rabiolabeled virus to BHK-21 cells in the presence or absence of DEAE-D (Figure 2). The binding-inhibitory fraction of the OG extract was resistant to proteases, boiling, and freeze-thaw cycles but sensitive to phospholipases. Unlike the result with whole cells, we found the OG extract inhibition to be destroyed by neuraminidase.

MEMBRANE FUSION BY RABIES VIRUS

To study adsorptive endocytosis, we have examined the fusion capabilities of rabies virus by exposing infected BHK-21 cells to mildly acidic pH (7). A suspension of 2.5 x 10^5 BHK-21 cells/ml in MEM containing 10% fetal calf serum and approximately 5 x 10^5 PFU/ml ERA virus (which caused the

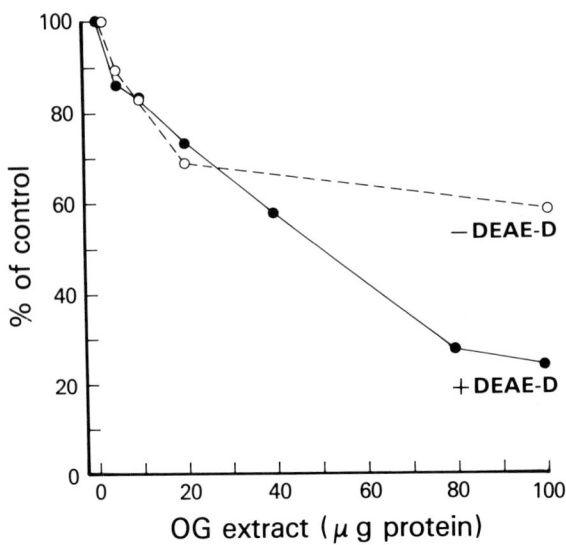

FIGURE 2. Inhibition of ERA virus binding by an OG extract from BHK-21 cells. BHK-21 cells were suspended in attachment medium with or without 50 µg/ml DEAE-D. Increasing amounts of dialyzed OG extract were added to the cell suspension followed by [^{35}S]methionine-labeled ERA virus. Virus bound after 3 hours at 4°C was determined and is expressed as a percentage of a control culture which received no OG extract.

development of rabies virus-specific antigen in 100% of the cells within 24 hours) was cultured on plates containing coverslips and incubated for 24 hours at 37°C. Cover-slips were removed, washed in PBS, and exposed for 60 seconds to warmed MEM buffered to pH 5.8 or 7.0 with 10 mM MES and 10 mM HEPES. Cells were replaced in warmed MEM containing 0.2% BSA, 50 mM HEPES, pH 7.4, and incubated for 30 minutes at 37°C, then fixed and stained. As shown in Figure 3, infected cells exposed to pH 5.8, but not pH 7.0, were extensively fused. This polykarion development which occurred only in infected cells could be inhibited by monoclonal antibodies directed to the rabies virus glycoprotein. The pH threshold for triggering fusion was 5.9.

DISCUSSION

These data suggest a specific, rhabdovirus-common CRU exists on the plasma membrane of susceptible cells in

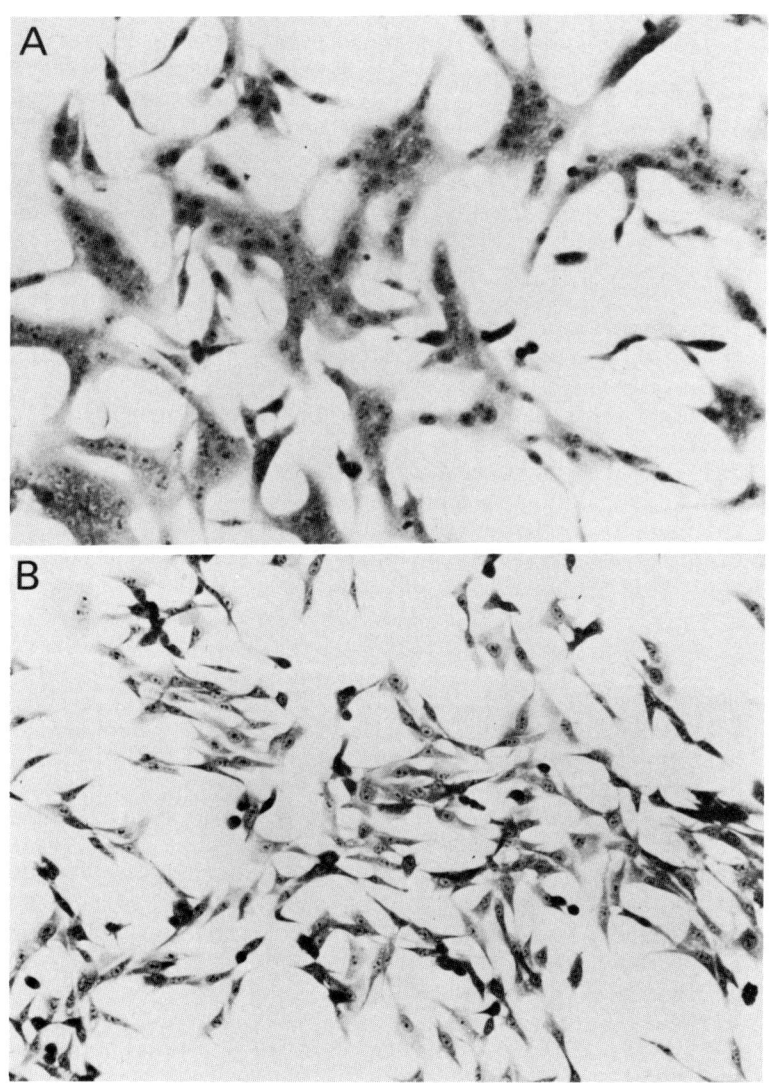

FIGURE 3. Fusion of ERA virus-infected BHK-21 cells by brief exposure to MEM, pH 5.8 (A). Replicate cultures exposed to MEM, pH 7.0 (B), failed to fuse.

culture. The areas of similarity between VSV and rabies virus are summarized below. VSV attaches to a limited number (approximately 4×10^3) of high affinity sites on Vero cells (8), a number which correlates well with the $3\text{-}15 \times 10^3$ sites for rabies virus on BHK cells. VSV and rabies virus attachment to cells was independent of pretreatment of cells with proteolytic enzymes or the ommission of Mg^{+2} or Ca^{+2} from the binding medium (4, 6, 9). In both systems, isolated viral G failed to block whole virus attachment (4, 9, 10), yet VSV was capable of competing with rabies for its receptor on BHK-21 and NA cells. An OG extract of susceptible cells, which was sensitive to phospholipases, could block specific attachment of either virus (6; Figure 2), suggesting that a lipid component had binding activity. We are currently investigating whether this component represents a functional receptor. Finally, both VSV and rabies virus promote cellular membrane fusion with a similar pH threshold (7; Figure 3).

The nonpathogenic variant of ERA strain rabies virus, RV194-2, recognized and competed for the same CRU as its virulent parent virus. This suggests that in cultured cells, an alteration of cellular receptor specificity does not represent the mechanism of attenuation.

REFERENCES

1. Murphy, F. A. (1977). Arch. Virol. 54, 279.
2. Wiktor, T. J., and Clark, H F. (1975). In "The Natural History of Rabies", Vol. 1 (G. M. Baer, ed.), p. 155. Academic Press, New York.
3. Clark, H F., Parks, N. F., and Wunner, W. H. (1981). J. gen. Virol. 52, 245.
4. Wunner, W. H., Reagan, K. J., and Koprowski, H. J. Virol., submitted.
5. Kaplan, M. M., Wiktor, J. J., Maes, R. F., Campbell, J. B., and Koprowski, H. (1967). J. Virol. 1, 145.
6. Schlegel, R., Tralka, T. S., Willingham, M. C., and Pastan, I. (1983). Cell 32, 639.
7. White, J., Matlin, K., and Helenius, A. (1981). J. Cell Biol. 89, 674.
8. Schlegel, R., Willingham, M. C., and Pastan, I. (1982). J. Virol. 43, 871.
9. Thimmig, R. L., Hughes, J. V., Kinders, R. J., Milenkovic, A. G., and Johnson, T. C. (1980). J. gen. Virol. 50, 279.
10. Perrin, P., Portnoi, D., and Sureau, P. (1982). Ann. Virol. (Inst. Pasteur) 133E, 403.

MICROINJECTION OF MONOCLONAL ANTIBODIES TO VESICULAR
STOMATITIS VIRUS NUCLEOCAPSID PROTEIN INTO HOST CELLS:
EFFECT ON VIRUS REPLICATION[1]

Heinz Arnheiter[2], Monique Dubois-Dalcq[2], Manfred Schubert[2],
Nancy Davis[3], John Patton[3], and Robert Lazzarini[2]

[2]Laboratory of Molecular Genetics, IRP
National Institute of Neurological and Communicative
Disorders and Stroke, NIH
Bethesda, Maryland 20205;
[3]Department of Bacteriology and Immunology
University of North Carolina
Chapel Hill, North Carolina 27514

MICROINJECTION OF ANTIBODIES INTO CELLS

Microinjection of antibodies into cells is a powerful technique to interfere with distinct viral (1) or cellular (2) functions. It is also attractive to use this technique to dissect viral replication and maturation at the molecular level: the living cell can be used as an experimental system, it is possible to inject large quantities of antibodies, the half life of the injected antibodies is usually longer than 24 hours (3), and injection of antibodies as such does not alter cellular ultrastructure (4) nor biofunction (5). In the present communication, we analyze aspects of the role of the vesicular stomatitis virus (VSV) nucleocapsid protein N in viral transcription and replication. We compare effects of monoclonal antibodies either introduced into cells by manual needle microinjection (6, 7) or mixed into cell-free transcription/replication systems.

MONOCLONAL ANTIBODIES THAT DISTINGUISH BETWEEN TWO POOLS OF CYTOPLASMIC N PROTEIN

Mouse monoclonal antibodies were obtained by standard procedures after immunization with a preparation of nucleocapsid free N protein obtained after high salt treatment of

---

[1]In vitro replication studies were supported by Public Health Service Grants AI 12464 and AI 15134 from the National Institute of Allergy and Infectious Diseases to the laboratory of Dr. Gail Wertz.

purified VSV nucleocapsids (serotype Indiana). Antibodies were purified to homogeneity by affinity chromatography on N protein attached to CNBr-activated Sepharose 4B, and binding specificity was tested on VSV proteins separated by poly-acrylamide electrophoresis and transferred to nitrocellulose papers.

Two anti-N antibodies were selected for these studies based on their pattern of immunofluorescent staining of N protein in VSV infected Madin Darby bovine kidney cells. Antibody 1 shows cytoplasmic clusters of N protein and a weak diffuse fluorescence (Figure 1a). Antibody 2 shows only the diffuse pattern of the N fluorescence (Figure 1b). These two antibodies recognize different epitopes (data not shown), and it is, therefore, conceivable that the epitope recognized by antibody 2 is not accessible on clustered N protein. Since electron microscopic studies show that clustered N protein may represent nucleocapsids (Shinishi Ohno, unpublished), we can postulate that antibody 1 recognizes nucleocapsids and free N protein and antibody 2 only free N protein.

EFFECTS OF MICROINJECTION OF ANTI-N ANTIBODIES ON VIRUS YIELDS

If the above hypothesis of a differential recognition of nucleocapsids and free N protein is correct, then microinjection of antibody 1 should protect cells against subsequent infection because antibody 1 would bind to infecting nucleo-

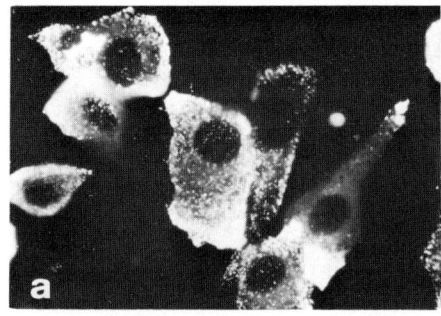
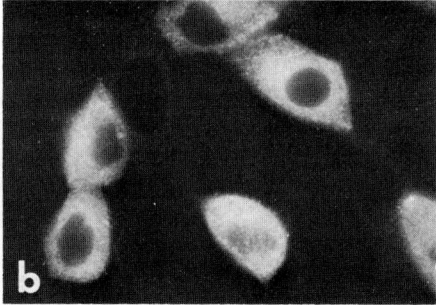

FIGURE 1. Immunofluorescent staining of N protein. Confluent monolayers of Madin Darby bovine kidney cells were fixed with 2% formaldehyde and permeabilized with 0.05% Triton X 100 5 hours after infection with VSV at a multiplicity of ≃ 0.5 pfu/cell. They were then incubated with mono-clonal antibodies (50 µg/ml), washed, and incubated with a goat anti-mouse IgG-fluorescein conjugate. a) monoclonal antibody 1, b) monoclonal antibody 2.

capsids and inhibit primary transcription. Microinjection of antibody 2 should not interfere with primary transcription, but the antibody should bind to newly synthesized N protein, thus making it unavailable for binding to viral RNA which may inhibit viral replication. We first injected antibodies at a concentration of 5 mg/ml (equivalent to $\simeq 2 \times 10^6$ molecules/cell) into bovine kidney cells which were subsequently infected with 10 pfu/cell of VSV. Immunofluorescent double staining for injected antibody and viral G protein 4 hours after infection showed that of 56 cells injected with antibody 1, none expressed G; of 55 cells injected with antibody 2, 42 expressed G at low levels and 12 showed no G; and of 42 cells injected with antibody 3, a control antibody made to an unrelated synthetic peptide (8), all except one expressed G at high levels. To test the effect of antibodies on virus yields, antibodies were injected at the same concentration into all of approximately 100 cells grown on coverslips. Cells were subsequently infected as above and supernatants were titrated for infectivity at several time points after infection. Results are shown in Figure 2. Cells injected with antibody 1 made negligible amounts of virus, cells injected with antibody 2 produced peak infectivities at

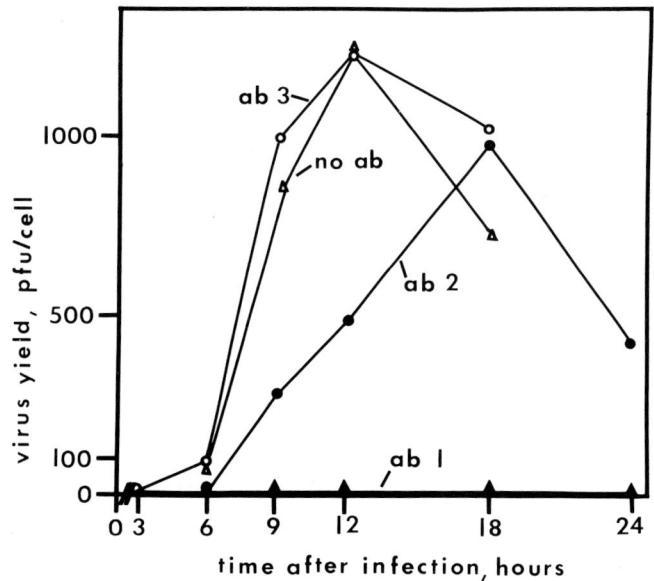

FIGURE 2. Cumulative virus yields in microinjected cells. Plaque titration of supernatant fluids was done in baby hamster kidney cells.

18 hours, and cells injected with control antibody 3 had a peak infectivity at 12 hours, similar to cells not injected with antibody. Thus, antibody 1 inhibited and antibody 2 delayed viral replication.

EFFECT OF ANTIBODIES ON IN VITRO TRANSCRIPTION OF VSV

The small number of cells that could be injected in reasonable time made biochemical studies in the living cell unfeasible. We, therefore, tested the effect of the two antibodies in standard in vitro transcription assays (9). Detergent disrupted VSV was mixed with various concentrations of antibodies and incubated with the four nucleotide triphosphates, one of which was radiolabeled (^{3}H-UTP, 2 Ci/mmol). Trichloroacetic acid precipitable transcripts were counted. Results are shown in Figure 3. Antibody 1 (15 µg/ml) inhibited transcription by more than 80%, whereas neither antibody 2 (17.5 µg/ml) nor antibody 3 (67.5 µg/ml) had any effect. Partially purified antibody 2 tested at concentrations of antibodies of up to 330 µg/ml had, likewise, no

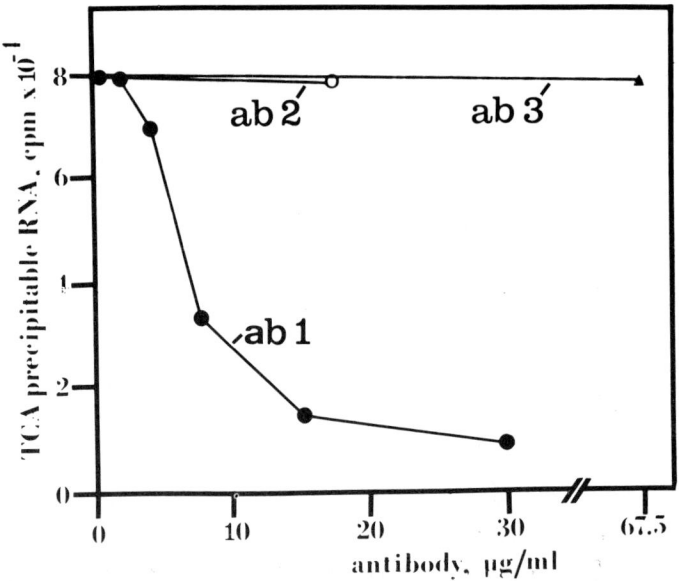

FIGURE 3. Effect of antibodies on in vitro transcription of VSV. Detergent disrupted virus (4 µg protein/100 µl) was incubated with antibodies and 10 U/100 µl of RNasin, 1 mM ATP, 200 µM CTP, 200 µM GTP and 50 µM ^{3}H-UTP for 90 minutes at 30°C. TCA precipitable RNA was counted in a Beckman scintillation counter.

effect (data not shown). Thus, the delay in viral growth in cells injected with antibody 2 cannot be explained by an effect on primary transcription.

EFFECT OF ANTIBODIES ON IN VITRO REPLICATION OF VSV

An in vitro system, including micrococcal nuclease treated rabbit reticulocyte lysate and intracellular VSV defective interfering (DI) nucleocapsid templates, supports the synthesis of DI genome length RNA when viral protein synthesis is programmed by added viral mRNAs (10). In the absence of viral mRNA, the intracellular DI nucleocapsids

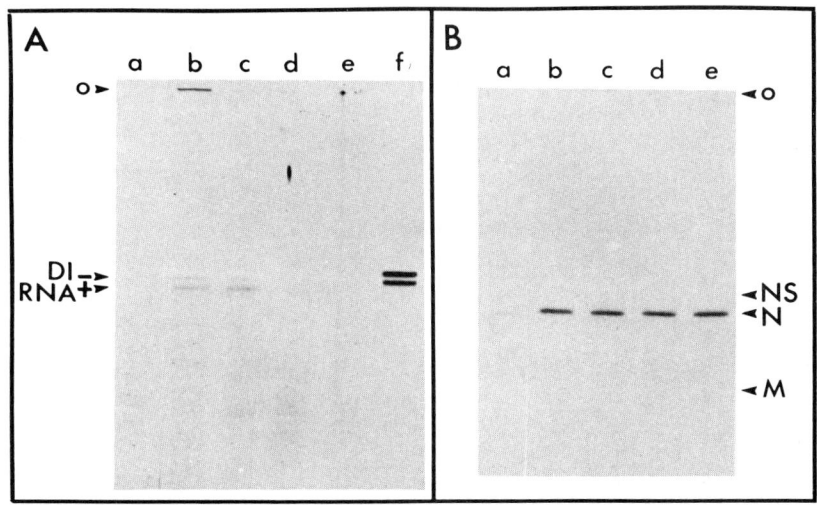

FIGURE 4. In vitro RNA replication and protein synthesis in the presence of monoclonal antibodies to N protein. Reaction mixtures (13) containing a constant amount of hybrid-selected N mRNA, ^{3}H-UTP and ^{35}S-methionine were incubated at 30°C for 45 min. Anisomycin (25 µM), the respective antibody, and purified intracellular DI nucleocapsids were added and the incubation was continued for 2¼ hours. Deproteinized RNA products (A) were electrophoresed in a 6 M urea agarose gel (10), and protein products (B) were analyzed by PAGE (14). (a) No added mRNA, no antibody, (b) N mRNA, no antibody, (c) N mRNA, antibody 3, (d) N mRNA, antibody 2, (e) N mRNA, antibody 1, (f) in vivo labeled DI intracellular nucleocapsid template RNA.

produce only a 46 base leader RNA. Hybrid-selected N mRNA was used to program protein synthesis in this *in vitro* system. The results showed that a source of N protein alone was sufficient to promote genome length RNA synthesis by the enzymatically active templates (11) (Figure 4, reactions a and b). These data and the finding of others (12) lead to the prediction that antibodies to the N protein would inhibit replication *in vitro*. In fact, both antibody 1 and 2 inhibited genome length RNA synthesis when added to reactions programmed with hybrid-selected mRNAs. Control antibody 3 had no effect (Figure 4, reactions c, d and e). A source of free N protein is required for RNA replication, but not for transcription; the N genome complex is the template for both reactions. Therefore, the finding that antibody 2 inhibits replication, but not transcription, is consistent with the hypothesis that this antibody recognizes only free N protein. The mechanism by which antibody 1 inhibits replication may involve interaction with both free and nucleocapsid-bound N protein.

ACKNOWLEDGMENTS

We thank Dr. Mark Willingham, NCI, NIH, for advice with the injection techniques.

REFERENCES

1. Antman, K. H. and Livingston, D. M. (1980). Cell 19, 627.
2. Wehland, J., Henkart, M., Klausner, R., and Sandoval, I. (1983). Proc. Natl. Acad. Sci. (USA) 80, 4286.
3. McGarry, T., Rottough, S., Rogers, S., and Rechsteiner, M. (1983). J. Cell. Biol. 96, 338.
4. Wehland, J., Willingham, M. C., Dickson, R., and Pastan, I. (1981). Cell 25, 105.
5. Klimkowski, M. W., Miller, R. H., and Lane, E. B. (1983). J. Cell Biol. 96, 494.
6. Graessmann, M., and Graessmann, A. (1976). Proc. Natl. Acad. Sci. (USA) 73, 366.
7. Wehland, J., Osborn, M., and Weber, K. (1977). Proc. Natl. Acad. Sci. (USA) 74, 5613.
8. Arnheiter, H., Ohno, M., Smith, M., Gutte, B., and Zoon, K. (1983). Proc. Natl. Acad. Sci. (USA) 80, 2539.
9. Colonno, R. J., and Banerjee, A. K. (1976). Cell 8, 197.
10. Wertz, G. (1983). J. Virol. 46, 513.
11. Patton, J., Davis, N., and Wertz, G. (1983). This volume.
12. Hill, V. M., and Summer, D. F. (1982). Virol. 123, 407.
13. Davis, N., and Wertz, G. (1982). J. Virol. 41, 821.
14. Laemmli, U. K., and Faure, M. (1973). J. Mol. Biol. 80, 575.

THE COILING OF VESICULAR STOMATITIS VIRUS NUCLEOCAPSIDS AT THE INNER SURFACE OF PLASMA MEMBRANES: IMMUNOLOCALIZATION OF THE MATRIX PROTEIN

Ward F. Odenwald, Heinz Arnheiter, Monique Dubois-Dalcq, and Robert Lazzarini

Laboratory of Molecular Genetics, IRP
National Institute of Neurological and Communicative Disorders and Stroke, NIH
Bethesda, Maryland 20205

Nucleocapsids of vesicular stomatitis virus (VSV) are loosely coiled or extended in the cytoplasm of infected cells and tightly coiled in the virion. Thin-section electron microscopy has shown that tight coiling occurs during viral budding from cellular membranes (1). However, the molecular mechanisms of this coiling are not yet understood, and little ultrastructural information on this process is available. Here, we present high resolution views of nucleocapsid coiling at the cytoplasmic surface of plasma membranes as it would be seen by an observer localized inside the cell.

PLATINUM-CARBON REPLICAS OF THE CYTOPLASMIC SURFACE OF PLASMA MEMBRANES

Cytoplasmic surfaces of plasma membranes of VSV infected baby hamster kidney cells were prepared by a modification of the technique originally described by Büechi and Büchi (2). Cells were grown on glass coverslips and incubated for six hours after infection at 37^0C. They were then vigorously rinsed in ice cold physiological buffer (100 mM KCl, 5mM $MgCl_2$, 2 mM EGTA and 30 mM Hepes, pH 7.1). This resulted in the removal of cell bodies from the coverslip, leaving behind basal plasma membranes with their cytoplasmic surfaces exposed. Fixation with 1% glutaraldehyde in the above buffer was followed by a brief rinse in 15% methanol in distilled water. The coverslips were then quickly hand-dipped into liquid propane/2-methyl butane (3:1) cooled by a liquid nitrogen bath. Etching and platinum-carbon rotary shadowing were performed in a Balzers 301 freeze-fracture device according to Heuser and Kirschner (3).

MORPHOLOGY OF VSV NUCLEOCAPSIDS ON THE INSIDE OF PLASMA MEMBRANES

Embedded between numerous clathrin sheets, clathrin baskets and cytoskeletal elements, we observed nucleocapsids in three different conformations: tightly coiled, tortuous and linear. These different conformations were either separate or attached to each other (Figures 1-2). The bullet-shaped, tightly coiled structures consisted of a nose cone of 5 to 6 turns of increasing diameter followed by a variable number of turns of 42.5 $\pm$ 2.5 nm diameter. These structures resembled nucleocapsid "skeletons" as derived from octylglucoside treated virus particles (4) and were never found in non-infected cells. The apex of the nose cone was observed free of uncoiled material and attached to the plasma membrane. In addition, the coils forming the nose cone were often separated from adjacent strands in the coil (Fig. 1 and 2), probably the result of physical stress induced during the coverslip rinsing. Stereo views demonstrated that many tight coils were embedded in the membranes. Surface subunits (33 $\pm$ 3 Å in diameter) formed a lattice along the helical coils which had a periodicity of approximately 5 nm. The distance between subunits along the cylindrical portion of the tight coils was approximately 17 Å. The nose cones appeared to have a tighter subunit packing. Uncoiled tortuous material was consistently found attached to the base of the tight coils and most likely represents nucleocapsid that was not yet tightly packed. Several tight coils were often found associated with one tortuous complex. Stereo views show that the tortuous complexes often projected away from the plane of the membrane. The surface underlining tortuous material was more granular than the surrounding cytoplasmic surfaces. In addition, linear nucleocapsids were observed traversing into or out of tortuous complexes. These linear structures were often in close association with the plasma membrane. The linear portions of the nucleocapsids had a morphology similar to that of negatively stained nucleocapsids prepared from cells or virions (5) and could easily be distinguished from the thicker microtubules and intermediate filaments. The distinction between microfilaments and linear nucleocapsids could be made on the basis of differences between subunit morphology and periodicity. Nucleocapsid subunit periodicity, approximately 37 Å, was less than the microfilament subunit spacing, approximately 55 Å. Depending on the angle of viewing, linear nucleocapsids had a width of 37 to 87 Å while the more symmetrical microfilaments were between 85 and 100 Å.

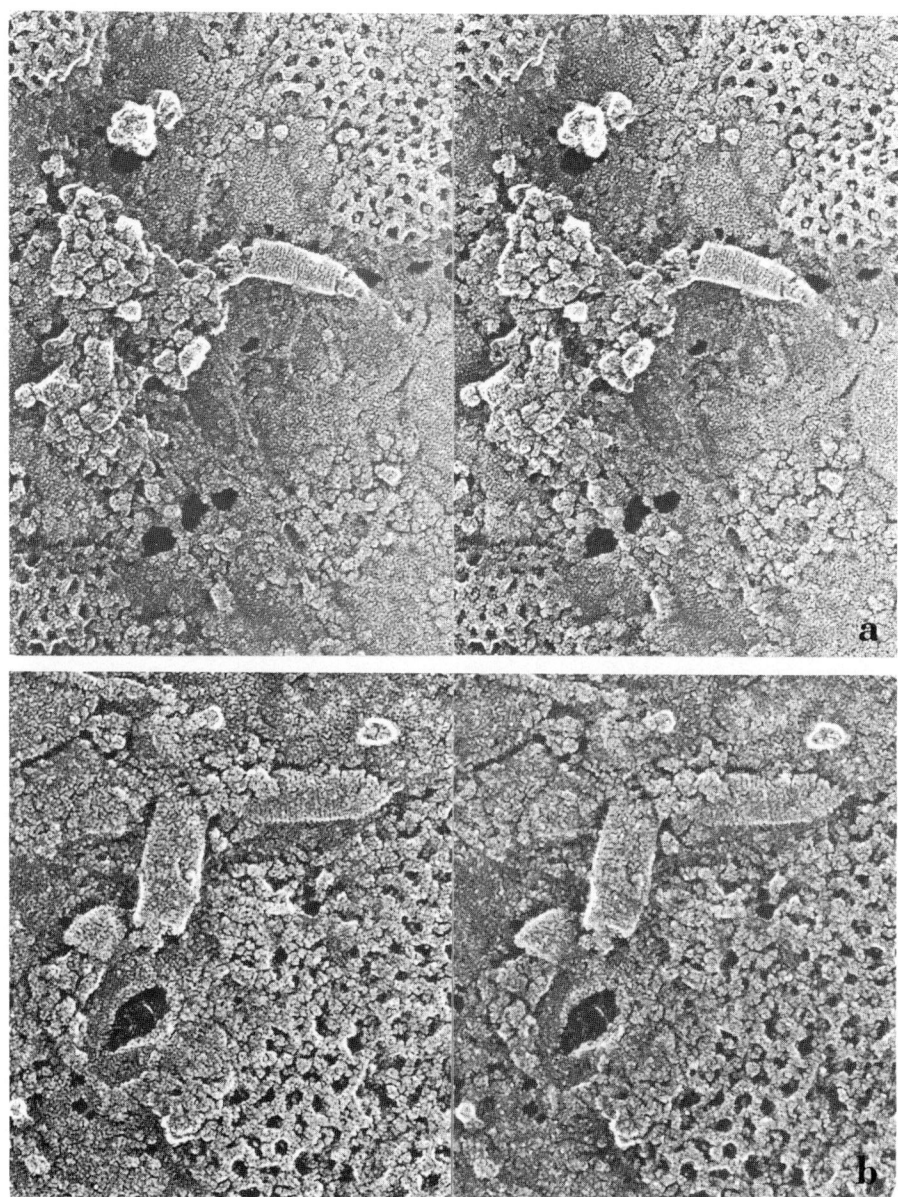

FIGURE 1. Stereo views of nucleocapsids on the cytoplasmic side of basal plasma membranes. (a) note the tortuous complex attached to the base of the tight coil whose apex is attached to the membrane (127,500x). (b) varying degrees of nucleocapsid coiling (172,500x).

IMMUNOLOCALIZATION OF VSV MATRIX PROTEIN

The exact localization of the viral matrix (M) protein in the virus and in the cell is still controversial. When viral membranes are solubilized by low ionic strength detergents, the M protein remains tightly associated with the nucleocapsid "skeletons" and is involved in maintaining the tightly coiled "skeleton" conformation (6). Other studies suggest that the M protein is located on the inner surface of the viral membrane (7, 8, 9) or that it even traverses the lipid bilayer (10). However, the amino acid sequence of the M protein does not indicate any long hydrophobic or nonpolar domains that would suggest membrane association (11). Therefore, we wanted to immunolocalize the M protein on the cytoplasmic side of the plasma membrane and the different nucleocapsid structures described above.

As a specific immunoreagent, we used polyclonal rabbit antibodies which were raised against M protein purified by SDS polyacrylamide gel electrophoresis and were affinity purified on M protein coupled to CNBr-activated Sepharose 4B. Purified antibodies were specific for M protein as tested by immunoblotting procedures (data not shown). They were coupled to 5 nm colloidal gold particles according to standard methods (12). This conjugate was used on plasma membrane fragments either before or after fixation with 1% glutaraldehyde. To enhance the observed frequency of nucleocapsid tight coils on the plasma membrane, infected cells were incubated with 15 μg/ml of Concanavalin A in the above buffer for 2 hours at 4°C and then rinsed as above. Figure 2 represents views from the immunolocalization. The colloidal gold particles which appear as white dots on these reversed image micrographs were only found on tortuous nucleocapsids and not on the surface of tight coils, linear nucleocapsids, cytoskeletal elements, or the plasma membrane.

CONCLUSIONS

Tortuous nucleocapsid was found at the bases of bullet-shaped tight coils of varying lengths while the apex of their nose cones, free of uncoiled nucleocapsid, were closely associated with the plasma membrane. This consistant observation suggests that the nucleocapsid coiling initiates at a membrane attachment site and then proceeds unidirectionally with nose cone formation followed by a helical cylinder. It further suggests that the tortuous conformation is necessary for the development of the tight coils or, conversely, is the result of tortional stress induced by the tight coiling.

The M protein is immunologically labeled only on the tortuous nucleocapsids. Two different scenarios of nucleocap-

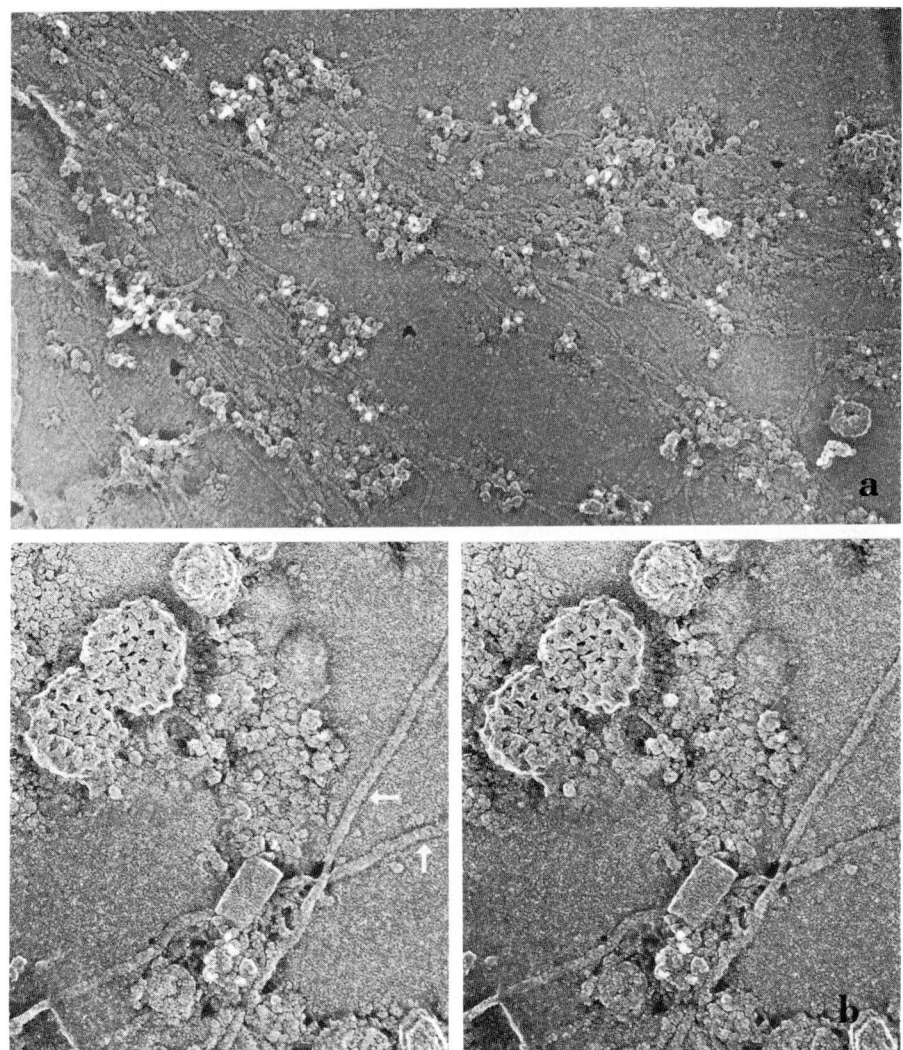

FIGURE 2. Immunolocalization of M protein on the cytoplasmic side of the plasma membrane. (a) low magnification view of linear and labeled tortuous nucleocapsids. Note the close association between the linear nucleocapsids and the membrane (72,000x). (b) stereo view of a tight coil and labeled tortuous nucleocapsid, arrows indicate fibers with the dimensions of intermediate filaments (120,000x).

sid coiling may have led to this M protein localization: twisting of nucleocapsids would expose sites on the nucleocapsid that enable the M protein to bind; alternatively, M protein could bind to membrane bound nucleocapsid and induce a twisted nucleocapsid conformation which would then favor tight coiling. Matrix protein could then cross-link adjacent turns in the helix. What molecular interaction would allow initial binding of M protein to nucleocapsid is not clear. We are currently investigating whether the cytoplasmic portion of the viral G protein has any role in binding of nucleocapsid to the plasma membrane. It is evident that our present results do not distinguish between these two scenarios. However, it is clear that under our experimental conditions the M protein cannot be localized on the inner side of the plasma membrane or on the surface of the tight coils and may therefore, not play a role in the association of nucleocapsids with membranes.

ACKNOWLEDGMENTS

The authors thank Dr. Paul Bridgeman for helpful discussions.

REFERENCES

1. Murphy, F. A., and Harrison, A. K. (1980). In "Rhabdoviruses Vol. 2" (D. H. L. Bishop, ed.), p. 65. CRC Press, Inc., Florida.
2. Büchi, M., and Büchi, T. (1979). J. Cell Biol. 83, 338.
3. Heuser, J. E., and Kirschner, M. W. (1980). J. Cell Biol. 86, 212.
4. Newcomb, W. W., and Brown, J. C. (1981). J. Virol. 39, 295.
5. Blumberg, B. M., Giorgi, C., and Kolakofsky, D. (1983). Cell 32, 559.
6. Newcomb, W. W., Tobin, G. J., McGowan, J. J., and Brown, J. C. (1982). J. Virol. 41, 1055.
7. Morrison, T. G., and McQuain, C. O. (1978). J. Virol. 26, 115.
8. Pepinsky, R. B., and Vogt, V. M. (1979). J. Mol. Biol. 131, 819.
9. Zakowski, J. J., and Wagner, R. R. (1980). J. Virol. 36, 93.
10. Walter, G., and Mudd, J. A., (1973). Virol. 52, 574.
11. Rose, J. K., and Gallione, C. J. (1981). J. Virol. 39, 519.
12. Slot, J. W., and Gruze, H. J. (1981). J. Cell Biol. 90, 533.

INTERACTIONS OF VIRAL PROTEINS WITH MURINE LYMPHOCYTES[1]

James J. McSharry, Gail Goodman-Snitkoff,[2] and Shinae Kizaka[3]

Department of Microbiology and Immunology
Albany Medical College of Union University
Albany, New York 12208

SYNTHETIC SUBUNIT VIRAL VACCINES

Progress in the development of synthetic subunit viral vaccines has been reported (1). Small peptides have been isolated from immunogenic proteins by enzymatic or chemical cleavage, tested for their ability to elicit antibodies that react with the isolated protein and the virion, their amino acid sequence determined, and the synthetic peptide made by the Merrifield technique (2). A number of investigators have used these procedures to synthesize biologically and chemically active peptides. For example, a synthetic peptide consisting of 20 amino acids corresponding to a portion of the MS-2 coat protein elicits antibodies which efficiently neutralize MS-2 infectivity (3); a number of synthetic peptides containing greater than 10 amino acids and corresponding to peptides of the surface protein of hepatitis B virus elicit antibodies in rabbits which precipitate the 23,000 and 28,000 dalton envelope proteins of Dane particles (4); synthetic peptides of the influenza virus hemagglutinin protein elicit antibodies that react with the complete hemagglutinin protein in an ELIZA test (5); and a 20 amino acid peptide of VP1 of foot and mouth disease virus will protect mice and guinea pigs against challenge with live foot and mouth disease virus (6). Thus, small synthetic peptides can be made which elicit antibodies that react with the complete antigen, and in some cases neutralize infectivity and protect animals against challenge with live virus. These synthetic immunogenic peptides should be useful as effective, inexpensive subunit vaccines.

---

[1]This work was supported by a National Science Foundation Grant No. PCM-8003126 and a New York State Health Research Council Grant No. 10-023.
[2]G.G.S. is the recipient of an NIH Postdoctoral fellowship No. F32-AI06060.
[3]S.K. is the recipient of an International Rotary Foundation Graduate Scholarship.

For the development of these synthetic subunit vaccines a rapid, inexpensive in vitro assay system for determining the ability of isolated viral proteins and their peptides to stimulate the cells of the immune system would be useful. Recently it has been demonstrated that viruses and their isolated antigens can stimulate lymphocytes in vitro. This has been shown for herpes simplex virus types 1 and 2 (7,8,9), influenza virus (10) and its hemagglutinin (11), vesicular stomatitis virus and its glycoprotein (12,13,14), Sindbis virus and its glycoproteins (15), and Sendai virus and its glycoproteins (16). In vitro mitogenesis is a rapid, inexpensive assay system for measuring the ability of viruses, their isolated proteins and peptides to activate lymphocytes. The requirements for various cell populations and the cooperation of B and T cells in the mitogenic response are easily measured in this in vitro assay. Thus, it would be useful for studying the interaction of viral proteins and peptides alone, chemical modifications of these antigens and combinations of antigens to determine their ability to elicit a mitogenic response in vitro. Proteins and smaller peptides that are active in this in vitro mitogenesis assay can be tested either alone or attached to carrier molecules for immunogenicity in vivo. Those mitogenic peptides that neutralize infectivity will be candidates for synthetic subunit viral vaccines.

In this communication, we present data that show that the immunogenic glycoprotein (G) of VSV, cyanogen bromide peptides and tryptic peptides of the G protein activate murine spleen cells in vitro. In addition, we demonstrate that the purified F glyco-protein of Sendai virus is the major mitogen associated with the virion. The use of an in vitro assay system and the mitogenic activity of these proteins and their peptides as model systems for the development of synthetic subunit vaccines are discussed.

PEPTIDES OF THE VSV GLYCOPROTEIN

Mitogenic Activity of Cyanogen Bromide Peptides of the G Protein. Since we have established that the G protein is the major mitogen associated with VSV (13), we attempted to determine if peptides of the G protein are also mitogens. The G protein was isolated from purified virus by Triton X-100 extraction as previously described (13) and cleaved with cyanogen bromide according to the procedure of Stanworth and Turner (17). Cleavage of the G protein was monitored by PAGE on 14% polyacrylamide gels in the presence of 1% SDS and 1% 2-mercaptoethanol (18). Treatment with cyanogen bromide removed essentially 80-90% of the band migrating as G protein and yielded 9 to 10 peptides smaller than native G protein. This result is consistant with the potential number of cyanogen bromide peptides present in the G protein as determined by its amino acid sequence (19).

Various concentrations of the G protein or cyanogen bromide-treated G protein, along with known mitogens LPS, Con A, and VSV, were incubated with murine spleen cells and then assayed for

TABLE 1
LYMPHOCYTE ACTIVATION BY CYANOGEN BROMIDE (CB)-TREATED G PROTEIN[a]

Mitogen	μg/Well	^{3}H-Thymidine Incorporation	
		Average CPM	S.I.[b]
None	--	5,122	--
LPS	10	69,285	12.2
Con A	0.12	109,265	21.3
HKCC-VSV	5	24,468	4.8
G Protein	1	17,418	3.4
	5	31,609	6.2
	10	55,972	10.9
CB-treated G Protein	1	21,020	4.1
	5	53,878	10.5
	10	37,875	7.4

[a] Spleen cells from BALB/c female mice were used; standard error of the mean was less than 10%.
[b] S.I.= stimulation index: cpm experimental/cpm control.

mitogenesis 48 hr after culture initiation. The data presented in Table 1 show that the mixture of cyanogen bromide peptides of the G protein is at least as active as the untreated G protein in the mitogen assay. At 1 and 5 μg/well, the mixture of cyanogen bromide peptides of G are more mitogenic than the same concentrations of untreated G protein, indicating that cleavage of the G protein changed the optimal concentration of G protein for mitogenesis from 10 to 5 μg/well. Both the G protein and the mixture of cyanogen bromide peptides of the G protein are more mitogenic than VSV, a result similar to our previous finding (13,14).

Mitogenic Activity of the Tryptic Peptides of the G Protein. The VSV glycoprotein, isolated as described above, was treated with an equal amount of trypsin. This treatment reduced all of the band migrating as the uncleaved G protein to peptides which were smaller than trypsin. This result is consistent with the size of tryptic peptides present in the G protein (19).

To determine if tryptic peptides of the G protein are mitogens, various concentrations of untreated or trypsin-treated G protein were incubated with murine spleen cells and then assayed for mitogenesis. The data in Table 2 show that at 0.5 and 1.0 μg/well, a mixture of tryptic peptides of the G protein are approximately 4 times as stimulatory as the same concentrations of untreated G protein, suggesting that cleavage has shifted the optimal concentration for mitogenesis from 10 to 1 µg/well. Alternatively, the lower mitogenic activity of trypsin-treated G protein at 5 and 10 µg/well may be due to the effect of trypsin on the lymphocytes. The data in the lower portion of Table 2 show that increasing concentrations of trypsin decrease mitogenesis, possibly by removing cell surface receptors for the mitogens.

The data from these experiments demonstrate that one or more of the cyanogen bromide and tryptic peptides of the G protein are mitogenic for murine spleen cells. We are presently attempting to isolate the cyanogen bromide and tryptic peptides in order to determine the mitogenic activity of individual peptides. On the basis of size and partial amino acid sequence data we should be able to determine which portions of the G protein have mitogenic activity. Mitogenic peptides will be injected into mice to determine their immunogenicity and their ability to elicit neutralizing antibodies for infectivity and mitogenicity of the virion.

SENDAI VIRUS GLYCOPROTEINS

We have demonstrated that a mixture of the Sendai virus glycoproteins HN and F are mitogenic for murine spleen cells (16). The mitogenic response of the glycoproteins has two components, a T cell-independent B cell response and a T cell dependent B cell response. To determine if these different mitogenic responses reside in one or both of these Sendai virus glycoproteins we have separated the two glycoproteins and begun to characterize their mitogenic activities.

Mitogenic Activity of the Sendai Virus HN Glycoprotein. The glycoproteins of Sendai viruses were isolated by solubilization in Triton X-100 and the HN glycoproteins separated from the F glycoprotein by adsorption to and elution from gluteraldehyde treated RBC (20). The adsorbed and eluted material contained only HN as judged by PAGE. Various concentrations of the purified HN glycoprotein, along with known mitogens, were incubated with lymphocytes from CBA/J mouse spleens and mitogenic activity was determined at various times after culture initiation. The data in Table 3 show that the isolated HN glycoprotein is mitogenic at 48 and 72 hr after culture initiation with 0.5 and 1.0 µg/well. However, the stimulation index was only slightly above 3.0, which is the lowest value considered significant.

Mitogenic Activity of the Sendai Virus F Glycoprotein. The F glycoprotein was isolated from Sendai virus as described above. PAGE of the unadsorbed material indicated that the F glycoprotein was the only Sendai virus protein present. Contaminants derived

TABLE 2
LYMPHOCYTE ACTIVATION BY THE TRYPSIN-TREATED G PROTEIN[a]

Mitogen	µg/Well	^{3}H-Thymidine Incorporation	
		Average CPM	S.I.[b]
None	--	1,079	--
LPS	10	33,691	31.2
Con A	0.12	133,580	123.8
G Protein	0.5	7,908	7.3
	1	10,982	10.2
	5	26,053	24.1
	10	41,156	38.1
Trypsinized G Protein	0.5	37,193	34.5
	1	41,521	38.5
	5	35,688	33.1
	10	25,785	24.0
Trypsin	10	3,136	2.9
	25	1,603	1.5
	50	814	0.8
	100	519	0.5

[a,b]Same as in Table 1.

from the gluteraldehyde-treated RBC were present, but were not mitogenic in our assay. Various concentrations of the F glycoprotein, along with standard mitogens, were tested for their mitogenic activity in mouse splenocytes. The data in Table 4 show that the purified F glycoprotein is mitogenic for murine splenocytes. Significant mitogenicity occurs at all concentrations and times tested. Optimal activity occurs at 72 hr after culture initiation with 0.5 and 1.0 µg/well. The stimulation index of approximately 14 is about 3 times greater than that for the HN glycoprotein and is comparable to that of the isolated VSV glycoprotein (14).

The data presented here suggest that the isolated F glycoprotein contains the majority of the mitogenic activity associated with a mixture of the HN and F glycoproteins reported earlier (16). We are currently attempting to determine if the isolated F glycoprotein stimulates B cells or T cells or both, and if T cell help is required as was demonstrated for the mixture of HN and F glycoproteins (16). In addition, we will prepare peptides of the F glycoprotein and determine its mitogenic potential. Any mitogenic

TABLE 3
LYMPHOCYTE ACTIVATION BY HN GLYCOPROTEIN

Mitogen	μg/well	CPM (^{3}H)THYMIDINE INCORPORATION[a]			
		48 hr	S.I.[b]	72 hr	S.I.
None	--	6,737	(-)	713	(-)
LPS	20	35,721	(5.3)	6,695	(9.4)
Con A	0.125	157,466	(23.4)	25,795	(36.2)
HN	1.0	24,780	(3.7)	3,417	(4.8)
	0.5	20,003	(3.0)	2,533	(3.6)
	0.25	17,410	(2.6)	2,045	(2.9)
	0.125	18,826	(2.8)	2,995	(4.2)

[a]Spleen cells from CBA/J female mice were used; standard error of three replicate samples was less than 10%.
[b]Same as in Table 1.

TABLE 4
LYMPHOCYTE ACTIVATION BY F GLYCOPROTEIN

Mitogen	μg/well	CPM (^{3}H)THYMIDINE INCORPORATION[a]			
		48 hr	S.I.[b]	72 hr	S.I.
None	--	1,094	(-)	581	(-)
LPS	20	15,106	(13.8)	18,618	(32.0)
Con A	0.125	125,466	(114.7)	150,795	(259.5)
F	1.0	7,535	(6.9)	8,319	(14.3)
	0.5	6,645	(6.1)	7,827	(13.5)
	0.25	6,216	(5.7)	6,569	(11.3)
	0.125	5,869	(5.4)	4,264	(7.3)

[a],[b]Same as in Table 3.

peptides will be tested for immunogenicity in vivo and their ability to induce antibodies that neutralize mitogenicity of the F glycoprotein and infectivity of Sendai viruses.

Isolation of mitogenic peptides and correlation with in vivo immunogenicity will be the first step in a rational approach to the development of synthetic viral subunit vaccines. The in vitro mitogen assay is rapid and inexpensive, and we have demonstrated that viral proteins and their peptides are active in the system. Those that are active in the in vitro system should be tested in vivo for immunogenicity. Those that are immunogenic, either alone or attached to a carrier, and are small enough to be synthesized, should be sequenced and then synthesized. With the rapid advances in peptide synthesis, the era of inexpensive, synthetic subunit vaccines is rapidly approaching.

REFERENCES

1. Arnon, R. (1980). Ann. Rev. Microbiol. 34, 593.
2. Merrifield, R.B. (1965). Science. 150, 178.
3. Langbeheim, H., et al. (1976). Proc. Natl. Acad. Sci. USA. 73, 4636.
4. Lerner, R.A., et al. (1981). Proc. Natl. Acad. Sci. USA. 78, 3403.
5. Green, N.H., et al. (1982). Cell. 28, 477.
6. Bittle, J.L., et al. (1982). Nature. 298, 30.
7. Kirchner, H.G., et al. (1976). J. Immunol. 117, 1753.
8. Kirchner, H.G., et al. (1978). J. Immunol. 120, 641.
9. Mochizuki, D., et al. (1977). J. Exp. Med. 146, 1500.
10. Butchko, G.M., et al. (1978). Nature. 271, 66.
11. Armstrong, R.B. (1981). Infect. Immunity. 34, 140.
12. Goodman-Snitkoff, G.W., and McSharry, J.J. (1980). J. Virol. 35, 757.
13. Goodman-Snitkoff, G.W., et al. (1981). J. Exp. Med. 153, 1489.
14. McSharry, J.J., and Goodman-Snitkoff, G.W. (1981). In "The Replication of Negative Strand Viruses" (D.H.L. Bishop and R.W. Compans, eds.), p. 929. Elsevier-North Holland Press, N.Y.
15. Goodman-Snitkoff, G.W., and McSharry, J.J. (1982). Infect. Immunity. 38, 1242.
16. Kizaka, S., et al. (1983). Infect. Immunity. 40, 592.
17. Stanworth, D.R., and Turner, M.W. (1978). In "Handbook of Experimental Immunology". (D.M. Weir, ed.), pp. 6.34-6.35. Blackwell Scientific Publications, Oxford.
18. Beckman, L.D., et al. (1976). Virology. 73, 216.
19. Rose, J., and Gallione, C.J. (1981). J. Virol. 39, 519.
20. Hosaka, Y. (1980). Infect. and Immunity. 30, 212.

HOST RANGE MUTANTS OF PIRY VIRUS: A NEW TYPE OF MUTANT IN DROSOPHILA

G. Brun

Laboratoire de Génétique des Virus
C.N.R.S.
91190 Gif sur Yvette (France)

I will recall the principles of selection of host range mutants in Drosophila. Flies were inoculated with one effective viral unit (cloning conditions). The observed frequency distribution of individual incubation times began as a Gaussian one and was extended by a tail due principally to physiological heterogeneity of flies but also to genetic heterogeneity of viruses. An attempt at selecting mutants in the head of incubation distribution has been unsuccessful. On the contrary selection in the tail gave a great variety of mutants(1).

As selection was based on incubation time, it seemed possible to select mutants specific of symptom production; in fact all mutants I have studied were mutants restricted in their capacity to invade drosophila fly. These mutants were called agD (affected growth in Drosophila). The majority seems to have abnormal cellular cycle in all drosophila cells, but tissular specificity is possible.

There are mutants restricted specifically in some drosophila strains. tsl (F1 clone) is one of them. tsl is submitted to a general restriction in all flies but is specifically highly restricted by an allele of ref(3)A gene present in the Paris drosophila strain. A secondary mutant, obtained after one passage of F1 in a standard fly (hybrid Oregon ♀ x ebony ♂), F1SS4, appeared to have lost general restriction but retained specific one. Study of F1SS4 showed that in permissive drosophila strains incubation times are shorter than those of wild type (wt) original Piry virus. The same event has been observed a second time after a passage in an ebony fly of an unspecifically restricted agD mutant. Lastly, when I have searched directly for mutants restricted in some drosophila strains other than Paris strain, by selecting the tail of incubation distribution, an important fraction of mutants I found were more rapid than wt in appearance of symptom. I have interpreted these mutants as double mutants: a second mutation which appeared and was selected during multiplication of a restricted mutant might have an enhancing effect which would not only rub out the effect of the first mutation but also go beyond.

We shall neglect the problem that these double mutants had been obtained in all drosophila strains but Paris one, and keep our attention on the fact that a new type of mutation does exist which enhances the invading fly capacity of virus. I shall call these mutants rgD (rapid growth in Drosophila).

Direct Search of Single rgD Mutants. In the $\mathcal{S}6$ population, multiplied in chick embryo fibroblasts (CEF), by selecting the head of Paris flies' incubation-time distribution, I found such a mutant out of 500 infective units (IU) approximately.

The figure shows the relation of Mean Incubation Time (MIT) in Paris flies to inoculum for the first rgD mutant PSP2 and the wt. Disposition of curves may be interpreted as resulting of more rapid or more productive cellular cycle in all cells (2).

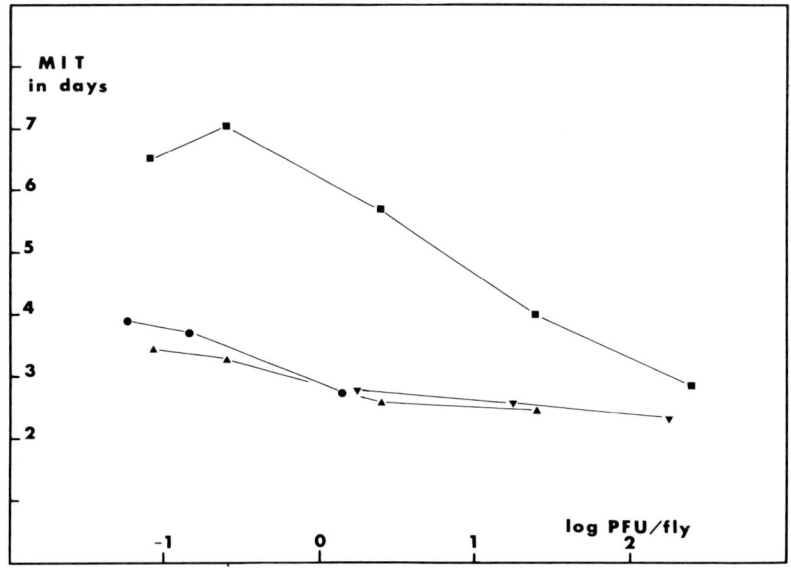

FIGURE. Comparison of MIT, at 20° in Paris flies from the same stock, for wt ■ , PSP2 ● and 20S3 ▲ mutants and for a ts$^+$ revertant of the last, 20S3r ▼ .

I noticed also a little increase (approximately 0.3 log unit) of the Relative Probability of Infection (RPI: ratio of the probability of infecting a fly to the probability of forming plaque in CEF) in Paris flies, but a great increase (1 log unit) in standard flies, so that RPI is the same for this mutant in Paris and standard flies (and in all strains used). I recall that Paris flies have been chosen as the most permissive for wt Piry virus, standard flies, hybrid Oregon x ebony, being the least. So, effects of diverse restrictive alleles (not still localized) on RPI are wholly rubbed out, but are still visible on incubation time.

The study of rgD mutants can be approached by two ways: genetic and physiologic; I have begun by genetical study.

GENETICAL STUDY OF rgD MUTANTS

Making up a Collection of Independent rgD Mutants. This collection obtained from 7 independent viral populations by selection in Paris and standard flies, is presented in table 1.

All nine rgD mutants have the same characteristics (MIT as a function of inoculum -see figure- and RPI) in Paris and standard flies, except F7S11 mutant. F7S11 shows peculiarities in Paris flies which may be interpreted as tissular specificity. Underlined mutants in table 1 are temperature sensitive (ts) in CEF (restrictive temperature 39°7); mutant 20S3 is leaky. Frequency of ts, 5/9, is high and exceeds random possibility (mean frequency of ts mutants in populations 5%), so that in some cases the two phenotypes rgD and ts should result from the same mutation. Reverting mutants for ts phenotype have been selected; for all ts,rgD mutants except 20S3 (see figure) this second mutation produces a reversion of the rgD phenotype.

TABLE 1
rgD MUTANTS SELECTED IN DIVERSE POPULATIONS MULTIPLIED IN CEF

Viral population	Selection in Paris	Selection in standard flies
S6	PSP2	S16
F7	–	F7S11 , F7S23
clone 12	–	–
clone 13	13P1	13S3
clone 14	–	–
clone 20	20P3	20S3
clone 21	–	21S3

So, for the four others, S16, F7S11, 13P1 and 13S3, the same mutation is probably responsible for the two phenotypes. This type of mutation is rare among spontaneous ts mutants since it has not been observed in more than 50 ts mutants.

Classification of ts Mutants in Complementation Groups. The chosen technique was complementation by UV irradiated virus (3). This technique is very dissymetric: a complemented ts virus is used at low multiplicity, the complementer, wt or ts, virus, used at high multiplicity, is UV irradiated; so this technique can be used with viral populations contaminated by defective interfering particles. Two mechanisms may occur. 1) functional survival of the gene, only effective for the first transcribed ones(N protein, group IV of VSV, and NS, group II) (4) and 2) reusing of the complementer virion's proteins. The first mechanism is very sensitive to UV dose; the second, less UV sensitive, has good efficiency for enzymatic proteins (L, group I, and NS), but for essentially structural proteins it is effective only for G (group V) (5,6). So, each complementation group has its peculiar relation of complementing efficiency by irradiated wt to a function of UV dose and of multiplicity. Such a preliminary classification was perfected by using complementer ts viruses; physiological properties of used mutants should be taken into account, for example group IV complementation requires complementer's transcription which seems defective for F4ts9 group I mutant. This work is in progress, the proposed classification (table 2) results only from formal study, but confirmation of physiological hypothesises

TABLE 2
COMPLEMENTATION GROUPS OF ts MUTANTS

	Ia(L)	Ib(L)	II(NS)	III(M)	IV(N)	V(G)	U.C.	Σ
ts,agD$^+$	18	2	0	1	2	0	2	25
ts,agD	17	4	0	0	3	0	0	24
Σ	35	6	0	1	5	0	2	49
ts,rgD	2 (13S3, 20S3)	3 (13P1, S16, F7S11)	0	0	0	0	0	5

Ia: uncomplemented, Ib: complemented by F4ts9
U.C.: uncertain classification

TABLE 3
INTRACISTRONIC COMPLEMENTATION IN GROUP I

Mutant	Complementation by F4ts9	Complementation by 13P1	Complementation by S16
13P1	35	N.S.[a]	N.S.
S16	15.5	1.6	N.S.
VXER6	5.4	16.5	3.7
21PR7	4.0	17.7	1.6
F7S11	3.0	11.0 ; 8.4	N.S.
F11ts5	1.9	5.9	2.0
ts1(F1A)	1.9	63	52
F11ts8	1.2	104	39
14PR1	0.2	69	28
F4ts9	no compl.	144 ; 180	86
wt	"	69	45
ts20	"	81	39
F2ts6	"	100	49
20S3	"	49	-
13S3	"	50 ; 127	47

[a] N.S.: non significative

Complementation by mutants is expressed as percent of complementation by wt in the same experiment.

(for F4ts9 for example) and direct confirmation of homology of established groups with VSV groups are in project.

Group frequencies among spontaneous ts mutants, selected as ts or agD, are similar to those observed for VSV Indiana (recapitulated data in (7)). Concerning ts mutants selected as rgD, difference in group frequencies has no signification because of the fewness of them; but the difference in frequency of group I mutants complemented by F4ts9 (3/4 if one excludes 20S3, probably double mutant) is highly significative. Moreover, complementation between F4ts9 and 13P1 or S16 is very effective in the two reciprocal ways (Table 3). Intracistronic complementation in group I of rhabdoviruses is classical (8,7) and high efficiencies have been observed with Chandipura (9).

Table 3 gives results of group I intracistronic complementation experiments. All the mutants complemented by F4ts9 are presented and ordered in decreasing efficiency of complementation by F4ts9. Complementation by F4ts9, 13P1 and S16 is expressed as percent of complementation by wt at comparable multiplicity in the same experiment (irradiation 4000erg/mm^2). We shall note that wt has a group I defect at 39°7 which is efficiently complemented by irradiated wt virions (autocomplementation). Four classes can be distinguished; 13P1 complements the majority of group I mutants as wt does (S16 in the same class, appeared less effective), but a middle class exists only partly complemented by both F4ts9 and 13P1 mutants.

Existence of this middle class is a second argument (the first is the relation of efficiency of complementation by irradiated wt to UV dose) to classify 13P1 and S16 in group I. But 13P1 and S16 define clearly an under-group which corresponds probably to a particular site of the L protein.

PHYSIOLOGICAL STUDY, PROJECTS.

For a well host-adapted virus as wt Piry, especially in Paris flies, there are many ways of losing its multiplying capacities, but certainly a few ones of becoming more efficient. In fact rgD mutants actually appeared phenotypically all alike (F7S11 seems to have a similar rapid multiplication but to be restricted in a specific tissue commanding access to symptom site).

The efficiency of rgD mutations to correct (in fact overcorrect) diverse agD mutations and to correct effects of diverse restrictive alleles (especially for RPI) suggests that mutations of this type enhance primary transcription. So, my project is to study the cellular cycle of rgD mutants in cultured drosophila cells and especially the primary transcription.

My provisional hypothesis (suggested by existence of F7S11 and of a double mutant which seems rgD cryosensitive) is that this enhancement is due to an association of L with an host protein which renders transcription more effective. This is only remaining to be proved.

REFERENCES

1. Brun, G. (1981). In "The Replication of negative Strand Viruses" (D.H.L. Bishop and R.W. Compans, eds.), p.921, Elsevier, North-Holland.
2. Brun, G. (1963). "Etude d'une association du virus sigma et de son hôte la drosophile". Thèse Paris XI, Orsay.
3. Deutsch, V., Muel, B. and Brun, G. (1977). Virology 77, 294.
4. Ball, L.A. and White, C.N. (1976). Proc. Nat. Acad. Sci. USA 73, 442.
5. Deutsch, V. (1975). J. Virol. 15, 798.
6. Deutsch, V. (1976). Virology 69, 607.

7. Pringle, C.R. (1977). In "Comprehensive Virology" (H. Fraenkel-Conrat and R.R. Wagner,eds.), Vol. 9, p.239, Plenum, New York.
8. Flamand, A. (1970). J. Gen. Virol., 8, 187.
9. Gadkari, D.A. and Pringle, C.R. (1980). J. Virol. 33,100.

EARLY APPEARANCE AND CO-LOCALIZATION OF INDIVIDUAL MEASLES VIRUS PROTEINS USING DOUBLE-LABEL FLUORESCENT ANTIBODY TECHNIQUES

R.N. Hogan, F. Rickaert, W.J. Bellini,
C. Richardson, M. Dubois-Dalcq and D.E. McFarlin[1]

Laboratory of Molecular Genetics
National Institute of Neurological and Communicative
Disorders and Stroke, NIH
Bethesda, Maryland 20205

In order to understand the sequence of events which lead to the assembly of measles virus, it is necessary to establish the precise time and location at which viral proteins first appear and how they interact with one another at the cell membrane during viral maturation. Norrby et al. (1) used monoclonal antibodies directed against the hemagglutinin protein (HA), phosphoprotein (P), nucleocapsid protein (NP), fusion protein (F) and matrix protein (M) of measles virus and found differential localizations of some of these proteins within infected cells. We have extended these studies using mouse monoclonal and rabbit polyclonal antibodies directed against individual viral proteins. Here we present a sequential immunofluorescence analysis of the emergence of viral proteins and, in some cases, their co-localization at sites of viral assembly.

Vero cells were infected with the Edmonston strain of measles virus at an MOI of 10 which resulted in infection of 90% of the cells and initiation of viral induced fusion at 18 hrs after infection. Cells were fixed at 2 hr intervals from 2 to 24 hrs after infection with periodate-lysine-paraformaldehyde fixative (2), a fixative which preserves both cell morphology and antigenic sites on proteins. To detect intracellular viral proteins, the cell membrane was permeabilized by treatment with 0.05% triton-x for 10 minutes. The mouse antibodies used in indirect immunofluorescent experiments were monoclonals to the HA, NP, P and M proteins (3, 4). Two polyclonal antibodies were raised in rabbits against: (a) purified HA protein and (b) a synthetic 20 amino acid peptide (p20) constructed from the nucleotide sequence of a region of the gene coding for

[1]Neuroimmunology Branch, NINCDS, NIH, Bethesda, MD 20205

the P protein (Bellini et al., this volume). The F protein was not investigated. Conjugated antibodies used for double-labeling experiments were goat anti-rabbit fluorescein and goat anti-mouse rhodamine. None of the antibodies or conjugates stained uninfected cells.

Staining of the HA protein was first detected at 4 hrs after infection (Fig. 1) in the form of a few bright granules and small dots scattered in the cytoplasm. Preliminary EM analysis suggest that the small dots correspond to HA associated with the membrane of smooth vesicles (Rickaert, unpublished observations). By 6 hrs, HA protein staining was predominantly found in the perinuclear area. At 8 hrs, HA was localized throughout the cytoplasm and at the cell surface where occasional punctate accumulation was seen. The HA glycoprotein was the only viral protein tested that could be stained in non-permeabilized cells or in living cells. NP was also first detected at 4 hrs in bright dots scattered over the cytoplasm (Fig. 1). With time, NP accumulated into larger and more abundant cytoplasmic dots.

Using monoclonal antibodies, P and M proteins were first detected at 8 hrs. P protein formed bright cytoplasmic dots in a pattern similar to NP, while M protein staining was weak and diffuse. After 8 hrs of infection, staining intensity of all proteins increased rapidly.

By 14 hrs the HA, NP, M and P proteins were co-localized into patches at the membrane, leaving surrounding membrane areas free of stain. Patches increased in number and extent until cell fusion began at 18 hrs, and gradually decreased thereafter.

Cytoplasmic inclusions known to contain nucleocapsids (5) were first detected at 12 hrs after infection. Inclusions could be stained with antibodies to NP and P proteins but not with four different monoclonal antibodies to the M protein (6-3 from W. Bohn; 16BB2, 16BB6 and 19CG6 from E. Norrby).

Double-labelling of viral proteins at 18 hrs is shown in Figs. 2 and 3 and the results are summarized in Table I.

Our results suggest that HA, a transmembrane glycoprotein, and NP protein, the major structural component of the nucleocapsid, emerge simultaneously in the cytoplasm at 4 hrs postinfection and that HA is transported from ER and Golgi regions to the surface within the next 4 hrs. P, another nucleocapsid associated protein, and M, the matrix protein, were found later. While it cannot be excluded that P and M proteins may be synthesized at only low levels early in infection, it seems more likely that their delayed appearance may be more apparent than real. Monoclonal antibodies bind to only one antigenic epitope and with

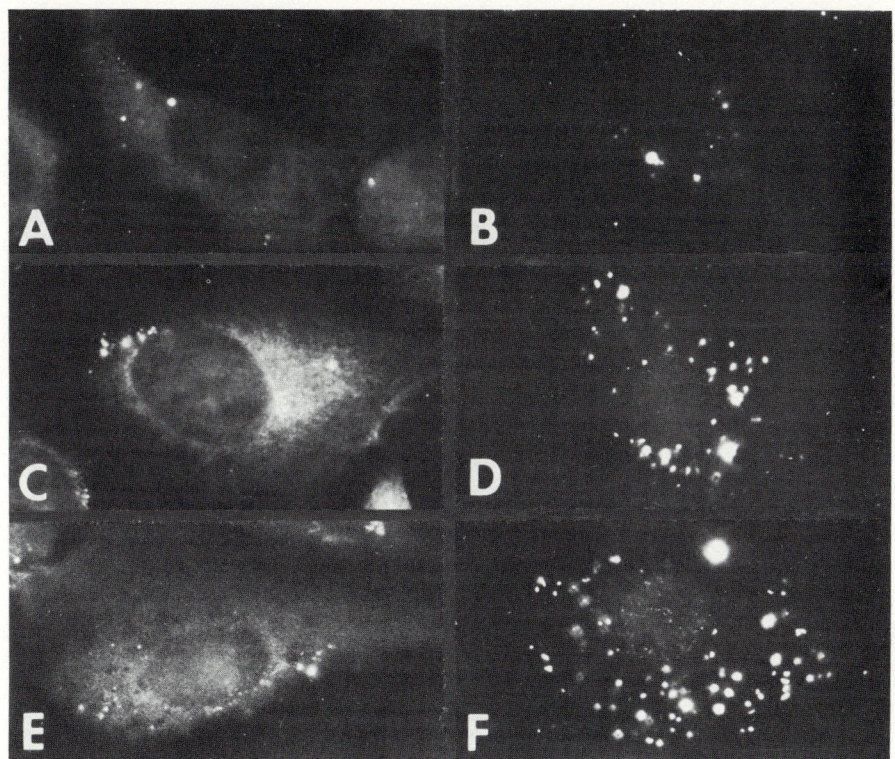

FIGURE 1. Immunofluorescence staining of HA and NP proteins of measles virus in Vero cells at 4 hrs (A,B), 6 hrs (C,D) and 8 hrs (E,F) after infection. (A,C,E): monoclonal anti-HA; (B,D,F): monoclonal anti-NP.

TABLE 1
VIRAL PROTEINS IN CELLS 18 HOURS AFTER INFECTION

Immunofluorescent localization	Genome associated proteins		Envelope associated proteins	
	N	P	M	HA
Cytoplasm				
Inclusions	+	+	−	−
Diffuse small granules	−	−	+	−
Membrane				
Patches	+	+	+	+
Surface stain	−	−	−	+

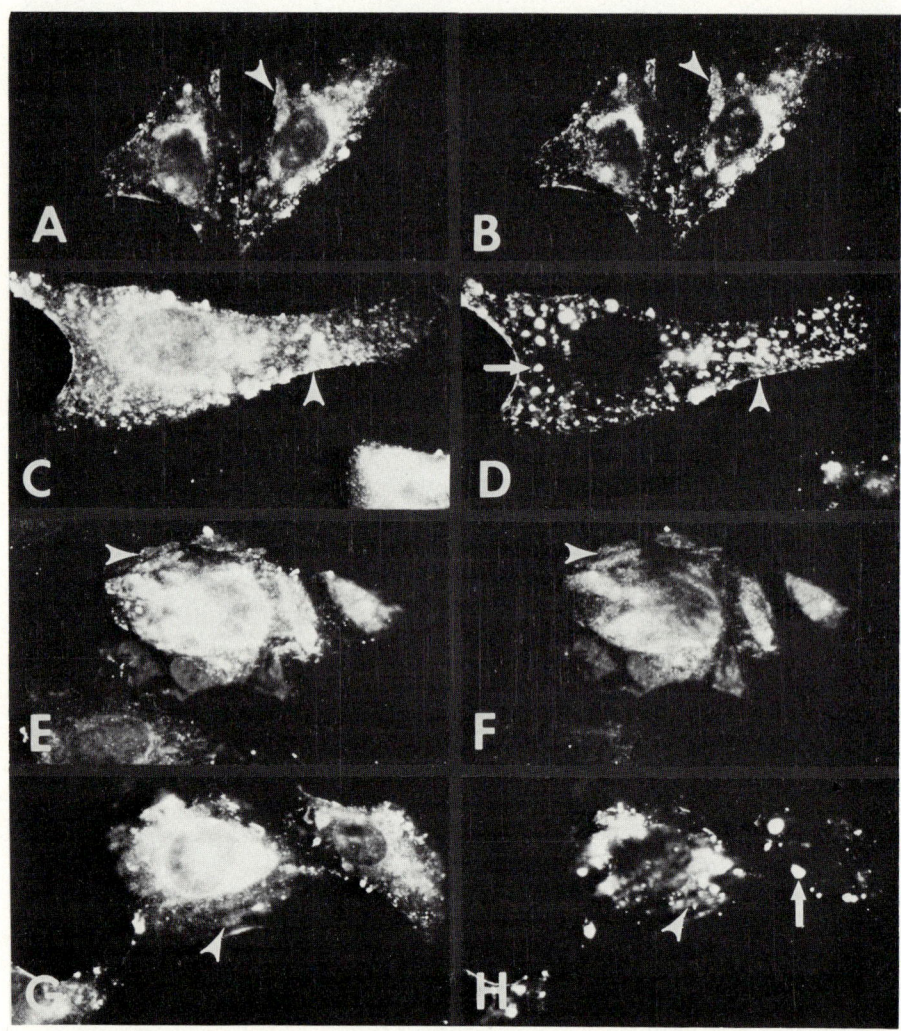

FIGURE 2. Double-label fluorescence in Vero cells 18 hrs after measles virus infection. Cells on left were stained with rabbit anti-HA (A, C, E, G) and identical cells on the right were stained with mouse monoclonal antibodies: (B) anti-HA, (D) anti-NP, (F) anti-M, (H) anti-P. Arrowheads indicate "patches" at membrane. Arrows indicate cytoplasmic inclusions. Note co-localization of all four proteins at membrane patches and lack of M staining of inclusions.

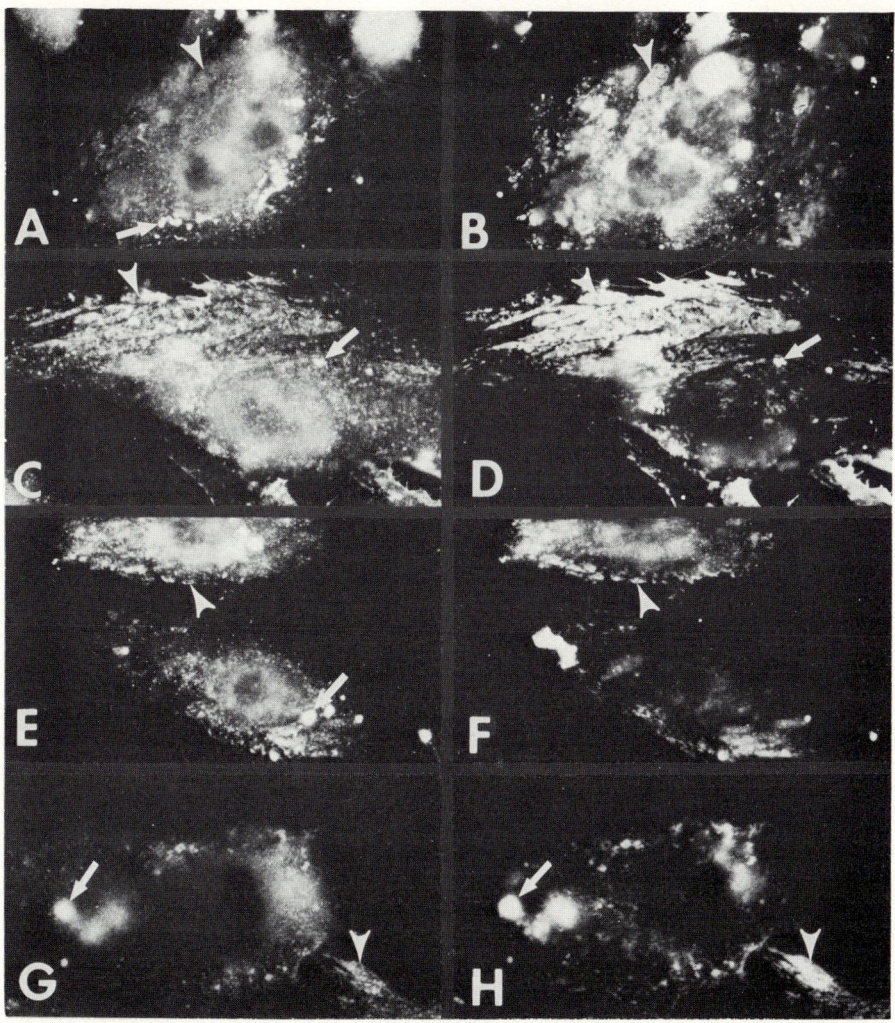

FIGURE 3. Double-label fluorescence in Vero cells 18 hrs after measles virus infection. Cells on left were stained with rabbit anti-p20 (A, C, E, G) and identical cells on right were stained with mouse monoclonal antibodies: (B) anti-HA, (D) anti-NP, (F) anti-M, (H) anti-P. Arrowheads indicate "patches" at membrane. Arrows indicate cytoplasmic inclusions. Note co-localization of all four proteins at membrane patches and lack of M staining of inclusions.

differing avidites. Monoclonal or polyclonal antibodies with high affinity may be more sensitive probes for detection of early forms of P and M proteins. The most striking observation at later stages of infection is the colocalization of viral proteins in two main cellular sites: (a) membrane patches containing NP, P, N and HA which probably represent active sites of transmembrane interactions between these components in preparation for viral budding (6); (b) cytoplasmic inclusions consisting of nucleocapsids not incorporated into virion and containing NP and P proteins but apparently not M. The interaction of M protein with nucleocapsids in inclusions in paramyxovirus infections is controversial (7,8,9). Since M protein is known to play a critical role in viral assembly (10), it is possible that binding of M protein to nucleocapsid occurs only at sites of viral budding at the plasma membrane. EM localization of M protein might elucidate this question.

ACKNOWLEDGMENTS

We thank W. Bohn, E. Norrby and K. Rammohan for antibody donations and Judy Hertler for editing the paper.

REFERENCES

1. Norrby, E., Chen, S., Togashi, T., Shesberadaran, H., and Johnson, K. (1982). Arch. Virol. 71, 1.
2. McLean, I.W., and Nakane, P.K. (1974) J. Histochem. Cytochem. 22, 1077.
3. Bohn, W., Rutter, G., and Mannweiler, K. (1982). Virol. 116, 368.
4. Bellini, W., Englund, G., Richardson, C., Hogan, N., Rozenblatt, S., Myers, C., and Lazzarini, R. (this volume).
5. Nakai, T., Shand, F.L., and Howatson, A.F. (1969). Virol. 38, 50.
6. Dubois-Dalcq, M. and Reese, T. (1975). J. Cell Biol. 67, 551.
7. Fraser, K., Gharpure, M., Shirodaria, P., Armstrong, M., Moore, A., and Dermott, E. (1978). In "Negative Strand Viruses and the Host Cell" (B. Mahy and R. Barry, eds.), p. 78. Academic Press, New York.
8. Yoshida, T., Yoshiyuki, N., Yoshii, S., Maeno, K., and Matsumoto, T. (1976). Virology 71, 143.
9. Johnson, K., Norrby, E., Swoveland, P., and Carrigan, D. (1981). J. Infect. Dis. 144, 161.
10. Choppin, P., and Compans, R. (1975). In "Comprehensive Virology" (H. Fraenkel-Conrat and R. Wagner, eds.), p. 95, Plenum Press, New York.

CROSS-REACTION OF MEASLES VIRUS PHOSPHOPROTEIN WITH A HUMAN INTERMEDIATE FILAMENT: MOLECULAR MIMICRY DURING VIRUS INFECTION[1]

Robert S. Fujinami and Michael B. A. Oldstone

Department of Immunology
Scripps Clinic and Research Foundation
La Jolla, California 92037

Several mechanisms have been proposed to explain the induction of autoimmunity in humans. One explanation is infection by viruses. Viruses could induce autoimmunity through shared determinants between virus structures and molecules normally present on host cells, by deregulating the host's immune system, or by causing the expression or liberation of "normally sequestered" self antigens. Auto-antibodies are commonly found in the circulation of virus-infected individuals, both during and post infection. For example, after infection with Epstein-Barr virus (1,2) antibodies reacting with intermediate filament proteins, immunoglobulin, or thyroglobulin could be detected. Similarly, humans infected with hepatitis, herpes, mumps, or measles viruses can produce antibodies to their own cytoskeleton components (3-5), although the antibodies that reacted with cytoskeletal proteins were not shown to bind to viruses (5).

Monoclonal antibodies (MAbs) provide useful reagents to analyze unique determinants on viruses and on self constituents. During the derivation of such reagents we noted that a large proportion of MAbs originally derived by immunization of mice with purified measles virus react with self components of human cells. We describe here one of these MAbs with dual specificity; one that recognizes the phosphoprotein of measles virus and an intermediate filament protein, M_r 52,000 found in uninfected cells.

---

[1]This work was supported by USPHS Grants NS-17214 and NS-12428 and a National Multiple Sclerosis Society grant. R. S. F. is a recipient of the Harry Weaver Award from the National Multiple Sclerosis Society.

Procedures. HeLa, Vero, L929, BHK$_{21}$, and human astrocyte cells were maintained in Eagles minimal essential medium (MEM) containing 10% fetal calf serum (FCS), 1% glutamine and antibotics. Astrocytes derived from a human brain biopsy sample were positive for glial fibrillary acidic protein by immunofluorescent staining.

P3X63Ag8 mouse plasmacytoma cells (P3) were grown in Eagle's medium containing 10% FCS and 0.1 mM 8-azaguanine. Cloned hybridomas producing antibody to the measles virus nucleocapsid were from E. Norrby (Karolinska Institute, Stockholm) and those producing antibodies to the hemagglutinin were from D. McFarlin, (National Institutes of Health, Bethesda, MD). All other measles MAbs were generated after immunization of BALB/c mice with measles virus (6,7). The preparation, purification and use of measles virus (Edmonston strain) and the subsequent immunofluorescence, have been reported (8,9).

A preparation of Triton-X100-insoluble proteins (intermediate filament enriched) of HeLa cells was obtained as described by Pruss et al. (10).

SDS-gels were analyzed as described (8,9). Proteins from gels were transferred to nitrocellulose paper (Schleicher and Schuell, Keene, NH) as reported by Towbin et al. (11). Strips were incubated with 5 ml of supernatant fluid from hybridoma or P3 cell culture (7) in 5 ml of 3% bovine serum albumin (BSA), 0.1% deoxycholic acid, 0.5% Nonidet P-40, 1% normal goat serum in PBS (binding buffer), or 20 µl of ascitic fluid derived from BALB/c mice injected with the hybridoma cells in 10 ml binding buffer. The strips were then incubated for 1 hr at 37°C, washed 3x with 3% BSA in PBS and incubated with 10 ml of binding buffer containing ^{125}I-labeled goat anti-mouse Ig. After 3 hr at 37°C, the strips were washed 3x with 10 ml of 1.5% BSA in PBS and dried overnight. Strips were mounted and exposed to Kodak X-RPI x-ray film the next day.

Detection of Dual-Reactivity by Immunofluorescence. The anti-measles virus MAb 2A-54-5, which reacted with measles virus infected and uninfected HeLa cells (Fig. 1), stained uninfected cells in a pattern similar to that reported by Franke et al. (12) for vimentin and cytokeratins, i.e., a network-like pattern in cells with a large cytoplasm-to-nucleus ratio (Fig. 1A, B). In contrast, a speckled pattern of labeling was observed with 2A-54-5 in large, rounded cells and staining was concentrated between dividing cells as shown in Fig. 2C and D. Similar patterns of reactivity with the MAb occurred in HeLa, Vero and the primary human astrocytes and was cell cycle dependent. No reactivity was observed with BHK_{21} or mouse L929 cells. Unlike the uninfected cells, cells infected with measles virus (Fig. 2A, B) showed a bright globular pattern that was more intense than the staining observed in uninfected cells. Many of the staining bodies could be seen by phase-contrast illumination as intracytoplasmic bodies (Fig. 2B). Immunofluorescent staining of uninfected cells with MAb 2A-54-5 was seen only when cells were fixed with acetone. This suggests that MAb 2A-54-5 reacts with a host specific determinant located inside the cell. The reactivity of MAb 2A-54-5 against infected cells could be removed by absorption with an intermediate filament preparation.

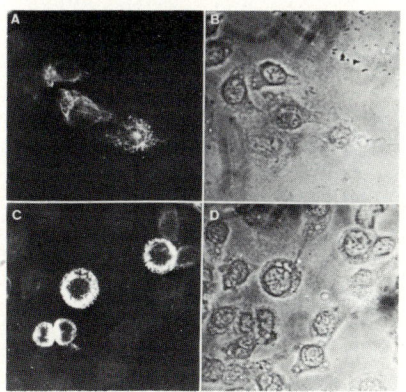

Figure 1. Network and speckled pattern of staining of uninfected HeLa cells (A-D) with antimeasles virus MAb 2A-54-5 (all x300). (A) Uninfected cells spread and stained in a network pattern. (B) Phase-contrast of A. (C) Uninfected cells in mitosis showing characteristic speckling. (D) Phase-contrast micrograph of cells in C.

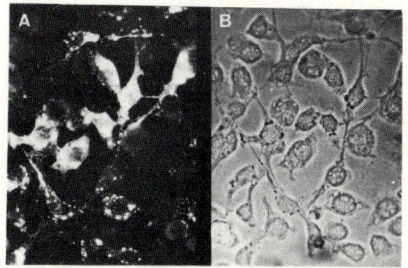

Figure 2. Globular staining of measles virus infected HeLa cells with MAb 2A-54-5. (A) Globular staining of the cytoplasm of measles virus infected cells. (B) Phase-contrast of the same field as in A. Note the cytoplasmic inclusion bodies.

Biochemical Characterization. We first identified the cytosol component of measles virus infected cells that MAb 2A-54-5 detected. The results are shown in Figure 3. A band at M_r 70,000 is visible in infected cell cytosol (lane 3) but not in uninfected preparations (lane 4). This band migrated with the phosphoprotein of measles virus (lane 5). Cytosols from uninfected (not shown) or infected cells failed to react with the second antibody alone or P3 supernatant fluid (lanes 1 and 2).

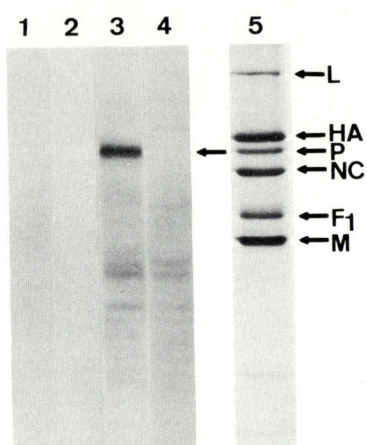

Figure 3. Immunoblots of reaction of measles virus P protein with MAb 2A-54-5. Lanes: 1, infected cell cytosol transferred to nitrocellulose paper incubated with labeled 2nd antibody only; 2, the same infected cell cytosol preparation incubated with P3 fluid and then with labeled 2nd antibody; 3 and 4, infected cell cytosol and uninfected cell cytosol, respectively, incubated with MAb 2A-54-5 and with labeled 2nd antibody; 5, marker purified measles virus.

Because cytosol preparations contain small amounts of intermediate filaments, Triton-insoluble material (intermediate filament-enriched) from HeLa cells was prepared and electrophoresed on SDS_4-gels. The proteins were then transferred to nitrocellulose paper and incubated with MAb specific for either measles virus phosphoprotein (3-60-5) or HeLa cytoskeleton (3-3-2), and MAb 2A-54-5. As detected by immunofluorescence, MAb 3-60-5 reacted with infected cells in a globular pattern but not with uninfected cells, whereas MAb against HeLa cytoskeleton (3-3-2) reacted with uninfected and infected cells in the network-like staining pattern indicative of intermediate filaments. MAb 3-60-5 did not react with the Triton-insoluble preparation (Fig. 4, lane 2). MAbs 3-3-2 and 2A-54-5 both reacted with a M_r 52,000 protein (Fig. 4, lanes 3 and 4). Thus, MAb 2A-54-5 (Fig. 4, lane 4) reacts with the same or similarly migrating intermediate filament protein as the 3-3-2 MAb (Fig. 4, lane 3). The MAb against measles virus phosphoprotein (Fig. 4, lane 2) does not react with the M_r 52,000 protein and thus does not recognize the same determinant as 2A-54-5.

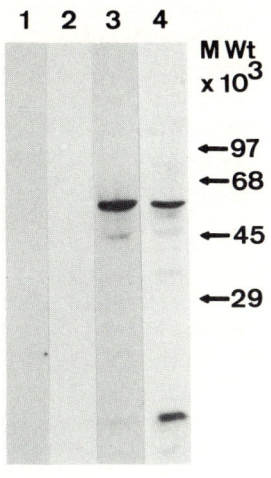

Figure 4. Immunoblots showing reaction of MAbs to intermediate filaments of HeLa cells. Lanes 1-4 represent strips to which intermediate filaments proteins were bound incubated with various MAbs followed by 2nd antibody. Lanes: 1, P3 fluid was incubated with the strip followed by labeled 2nd antibody; 2, strip incubated with MAb specific for measles virus P protein (3-60-5); 3, strip incubated with MAb to intermediate filaments (3-3-2); 4, strip incubated with MAb 2A-54-5. The prominant band is M_r 52,000.

To determine whether the phosphoprotein of measles virus detected by MAb 2A-54-5 was present both in the cytoplasm of infected cells and in virions, purified measles virus was disrupted, electrophophoresed on SDS gels, and the proteins were transferred to nitrocellulose paper and blotted (Fig. 5). No reaction was seen when the strip was incubated with labeled second antibody alone (lane 1), with P3 supernatant fluid (lane 2), or with MAb to intermediate filament (lane 3). MAb to measles virus nucleocapsid reacted with the nucleocapsid protein (lane 4). Reaction with measles virus phosphoprotein was seen when 2A-54-5 was incubated with the strip (lane 5). MAb to measles virus hemagglutinin reacted specifically with the viral hemagglutinin protein (lane 6). Lanes 7 and 8 are MAb 2A-54-5 and 3-60-5, indicating recognition of the same polypeptide.

In attempting to produce MAbs to measles virus, we obtained stable clones that secreted antibody reactive both with intermediate filaments of normal cells and with the phosphoprotein of measles virus. The M_r 52,000 intermediate filament protein was obtained as insoluble material from cells treated with high salt/Triton X-100 buffers and is probably vimentin or a cytokeratin (12). Our staining patterns with 2A-54-5 are similar to those presented by Franke et al. (12) who described the rearrangement of cytokeratins and vimentin during the cell cycle. The cross-reaction of the intermediate filament protein with the phospho-

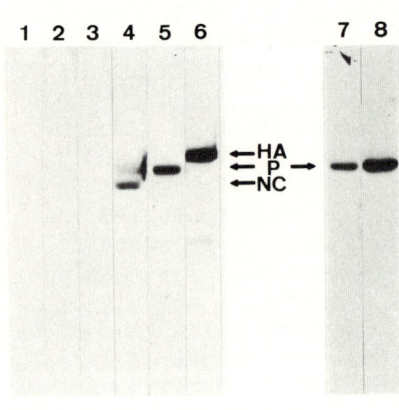

Figure 5. Immunoblots showing reaction of MAbs with purified measles virus. Lanes 1-8 are strips to which purified measles virus proteins are bound. Lane 1, reaction with the second antibody alone (affinity-purified ^{125}I-labeled goat anti-mouse Ig); the remaining strips were incubated with various MAbs and then the second antibody. Lanes 2, reaction with P3 supernatants; 3 MAb to intermediate filaments (3-3-2); 4, measles virus nucleocapsid MAb; 5, MAb 2A-54-5; 6, MAb to measles virus hemagglutinin; 7 and 8, MAb 2A-54-5 and MAb to measles virus P protein (3-60-5), respectively.

protein of measles virus was detected using immunofluorescence and immunoblotting. The 2A-54-5 MAb reacted with the phosphoprotein of measles virus both from infected cells and from purified virus. It is interesting that many of the intermediate filament proteins are phosphorylated as well as the phosphoprotein of measles virus.

There are two alternative explanations for the observed antigenic relationship between vimentin or cytokeratin and viral proteins. First, these cellular proteins may have certain sequences (or surface configurations) of amino acids in common with measles virus phosphoprotein. This relationship may be entirely fortuitous. Alternatively, parts of intermediate filament proteins may become incorporated into viral proteins in the course of virus replication. The observations of Blase et al. (13) and of Dulbecco et al. (14) on cross-reactive determinants between vimentin and tropomyosin (13) as well as vimentin and Thy-1 (14) and of Pruss et al. (10) of common antigenic determinant shared by all classes of intermediate filaments support the concept of molecular mimicry. This notion of molecular mimicry discussed by Lane and Koprowski (15) has several important biological implications. Recently Wood et al. (16) described an antibody that reacts with neurons, cardiac muscle cells, and Trypanosoma cruzi. Those investigators suggested that a common antigen seen by this antibody may be important in the pathogenesis of trypanosomiasis. A similar scenario may occur with viruses that share epitopes with cellular proteins. In this instance, an immune response against virus may result in formation of cross-reacting antibodies that bind to "normal" cell surface determinants and result in the modulation of cellular function (17,18) or lysis of the cell with complement or lymphocytes bearing an Fc receptor. Cells lysed by virus infection or immune attack expose intracellular determinants. The reaction of autoantibody with intracellular components might yield secondary pathogenic sequalae during viral infection. Hence the reaction of autoantibody with self molecules may increase injury at the site of viral infection or at distant sites due to trapped immune complexes (19).

Autoimmunity associated with virus infection could occur by several mechanisms in addition to "molecular mimicry". Viruses may act as polyclonal B cell activators (20-26), thereby expanding a preexisting clone that reacts with "self". Viruses may also cause the shedding of antigens not ordinarily seen by the host immune system (27). Similarly, viruses can potentiate or augment the response to many antigens (28,29). Any of these mechanisms may explain the appearance of autoantibodies during and after virus infections (1-5, 30) and perhaps the origin of many so-called "natural" antibodies (31-33).

ACKNOWLEDGEMENTS

This is Publication No. 3198-IMM from the Department of Immunology, Scripps Clinic and Research Foundation, La Jolla, California 92037. The authors express their graditude to Dr. Michael J. Buchmeier, Hanna Lewicki and Janet Anderson for their help in the production of the MAbs. We acknowledge Ms. Callie-Jane Mack for help with the technical aspects of the project and Ms. Lisa Ann Flores and Ms. Gay L. Wilkins for assistance in the preparation of this manuscript.

REFERENCES

1. Linder, E., Kurki, P. and Andersson, L.C. (1979). Clin. Immunol. Immunopathol. 14, 411.
2. Fong, S., Tsoukas, C.D., Frincke, L.A., Lawrance, S.K., Holbrook, T.L., Vaughan, J.H. and Carson, D.A. (1981). J. Immunol. 126, 910.
3. Kurki, P., Virtanen, I., Stenman, S. and Linder, E. (1978). Clin. Immmunol. Immunopathol. 11, 379.
4. Toh, H., Yildiz, A., Sotelo, J., Osung, O., Holborow, E.J., Kanakoudi, F. and Small, J.V. (1979). Clin. Exp. Immunol. 37, 76.
5. Haire, M. (1972). Clin. Exp. Immunol. 12, 335.
6. Kohler, G. and Milstein, C. (1975). Nature 256, 459.

7. Buchmeier, M.J., Lewicki, H.A., Tomori, O. and Oldstone, M.B.A. (1981). Virology 113, 73.
8. Fujinami, R.S., Sissons, J.G.P. and Oldstone, M.B.A. (1981). J. Immunol. 127, 936.
9. Fujinami, R.S. and Oldstone, M.B.A. (1981). J. Exp. Med. 154, 1489.
10. Pruss, R.M., Mirsky, R., Raff, M.C., Thorpe, R., Dowding, A.J. and Anderton, B.H. (1981). Cell 47, 419.
11. Towbin, H., Stachelin, T. and Gordon, J. (1979). Proc. Natl. Acad. Sci. USA 76, 4350.
12. Franke, W.W., Schmid, E. Grund, C. and Geiger, B. (1982). Cell 30, 103.
13. Blase, S.H., Matsumura, F. and Lin, J.J-C. (1982). Cold Spring Harbor Symp. Quant. Biol. 46, 455.
14. Dulbecco, R., Unger, M., Bolgna M., Battifora, H., Syka, P. and Okada, S., (1981). Nature 292, 772.
15. Lane, D. and Koprowski, H. (1982). Nature 296, 200.
16. Wood, J.N., Hudson, L., Jessel, T.M. and Yamamoto, M. (1982). Nature 296, 34.
17. Kahn, C.R., (1979). Fed. Proc. 38, 2607.
18. Oldstone, M.B.A., Fujinami, R.S. and Lampert, P.W. (1980). Prog. Med. Virol. 26, 45.
19. Oldstone, M.B.A. (1975). Prog. Med. Virol. 19, 84.
20. Butchko, G.M., Armstrong, R.B., Martin, W.J. and Ennis, F.A. (1978). Nature 271, 66.
21. Bird, A.G and Britton, S. (1979). Immunol. Rev. 45, 41.
22. Mochizuki, D., Hedrick, S., Watson, J. and Kingsbury, D.T. (1977). J. Exp. Med. 146, 1500.
23. Kirchner, H., Darai, G., Hirt, H.M., Keyssner, K. and Munk, K. (1978). J. Immunol. 120, 641.
24. Goodman-Snitkoff, G.W. and McSharry, J.J. (1980). J. Virol. 35, 757.
25. Goodman-Snitkoff, G., Mannino, R.J. and McSharry, J.J. (1981). J. Exp. Med. 153, 1489.
26. Cafruny, W.A. and Plagemann, P.G.W. (1982). Infect. Immun. 37, 1001.
27. Haspel, M.V., Onodera, T., Prabhakar, B.S., Horita, M., Suzuki, H. and Notkins, A.L. (1982). Science, in press.
28. Lindenmann, J. and Klein, P.A.(1967). J. Exp. Med. 126, 93.
29. Bromberg, J. S., Lake, P. and Brunswick, M. (1982). J. Immunol. 129, 683.
30. Sotelo, J., Gibbs, C.J. and Gajdusek, D.C. (1980). Science 210, 190.
31. Koprowski, H. (1946). J. Immunol. 54, 387.
32. Guilbert, B., Dighiero, G. and Avrameas, S. (1982). J. Immunol. 128, 2779.
33. Dighiero, G., Guilbert, B. and Avrameas, S. (1982). J. Immunol. 128, 2788.

MEASLES VIRUS INFECTION OF HUMAN PERIPHERAL BLOOD LYMPHOCYTES: IMPORTANCE OF THE OKT4$^+$ T-CELL SUBSET[1]

Steven Jacobson and Henry F. McFarland

Neuroimmunology Branch
National Institutes of Health
Bethesda, MD 20205

INTRODUCTION

Measles virus has been know to infect human lymphocytes and studies employing mainly lymphoblastoid cell lines have indicated that both T and B cells can be infected (1,2,3,4,5). Lucas (6) has shown that measles virus can produce a silent infection in unstimulated peripheral blood lymphocytes (PBLs). We have reported previously, that infection of unstimulated PBLs results in the production of high levels of virus-induced alpha interferon associated with low levels of detectable virus (5). Expression of viral antigens on cell surface membranes was induced by either incubation of these infected cells with a mitogen or with a non-mitogenic anti-leukocyte interferon serum that neutralized the virus-induced interferon. These treatments produced a concommitant increase in the production of infectious virus. To extend our previous observations we have investigated the replication of measles virus in T-cell subsets. We will show that measles virus preferentially infects the T-cell subset with an OKT4 (helper/inducer) phenotype. Infection of purified T-cell subsets results in higher viral yields form OKT4$^+$ cells compared to simarlarly infected OKT8$^+$ lymphocytes (suppressor/cytotoxic subpopulation). The decreased measles virus titers from infected OKT8$^+$ cells is associated with an increase in the production of virus-induced alpha interferon. Finally, the observation that measles virus preferentially replicates in the OKT4$^+$ T-cell subset is discussed in terms of the importance of this T cell population in measles virus-lymphoctye regulation with particular emphasis to Multiple Sclerosis.

---

1. This work was supported by a postdoctoral fellowship from the National Multiple Sclerosis Society.

RESULTS

Human PBLs isolated on Ficoll-Hypaque gradients and infected with the Edmonston strain of measles virus at an MOI of 1.0 were incubated with a 1:500 dilution of a potent anti-leukocyte interferon serum for 6 days. At that time cells were stained for the presence of cell associated

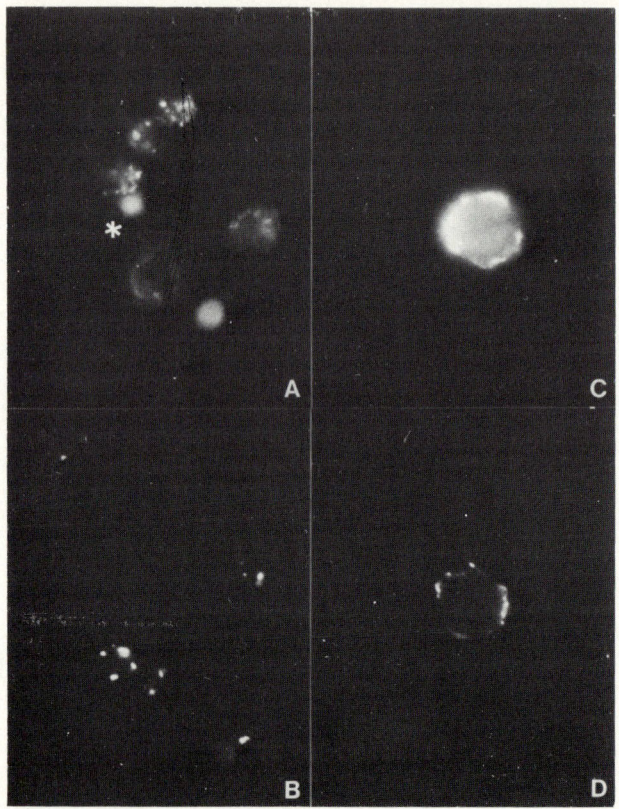

Figure 1. a) RITC-human anti-measles virus, b) same cells, FITC-OKT8, c) RITC-human anti-measles virus, d) same cell, FITC-OKT4

measles virus antigens (human SSPE serum coupled to rhodamine (RITC) conjugated goat anti-human IgG) and phenotypic expression of either OKT4 or OKT8 determinants (mouse OKT4 or OKT8 antisera coupled to fluorescein (FITC) conjugated to sheep anti-mouse IgG). Using immunofluorescence procedures any cell expressing measles virus antigens stained red and could be classified as an OKT4 or OKT8 cell (staining green). These results are shown in Figure 1.

The use of the anti-interferon serum enabled us to detect measles virus antigens on infected PBL cell membranes (5). In Figure 1, the cell infected with measles virus (Panel C) was the same cell reacting with the OKT4 antibody (Panel D). Infected PBLs were rarely of the OKT8 phenotype. The asterisk in Figure 1, Panel A denotes the presence of the OKT8$^+$ cell in Panel B. None of the infected cells in Panel A were stained by the OKT8 antibody (Panel B). These data would suggest that measles virus infection of human PBLs was preferential for the OKT4$^+$ T-cell subset relative to the OKT8 population.

To confirm this T cell preference for measles virus replication we isolated purified populations of OKT4 and OKT8 cells. We employed a lymphocyte panning procedure whereby sheep resette forming cells (E$^+$) were positively selected for either OKT4 or OKT8 by incubation with the appropriate monoclonal antibody and subsequently plated on plastic plates coated with an F(ab')$_2$ rabbit anti-mouse IgG. Adherent cells were removed and the surface phenotype of these lymphocytes were determined on the FACS-IV Fluorescence-Activated Cell Sorter. The purity of these T-cell populations are shown in Table 1.

TABLE 1
PURITY OF HUMAN T-CELL SUBSETS: ANALYSIS BY FLUORESCENCE-ACTIVATED CELL SORTER

	OKT8+	OKT4+
OKT3[a]	96%	95%
OKT4	5%	94%
OKT8	93%	5%

[a] All determinations are relative to a negative control of mouse ascites fluid produced by the P3X63 myeloma cell line.

TABLE 2
MEASLES VIRUS PRODUCTION IN HUMAN T-CELL SUBSETS (PFU/ml)

	OKT8+	OKT4+
Unstimulated [a]	8.0×10^3	2.7×10^4
Pokeweed mitogen	3.5×10^4	4.0×10^5
Anti-leukocyte interferon	8.0×10^3	2.3×10^5
Mock anti interferon	8.3×10^3	2.1×10^4

[a] Lymphocyte subsets cultured at a concentration of 2×10^6 cells/ml. Virus harvested at 4 days post infection. PWM - final concentration 1:50; Anti-interferon - final concentration 1:500; mock anti-interferon - final concentration 1:100.

To determine if measles virus can preferentially replicate in one T-cell subset, purified T cell populations were infected at an MOI of 1.0 and incubated with either PWM, an anti-interferon serum, a mock anti-interferon preparation or left unstimulated. Four days later supernatants were harvested and assayed for infectious virus. Unstimulated, "silently" infected T-cell subsets produced low levels of virus although even in this group there was an indication that more virus was produced in the OKT4+ population (Table 2). Stimulation of infected PBLs with PWM had been shown to augment the production of measles virus (5,6). The stimulatory effect of PWM on these infected T-cell subsets was greatest for the OKT4+ cells compared to the OKT8+ population. The addition of a non-mitogenic anti-leukocyte interferon serum to "silently" infected PBLs also activated these lymphocytes to produce large amounts of virus (5). Incubation of infected T cell subsets with this anti-interferon serum resulted in no increase in virus production from infected OKT8+ cells and an approximately one log increase in infectious measles virus from the infected OKT4+ population (Table 2). A control serum with no antibody to interferon had no effect on virus production and virus titers from this group were similar to unstimulated, infected T-cell subsets. Again, even in this latter group, more virus was produced from infection of OKT4+ cells. These data coupled with the immunofluorescent pattern seen with infected PBLs (Figure 1) strongly suggest that measles virus appears to replicate preferentially in the OKT4+ T-cell subset relative to the OKT8+ population.

TABLE 3
ALPHA INTERFERON PRODUCTION IN MEASLES VIRUS INFECTED
HUMAN T CELL SUBSETS (INTERFERON UNITS/ml)[a]

	OKT8+	OKT4+
Unstimulated [b]	320	160
Anti-leukocyte interferon	0	0
Mock anti-interferon	320	160
Uninfected	0	0

[a] One laboratory unit of interferon is equal to .05 international units.

[b] Lymphocytes cultured at a concentration of 2×10^6 cells/ml; supernatants assayed for interferon at 4 days post infection.

We had shown that low levels of infectious measles virus produced by unstimulated, infected PBLs was associated with the production of alpha interferon (5). We wished to determine if increased levels of interferon of infected OKT8+ cells could also be associated with low yields of measles virus from this T-cell subset. Purified T-cell subpopulations were infected with measles virus at an MOI of 1.0 and supernatants were harvested 4 days later to be assayed for interferon. Interferon was determined by CPE reduction of an EMC virus challenge on GM2504 cells. The results are shown in Table 3. Unstimulated, measles virus infected OKT8+ cells consistently produced twice as much alpha interferon as comparably treated OKT4+ cells. Incubation of these lymphocytes with an anti-leukocyte interferon preparation (which augmented the production of infectious measles virus from infected OKT4+ cells, Table 2), completely neutralized this virus-induced interferon. A control antisera had no inhibitory effect. Unstimulated cells produced no interferon (Table 3).

DISCUSSION

The purpose of this study was to examine measles virus replication in lymphocyte subpopulations. Numerous reports have indicated that measles virus can persist in lymphoblastoid cell lines with either B or T cell characteristics (1,2,3,4). Infection of peripheral blood lymphocytes (PBLs) may represent a more relevent model for the study of measles virus replication in vivo. Studies have shown that measles virus can replicate in human PBLs (1,2,5,6,7). Unstimulated, infected PBLs produced low levels of infectious virus and failed to express viral antigens on cell surface membranes. These cells produced high levels of virus-induced alpha interferon (5). Treatment of these silently infected PBLs with mitogen or an anti-leukocyte interferon serum produced a productive infection.

To extend these findings the present studies have examined measles virus replication in T-cell subsets. Our results would suggest that measles virus preferentially replicates in T lymphocytes with an $OKT4^+$ phenotype. This is based on two observations. First, by employing a dual-labelling immunofluorescent procedure we were able to detect measles virus antigens associated predominantly in those infected PBLs co-staining with the OKT4 monoclonal antibody and not OKT8 (Figure 1). Secondly, infection of purified T cell subsets (Table 1) resulted in higher virus yields from $OKT4^+$ cells compared to infected $OKT8^+$ cells (Table 2). Taken together, these data strongly suggest that the preferred lymphocyte population for measles virus replication is the $OKT4^+$ T cell subset.

We had demonstrated previously that virus-induced alpha interferon was associated with measles virus persistence in PBLs (5). Infection of purified T-cell subsets also induced alpha interferon (Table 3), although considerably less than infected PBLs. In the percent study, separated T cell populations were greater than 90% pure when stained by the monoclonal antibody to which it had been positively selected (Table 1). Less than 1% of the cells in each purified population reacted with monoclonal antibodies that detect macrophages (data not shown). It is known that macrophage and NK cells are important co-factors in the production of interferon in PBLs (8). This may account for the higher yields of alpha interferon from infected PBLs (5) compared to similarly infected T cell subsets (Table 3). Our data also indicated that $OKT8^+$ cells produced twice as much virus-induced alpha interferon as infected $OKT4^+$ cells (Table 3). This may be associated with lower virus yields from $OKT8^+$ cells (Table 2), however it could not fully account for this phenomena since an anti-leukocyte interferon serum which augmented the production of measles virus from infected-PBLs (5) and purified $OKT4^+$ cells had no effect on infected $OKT8^+$ cells (Table 2).

The ability of measles virus to replicate in $OKT4^+$ T cells can permit some interesting speculation. Firstly, this may suggest that measles virus is capable of persistence in T cells with an OKT4 phenotype. Immunological activation of these cells might then specifically augment the immune response to this virus. Secondly, antigen presentation may occur on the surface of OKT4 cells allowing recognition of viral antigen in association with histocompatability antigens that occur on this subset. Lastly, it has long been suggested that measles virus may play a role in the pathogenesis of Multiple Sclerosis (9,10) a disease in which abnormalities of T-cell subsets have been reported (11,12,13,14,15,16,17). With regards to this present study, OKT4$^+$ cells have been implicated in MS both with augmented amplifier activity for specific antibody production (9,18) and possibly with lesion progression (19,20). The significance of these observations to our findings is uncertain at this point.

ACKNOWLEDGMENTS:

The authors wish to acknowledge Mrs. Pamela Danner for her excellent secretarial assistance.

REFERENCES

1. Joseph, B.S., Lampert, P.W., and Oldstone, G.B.A. (1975). J. Virol. 16, 1638.
2. Sullivan, J.L., Barry, D.W., Lucas, S.J., and Albrecht, P. (1975). J. Exp. Med. 142, 773.
3. Barry, D.W., Sullivan, J.L., Lucas, S.J., Dunlap, R.C., and Albrecht, P. (1976). J. Immunol. 116, 89.
4. Fagraeus, A., Bottiger, M., Heller, L., and Norrby, E. (1981). Arch. Virol. 69, 229.
5. Jacobson, S., and McFarland, H.F. (1982). J. Gen. Virol. 63, 351.
6. Lucas, C.J., Ubels-Postma, J.C., Rezee, A., and Galama, J.M.D. (1978). J. Exp. Med. 148,940.
7. Huddlestone, J.R., Lampert, P.W., and Oldstone, M.B.A. (1980). Clin. Immunol. Immunopath. 15, 502.
8. Epstein, L.B. (1977). In "Interferons and Their Actions" (W.E. Stewart, ed.) p. 91. CRC Press, Florida.
9. Adams, J.M., and Imagawa, D.T. (1962) Proc. Soc. Exp. Biol. Med. 3, 562.
10. Haase, A.T. Ventura, P., Gibbs, C.J., and Tourtellotte, W.W. (1981). Science. 212, 672.
11. Antel, J.P., Arnason, B.G.W., and Medof, M.E. (1979). Ann. Neurol. 5, 338.

12. Huddlestone, J.R., and Oldstone, M.B.A. (1979) J. Immunol. 123, 1615.
13. Santoli, D., Moretta, J., Lisak, R., Gilder, D., and Koprowski, M. (1979). J. Immunol. 120, 1369.
14. Bach, M.A., Phan-Dinh-Tuy, F., Tournier, E., Chatenoud, L., and Bach, J-F. (1980). Lancet. 2, 1221.
15. Reinherz, E.L., Weiner, H.L., Hauser, S.L., Cohen, J.P., Distago, J.A., and Schlossman, S.F. (1980). N. England, J. Med. 303, 125.
16. Hauser, S.L., Reinherz, E.L., Hoban, C.J., Schlossman, S.F., and Weiner, H.L. (1983). Ann. Neurol. 13, 418.
17. Hauser, S.L., Reinherz, E.L., Hoban, C.J., Schlossman, S.F., and Weiner, H.L. (1983). Neurol. (Cleaveland). 33, 575.
18. Albrecht, P., Tourtellotte, W.W., Hicks, J.T., Sato, H., Boone, E.J., and Potvin, A.R. (1983). Neurol. (NY). 33,45.
19. Traugott, U., Reinherz, E.L., and Raine, C.S. (1982). Science. 219, 308.
20. Traugott, U., Reinherz, E.L., and Raine, C.S. (1983). J. Neuroimmunol. 4, 201.

TREATMENT OF EXPERIMENTAL MUMPS VIRUS MENINGOENCEPHALITIS
USING MONOCLONAL ANTIBODIES

Jerry S. Wolinsky, M.D., M. Neal Waxham,
Alfred C. Server, M.D., Ph.D., David C. Merz, M.D., Ph.D.[1]

Department of Neurology
University of Texas Health Science Center
P.O. Box 20708
Houston, Texas 77025

INTRODUCTION

Viral invasion of the central nervous system (CNS) is common in the course of mumps. While this is usually subclinical, mild signs and symptoms of meningeal involvement are not infrequent and encephalitis with severe CNS damage does occur (1). There is some evidence that the administration of hyperimmune immunoglobulin (Ig) beneficially modifies the subsequent course of human mumps (2), and in a newborn hamster model of mumps menigoencephalitis, the development of specific humoral immune responses to mumps virus correlates well with the fall of free virus content in brain (3). Given the availability of monoclonal antibodies to mumps virus (4), we sought to determine whether or not the passive administration of such defined Ig could exert a protective effect on suckling hamsters previously infected with an otherwise lethal dose of a neurovirulent mumps virus strain, and to determine the mechanism of any observed protection. Our data show that some monoclonal antibodies have therapeutic potential against otherwise fatal viral infections of the CNS.

MATERIALS AND METHODS

Virus. The Kilham strain of mumps virus was plaque purified three times, expanded in CV-1 cells and aliquots stored at -70°C prior to use. Stock virus contained $10^{7.1}$ plaque forming units (PFU)/ml when assayed on CV-1 cell monolayers as previously detailed (5). The virus inoculum used contained $128 LD_{50}$ as determined from survival data at day 14 after inoculation of newborn hamsters with serial decimal dilutions of

[1]Current Address: Department of Medicine, University of Michigan, Ann Arbor.

this virus stock.

Animals. Pregnant multiparous Syrian hamsters were obtained from Charles River (Wilmington, MA) during the second week of gestation and housed separately. All pups were under 24 hrs. of age when inoculated with 25ul virus into the parietal region of the right hemisphere. Treatments were administered in 0.1ml volumes by intraperitoneal (IP) injection at the various times indicated. The treatment protocol was not varied within a given litter, only between litters which were randomly selected for a given treatment. Census was taken daily. Deaths occuring within the first 72 hrs. of life were attributed to non-specific causes and excluded from both experimental and control groups from further analysis. Animals removed from the litters for histologic or virologic analyses were handled as censored data at the time of harvest. All censored animals were randomly selected appropriate with the experimental protocol. For most experiments, observations were terminated between 40 and 50 days after infection, at which time all surviving animals were censored.

Monoclonal Antibodies. Mouse monoclonal antibodies were generated and characterized as previously detailed (4), and are designated by the specific virus polypeptide which they precipitate, their order of isolation, and IgG isotype. The antibodies were either purified from tissue culture supernatants with protein A sepharose or ascitic fluids from Balb/C mice were used directly.

Virus Titers. Clarified 10% homogenates (wt/wt) of individual hamster brains were prepared in Hank's basic salt solution (HBSS) from four animals at each of the indicated time intervals as previously detailed (3) and stored at $-70°C$ until titered. All plaque assays were performed in triplicate on CV-1 monolayers (5).

Morphology. Eight experimental and two control animals were killed by perfusion-fixation with 4% paraformaldehyde solution on days 4, 6 and 10 after infection. The brains were sectioned coronally and embedded in paraffin for standard histology and immunohistochemical localization of virus antigen. Mumps virus nucleocapsid protein was localized on deparaffinized sections using NP4-1k ascitic fluid as primary antibody at a 10^{-5} dilution. Goat anti-mouse IgG conjugated to horseradish peroxidase was used at a 10^{-2} dilution as seeking antibody in an indirect technique. The reaction product was generated with diaminobenzidine as previously detailed (6).

Statistics. Survival curves were constructed according to the method of Kaplan and Meier and treatment dependant differences between the survival curves sought using a nonparametric comparison test for multiple samples with censored observations (7). Group contrasts were performed using an overall level of significance ≤ 0.01 in all analyses.

Distribution of Labeled Monoclonal Antibody. HN2-2ak purified on protein A sepharose was radiolabeled with ^{125}I using iodogen, and free iodine removed by a second purification on protein A sepharose with extensive washing of the column prior to elution of the iodinated IgG. Two infected pups each were injected with equal amounts of iodinated antibody, either 24 or 80 hrs. after infection and killed 24 hrs. later by perfusion-fixation with 4% paraformaldehyde solution. Brains and other organs were then removed and autoradiographs prepared (8).

RESULTS

Newborn hamsters inoculated with an intracerebral dose of $10^{5.5}$ PFU of plaque purified Kilham strain mumps virus and treated with a single 0.1ml IP injection of HBSS alone developed uniformly fatal disease (Fig. 1). The slope of the survival curve for control animals was both reproducible and quite steep. Fifty percent of the control animals were dead by day 11 of the infection and there were no survivors beyond day 12 in the cumulative control group of 127 animals, 80 of which were uncensored. However, the single IP inoculation of relatively small amounts of selected monoclonal antibodies was shown to shift the survival curves markedly to the right, while other monoclonal antibodies showed little or no beneficial effect. The monoclonal antibody HN-2ak has hemagglutination inhibition (HAI) and neutralization (Nt) activities when measured _in vitro_ and immunoprecipitates the 78K mol.

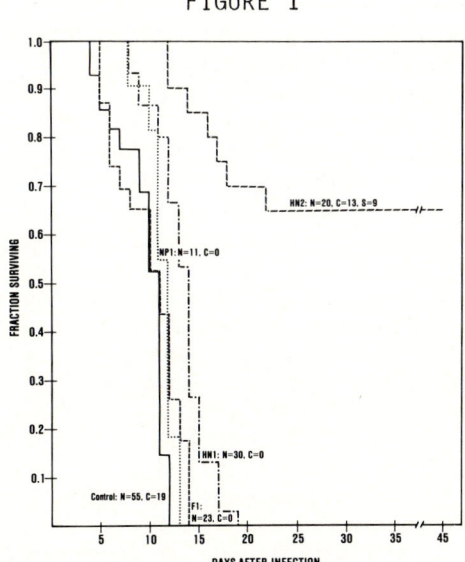

FIGURE 1

wt. glycoprotein of mumps virus. In vivo, a single IP inoculation of 100ug of HN2-2ak monoclonal antibody 12 hrs. after infection showed considerable protective effect (p $\leq$ 0.001) with 65% of treated animals surviving to 45 days after infection (Fig. 1). Some animals surviving the acute infection appeared to die from obstructive hydrocephalus. This was especially common at or about the time of weaning (day 22-24). However, most animals which survived to day 40 showed neither altered appearance or behavior. In other experiments, selected animals were followed for more than three months. Some of these proved endrocrinologically intact as evidenced by pregnancy and subsequent healthy litters.

In contrast, a single treatment with 100ug of the monoclonal antibody HN1-3k which also precipitates the 78k mol. wt. glycoprotein and has measurable HAI and Nt activity in vitro exerted a statistically significant protective effect (p $\leq$ 0.001) but did not produce long term survivors. Finally, single 100ug injections of the monoclones F1-2ak, directed to the 60K mol. wt. fusion glycoprotein, and NP1-2ak, directed against the 68k mol. wt. nucleocapsid protein, both of which lack measurable biologic activity in vitro, failed to beneficially effect the survival curve (Fig. 1).

The effect of a single IP treatment with 100ug of HN2-2ak, $10^{3.5}$ units of antibody as determined by end dilutional ELISA, given at different times after the initial infection was then explored. The most dramatic effects were seen with treatment 12 hrs. after infection with 65% of the animals surviving throughout the period of observation. However, treatment at 24, 36 and 48 hrs. after infection also prolonged survival over the HBSS treated control group (p $\leq$ 0.001). It therefore appeared that the course of an established virus encephalitis could be substantially modified with selected monoclonal antibodies.

The above studies were all undertaken with mouse monoclonal antibody purified from tissue culture super-

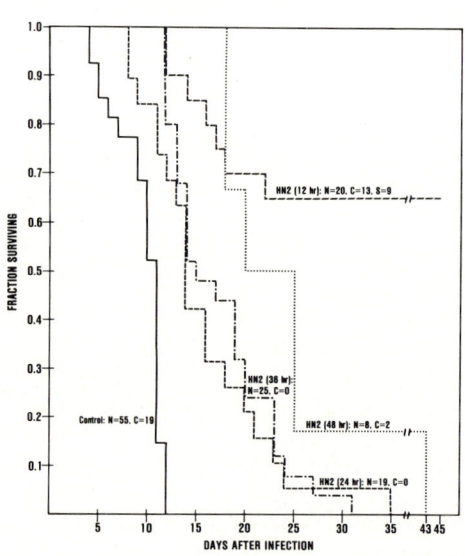

FIGURE 2

natants. This limited the quantity of antibody available for study. Neither HN1-3k nor HN2-2ak grow well in the peritoneal cavity of Balb/C mice. Therefore, a new monoclone was selected for the remaining studies. This monoclone, HN8-1k also has HAI and Nt activity in vitro. The effect of varying treatment dosage was explored with a single IP injection given 24 hrs. after the standard infection. Animals received either (A) $10^{3.8}$, (B) $10^{2.8}$ or (C) $10^{1.8}$ units of antibody (Figure 3). A dose response relationship was apparent with substantially improved survival at all doses ($p \leq 0.05$) and an increased fraction of long term survivors with the more concentrated doses.

In order to further explore the mechanism of protection by HN8-1k, infected animals were treated with a single inoculation of HN8-1k ($10^{3.6}$ units of antibody) or HBSS, 24 hrs. after infection and followed for survival, virus growth curves, histology and the distribution of viral antigen in brain. As anticipated, HN8-1k again afforded significant protection ($p \leq 0.001$) with only 30% overall mortality at 1.5 months (Fig. 4).

Free virus titers were substantial in these animals with $10^{4.37}$ PFU/gm recoverable from brain tissue at 24 hrs. after infection. (Fig. 5). Viral titers rapidly peaked in HBSS treated control animals (filled squares and fitted curve) and remained high until death. In contrast, few animals treated with HN8-1k (filled circles)

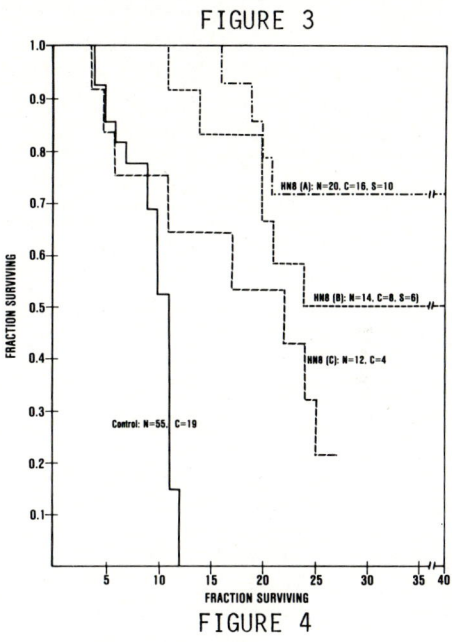

FIGURE 3

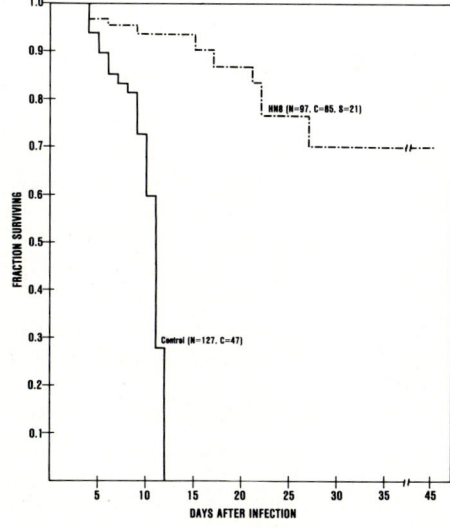

FIGURE 4

had virus recoverable at the dilutions assayed (11/28) and when virus was found it was generally at lower titer than from that recovered from controls. Similarly, all of the control animals (6/6) showed viral nucleocapsid antigen widely disseminated throughout brain by day 4 which subsequently further increased in distribution. Most HN8-1k treated animals (13/18) had only limited or no detectable mumps virus nucleocapsid antigen in brain at any time, and when present it was restricted to scattered ependymal cells. Of those HN8-1k treated animals with viral antigen in brain parechyma (5/18) only two had morphologic evidence of infection comparable to matched controls. Finally, the distribution of injected monoclonal antibody was sought in a pilot study. Two infected animals were inocculated on each of days 1 and 5 post-infection, with ^{125}I HN2-2ak and killed by perfusion-fixation 24 hrs. later. Autoradiography localized activity in well perfused brains primarily to meningeal and parenchymal vasculature of the choroid plexus.

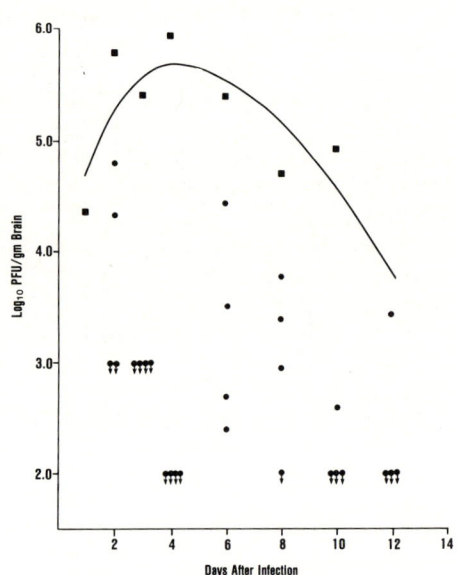

FIGURE 5

DISCUSSION

Monoclonal antibodies have recently been used to protect animals from the lethal effects of a variety of experimental viral infections of the CNS including herpes simplex virus (9,10), Sindbis virus (11), Venezuelan equine encephalitis virus (12), and Semliki Forest virus (13) infections. In these model infections, monoclonal antibody has been administered prior to establishing the infection. While extremely useful studies in delineating protective effects of antibodies which exist in vivo that were unexpected from the known functional activities of the antibodies as measured in vitro, they provide only limited insight into how such molecules might modify the course of natural infections. The present studies, like these using monoclonals to influenza virus (14) and measles virus (15) provide preliminary evidence that selected monoclonal antibodies may beneficially modify the course of established CNS infections.

In the mumps virus-hamster meningoencephalitis model, heterologous mouse monoclonal Ig directed against one or more epitopes on the HN glycoprotein of mumps virus would appear to diffuse into the CNS across vascular endothelium of the choroidal, meningeal, and parenchymal vasculature to reach cells within the CNS. At a minimum, these defined Ig prevent the spread of infection by neutralizing progeny virus shed from ependymal and choroidal epithelium infected up to 48 hrs. previously. These conclusions are based on the preliminary autoradiographic studies which localize labelled Ig to the above anatomic sites within the CNS, the effects on survival of appropriate monoclonal Ig administered as late as 48 hrs. after infection, the restricted sites of viral antigen in the brains of treated animals, and the markedly diminished viral titers in the brains of most treated hamsters. Which domains of the polypeptides of mumps virus are most important for inducing humoral responses that are protective from infection and reinfection, what restrictions for Ig entry and spread within the CNS may exist during the course of encephalitis, and the role of complement and antibody dependent cell mediated cytotoxicity mechanisms in the therapeutic effects seen with these monoclonal Ig are now approachable problems with this model system.

REFERENCES

1. Wolinsky, J., and Server, A. (in press). In "Virology" (B. Fields, J. Melnick, R. Chanock, R. Shope, and B. Roizman, eds.) Raven Press, New York.
2. Gellis, S., McGuiness, A., and Peters, M. (1945). Amer. J. Med. Sci. 210, 661.
3. Wolinsky, J., Klassen, T., and Baringer, J. (1976). J. Infect. Dis. 133, 260.
4. Server, A., Merz, D., Waxham, M. and Wolinsky, J. (1982). Infect. Immun. 35, 179.
5. Merz, D., Server, A., Waxham, M., and Wolinsky, J. (1983). J. Gen. Virol. 64, 1457.
6. Schwendemann, G., Wolinsky, J., Hatzidimitriou, G., Merz, D., and Waxham, M. (1982) J. Histochem. Cytochem. 30, 1313.
7. Lee, E. (1980). "Statistical Methods for Survival Analysis." Lifetime Learning Publications, Belmont.
8. Wolinsky, J., Jubelt, B., Burke, S. and Narayan, O. (1982). Ann. Neurol. 11, 59.

9. Balachandran, N., Bacchetti, S., and Rawls, W. (1982). Infect. Immun. 37, 1132.
10. Dix, R., Pereira, L., and Baringer, J. (1981). Infect. Immun. 34, 192.
11. Schmaljohn, A., Johnson, E., Dalrymple., J. and Cole, G. (1982). Nature 297, 70.
12. Mathews, J., and Roehrig, J. (1982). J. Immunol. 129, 2763.
13. Boere, W., Benaissatrouw, B., Harmsen, M., Kraaijeveld, C., and Snippe, H. (1983). J. Gen Virol. 64, 1405.
14. Doherty, P., and Gerhard, W. (1981). J. Neuroimmunol. 1,227.
15. Rammohan, K., McFarland, H., and McFarlin, D. (1981). Nature 290,588.

BIOCHEMICAL ASPECTS OF CHEMILUMINESCENCE INDUCED BY SENDAI VIRUS IN MOUSE SPLEEN CELLS[1]

Bernard Semadeni[2], Maurice J. Weidemann[3], & Ernst Peterhans[2]

[2] Institute of Virology, University of Zurich, Switzerland
[3] Department of Biochemistry, Faculty of Science, A.N.U., Canberra, A.C.T., Australia

Sendai and influenza viruses induce an immediate, short-lived burst of luminol-dependent chemiluminescence (CL) in mouse spleen cells that is triggered by the interaction of the envelope glycoproteins with the membranes of the responsive subpopulation of cells (1,2,3). In the case of Sendai virus, the chemiluminescent response depends upon both the hemagglutinin-neuraminidase (HN) and fusion (F) glycoproteins, the bulk of the light emission being associated with the action of the latter (2,3). Chemiluminescence reflects the generation by the cells of unstable oxygen species (e.g. O_2^-, H_2O_2, OH·) that react with the easily oxidized chemiluminogenic probe luminol (5-amino-2,3-dihydro-1,4-phthalazinedione) (reviewed in 4). The emission of photons from this reaction can be monitored in a conventional liquid scintillation counter (LSC) operated in the "out-of-coincidence" mode. An understanding of virus-induced CL is significant for several reasons. Firstly, being an event associated with the action of HN and F, an understanding of its origin and mechanism would help to elucidate biochemical events associated with adsorption and penetration of the virus into its host cell. Secondly, if it reflects the interaction of the virus with phagocytic cells, the biochemistry of CL might help to elucidate some aspects of the pathogenesis of virus infection such as the impairment of immune responsiveness and the appearance of systemic symptoms associated with locally restricted virus multiplication (5).

In this paper, we report the bivalent cation dependence of virus-induced CL and present evidence that indicates a prominent role for arachidonic acid metabolism in this process.

---

[1] This work was supported by Grant no. 3.649.0.80. from the Swiss National Science Fund and Grant no. D 27915664 from the Australian Research Grants Scheme.

CL Induced by Sendai Virus. Fig. 1 shows a typical CL response of mouse spleen cells to Sendai virus. CL increased rapidly on addition of virus to the cells, reaching a peak at approx. 2-3 min. post-infection and declining thereafter to prestimulation levels at 20-40 min. post-infection.

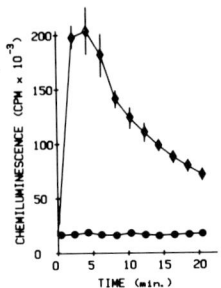

FIGURE 1. The CL response of mouse spleen cells to Sendai virus. Spleen cells and virus were prepared as described previously (3). CL was measured in a modified Kontron MR 300 DPM liquid scintillation counter using the settings for tritium with the dead time correction and coincidence circuits switched off. Data were recorded and analyzed as described (3). To 3×10^6 cells suspended in 750 µl Hanks'balanced salt solution (HBSS) were added 390 hemagglutinating units of virus. Zero time represents the addition of virus to the spleen cells. Three samples of virus-stimulated cells and one sample mock-stimulated with PBS were studied. The points represent the means and the bars standard deviations. ♦——♦ Virus-stimulated cells; ●——● mock-stimulated cells.

The Role of Bivalent Cations in the Generation of CL. To investigate the role of bivalent cations, cells were suspended in HBSS lacking either Ca^{2+} or Mg^{2+}, or both. Fig. 2 shows that omission of Mg^{2+} from the medium did not decrease CL stimulated by Sendai virus; it was inhibited more than 75%, however, in medium lacking Ca^{2+} but supplemented with Mg^{2+}. When EGTA (0.1-4 mM) was included in the Ca^{2+}-free assay medium, CL was not decreased further, suggesting that the residual CL was generated independently of trace amounts of Ca^{2+} present in the assay medium. In medium lacking both Ca^{2+} and Mg^{2+} and containing EGTA and EDTA (1.0 mM) to bind trace amounts of bivalent cations, CL was similar to that observed in the medium lacking Ca^{2+} alone. To compare virus-induced CL with that evoked by a stimulus known to be Ca^{2+}-dependent, similar experiments were carried out using the Ca^{2+} ionophore A23187 at a final concentration of 1 μM. The ionophore stimulated CL with a time course similar to that of the virus-induced response (results not shown). However, in contrast to virus-induced CL, the response evoked by the ionophore was completely abrogated in the absence of extracellular Ca^{2+}. Fig. 2 also shows that Mg^{2+} exerted an inhibitory effect on CL induced by the ionophore; the light emission in the presence of Ca^{2+} alone exceeded that generated when both Ca^{2+} and Mg^{2+} were present.

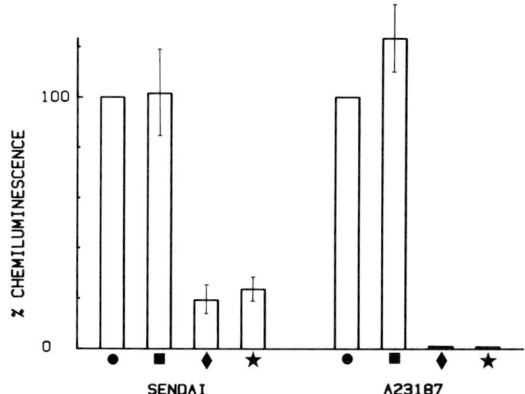

FIGURE 2. The effects of Ca^{2+} and Mg^{2+} on Sendai virus-induced CL. Spleen cells were washed twice in Ca^{2+}/Mg^{2+}-free HBSS. Ca^{2+} or Mg^{2+} (final concentration 1 mM) were added to the cell suspension at 30 min. and EGTA or EDTA at 10 min. before stimulation with the virus. The different ionic conditions did not alter the kinetics of the CL response and integrated light emission over 20 min. was compared in each case. CL in the presence of Ca^{2+} and Mg^{2+} is taken as 100%. Columns represent integrated light emission and bars represent the standard errors of the respective means. Values from 6 and 5 separate experiments are compared for virus- and ionophore-stimulated CL, respectively. The response observed in the Ca^{2+}-depleted buffers was significantly different at the $p < 0.05$ level. ● control, Ca^{2+} and Mg^{2+} at 1 mM; ■ Ca^{2+} alone; ◆ Mg^{2+} alone plus EGTA; ★ no Ca^{2+} and Mg^{2+}, but EGTA and EDTA present.

Glucose Requirement of the CL Response. CL associated with the respiratory burst in phagocytic cells is thought to originate from a membrane-spanning NAD(P)H oxidase located within the plasma membrane (6,7). The electrons required for the reduction of oxygen originate mainly from the oxidation of glucose via the oxidative segment of the pentose phosphate pathway. To investigate the glucose dependence of virus-induced

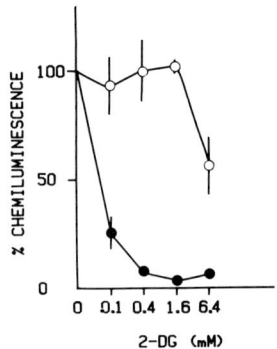

FIGURE 3. The effect of 2 deoxy-D-glucose on virus-induced CL. Points represent the means of integrated light emission curves (20 min) of 6 separate experiments and bars represent the standard errors of the means. ●——● No D-glucose; ○——○ D-glucose present (5 mM).

CL, we compared the responses of cells incubated in the presence and absence of glucose. Preliminary experiments had shown that the extent of CL is dependent upon the time of preincubation of cells in glucose-free medium; it begins to decrease when the period of preincubation exceeds 1 h. The decrease in CL is not accompanied by a decrease in cell viability as assessed by the Trypan Blue exclusion test (results not shown).

These observations suggested that cellular light emission has a requirement for glucose that can be met partly by mobilization of glycogenolytic intermediates from endogenous stores. To investigate this possibility, cells were prepared in medium containing glucose, further incubated in this medium for 30 min at $20°C$, washed twice and incubated thereafter for 60 min at $37°C$ in glucose-free medium. Under these conditions, the CL produced was similar to that observed in control cells incubated in a glucose-supplemented medium. However, when the glucose analogue 2-deoxy-D-glucose (2-DG) was added to the cells 10 min before virus (i.e. after 50 min incubation in the glucose-free medium), CL was decreased in a concentration-dependent manner (Fig. 3). The inhibitory effect of 2-DG was practically eliminated when D-glucose was present in the assay medium (Fig. 3), suggesting that the inhibition by 2-DG was due to interference with D-glucose metabolism. In contrast to 2-DG, the transportable but non-metabolizable glucose analogue 3-O-methyl-D-glucose did not affect CL when tested over the same concentration range (result not shown). Since this analogue, unlike 2-DG, is not phosphorylated by hexokinase, it would be unlikely to compete with the metabolism of glycogenolytically-derived intermediates.

The Role of Phospholipase Metabolism. The release of arachidonic acid from membrane phospholipids and its subsequent metabolism via the cyclooxygenase and lipoxygenase pathways represents a major regulatory network in the inflammatory response to a wide variety of pathogens. The former pathway leads to the production of prostaglandins and thromboxanes and the latter to leukotrienes and HETE's. In both pathways, highly unstable epoxy- and hydroperoxy-intermediates are formed, respectively, and it has been suggested that these intermediates may play a role in the generation of CL in various systems (8,9). To investigate the role of arachidonic acid metabolism in virus-induced CL, we used a variety of metabolic inhibitors (see Fig. 4).

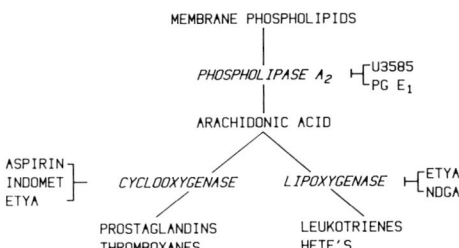

FIGURE 4. The metabolism of arachidonic acid and the sites of action of the inhibitors used.

Of these inhibitors, U3585, prostaglandin E_1, ETYA and NDGA inhibited CL in a concentration-dependent manner, whilst aspirin had no significant effect and indomethacin at high concentration only affected CL (Fig. 5). Control experiments suggested that none of the inhibitors used interfered with the binding of ^3H-uridine-labelled Sendai virus to mouse spleen cells, nor did they quench the chemical reaction between H_2O_2 and luminol (except for NDGA, which quenched H_2O_2 stimulated luminol CL by 50% whilst decreasing cellular CL by 97% when used at 10 µM).

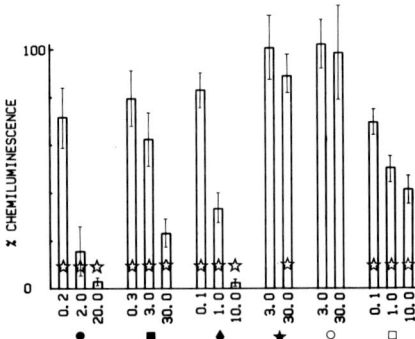

FIGURE 5. Effect of inhibitors of arachidonic acid metabolism on Sendai virus-induced CL. The columns show light emission integrated over 20 min in the presence of inhibitors at the concentrations indicated in μM. Values represent the means ± SEM of 5 separate experiments with U3585 and 6 separate experiments with the other inhibitors. ☆ Indicates significant difference at the $p \leq 0.05$ level (t-test). ● U3585, ■ ETYA, ♦ NDGA, ★ Indomethacin, ○ Aspirin, □ Prostaglandin E_1.

DISCUSSION

In this paper we have characterized the metabolic requirements of Sendai virus-induced CL in mouse spleen cells. The results show that CL depends on: (a) glucose metabolism and (b) the presence of external Ca^{2+} but not Mg^{2+}; and (c) that there is a correlation between the magnitude of the CL response and phospholipase A_2 and lipoxygenase activities.

Within this conceptual framework, two reactions have been shown to lead to CL. An NAD(P)H oxidase located at the surface of phagocytic cells is known to be associated with the production of reactive oxygen species that generate CL in the presence of luminol (4,6,7). More recent evidence indicates that lipoxygenase activity may also lead - either directly or indirectly - to CL, most likely through the decomposition of highly unstable hydro-peroxy-intermediates generated during the formation of HETE's and leukotrienes (8,9) or through the activation of NAD(P)H oxidase by one of these metabolic intermediates and/or products. To definitely attribute the source of reactive oxygen species to one or other or both of these pathways, it will be necessary to carry out additional experiments. The evidence for the involvement of the lipoxygenase pathway and glucose catabolism in Sendai virus-induced CL is indirect and the specificities of the inhibitors used in the present study must be checked by assaying for product formation using independent methods such as ion-pair HPLC (10). Moreover, although we have checked the effects of all the inhibitors used in the present work on a luminol-H_2O_2 reaction in a cell-free system, the possibility that they might cause interference with the reaction between luminol and the (unknown) reactive oxygen species responsible for virus-induced CL cannot be excluded with certainty. In spite of these limitations, the results of the present study are of interest in several respects. Firstly, two recent papers have shown that a correlation exists between glucose metabolism and lipoxygenase activity (as measured by product formation) similar to that observed in the present experiments (10,11), and it was suggested that lipoxygenase activity and the respiratory burst are interrelated phenomena with mutual regulatory activity (10). Secondly, the dependence of CL on exogenous Ca^{2+} can be explained in terms of the requirement for this cation of phospholipase A_2 (12). Several observations suggest that the reactions underlying CL favour membrane-membrane fusion. Thus, lyso-phospholipid products that result from the activation of phospholipase A_2 are known to be fusogenic and the reincorporation into the membrane of lipoxygenase products (e.g. HETE's) (13) is likely to lead to increased membrane fluidity. However, Sendai virus has been reported to fuse with artificial, enzyme-free membranes (14), suggesting that factors in addition to those described in this paper may contribute to envelope-cell membrane fusion.

Finally, it seems likely that the biochemical reactions responsible for generating chemiluminescence may play a role in the pathogenesis of virus infection. The metabolism of arachidonic acid leads to the production of several highly potent mediators of inflammation (e.g. leukotrienes, HETE's), and reactive oxygen metabolites released from phagocytic cells on activation by the virus might contribute significantly to tissue damage.

ACKNOWLEDGEMENTS

We thank Dr. Wallach, The Upjohn Company, Kalamazoo, USA for a gift of the phospholipase A_2 inhibitor U3585 and Ms Christina Gerber for typing the manuscript.

REFERENCES

1. Peterhans, E. (1979). Biochem.Biophys.Res.Commun. 91, 383.
2. Peterhans, E. (1980). Virology. 105, 445.
3. Peterhans, E., Bächi, T., and Yewdell, J. (1983). Virology. 128, 366.
4. Allen, R.C. (1980). In "The Reticuloendothelial System". (J. Sbarra and R.R. Strauss, eds.) p.309. Plenum Press, New York.
5. Mims, C.A. (1982). "The Pathogenesis of Infectious Disease." Academic Press, New York.
6. Rossi, F., Romeo, D., and Patriarca, P. (1972). J. Reticuloendothel. Soc. 12, 127.
7. Salin, M.L., and McCord, J.M. (1974). J.Clin.Invest.54, 1005.
8. Smith, R.L., and Weidemann, M.J. (1980). Biochem.Biophys. Res.Commun. 97, 973.
9. Cheung, K., Archibald, A.C., and Robinson, M.F. (1983). J. Immunol. 130, 2324.
10. Ziltener, H.J., Chavaillaz, P.-A., and Jörg, A. (1983). Hoppe-Seyler's Z. Physiol. Chem. 364, 1029.
11. Walker, J.R., and Parish, H.A. (1981). Inter.Archs.Appl. Immun. 66, 83.
12. Wightman, P.D., Humes, J.L., Davies, P., and Bonney, R.J. (1981). Biochem.J. 195, 427.
13. Stenson, W.F., and Parker, C.W. (1979). J.Clin.Invest. 64, 1457.
14. Hsu, M.-C., Scheid, A., and Choppin, P.W. (1983). Virology. 126, 361.

Defective Viruses
and Virus Persistence

TRANSCRIBING VSV$_{NJ}$ DI PARTICLE AND ITS BIOLOGICAL ACTIVITIES

C. Yong Kang, Ruth Park, John McCulloch and Jeong Sun Seo [1]

Department of Microbiology and Immunology
University of Ottawa School of Medicine
Ottawa, Ontario, Canada K1H 8M5

Defective interfering (DI) particles contain only a portion of the genetic information of the parental standard virus [1,2]. DI particles are not capable of self replication in infected host cells, however, they can be amplified when they are coinfected with the parental standard virus. Coinfection of cells with the standard virus and DI particles results in suppression of standard virus production. This phenomenon is known as viral interference [1,2]. The molecular mechanisms of DI particle mediated viral interference is not yet known.

The genetic contents of a substantial number of VSV and Sendai virus DI particles have been characterized [2,3,4]. It is clear that the biological activities of DI particles are related to the sequences present in the DI genome [2,5].

DI particle genomes represent true deletions of the standard virus genome at either the 3' or 5' terminus of the genome. The most abundant DI particles of RNA viruses represent the 5' terminus of the standard virus. A recent report has described the isolation of 3' DI particles of Sendai virus [4] and we have isolated several 3' DI particles of VSV$_{IND}$ [6]. One well defined DI particle of VSV$_{IND}$, designated as DI-LT, represents the 3' half of the genome [7,8,9]. The genome of this DI particle contains genes for N, NS, M and G proteins including the leader RNA sequences. It can be transcribed by the endogenous RNA polymerase activity yielding 4 functional messenger RNAs and leader RNA [10,11]. The DI-LT interferes with replication of standard virus from both Indiana serotype (homotypic interference) and New Jersey serotype (heterotypic interference) of VSV [5]. In contrast, all 5' DI particles are capable of interfering only with the replication of the standard virus of their own serotype [1,2,5]. This paper deals with the isolation and partial characterization of a New Jersey serotype DI particle which interferes with the replication of the standard virus of both New Jersey and Indiana serotypes of VSV.

---

[1] Present address, Department of Biochemistry, Seoul National University College of Medicine, Seoul, Korea.

Generation of DI 121 from VSV_{NJ}. The prototype of 3' DI particles, DI-LT, has been isolated from a heat resistant strain of VSV_{IND}[12]. We used the same strategy to isolate 3' DI particles of New Jersey serotype. We chose VSV_{NJ} (Ogden) because the DI-LT which originated from VSV_{IND} interferes with the replication of the standard virus of heterotypic VSV_{NJ} (Ogden) [5]. The virus was heated at 45°C for 30 mins, appropriate dilutions were made and used to infect rat cells transformed with B77 strain of avian sarcoma virus (R(B77)). The cells were overlayed with 0.9% agar and incubated at 40.5°C overnight. Plaques arising from the heat treatment were selected and made subsequent plaque isolations were made by repeating the same procedure. After three consecutive plaque isolations, the virus was propagated in R(B77) cells. We made subsequent high multiplicity of infection passages of the virus until we detected an appropriate concentration of DI particles. It took approximately five 15-hour passages to obtain adequate quantities of DI particles. Fig. 1 illustrates the sucrose gradient purification of DI particles from both VSV_{IND} and VSV_{NJ}. The DI particle from New Jersey serotype, DI-121, sedimented close to the standard virion and relatively free of other DI particles. In contrast, DI-011 and DI-LT preparations had some minor contamination of other DI particles. The DI-121 of VSV_{NJ} sedimented at the same position as DI-LT, thus, DI-121 may have genomic RNA of approximately 50% that of the standard virus genome since DI-LT represents 50% of the standard viral genome.

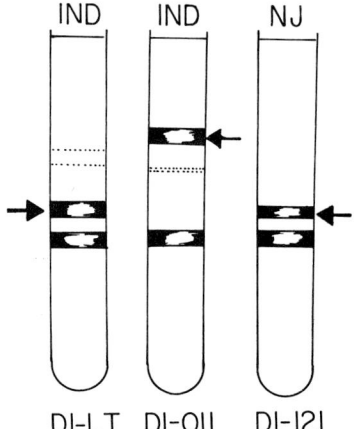

FIGURE 1. Sucrose gradient purification of DI particles and the standard virions. Samples of lysates were centrifuged at 81,000 x g for 75 minutes, resuspended in 0.2-0.3 ml of PBS containing 1M NaCl and layered on a linear 5-30% sucrose gradient made in PBS. After centrifugation at 110,000 x g for 35 minutes at 5°C in a Spinco SW41 rotor, the visible DI particles (with arrows) and the standard virions (the bottom bands) were collected.

Endogenous RNA transcriptase activity of DI-121. DI particles representing the 5' terminus of the standard VSV$_{IND}$ synthesizes RNA products of 46 nucleotides in length when the endogenous RNA polymerase reaction is carried out in vitro. There is a strong termination signal which prevents polymerase readthrough (2). In contrast, the 3' DI particle, DI-LT, synthesizes the leader RNA and four functional messenger RNAs (10,11).

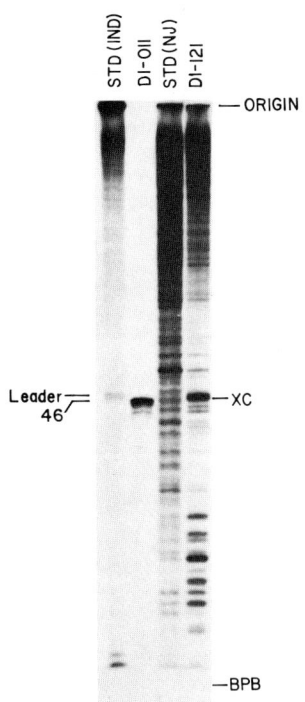

FIGURE 2. Polyacrylamide gel electrophoresis of RNA transcripts. RNAs were synthesized in vitro in the presence of -^{32}P GTP by DI-011, DI-121, and the standard virions of both VSV$_{IND}$ and VSV$_{NJ}$. The resulting RNA products were purified by Sephadex G-50 chromatogrpahy followed by extraction with phenol and chloroform. The ^{32}P GTP-labeled RNA products were analyzed by electrophoresis in 12% polycacrylamide gel containing 8M urea.

Figure 2 shows the analyses of RNA transcripts from in vitro polymerase reaction. VSV$_{IND}$ standard virus synthesizes 47 nucleotide leader RNA and large transcripts whereas, DI-011 synthesizes predominantly the 46 nucleotide products without larger RNAs. In contrast, DI-121 synthesizes large transcripts similar to that of VSV$_{NJ}$ standard virus (Fig. 2). There are numerous RNA products which are much smaller than messenger RNAs. It is unlikely tht these smaller products are the result of post-transcritional degradation since VSV_{IND} standard virus shows predomintly large RNA transcripts under the same conditions of preparation. It is possible that VSV_{NJ} polymerase terminates RNA transcription much more frequently. The 46 nucleotide DI products yields 21 and 13 oligonucleotides after ribonuclease T_1 digestion. In contrast, the leader RNA of VSV_{IND} has 28 nucleotides which can be identified easily in 20% polyacrylamide-urea gels after ribonuclease T_1 digestion (Fig.3). We have digested RNA transcripts of both VSV_{NJ} (Ogden) and DI-121 with ribonuclease T_1 and analyzed in the 20% polyacrylamide gel. A comparison of the digests of the RNA transcripts of DI-121 and the standard virus shows no significant differences betwen the standard virus of VSV_{NJ} and DI-121 particles (Fig. 3).

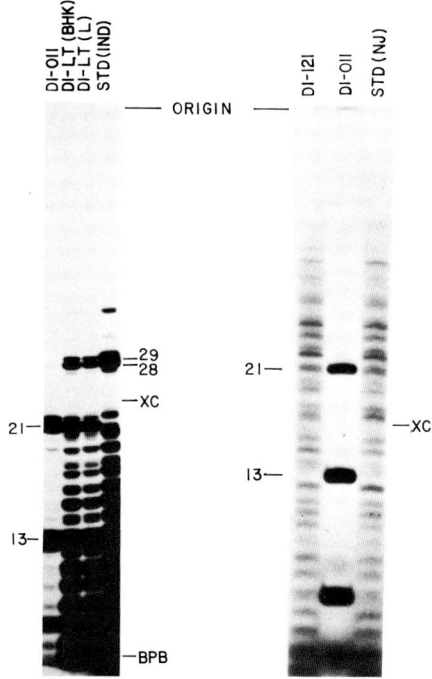

FIGURE 3. Polyacrylamide gel electrophoresis of RNA transcripts after ribonuclease T_1 digestion. The ^{32}P GTP-labeled RNA products from Fig. 2 were digested with ribonuclease T_1 and analyzed in 20% polyacrylamide gel containing 8M urea.

Homotypic and heterotypic interference by DI-121. The DI-LT, the 3' DI particle isolated from the heat resistant strain of VSV_{IND}, interferes the replication of the standard viruses of both Indiana and New Jersey serotypes [5]. The biological activities of DI-121 have been investigated to determine whether this DI particle also interferes with the replication of standard virus of both

TABLE 1.
Interference by DI-011 and DI-121 on production of the standard virus of VSV_{IND} and VSV_{NJ}

	VSV_{IND} PFU/ml (% control)	VSV_{NJ} PFU/ml (% control)
Standard virus alone	3.2×10^9 (100%)	1.3×10^9 (100%)
Standard virus + DI 011	1.5×10^7 (0.5%)	1.0×10^9 (76.9%)
Standard virus + DI 011 (Anti-VSV_{IND}-Ab)	1.3×10^9 (40.6%)	9.0×10^8 (69.2%)
Standard virus + DI 011 (Anti-VSV_{NJ}-Ab)	1.2×10^7 (0.4%)	8.5×10^8 (65.4%)
Standard virus + DI 121	5.6×10^7 (1.8%)	2.7×10^7 (2.1%)
Standard virus + DI 121 (Anti-VSV_{IND}-Ab)	1.5×10^8 (4.7%)	4.2×10^7 (3.2%)
Standard virus + DI 121 (Anti-VSV_{NJ}-Ab)	1.8×10^9 (56.3%)	8.0×10^8 (61.5%)

The gradient purified DI particles were treated with either anti-VSV_{IND}-antibody or anti-VSV_{NJ}-antibody. The antibody treated DI particles were layered on a 10% sucrose made in PBS and 0.5 ml of 65% sucrose cushion in Spinco SW50.1 rotor and centrifuged at 189,000 x g for 90 minutes at 5°C. The pelleted antibody treated DI particles were mixed with standard virus and infected with multiplicity of infection of 10 PFU/cell. The progeny virus was harvested after 8 hours of infection and tested for infectivity by plaque assay.

serotypes since DI-121 resembles the DI-LT both in terms of its size and biochemical properties. It is clear from the data presented in Table I that over 95% of infectious virus production of standard viruses of both Indiana and New Jersey serotypes was inhibited by DI-121. Furthermore, DI-121 interferes with replication of the standard virus of VSV_{IND} and VSV_{NJ} at about the same level (Table I). Thus, the DI-121 is capable of both homotypic and heterotypic interference.

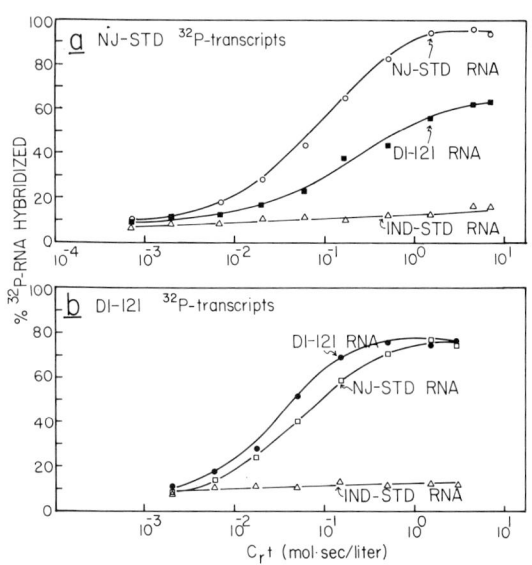

FIGURE 4. Kinetics of hybridization of in vitro transcripts of VSV_{NJ} standard virion and DI-121 to different VSV RNAs. ^{32}P GTP-labeled transcripts from in vitro transcription of VSV_{NJ} standard virion (4000 cpm/sample) and DI-121 (3500 cpm/sample) were incubated at 63°C in 25 u liters of annealing buffer containing 0.3M salt sealed in 25 u liter micropipettes with 300ng of RNAs of VSV_{IND} standard virion, VSV_{NJ} standard virion, DI-011, or DI-121. At different times, samples were withdrawn and were frozen. At the end of all incubations, the extent of hybridization was determined by ribonucleases A and T_1 digestion in 0.3M salt. The results are expressed as the proportion of total ^{32}P RNA hybridized at a given $C_r t$ value in 0.3M salt (RNA concentration in moles/liters × time of incubation in seconds).

The origin of DI-121. Since the size and biological activities of DI-121 resembles the DI-LT, it was necessary to prove that this DI particle originated from VSV_{NJ} rather than VSV_{IND}. The gradient purified DI particles were treated with either anti-VSV_{IND}-antibody or anti-VSV_{NJ}-antibody, centrifuged through 10% sucrose and pelleted on 65% sucrose cushion to remove unbound antibodies. The pelleted antibody treated DI particles on the 65% sucrose cushion were mixed with known amounts of the standard virus and infected. The progeny virus was harvested after 8 hours of infection and tested for infectivity by plaque assay. As can be seen in Table I, the DI-011 activity for interference can be neutralized by anti-VSV_{IND}-antibody but not by the anti-VSV_{NJ}-antibody. In contrast, DI-121 activity for interference was neutralized by anti-VSV_{NJ}-antibody only. Thus, DI-121 contains surface antigens of VSV_{NJ}.

To further ensure that DI-121 is indeed originated from VSV_{NJ} rather than being a pseudotype representing genomic RNA from Indiana serotype and surface antigens derived from VSV_{NJ}, hybridization experiments have been carried out. A significant amount of ^{32}P-labeled RNA transcripts of VSV_{NJ} hybridized with RNAs from the standard virus of VSV_{NJ} and from DI-121 (Fig. 4a). A reciprocal experiment was also carried out (Fig. 4b). The transcripts of DI-121 hybridized with RNAs from the standard virions of VSV_{NJ} and from DI-121. However, the transcripts failed to hybridize to RNAs of VSV_{IND}. These results demonstrate that DI-121 genome is derived from VSV_{NJ}.

Summary. The DI-121 isolated from heat resistant strain of VSV_{NJ} inteferes with the replication of standard viruses of both Indiana and New Jersey serotypes of VSV. It synthesizes RNA transcripts similar to that of the standard virus of VSV_{NJ}. We are currently analyzing the nucleotide sequences of DI-121 genome in order to determine the relationship between the biological activities and nucleotide sequences.

ACKNOWLEDGEMENTS

This study was supported by grant MA-7696 from the Medical Research Council of Canada.

REFERENCES

1. Huang, A. S., and Baltimore D. (1977). In "Comprehensive Virology" Vol. 10 (H. Fraenkel-Conrat and R.R. Wagner, eds). p. 73. Plenum Press, New York.
2. Perraut, J. (1981). In "Current Topics in Microbiology and Immunology", Vol. 93 (A. J. Shatkin, ed.), P. 151. Springer-Verlag, New York.
3. Keene, J.D., Chien, I. M., and Lazzarini, R. A. (1981). Proc. Natl. Acad. Sci. U.S.A. 78, 2090.
4. Re, G. G., Gupta, K. C., and Kingsbury, D. W. (1983). J. Virol. 45, 659.

5. Prevec, L., and Kang, C. Y. (1970). Nature 228, 25.
6. Kang, C. Y., Schubert, M., Rose, J., and Lazzarini, R.A. (1983). Manuscript in preparation.
7. Stamminger, G.M., and Lazzarini, R.A. (1974). Cell 3, 85.
8. Leamnson, R. N., and Reichmann, M.E. (1974). J. Mol. Biol. 85, 551.
9. Epstein, D. A., Herman, R.C., Cien, I., and Lazzarini, R. A. (1980). J. Virol. 33, 818.
10. Colono, R. J., Lazzarini, R. A., Keene, J. D., and Banerjee, A. K. (1977). Proc. Natl. Acad. Sci. U.S.A. 74, 1884.
11. Johnson, L.D., and Lazzarini, R. A. (1977). Proc. Natl. Acad. Sci. U.S.A. 74, 4387.
12. Petric, M., and Prevec, L. (1970). Virol. 41, 615.

RECOMBINATION EVENTS DURING THE GENERATION
OF DI RNAS OF VSV

Ellen Meier, George G. Harmison, Jack D. Keene[1],
and Manfred Schubert

Laboratory of Molecular Genetics, IRP
National Institute of Neurological and Communicative
Disorders and Stroke, NIH
Bethesda, Maryland 20205

The model for the generation of defective interfering (DI) particles of Vesicular stomatitis virus (VSV) proposes aberrant replication events rather than breakage and reunion of RNA molecules (1, 2). The central feature of this copy choice model is that the polymerase interrupts RNA synthesis, detaches from the template and moves with the nascent daughter strand attached to it to another position on the same or a different template where RNA synthesis resumes, thereby extending the nascent chain. With respect to genomic sequences retained and the degree of self-complementarity, four classes of DI particles can be distinguished: deletion, panhandle, snapback and compound DIs. The nucleotide sequence analysis of the L gene (Harmison et al., this volume) enabled us to identify the sites of rearrangement involved in the generation of DI genomes. In this communication, we describe their structural organization and discuss whether termination and/or resumption of RNA synthesis are specified by the primary sequence of the RNA template.

STRUCTURAL ORGANIZATION OF DI GENOMES

Figure 1 shows the structure of DI-LT, 011, T, T(L) and 611 as compared to VSV and the positions where the DI and L gene sequences diverge.

Deletion DI RNA. Hybridization studies revealed that DI-LT is a true deletion mutant in which most of the L gene is deleted, while the N, NS, M and G genes are retained (3). The site of the deletion was identified by sequencing cDNA clones of the relevant regions of the DI and the parental

[1]Department of Microbiology and Immunology, Duke University Medical Center, Durham, North Carolina.

genome (4). Comparison with the complete sequence of the L gene demonstrates that the deletion spans from positions 341 to 6190 of the genome involving 5848 nucleotides. It is important to note, that DI-LT represents a mixture of at least 2 DI particles of similar size, DI-LT$_1$ and DI-LT$_2$ (3).

Snapback DI RNA. Earlier studies established that the genome of the snapback DI 011 consists of a region which is homologous to the genomic 5' end for about 1 kb and which is covalently linked to its exact complement (5). Comparison of the published sequence of about 70 bases before and behind the turnaround point (5) with the sequence of the cDNA clones of the L gene reveals, that the last base of the (-) sense strand of DI 011 corresponds to position 1167 of the VSV genome.

Panhandle DI RNA. The panhandle DI particles T, T(L) and 611 have retained different amounts of genetic information from the 5' half of the genome and share at their 3' end a 45 (DI-T) or 48 (DI-T(L) and 611) nucleotides long sequence which is complementary to the 5' end (6) -- the panhandle region (Figure 1, a'). The sequences beyond the panhandle regions were determined for about 80-100 bases from the 3' end of the DI RNAs using the chemical sequencing method described by Peattie (7). Homology to the L gene sequence was found to extend to positions 2163 with DI-T, 3641 with DI-T(L) and 4417 with DI 611 RNA.

The locations of the recombination sites within the L gene of all the DI particle RNAs described above are consistent with the overall size of the DI particle genomes, based on hybridization data and oligonucleotide fingerprint mapping. We, therefore, assume that no additional severe

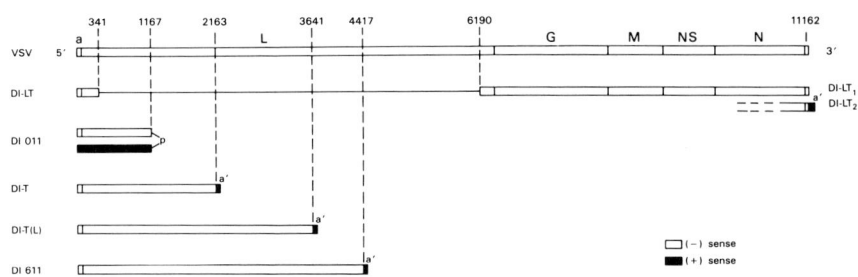

FIGURE 1. Maps of the genomes of VSV, DI-LT, DI 011, DI-T, DI-T(L) and DI 611. Black bars represent sequences of DI particle genomes not present in the VSV (-) sense sequence.

rearrangements have occurred in the RNA of these DI particles. The complete sequences of the genomes of DI 011, T, T(L), and 611 can, therefore, be deduced from the primary structure of the L gene.

SEQUENCE ANALYSIS OF THE RECOMBINATION SITES

The sequences of the genomic recombination sites involved in the generation of DI-LT, 011, T, T(L) and 611 are shown in Figure 2a-c. By comparing these data, we did not detect the existence of a general termination or resumption signal. This result, however, does not mean that the construction of DI particle genomes occurs in all cases independently of the primary structure of the VSV genome. In fact, in the following part, we present evidence that a particular sequence may be involved in the origin of the deletion DI-LT and another sequence may specify the events leading to the panhandle DIs T, T(L) and 611. We propose that different sequences play a role in the generation of some DI particle genomes by exerting their influence through different mechanisms.

Origin of Deletion DI RNA. The sequence immediately before the start of the deletion in DI-LT contains two hexanucleotides ATCTGA and GATTGG, marked by boxes in Figure 2a. Interestingly, these groups are also found about 6000 bases downstream at the end of the deletion. In addition, the spacing of the hexanucleotides in both regions is very similar: the number of bases between the hexanucleotides is in the same range (10 and 13 bases) and also the distance between the sequence GATTGG and the base flanking the deletion at each site differs only little (5 and 7 bases).

This arrangement suggests a model for the generation of DI-LT which may involve annealing of the nascent RNA strand to the template before synthesis resumes (Figure 2a). Unique for this scheme is that a sequence at the end of the nascent RNA chain determines the resumption site of RNA synthesis, while the termination cannot be addressed to a specific sequence at this time.

Origin of Snapback DI RNA. The recombination site involved in the generation of DI 011 does not reveal that a particular sequence may play a role in the origin of this DI particle (Figure 2b).

In order to explain the exact complementarity of the DI 011 genome, two models have been discussed (5). One of these models depends on the presence of a small region of dyad symmetry in the RNA template. It was suggested that an unencapsidated transcript of this region could potentially self-anneal and that the polymerase then copies back the nascent chain. Our sequencing data clearly rule out this model, because the sequence in the template before and behind the position corresponding to the turnaround point are not self-complementary. Our data are, however, consistent with the second model, which suggests replication across a replication fork (5).

Origin of Panhandle DI RNA. The similarity in the length of the stem regions of DI-T, T(L) and 611 was explained by involvement of a specific polymerase recognition site with the sequence GGUCUU which is located 43-48 bases from the 5' end of the VSV genome (6). Based on this copy back model (8, 9) the recombination sites within the L gene are presumptive termination sites. There is no evidence that termination may be specified by sequence information (Figure 2c).

Because of a repeat of two bases (AA) at the termination and resumption sites involved in the generation of DI-T(L) and 611, the 48th and 47th base from the 3' terminus of these DI particle genomes could also be encoded in the case of DI-T(L) by positions 3640 and 3641 of the L gene and by positions 4416 and 4417 in the case of DI 611. Thus, the added 3' terminal regions of these DI particle RNAs are identical in size as the predominant 46 nucleotides long (-) strand leader RNA (10). The same region of DI-T consists of 45 nucleotides.

A model for the origin of these DI particles can be proposed which involves as a first step the synthesis of (-) leader RNA. Instead of releasing the transcript the polymerase detaches with it from the (+) strand template and extends it by several thousand bases after resuming synthesis at a (-) strand nucleocapsid. In this model, the termination of RNA synthesis presumably can be addressed to the sequence 3'-GAA(A)-5' which is considered to function as part of a stop signal in VSV (11), while the resumption seems to occur randomly. This scheme is similar to the proposed model for the generation of the compound DI-LT$_2$ (12). Although the copy back model (6, 8, 9) and the model suggested here are contrary, they have in common that a sequence signal plays a role, specifying either the resumption or the termination of RNA synthesis.

In a modified form both models may also be applicable for DIs with larger panhandle regions (13). Based on the first model, the hexanucleotide GGUCUU which can be found at positions 43-48 and also at positions 156-161 from the 5' end of the genome is a necessary prerequisite for resumption of RNA synthesis, but the resumption site is less localized than assumed. Based on the second model, occasional readthroughs

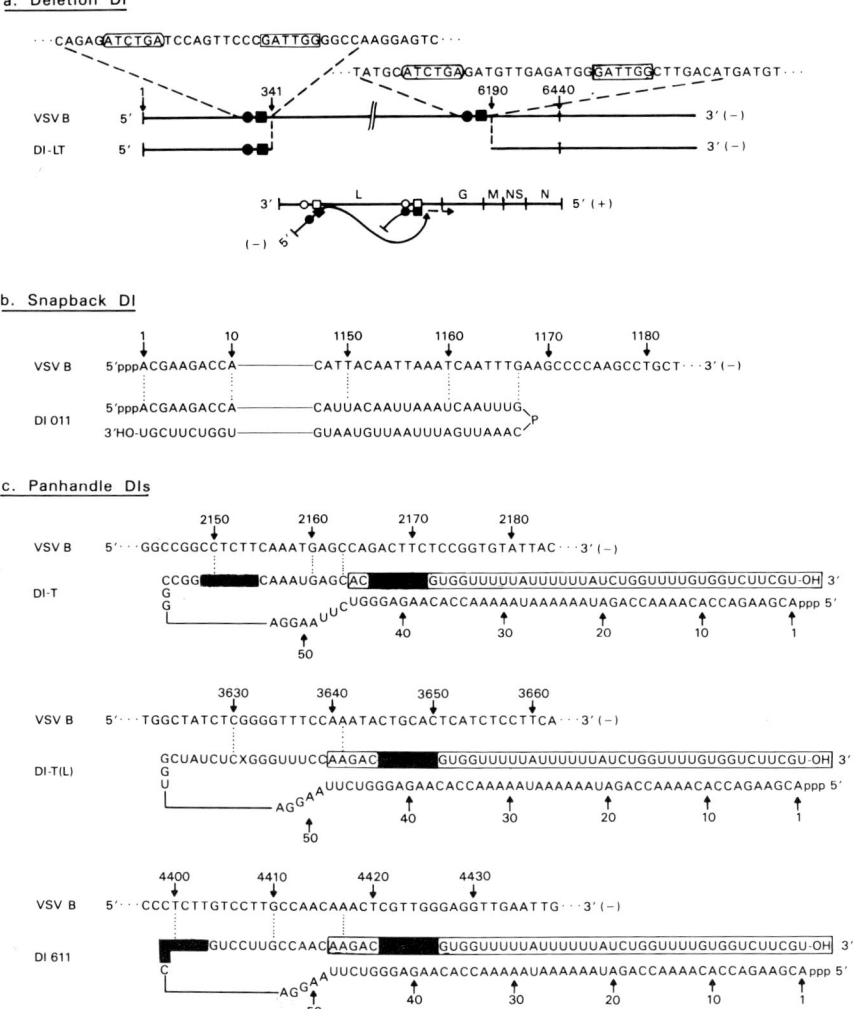

FIGURE 2. Recombination sites involved in the generation of DI particle genomes.

through the stop signal at the end of (−) leader RNA region could account for larger panhandles.

At the end of the 5' region, we found in the case of DI-T and DI 611, but not DI-T(L), the hexanucleotide CCUCUU which is also repeated at the beginning of the stem region in all three DI RNAs. If this sequence exerts an effect then the origin of DI-T(L) follows a different pattern than DI-T and DI 611.

The recombination sites reveal that in some, but not all cases, the termination and/or resumption of RNA synthesis which lead to DI particle generation may be due to the primary structure of the VSV genome. In addition, if sequences are involved they share no homology and operate by different mechanisms. Thus, it seems that the generation of DI particles of VSV has no strict requirement for sequence information. We propose that each site in the genome has the potential to give rise to DI particles through a copy choice mechanism, unless other yet unknown factors are competing. However, we also propose, that sequences play a role by modulating the process of DI particle generation in a quantitative manner in that they favor the construction of a particular type of DI particle. The large number of panhandle particles with stem regions of 45-48 bases, for example, may be explained by this model.

ACKNOWLEDGMENTS

We thank Dr. Robert A. Lazzarini for constructive criticism of this work and Charlene French for the computer analysis of the sequences and for typing the manuscript.

REFERENCES

1. Lazzarini, R. A., Keene, J. D., and Schubert, M. (1981). Cell 26, 145.
2. Perrault, J. (1981). In "Current Topics in Microbiology and Immunology" p. 151. Springer.
3. Epstein, D. A., Herman, R. C., Chien, J., and Lazzarini, R. A. (1980). J. Virol. 33, 818.
4. Yang, F., and Lazzarini, R. A. (1983). J. Virol. 45, 766.
5. Schubert, M., and Lazzarini, R. A. (1981). J. Virol. 37, 661.
6. Schubert, M., Keene, J. D., and Lazzarini, R. A. (1979). Cell 18, 749.
7. Peattie, D. A. (1979). Proc. Natl. Acad. Sci. (USA) 71, 760.
8. Leppert, M., Kort, L., and Kolakofsky, D. (1977). Cell 12, 539.
9. Huang, A. S. (1977). Bacteriol. Rev. 41, 811.
10. Schubert, M., Keene, J. D., Lazzarini, R. A., and Emerson, S. U. (1978). Cell 15, 103.
11. Keene, J. D., Schubert, M., and Lazzarini, R. A. (1980). J. Virol. 33, 789.
12. Keene, J. D., Chien, I. M., and Lazzarini, R. A. (1981). Proc. Natl. Acad. Sci. (USA) 78, 2090.
13. Kolakofsky, D. (1982). J. Virol. 41, 566.

STRUCTURE AND GENERATION OF DELETION MUTANTS OF VESICULAR STOMATITIS VIRUS[1]

Ronald C. Herman

Virology Laboratory
Center for Laboratories and Research
New York State Department of Health
Albany, New York 12201

One of the defective interfering (DI) particles generated by the heat-resistant strain of vesicular stomatitis virus (VSV-HR) contains genetic information from the 3' half of the genome (1-3). This DI particle (DI-LT; DI 0.46) is the only true internal deletion mutant of VSV to be identified (4-7). DI-LT retains both parental termini, whereas other VSV DI particles contain a short nonparental sequence at the 3' end which is the complement of the parental 5'-terminal sequence (8). Our previous analyses indicate that DI-LT contains 320-350 nucleotides of the 5'-terminal parental sequence (5).

Because DI-LT contains genetic information from the 3' half of the genome and retains the 3'-terminal sequence it can synthesize the leader and mRNAs for the nucleocapsid (N), phospho- (NS), matrix (M), and glyco- (G) proteins (9, 10). In contrast, DI-LT2 (DI 0.50) is also derived from the 3' half of the VSV genome but does not retain the 3'-terminal sequence and does not synthesize mRNA (11).

When the DI-LT mRNAs are resolved in a 1.5% agarose gel containing 6 M urea at pH 3.0, the messages for the NS, M, and N proteins comigrate with the corresponding mRNAs synthesized by VSV-HR. Surprisingly, only very little of the polyadenylated DI-LT mRNA comigrates with authentic G message. Instead, a new larger and relatively abundant message is observed (G*) (12). Structural analyses of the G* RNA indicate that it contains an estimated 250-nucleotide transcript of the remnant polymerase (L) gene covalently linked to the 3' end of a transcript of the G gene. This

---

[1]This work was supported by Public Health Service Grant AI17699 from the National Institute of Allergy and Infectious Diseases.

message is not interrupted by an intervening poly(A) (13), but contains a 3' poly(A) tail most probably encoded by the remnant L gene polyadenylation signal (12).

SEQUENCING OF THE G* TRANSCRIPT

In order to understand the underlying reasons for the synthesis of the G* message, I have cloned and sequenced a complementary DNA (cDNA) copy of this abnormal polycistronic RNA. These experiments have pinpointed the location of the deletion within the genome of the transcriptionally-active DI-LT particle. A comparison of this sequence with previously published sequences for the transcriptionally-inactive DI-LT2 (14, 15), for the wild type virus (16-20), and for other unrelated DI particles derived from the 5' half of the genome (21) has revealed some structural features shared by all DI particles. These observations suggest a single unified model for the generation of DI particles by VSV. This model may have broader implications for the regulation of RNA synthesis by the negative strand viruses.

cDNA was prepared from total intracellular poly(A)$^+$ mRNA that had been extracted from BHK cells at 4.5 h after infection with DI-LT. Since most, if not all, DI-LT stocks are mixtures of the transcriptionally-active DI-LT and the -inactive DI-LT2 (4, 5), the cloning and sequencing of cDNA prepared from poly(A)$^+$ mRNA ensured that only sequences from the biologically expressed cistron were examined.

After cloning the message cDNA by standard techniques into the Pst 1 site of the E. coli plasmid pBR322, recombinant colonies were identified by hybridization to radioactive plasmid p011 DNA. Plasmid p011 contains cloned sequences equivalent to the 368 nucleotides at the precise 5' end of the (-) sense VSV genome (14). Of several clones obtained, the largest, containing approximately 450 nucleotides of DI-LT-specific sequences, was chosen for further analysis.

The viral-specific insert in this plasmid was sequenced by the chemical method of Maxam and Gilbert (22). Examination of the sequence indicated that it could be aligned with the sequence of the VSV-San Juan G message (17) beginning at nucleotide 1429 (numbered from the 5' end of the normal message). The sequence of the cloned G* cDNA was homologous with that of G message, with the exception of four single base changes, up to nucleotide 1618; the two sequences diverged after that position (not shown).

The divergent sequence aligned with the sequence beginning at position 259 from the 3' end of the L mRNA (318 from the 5' end of the genome). Since the nucleotides at position 1618

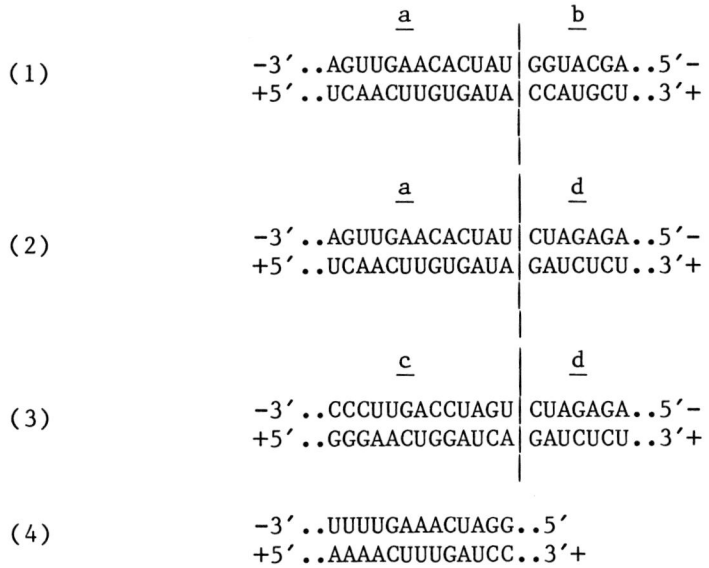

FIGURE 1. Sequences surrounding the deletion in DI-LT. The sequences in the (1) G, (2) G*, and (3) L cistrons surrounding the deletion point (|) have been aligned. The sequences a and d are conserved in DI-LT; the segment of the genome bounded by sequences b and c is deleted. (4) The sequence at the 5' end of the L gene which terminates with a U at position 60 from the 5' end of the genome.

from the 5' end of G message and at position 259 from the 3' end of L message are both adenosine residues, the base at position 1618 of G* could have been derived from either side of the deletion.

As a result of the large deletion in the DI-LT genome, the last 54 nucleotides of the G cistron including the polyadenylation signal, the intercistronic dinucleotide, and all but the last 258 nucleotides of the polymerase gene have been removed. Because the normal regulatory signals are no longer present at the junction between the G and L cistrons, an open transcription unit has been created in DI-LT which extends from the 3' end of the G cistron up to the remnant L gene polyadenylation signal. However, since the deletion is downstream from the normal translation termination signal, the protein encoded by G* is not affected (not shown).

Comparison of the sequences present at the borders of the deletion in the G and L cistrons reveals some elements of similarity (Figure 1). The sequence 3'..UUGA..5' is found in both the conserved region of the G cistron (a) and in the

```
                            a               d
                            ‾               ‾
DI-LT           -3'..AGUUGAACACUAU|CUAGAGA..5'-

DI-LT2          -3'..CUACAAUUGUAGU|ACCGGGG..5'-

DI-0.50         -3'..ACAAUUGUAGUAC|CXXGGUU..5'-
```

FIGURE 2. Comparison of deletion mutants. The sequences surrounding the deletion in the DI particles have been aligned at that point (|). DI-LT2 sequence from ref. 14; DI-0.50 sequence from ref. 15. X̲, unidentified nucleotide.

deleted region c̲ of the L cistron. A similar sequence has also been shown to be present at the 3' end of both the leader and L genes (16, 20) where it may play a role in transcription termination. Several other elements of homology are also observed at the borders of the deletion (Figure 1).

Yang and Lazzarini (14) and De and Perrault (15) have sequenced an internal deletion mutant(s) containing a deletion whose boundaries are entirely within the polymerase gene. By virtue of the techniques used to obtain those sequences (14, 15) it is likely that they are derived from the transcriptionally-inactive DI-LT2 (DI 0.50).

MODEL FOR THE GENERATION OF DI PARTICLES

A comparison of the three sequences at the boundaries of the internal deletions reveals no obvious elements of homology in the primary structure (Figure 2). Note, however, that De and Perrault (15) have located their deletion two bases downstream from the position at which Yang and Lazzarini have placed it. Although the primary sequences show no similarities, homology is detected between the potential secondary structures of these regions (Figure 3). Assuming only standard A:U, G:C base pairing, the conserved a̲ sequence in both DI-LT and DI-LT2 can be drawn as a short base-paired hairpin which is located just up stream from the deletion point (Figure 3). (If G:U pairing is also permitted, then slightly different hairpins could be drawn which encompass the deletion points.) The conserved d̲ regions can also be formed into small hairpins. The resulting DI genomes can, thus, be folded into very similar secondary structures.

A. DI-LT

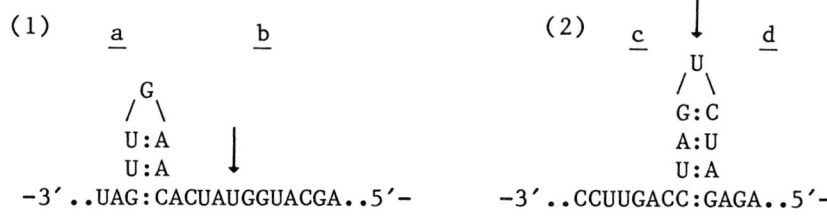

(1)
```
    a              b
    G
   / \
  U:A         ↓
  U:A
-3'..UAG:CACUAUGGUACGA..5'-
```

(2)
```
         c    ↓    d
              U
             / \
            G:C
            A:U
            U:A
     -3'..CCUUGACC:GAGA..5'-
```

(3)
```
        a         ↓ d
         G
        / \  U-C
        U:A A:U
        U:A U:A
      -3'..UAG:CAC:GAGA..5'-
```

B. DI-LT2

(1)
```
    a              b
    A-U
    A:U
    C:G
    A:U      ↓
    U:A
-3'..CUC:GUACAGUU..5'-
```

(2)
```
         c              d
                       G-G
                       C:G
                       C:G
                    ↘  A:U
     -3'..UCUGAGGA:UAG..5'-
```

(3)
```
        a         d
        A-U
        A:U
        C:G  G-G
        A:U  C:G
        U:A  C:G
      -3'..CUC:GUA:UUAG..5'-
```

FIGURE 3. Secondary structure at deletion sites. The sequences of the (-) sense 42S genome surrounding the deletion points have been folded into hairpins using standard base pairing (:). Structures (1) and (2) are the upstream and downstream regions, respectively; structure (3) is the (-) sense sequence of the resulting DI particle. Sequences for DI-LT2 from ref. 14. <u>Arrow</u>, deletion point.

Thus, for the first time, two independently generated DI particles can be shown to be related, not by primary sequence, but by the potential to form very similar secondary structures. A model based on these observations suggests that

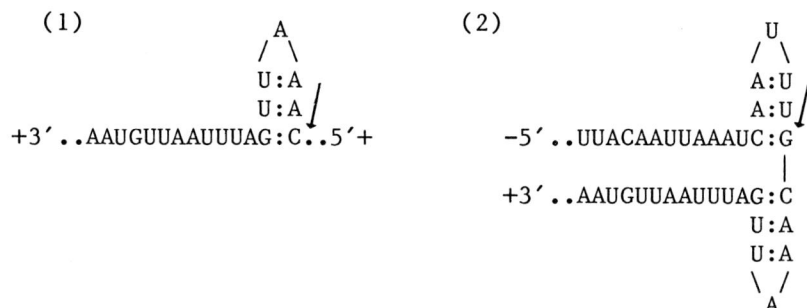

FIGURE 4. Structure of DI-011. (1) The sequence of the (+) sense 42S template at the site of premature termination (19) with potential hairpin. (2) The structure of the resulting DI-011. <u>Arrow</u>, site of premature termination.

secondary structure elements may play an important role in the generation of DI particles. The model predicts that the polymerase with attached nascent chain may be prematurely ejected from the template at or very close to sequences having the potential to form a small base-paired hairpin. The polymerase may subsequently reattach to a template at a another site possessing a hairpin sequence and then resume RNA chain elongation.

This model is also applicable to snapback type DI particles such as DI-011 which contains approximately 1700 nucleotides of covalently linked and complementary sequences from the 5' end of the VSV genome (19). The sequence of the (+) sense strand at the site where the polymerase prematurely detached from the template can be written as a small base-paired hairpin which is similar in structure to that found at the boundaries of the deletion in DI-LT (Figure 4). However, instead of reattaching to a template RNA, the polymerase appears to have recognized a hairpin structure at the 3' end of the nascent chain to resume RNA synthesis. The resulting DI-011 genome contains two small hairpin sequences which are tandomly arranged in virtually same way as they are in the internal deletion mutants. Thus, a single model accurately predicts the generation of DI particles with structures as diverse as those found in DI-LT and DI-011.

Since panhandle type DI particles such as DI-T differ from snapback DIs only in the length of the sequence that was copied from the nascent chain, it is likely that the generation of these DI particles is also predicted by the model. Most panhandle DI particles are generated when the polymerase resumes synthesis at a position 45-48 nucleotides

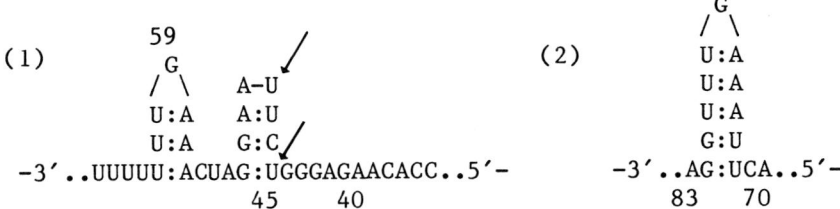

FIGURE 5. Structures at resumption points. (1) The most common "fall-back" site (21) results in 45-48-nucleotide panhandles. (2) Possible site from which 70-nucleotide panhandles (23) would be generated. Numbering is from the 5' end of the 42S RNA. <u>Arrows</u>, sites of resumed synthesis.

from the 5' end of the nascent genome (21). Schubert et al. (21) suggest that the sequence in this region resembles a normal polymerase binding site and this implies that the generation of DI particles is sequence-specific. The model presented here suggests that this sequence may be more relevant to DI formation because it forms one arm of a small hairpin (Figure 5). As shown by Kolakofsky (23), DI particles containing panhandle sequences ranging from perhaps 70 to 200 base pairs can be isolated. While the primary genomic sequences at these positions show no direct relationship to the sequence at position 45-48, they are related by the ability to form similar secondary structures (Figure 5). Since only limited sequencing data is available for panhandle type DIs at the sites of premature termination (21), it is not presently possible to analyze these sequences directly. Nevertheless, the model predicts that these sequences also contain small potential hairpins.

The model presented here may have broader implications for the regulation of viral RNA synthesis. Hairpin sequences are located at both the end of the leader and L genes, positions at which the polymerase is known to terminate synthesis, and at each of the intergenic junctions in the VSV genome (not shown). While it is has not been demonstrated that such hairpins can form in RNA in the environment of a nucleocapsid, the presence of these structures, even if only transient, may be sufficient to regulate RNA synthesis. Consequently, elements of secondary structure may punctuate the viral genome and serve as attenuators and enhancers of RNA synthesis by the negative strand viruses.

ACKNOWLEDGMENTS

I thank R. A. Lazzarini for providing plasmid p011-2.

REFERENCES

1. Leamson, R. D., and Reichmann, M. E. (1974). J. Mol. Biol. 85, 551.
2. Stamminger, G. and Lazzarini, R. A. (1974). Cell 3, 85.
3. Schnitzlein, W. M., and Reichmann, M. E. (1976). J. Mol. Biol. 101, 307.
4. Perrault, J. and Semler, B. (1979). Proc. Natl. Acad. Sci. USA 76, 6191.
5. Epstein, D. A., Herman, R. C., Chien, I. M., and Lazzarini, R. A. (1980). J. Virol. 33, 818.
6. Chanda, P. K., Kang, C. Y., and Banerjee, A. K. (1980). Proc. Natl. Acad. Sci. USA 77, 3927.
7. Clerx-Van Haaster, C, Clewley, J. P., and Bishop, D. H. L. (1980). J. Virol. 33, 807.
8. Perrault, J., and Leavitt, R. W. (1977). J. Gen. Virol. 38, 35.
9. Colonno, R. J., Lazzarini, R. A., Keene, J. D., and Banerjee, A. K. (1977). Proc. Natl. Acad. Sci. USA 74, 1884.
10. Johnson, L. D. , Bender, M., and Lazzarini, R. A. (1979). Virology 99, 203.
11. Keene, J. D., Chien, I. M., and Lazzarini, R. A. (1981). Proc Natl. Acad. Sci. USA 78, 7090.
12. Herman, R. C., and Lazzarini, R. A. (1981). J. Virol. 40, 78.
13. Herman, R. C., Adler, S., Lazzarini, R. A., Colonno, R. J., Banerjee, A. K., and Westphal, H. (1978). Cell 15, 587.
14. Yang, F., and Lazzarini, R. A. (1983). J. Virol. 45, 766.
15. De, B. K., and Perrault, J. (1982) Nuc. Acids. Res. 10, 6919.
16. Rowlands, D. (1979). Proc. Natl. Acad. Sci. USA 76, 4793.
17. Rose, J. K. (1980). Cell 19, 415.
18. Rose, J. K., and Gallione, C. J. (1981). J. Virol. 39, 519.
19. Schubert, M. and Lazzarini, R. A. (1981). J. Virol. 37, 661.
20. Schubert, M., Keene, J. D., Herman, R. C., and Lazzarini, R. A. (1980). J. Virol. 34, 550.
21. Schubert, M., Keene, J. D., and Lazzarini, R. A. (1979). Cell 18, 749.
22. Maxam, A., and Gilbert, W. (1980). Meth. Enzymol. 65, 501.
23. Kolakofsky, D. (1982). J. Virol. 41, 566.

SENDAI VIRUS DI RNA SPECIES CONTAINING 3'-TERMINAL GENOME FRAGMENTS[1]

G.G. Re, E. Morgan, K.C. Gupta, and D.W. Kingsbury

Division of Virology and Molecular Biology
St. Jude Children's Research Hospital
P.O. Box 318, Memphis, Tennessee 38101

VARIETIES OF DI GENOMES

Defective interfering (DI) virions are deletion mutants. To survive, they require the replicative machinery provided by a co-infecting standard virus whose replication they, in turn, depress. Determining the mechanisms by which DI virions are generated, survive and interfere with replication of the standard virus are matters of special interest.

A variety of DI RNA genomes have been derived from nonsegmented negative strand RNA viruses. The most common species found in vesicular stomatitis virus (VSV) and also found in Sendai virus is the copy-back type (1,2), in which only a 5'-terminal portion of the genome is conserved, the 3' end being replaced by a complementary copy of a portion of the 5' terminus. However, both viruses have also yielded DI genomes that retain both parental termini; examples are the LT RNA of vesicular stomatitis virus (VSV) (3) and the fusion DI RNAs of Sendai virus (4). LT RNA is deleted exclusively in a major 3' portion of the L gene, whereas Sendai virus fusion DI RNAs have lost all of their internal genes and intergenic sequences. These fusion DI RNAs retain a 3' fragment of the NP gene attached to the 3'-terminal leader RNA template on one side of the deletion and a 5'-terminal portion of the L gene on the other side.

---

[1]Supported by research grants RG 1142 from the National Multiple Sclerosis Society and AI 05343 from the National Institute of Allergy and Infectious Diseases, by Cancer Center Support Grant CA 21765 from the National Cancer Institute, and by ALSAC.

An interesting fusion DI strain of Sendai virus is our strain 7, which contains three subgenomic RNA species, 7a, 7b, and 7c, whose electrophoretic mobilities indicate molecular weights of 1.24, 0.70, and 0.55 x 10^6 (about 26%, 15% and 11% of the standard virus genome), respectively (5). Oligonucleotide mapping had indicated that RNA species 7b retained at least part of the leader RNA template, but revealed no sequences characteristic of the adjacent NP gene (4). Direct RNA sequencing later showed that all three DI RNAs in strain 7 had genomic 3' termini (5). Here, we report that the minimum amount of 3' terminus retained by any member of this DI RNA family is 100 nucleotides.

SEQUENCES OF SEPARATED FUSION DI RNA SPECIES

By direct RNA sequencing (6) and by dideoxynucleotide sequencing with a synthetic primer (7), we determined the 3'-terminal 70 bases of RNA 7a and 100 bases of RNAs 7b and 7c. These were identical to the genomic 3'-terminal sequence (Fig. 1; Gupta, Re and Kingsbury; Morgan, Gupta, Re and Kingsbury, this volume). Guanine residues at positions 19 and 46 define a sequence whose first 26 bases are identical to the sequence of T1-oligonucleotide 7 (4) extracted from a 2-dimensional mapping gel (our unpublished data). Indirect evidence had indicated that oligonucleotide 7 was part of the positive-strand leader template. Our sequencing data have now confirmed this deduction.

At bases 56 to 65, there is 3'-UCCCAGUUUC (Fig. 1), homologous (where underlined) to the consensus sequence that starts each of the next 4 genes (Gupta, Re and Kingsbury, this volume). This is the start of the NP gene, according to sequence data on a complete NP gene clone (Morgan, Gupta, Re and Kingsbury, this volume) and data of D. Kolakofsky (personal communication).

Because the Sendai virus positive strand leader RNA has not yet been sequenced, the 5' terminus of the leader RNA template is not known. However, considering the many homologies in the organization of the Sendai virus and VSV genomes (5,8,10; Gupta, Re and Kingsbury, this volume), it may be base 51, since bases 52 through 55 are $(A)_4$ and a four- to five-base A-rich sequence intervenes between the leader RNA template and the start of the N gene in two VSV serotypes (9).

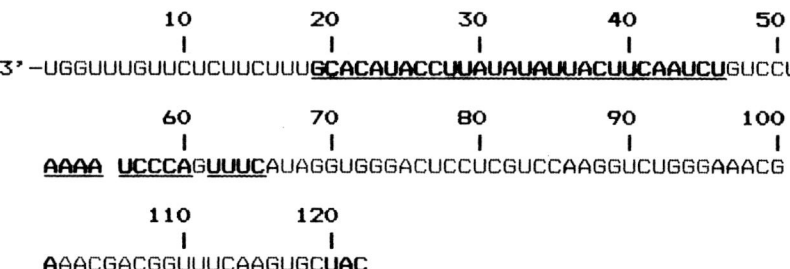

FIGURE 1. Sendai virus 3'-terminus. The positive strand leader template is thought to end at base 51. The highlighted sequence at bases 19 to 46 is T1-oligonucleotide 7 (4). The presumed intervening sequence $(A)_4$ is at bases 52-55 and the NP gene starts at bases 56-65. The A residue at base 101 marks the first deletion site in a DI RNA species of strain 7. The UAC at position 120 is the first translational anticodon.

FURTHER SEQUENCING OF MIXED FUSION DI RNAs

To expedite the task of learning the extent of 3' terminus conservation in the family of strain 7 fusion DI RNAs, we used the mixture as templates for dideoxynucleotide sequencing. Our primer was a ^{32}P-labeled dodecamer complementary to bases 13 to 24. We expected the sequence of the mixture to be homologous to the virus genome up to the point where one of the DI RNAs contained a mutation, either by base substitution or at the site of deletion. As shown in Fig. 2, the mixed DI RNA sequence matched the virus genome to position 100. This is evidently a deletion site, since subsequent positions in the sequencing ladders contain extra bands, missing bands, and anomalous band spacing, marked in Fig. 2 starting with the asterisk at position 101 and continuing with arrowheads.

Thus, at least one strain 7 DI RNA appears to retain only 100 nucleotides of the genomic 3' terminus. This may be RNA 7b, since 7b did not contain any NP-specific oligonucleotides (4). However, we have not examined 7a or 7c by oligonucleotide mapping. Taking the start of the NP gene as the U in position 56, this DI RNA species contains only the first 44 bases of the NP gene, falling 20 bases short of the first UAC initiation anticodon at position 120 (Fig. 1). The other two DI RNAs in the mixture may possess considerably more of the NP gene, since the sequencing gels contained traces of the genomic sequence up to about 300 bases from its 3' end (Fig. 2; other data, not shown).

FUSION DI GENOMES, TRANSCRIPTS, PROTEINS

Our results reveal that all of the DI RNAs of strain 7 include the signal for initiation of NP mRNA transcription. T1-oligonucleotide mapping studies had shown that RNA 7b also contains 5'-terminal L gene sequences and it seems likely that all three of these RNAs possess the gemonic 5' terminus, since no DI RNA lacking the 5' terminus has been isolated from a nonsegmented negative strand RNA virus (1). We have shown that the L gene termination signal lies in positions 58 to 66 from the 5' end of the virus genome (10). Since the molecular weights of all three strain 7 DI RNAs are great enough to include much more of the L gene, we expect that they all contain L gene termination signals.

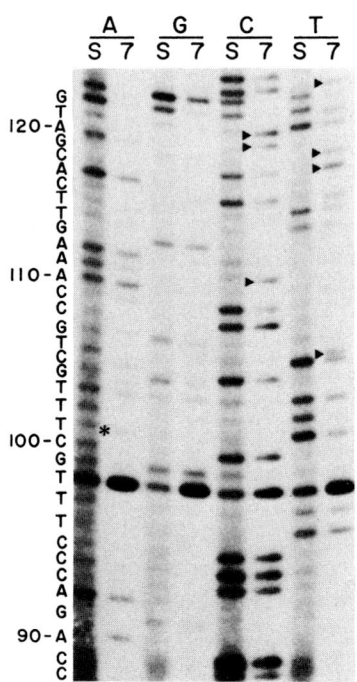

FIGURE 2. 3'terminal sequences of the Sendia virus genome and mixed strain 7 DI RNAs. S: genome; 7: DI RNAs. The asterisk and arrowheads denote loci that differ from the genomic sequence.

Possessing both an NP gene start signal and an L gene stop signal, a fusion DI RNA species would have all of the sequence information necessary to direct synthesis of a fusion transcript. If such transcripts appear in cells infected with strain 7, their impact on the replication of the standard virus genome can be determined. Furthermore, depending on the availability of an NP initiation codon and the reading frame of the L gene fragment, fusion DI transcripts might specify fusion proteins containing N-terminal fragments of the NP protein and C-terminal fragments of the L protein. (Other reading frames of the L gene fragment might also be translatable, yielding fusion proteins with missense C-terminal regions.) It seems unlikely that any such chimeric proteins could influence the outcome of infection, but the possibility is worth examining. We note in this context that members of the major class of DI RNAs of influenza virus, representing internally deleted RNA segments 1 to 3 (specifying the P proteins), possess transcription start and stop signals and have been transcribed in vitro (11), so a similar chain of events is conceivable in that case, as well.

ACKNOWLEDGMENTS

We appreciate the help of C. Naeve in the synthesis of deoxynucleotide primers and the technical assistance of K. Rakestraw and M.A. Oullette.

REFERENCES

1. Perrault, J. (1981). Curr. Top. Microbiol. Immunol. 93, 151.
2. Leppert, M., Kort, L., and Kolakofsky, D. (1977). Cell 12, 539.
3. Epstein, D.A., Herman, R.C., Chien, I., and Lazzarini, R.A. (1980). J. Virol. 33, 818.
4. Amesse, L.S., Pridgen, C.L., and Kingsbury, D.W. (1982). Virology 118, 17.
5. Re, G.G., Gupta, K.C., and Kingsbury, D.W. (1983). J. Virol. 45, 659.
6. Donis-Keller, H., Maxam, A.M., and Gilbert, W. (1977). Nucleic Acids Res. 4, 2527.
7. Sanger, F., Nicklen, S., and Coulson, A.R. (1977). Proc. Natl. Acad. Sci. U.S.A. 74, 5463.
8. Gupta, K.C. and Kingsbury, D.W. (1982). Virology 120, 518.
9. Keene, J.D., Schubert, M., and Lazzarini, R.A. (1980). J. Virol. 33, 789.

10. Re, G.G., Gupta, K.C., and Kingsbury, D.W. (1983). Virology, in press.
11. Chanda, P.K., Chambers, T.M., and Nayak, D.P. (1983). J. Virol. 45, 55.

LONG-TERM PERSISTENCE BY VESICULAR STOMATITIS VIRUS
IN HAMSTERS[1]

Patricia N. Fultz and John J. Holland

Department of Biology
University of California, San Diego
La Jolla, California 92093

Robert Knobler and M.B.A. Oldstone

Department of Immunology
Scripps Clinic and Research Foundation
La Jolla, California 92037

RNA and DNA viruses are being implicated with increasing frequency in pathological changes associated with chronic degenerative diseases. Since in many instances disease becomes apparent only many years after the initial virus infection, studies of viral persistence in animals are needed. As a model system of long-term viral persistence, we have been studying persistent infection of Syrian hamsters by vesicular stomatitis virus (VSV).

Acute infection of inbred LSH hamsters, following intraperitoneal (i.p.) injection of 100 plaque-forming units (pfu) of VSV, Indiana serotype, results in death of 90% of the animals (1); however, lethality can be attenuated by coinjection of defective interfering (DI) particles or polyI:polyC (2). We previously have shown (3) that adult hamsters that survive an acute infection of VSV become persistently infected and that VSV could be recovered from brain, spleen, and liver tissue homogenates as long as 8 months post-infection. All of the hamsters from which virus was recovered had high titers of anti-VSV neutralizing antibodies at the time of sacrifice and some hamsters became transiently or permanently paralyzed in one limb within two weeks of VSV injection. The present report summarizes additional observations concerning long-term persistence of VSV in hamsters.

---

[1]This work was supported by NIH grants AI14627, NS12428, and NS00803, a Teacher Investigator Development Award to R. K., and a postdoctoral fellowship from the National Multiple Sclerosis Society to P. N. F.

LONGEVITY AND CLINICAL STATUS

In our original study (3) only 5 hamsters were kept for an extended period following VSV infection and, aside from those that were partially paralyzed, no changes in the health of the infected animals were noted. It was of interest, therefore, to follow a larger number of animals to determine if VSV infection would alter the life expectancy of hamsters and/or result in any clinical manifestations of disease. At $2\frac{1}{2}$ months of age, 42 LSH hamsters received an i.p. injection of 10^4 pfu of VSV plus either 10^{10} DI particles or 100 μg of polyI:polyC. Figure 1 shows that at 14 months post-infection (age $16\frac{1}{2}$ months) only 48% of the hamsters were living. Although it was impractical to follow a large number of uninfected hamsters during this same time period, we know that hamsters can live to be 2 to 3 years old in captivity. It appeared, therefore, that VSV infection during early adulthood shortened the life expectancy of LSH hamsters and that an increase in the rate of death occurred at approximately 9 months post-infection.

In this same group of 42 hamsters we observed: (i) hemorrhage from the nose and anus prior to death--a feature of death due to acute systemic VSV infection in hamsters; (ii) kidney disease, as evidenced by dramatic fluctuations in urine output; and (iii) neurological disease, as evidenced by peri-

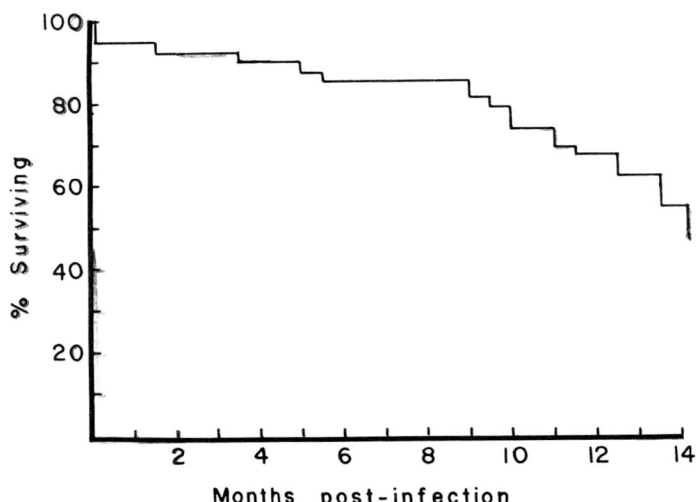

Figure 1. Mortality of LSH hamsters persistently infected with VSV.

odic seizures and tremors with accompanying changes in social behavior.

HISTOPATHOLOGY

Histologic examination of tissues taken from animals 4 to 16 months post VSV infection have revealed several pathologic changes. Changes seen in livers and kidneys included inflammatory cell infiltrates (Figure 2A), foci of hepatocellular necrosis, and dilated or congested tubules in kidneys. In liver tissue from an animal sacrificed 13½ months after VSV infection, foci of necrotic cells that resembled active sites of infection were positive for VSV antigens by indirect immunofluorescence assays with a hyperimmune rabbit anti-VSV antiserum. The pathology of the liver and kidney sections that were examined was consistent with immune complex disease, but this has not been verified.

Lesions in the central nervous system (CNS) were seen only in the gray matter where the most prominent evidence of disease was a marked cellular drop-out in both the brain and spinal cord. Spongiform lesions were seen most often in the hippocampus (Figure 2B), but in some animals other areas of the brain also were affected. There was not only neuronal cell drop-out but also neuronophagia with round cells adjacent to neurons in the pons of the brain of an LSH hamster sacrificed one year post-infection. The subsequent demonstration by indirect immunofluorescence of VSV antigens in the pons of this same animal indicated that these lesions probably were a result of VSV infection. These studies showed that pathologic lesions that are associated with VSV antigens are prominent in liver, kidneys, and the CNS of hamsters persistently infected with VSV.

IMMUNOCOMPETENCE

We showed above that VSV establishes persistent infections in hamsters and that many months after infection various organs contain cells that express VSV antigens. It is important to identify ways by which VSV can escape host immunosurveillance mechanisms. Escape from immune elimination could be accomplished either by specific failure or inactivation of components of the immune system or by viral mutation to avoid immune surveillance. As stated above, all hamsters from which VSV was isolated after long-term infections had titers of VSV neutralizing antibodies comparable to those found in animals 2 to 4 weeks post-infection. To determine if anti-VSV antibody titers decreased or fluctuated following initial VSV infection, we periodically obtained serum from

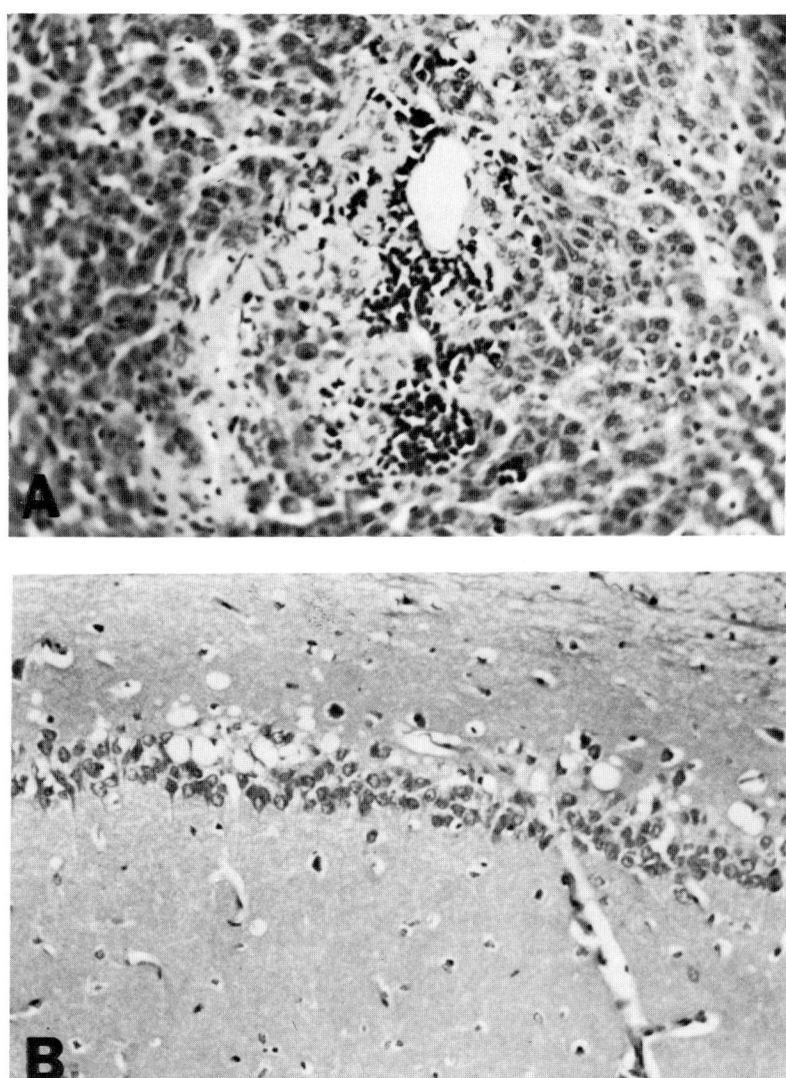

Figure 2. (A) Inflammatory cell infiltrate in liver parenchyma near a portal vein, found in an LSH hamster sacrificed 6 months post-infection. (B) Neuronal cell drop-out in the hippocampus of the same LSH hamster as in A. This animal had experienced at least 3 seizures in the month preceding sacrifice. Hemotoxylin and eosin, x400.

animals via the retro-orbital plexus and assayed the sera for neutralizing antibodies in plaque-reduction assays (3). Figure 3 shows antibody titers in 4 animals from one month to 16 months post-infection. It is obvious that following VSV infection neutralizing antibodies to VSV remain at high levels for the remainder of the animals' lives. These data indicate that not only are VSV-specific B cells synthesizing antibodies but also that helper T cells are functional during VSV persistence.

To determine if hamsters persistently infected with VSV can eliminate VSV-infected cells, we injected 10^7 BHK cells or VSV-persistently infected BHK cells (BHK-VSV) subcutaneously into inbred LSH and random-bred LVG hamsters that had received an i.p. injection of VSV 6 to 11 months earlier. Table 1 shows that irrespective of the status of the animal (normal or VSV-infected), hamsters cannot reject large numbers of BHK tumor cells, whether normal or VSV-infected. The only exception was an uninfected LVG control that did not allow tumor formation, presumably due to killing of the BHK-VSV cells by cytotoxic cells. It is assumed that inbred LSH hamsters are unable to reject BHK tumor cells (derived from random-bred hamsters) because LSH hamsters and all random-bred hamsters tested (4) have the same major histocompatibility haplotype.

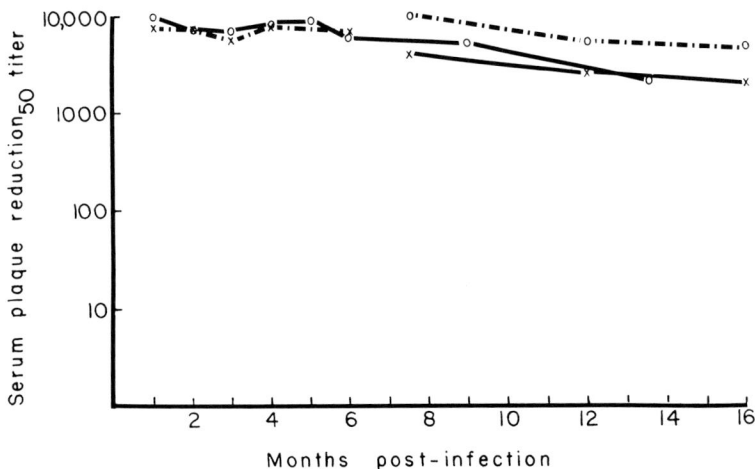

Figure 3. VSV neutralizing antibody titers in sera from hamsters infected with 100 pfu of VSV. The different combinations of symbols and lines represent serum titers in 4 animals at different times post-infection. Titers are the dilution of serum that reduced VSV plaques by 50%.

TABLE 1
GROWTH OF TUMOR CELL LINES IN NORMAL AND
VSV-INFECTED HAMSTERS[a]

Tumor cells	Hamster strain	Hamsters infected	Total no.	No. with tumors
BHK	LSH	-	3	3
		VSV	6	6[b]
	LVG	-	1	1
		VSV	4	4
Total			14	14
BHK-VSV	LSH	-	4	4[b]
		VSV	6	6[b]
	LVG	-	5	4
		VSV	5	5
Total			20	19

[a] See text.
[b] One tumor in this group grew to a diameter of approximately 2 cm and then regressed.

To determine whether the tumors that grew from the inocula of BHK-VSV cells contained VSV, cell lines were established from 19 excised tumors; 7 of the tumors were from animals that were normal at the time of tumor cell inoculation and 12 were from VSV-persistently infected animals. Indirect immunofluorescence assays and superinfection experiments with the 19 cell lines revealed that essentially all of the VSV infected cells in the inoculum had been eliminated. Of the 19 cell lines tested by immunofluorescence for VSV antigens, 16 were negative and 3 had less than 0.1% positive cells. In superinfection experiments all of the cell lines gave yields of VSV comparable to the yield of VSV on uninfected BHK cells (data not shown). Thus, host defense systems of VSV-persistently infected hamsters apparently can recognize and lyse VSV-infected cells, suggesting that the persistent state is not maintained via VSV-specific immunosuppressive mechanisms. We conclude that VSV-persistently infected hamsters are immunocompetent with respect to VSV.

CHARACTERIZATION OF RECOVERED VIRUSES

Since there appeared to be no obvious defect in immune reactivity to VSV in VSV-persistently infected hamsters, we looked for changes in the recovered viruses. VSV, serologically identical to the inoculated virus (VSV_O), was cloned by serial plaque isolations. The clones were analyzed by T1 oligonucleotide mapping, for temperature sensitivity (ts), and for alterations in plaque morphology. Of 18 clones tested for genomic mutations by T1 mapping, 11 clones had maps identical to VSV_O and 7 had one or two spot differences. Phenotypic analyses of the clones showed that greater than 90% were ts at 39.5°C and 4 clones with VSV_O T1 maps were small-plaque mutants (data not shown). The fact that T1 mapping did not reveal mutations in virus clones that were phenotypically different from VSV_O is not surprising since T1 mapping identifies only 10% of VSV nucleotide changes. We conclude that VSV does accumulate genomic mutations during in vivo persistence, analogous to earlier findings (5,6).

DISCUSSION

We have shown that VSV persists in various organs of inbred and random-bred hamsters following i.p. injection of the virus. Whether one or all of the resulting disease states are caused directly by VSV cytopathic effects or are immunopathological is not yet known. We do have evidence that autoantibodies are present in sera from some persistently infected hamsters. It will be interesting to see if autoimmunity is elicited in hamsters due to the polyclonal B-cell activation property of VSV (7) or to crossreaction of antibodies to VSV with cellular components (8). The VSV persistent state exists in hamsters despite continual production of VSV neutralizing antibodies. VSV apparently does not undergo major antigenic changes to escape immunosurveillance since antiserum to VSV_O neutralizes recovered viruses (3) and gives positive immunofluorescence in VSV-persistently infected tissues. The continued high production of VSV neutralizing antibodies and the apparent lack of VSV-specific immunosuppression in these animals could be explained if VSV replicates in and inactivates a subset of hamster suppressor T cells as has been demonstrated for VSV in the murine system (9). The persistence of VSV despite a functional immune system and the failure to accumulate antigenic mutations suggests that VSV may escape immune attack through viral mutations that alter replication and/or maturation. Perhaps VSV does accumulate debilitating mutations during long-term persistence in hamsters since we have failed to isolate virus from tissues taken 13 months post-infection that were positive for VSV antigens by immunofluorescence. The VSV-hamster system provides a unique model for studying chronic degenerative disease and the effects of persistent infection on the immune system in adult immunocompetent animals with a cloned, biochemically well-characterized virus.

REFERENCES

1. Fultz, P. N., Shadduck, J. A., Kang, C. Y., and Streilein, J. W. (1981). Infect. Immun. 32, 1007.
2. Fultz, P. N., Shadduck, J. A., Kang, C. Y., and Streilein, J. W. (1982). Infect. Immun. 37, 679.
3. Fultz, P. N., Shadduck, J. A., Kang, C. Y., and Streilein, J. W. (1982). J. Gen. Virol. 63, 493.
4. Streilein, J. W., and Duncan, W. R. (1979). Immunogenetics 9, 563.
5. Holland. J. J., Grabau, E. A., Jones, C. L., and Semler, B. L. (1979). Cell 16, 495.
6. Holland, J. J., Spindler, K., Horodyski, F., Grabau, E., Nichol, S., and VandePol, S. (1982). Science 215, 1577.
7. Goodman-Snitkoff, G. W., and McSharry, J. J. (1980). J. Virol. 35, 757.
8. Fujinami, R. S., Oldstone, M. B. A., Wroblewska, Z., Frankel, M. E., and Koprowski, H. (1983). Proc. Natl. Acad. Sci., USA 80, 2346.
9. Sy, M.-S., Tsurufuji, M., Finberg, R., and Benacerraf, B. (1983). J. Immunol. 131, 30.

PERSISTENT INFECTIONS OF BHK-21
CELLS WITH RABIES VIRUS[1]

Christine Tuffereau, Florence Lafay and Anne Flamand

Laboratoire de Génétique 2
Université de Paris-Sud, Bât 400
91405 Orsay Cedex - France

Three BHK-21 cell lines persistently infected with the CVS strain of rabies virus have been established from independent cell clones. They have been maintained at 37°C for 2 years. Production of plaque forming units (PFU) was low, irregular and heterogeneous (small plaque variants, temperature sensitive and semi lethal mutants). It ceased definitively between the thirthieth and the sixthieth passage depending on the cell line.

After that period, cells continued to release variable amounts of slowly replicating non-plaque forming units and non-infectious particles. The presence of defective interfering (DI) particles and interferon in the supernatant of the cell cultures was not evidenced using a technique which allowed to detect at least 10^4 to 10^5 DI particles or 1 interferon unit/ml. Immunofluorescence staining with fluorescein conjugated anti-nucleocapsid antibodies revealed that nearly every cell was positive. The fluoresceing pattern was identical to that of CVS acutely infected (AI) cells suggesting that the N protein was still preferentially associated into nucleocapsids.

Intracellular viral proteins and RNA syntheses have been studied and results are presented hereunder.

INTRACELLULAR VIRAL PROTEINS

Immunoprecipitation of intracellular viral proteins were performed with serum from CVS immune mice (Fig. 1). From AI cell lysates, the two forms of the glycoprotein, G_1 and G_2, were precipitated as well as the nucleocapsid protein N and the matrix protein M_1. From PI cell lysates only N and M_1 proteins were precipitated. The ratio between the N and M_1 proteins was the same in AI cells and in two PI cell lines (D, F). In PI

---

[1]This work was supported by the Centre National de la Recherche Scientifique through the L.A. 040086 and by le Commissariat à l'Energie Atomique.

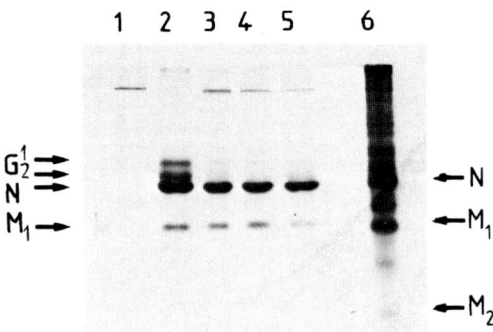

FIGURE 1. Immunoprecipitation with mouse antiserum of lysates from uninfected cells (lane 1), 44 hours-CVS-infected cells (lane 2) and PI cells D, F, J (lanes 3, 4, 5). 10^5 cells were labelled for 20 hours with 40 μCi of (^{3}H)-lysine. They were the disrupted in lysis buffer (0.5 % NP_{40}, 0.15 M NaCl, 50 mM Tris HCl pH 7.4) and sonicated for 15 sec. Cell debris were removed by centrifugation at 10,000 rpm in an Eppendorf centrifuge. Cell lysates were incubated overnight at 4°C with serum from adult mice inoculated with inactivated CVS. Then anti-mouse globulins were added for 4 hours. The complex was spun down, washed twice with buffer (0.15 M NaCl, 15 mM Tris HCl pH 7.4, 5 mM EDTA, 5 % sucrose, 1 % NP_{40}), disrupted in Laemmli buffer, and analyzed in a 10 % polyacrylamide SDS gel. Lane 6 shows viral proteins in AI cells .

cell line J it was two times lower (Fig. 1, lane 5). In the three PI cell lines, the G protein was undetectable and so represented less than 10 % of the amount present in AI cells.

Immunoenzymatic tests with monoclonal antibodies n° 101-1 and 509-6 (1) recognizing sites I and II of the glycoprotein (2) also showed a restriction of this protein. In order to point out a low amount of G protein in PI cells, (^{3}H) glucosamine labelling was performed for 20 hours. Cell extracts were analyzed on a 8-12 % polyacrylamide SDS-gel (Fig. 2). In uninfected cells no glycosylated protein migrated at the same place as the viral glycoprotein. Therefore minor amounts of G protein should have been detectable in PI cells. However no significant amount of the glycoprotein could be detected.

The L and M_2 proteins were visible in AI cells on longer exposed autoradiograms. In PI cells only the L protein was seen.

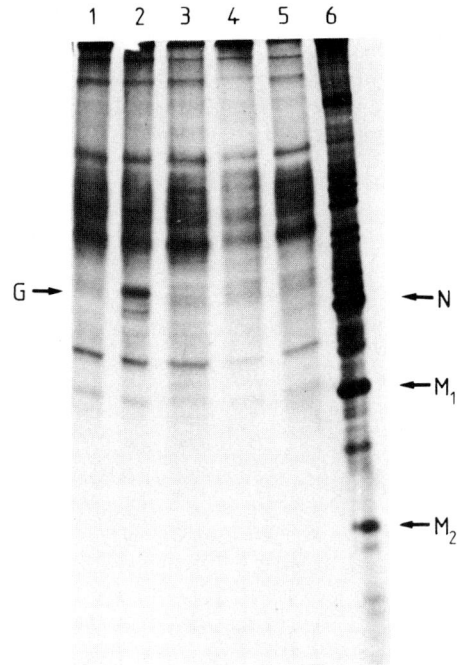

FIGURE 2. (^{3}H)-glucosamine labelled proteins
Uninfected cells (lane 1), 6 hours CVS-infected cells (lane 2) and PI cells D, F, J (lanes 3, 4, 5) were incubated for 20 hours in the presence of 50 µCi of (^{3}H) glucosamine. Cell were disrupted in Laemmli buffer and analyzed in an 8-12 % polyacrylamide SDS-gel. Lane 6 : marker proteins.

TRYPTIC MAP OF N AND M_1 PROTEINS

Whether or not N and M_1 underwent genetical evolution in PI cells was studied by establishing the tryptic peptide map of the proteins at the eightieth passage. Since it is possible that viral material evolves independently in each cell, a global extract may contain an heterogeneous viral protein popula-

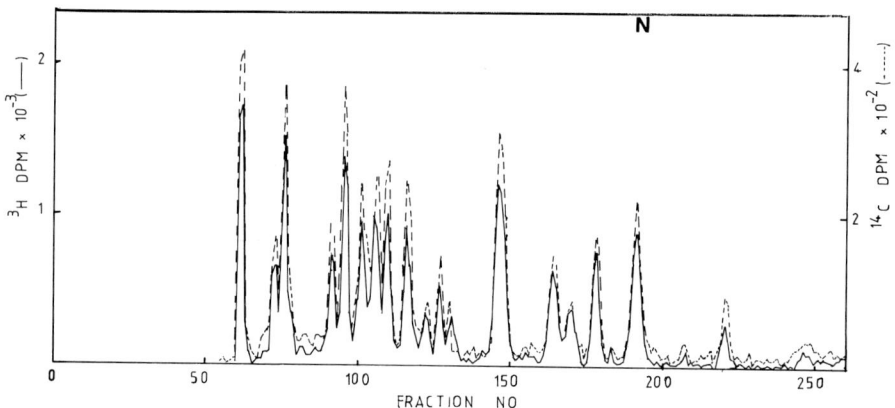

FIGURE 3. Tryptic peptide map of N protein extracted from PI cell line J.

(^{3}H) lysine labelled N protein was prepared from PI cells by immunoprecipitation with mouse antiserum. (^{14}C) lysine labelled N protein from purified virions was eluted from a 10 % polyacrylamide SDS gel. (^{3}H) and (^{14}C) proteins were mixed, reduced with dithiothreitol (0.1 M), alkylated with iodoacetamide (0.25 M) and incubated with trypsin (TPCK, Worthington). Tryptic peptides were separated by a pH gradient from 2.4 (0.1 M pyridine acetate) to 5.2 (2 M pyridine acetate) on a chromobead P (Technicon) column (3,4).

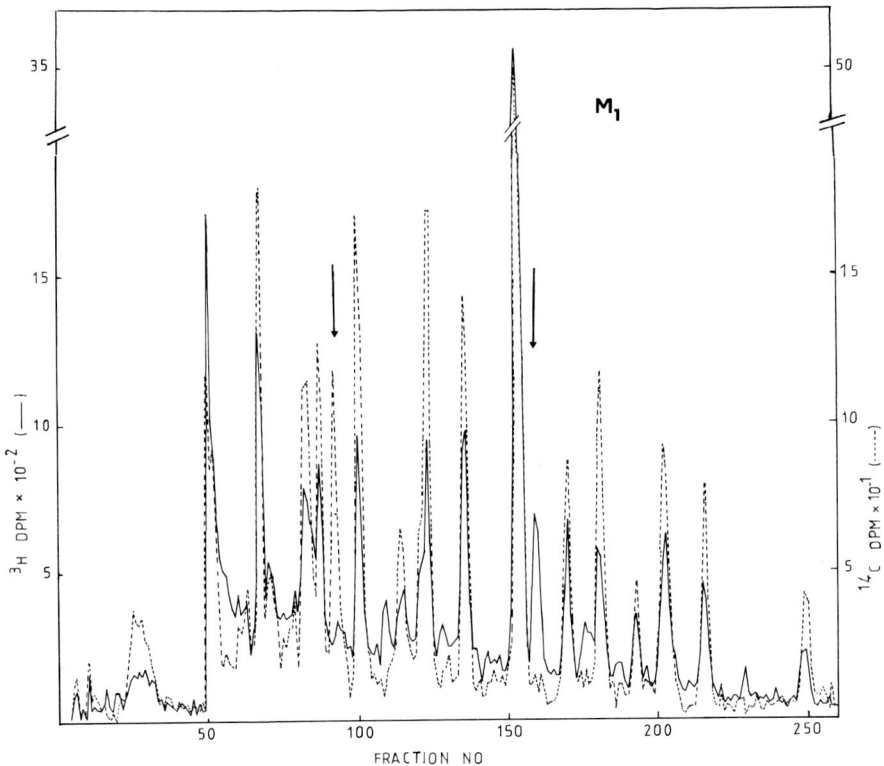

FIGURE 4. Tryptic peptide map of M_1 protein extracted from PI cell line J.
(^{3}H) lysine labelled protein from PI cell and (^{14}C) lysine labelled protein from purified virions were prepared and digested as explained in Fig. 3.

tion. To avoid this problem, cell extracts were performed from recently cloned PI cells. (^{3}H) labelled proteins were obtained from PI cell lysates by immunoprecipitation with mouse antiserum. (^{14}C) labelled proteins were prepared from purified CVS virions.

Fig. 3 presents the peptide map of N protein extracted from PI cell line J or from virions. No difference appeared between the two patterns. The same results were obtained with N proteins extracted from the other two PI cell lines.

On the contrary peptide pattern of M_1 protein from PI cell line J exhibited several modifications indicated by arrows on fig. 4. Tryptic maps performed with M_1 purified from other PI cell lines also revealed several modifications, although in other peaks (data not shown).

RNA CONTENT OF INTRACELLULAR VIRAL NUCLEOCAPSIDS

We have analyzed RNA from intracellular viral nucleocapsids in order to see if it was full size (42S) or defective. (^{3}H) uridine or (^{32}P) Na_3PO_4 labelled intracellular nucleocapsids from PI or AI cells were purified on a CsCl gradient and their RNA were analyzed on a sucrose gradient (fig. 5). PI nucleocapsids from AI cells contained an RNA which sedimented as a single peak. In the three PI cell lines, a major peak of (^{3}H) RNA sedimented as the single peak of (^{32}P). In two cases (PI cell lines D and J) additional RNA of smaller and heterogeneous size was also detected (fig. 5 (2) and (3)). (^{32}P) radioactivity at the top of the gradient probably corresponds to the N protein which is phosphorylated in vivo (5).

DISCUSSION

Long-term persistent infection with rabies virus was characterized by a relatively stable amount of N protein (detected with fluorescein-conjugated antinucleocapsid antibodies) in PI cells in the course of several passages. Analysis by tryptic peptide maps revealed that this protein did not undergo a great evolution. On the contrary M_1 protein presented several modifications in its peptide pattern indicating that some evolution of the protein was compatible with the maintenance of the PI state. L protein was also detectable in PI cells by radiolabelling but in too low an amount to be further analyzed. Persistence of L, N, and M_1 proteins is not surprising since those proteins are thought to be necessary for viral transcription and replication. G protein could never be detected either by immunoprecipitation, by immunoenzymatic test or by specific labelling with (^{3}H) glucosamine, and thus represented less than

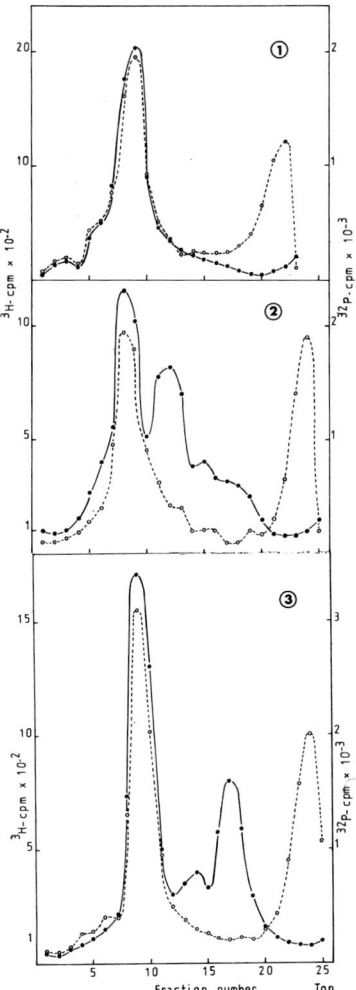

FIGURE 5. Nucleocapsid RNA from PI cells.

PI cells were labelled in presence of 2 μg/ml of Actinomycin D with 100 μCi of (^{3}H) uridine for 20 hours.

Eight hours-CVS infected cells were labelled in presence of 2 μg/ml of Actinomycin D with 700 μCi of (^{32}P) Na_3PO_4 for 20 hours in a medium without phosphate.

Intracellular nucleocapsids were purified on a CsCl gradient (20-40 %) (13) and pelleted. Nucleocapsids from PI cells (●——●) and from AI cells (●-----●) were mixed, treated with 0.5 % SDS and analyzed on a 5-23 % sucrose gradient.

(1),(2), (3) : viral RNA from PI cells F, D, J.

10 % of what is found in AI cells. The amount of M_2 protein seemed also to be restricted in PI cells (as shown by labelling and immunoprecipitation).

The requirement of functional G and M_2 does not seem to be necessary for the maintenance of PI state. Several authors also reported modification (or restriction) of maturation proteins in PI cells (6, 7, 8, 9). In SSPE disease the disappearance of a membrane protein have also been reported (10, 11, 12).

ACKNOWLEDGMENTS

The authors wish to thank Dr. Ph. Vigier for helpful discussion. The excellent technical assistance of J. Benejean, Ch. Thiers, J. Gagnat and J. Alexandre is gratefully acknowledged. Monoclonal antibodies n° 101-1 and 509-6 were a generous gift from Dr. Wiktor.

REFERENCES

1. Flamand, A., Wiktor, T.J., and Koprowski, H. (1980). J. Gen. Virol. 48, 105-109.
2. Lafon, M., Wiktor, T.J., and Mac Farlan, R.I. (1983). J. Gen. Virol. 64, 843-851.
3. Lafay, L., and Benejean, J. (1981). Virology. 111, 93-102.
4. Ewald, S.J., Wlein, J., and Hood, L.E. (1979). Immunogenetics. 8, 551-559.
5. Dietzschold, B., Cox, J.H., and Schneider, L.G. (1979). Virology. 98, 63-75.
6. Oldstone, M.B.A., and Buchmeier, M.J. (1982). Nature, 300, 360-362.
7. Nishiyama, Y., Ito, Y., Shimokata, K., Kimura, Y., and Nagata, I. (1976). J. Gen. Virol. 32, 73-83.
8. Eaton, B.T. (1982). Virology. 122, 486-491.
9. Roux, L., and Waldvogel, F.A. (1981). Cell. 28, 293-302.
10. Hall, W.W., and Choppin, P.W. (1979). Virology. 99, 443-447.
11. Lin, F.H., and Thormar, H. (1980). Nature. 285, 400-492.
12. Machamer, C.E., Hayes, E.C., and Zweerink, H.T. (1981). Virology. 108, 505-510.
13. Roux, L., and Waldvogel, F.A. (1981). Virology. 112, 400-410.

MEASLES VIRUS PERSISTENT INFECTIONS :
MODIFICATION OF FATTY ACID METABOLISM

T.F. Wild, P. Giraudon, P. Anderton and G. Zwingelstein

Unité de Virologie, INSERM,
1 Place Prof. J. Renaut
69008 Lyon, France.

INTRODUCTION

Measles virus belongs to the paramyxovirus group and in common with several of the other members of this group can give rise to persistent infections both in vitro and in vivo. Several years ago, we established a measles virus (Hallé strain) persistent infection in BGM (African green monkey kidney) cells (1). Initially the cultures produced high yields of infectious virus, but after 2 years of culture (150 passages), no infectious virus could be detected. Small quantities of virus particles continued to be released and these contained the normal complement of virus structural proteins (2).

Examination of the persistently infected cells (BGM-P) by immunofluorescence with polyclonal serum revealed that $\geq$ 95 % of the cells contained virus antigens, although the intensity of the staining varied with either the cell passage or time after trypsinisation. ^{125}I-iodination of the cell surface proteins and subsequent analysis by immunoprecipitation and polyacrylamide gel electrophoresis showed that both of the virus envelope antigens, hemagglutinin (HA) and the hemolysin (F) were expressed at the cell surface. When BGM-P cells were radiolabelled with ^{35}S-methionine and the virus-induced proteins analysed by the same techniques, the same spectrum of virus-induced proteins was found as in lytically infected cells (2). No modification in their electrophoretic mobility was noted.

MONOCLONAL ANTIBODY

The availability of measles virus monoclonal antibody to 5 of the 6 virus structural proteins (3) enabled us to probe further the comparison of the distribution of virus antigens in lytically (BGM-L) and BGM-P cells. The main differences observed in immunofluorescence studies were seen with the antinuclear protein (NP) and L-monoclonal antibodies, FIGURE 1. As many as 50 % of the BGM-P cells contained intranuclear virus inclusion bodies and these could only be stained with anti NP-monoclonal antibody. After cell passage (trypsinisation), the intranuclear inclusions were not found for a further 24 h, after which they accumulated, reaching a maximum by the third day. It is presumed that cells containing intranuclear virus antigens are in a terminal state and are lost during cell passage. In BGM-L cells, intranuclear antigens were observed only during late infection (28 h) and even then were limited to not more than 5 % of the cells.

Immunofluorescence studies on BGM-P cells with a single anti-L monoclonal antibody indicated that during the evolution of the persistent infection there was a gradual reduction of the protein detected. It was not established if there was less L-protein synthesised or if there was a loss of the epitope.

LIPID COMPOSITION

We were perplexed by our observations that similar quantities of virus envelope antigens were synthesised and expressed at the cell surface of BGM-P cells as in BGM-L cells without leading to cell lysis. As lysis is also intimately associated with the lipid composition, we examined the plasmic membranes for lipid changes.

The total lipids were extracted from monolayer cultures of uninfected (BGM), BGM-L and BGM-P cells and the individual phospholipids separated by thin layer chromatography (4). There were no major differences in the phospholipid composition of the 3 systems examined.

The fatty acid composition of the total phospholipid was examined by gas-liquid chromatography. There was a 25 % increase in the palmitic acid ($C_{16:0}$) content and a 30 % decrease in oleic acid ($C_{18:1}$) in BGM-P compared to BGM and BGM-L cells. When the individual phospholipids were separated, the increase of $C_{16:0}$ was found to be associated with phosphatidylcholine (PC). The decrease in $C_{18:1}$ was associated with both major phospholipids, PC and phosphatidylethanolamine (PE). There was also an accumulation of arachidonic acid ($C_{20:4}$) in the PE fraction (50 % more than BGM) and a decrease in BGM-L cells.

Typical distribution of virus proteins

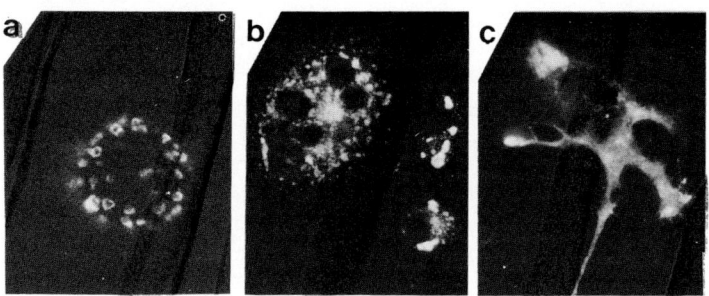

Modification of antigen distribution

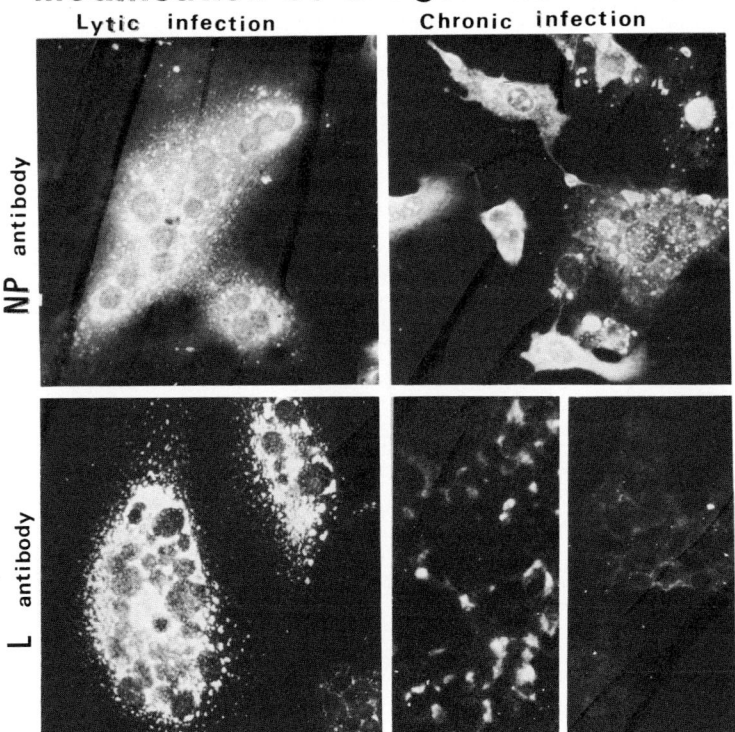

FIGURE 1. Immunofluorescence studies with measles virus monoclonal antibody. a. Membrane fluorescence with anti-HA (magnification X2000). b. Anti-NP on fixed cells. c. Anti-L fixed cells.

LIPID METABOLISM

To study if the modification of the fatty acid composition of the cell membrane of BGM-P cells was due to changes in the fatty acid metabolism we radiolabelled the cells with two fatty acids. (a) ^3H-$C_{20:4}$ was chosen because its composition is altered in BGM-P cells and is incorporated into the C-2 position of the glycerol moiety. (b) ^{14}C-stearic ($C_{18:0}$) whose composition is unchanged in BGM-P and is incorporated in the C-1 position of glycerol.

Monolayer cultures of BGM, BGM-L and BGM-P cells were labelled with ^3H-$C_{20:4}$ and ^{14}C-$C_{18:0}$ for 2, 4 and 6 h. The lipid was extracted and the incorporation of radioactivity determined. There was no difference in the uptake of the two isotopes into the different systems examined. The lipid extracts were resolved into their neutral and polar species by one dimensional thin layer chromatography and the partition of the ^3H and ^{14}C isotopes determined, FIGURE 2. ^3H-$C_{20:4}$ and ^{14}C-$C_{18:0}$ were incorporated into the neutral lipids of BGM-P, 10 times and 2 times respectively, greater than into BGM or BGM-L cells. To locate the increase in incorporation of the isotopes, the neutral lipids were fractionated and the distribution of radioacivity determined, TABLE 1.

TABLE 1
DISTRIBUTION OF ^3H-$C_{20:4}$ AND ^{14}C-$C_{18:0}$
IN NEUTRAL LIPIDS

	^3H -arachidonic acid		
	BGM	BGM-L	BGM-P
Cholesterol esters	0.38+0.09	0.41+0.10	3.05+0.40(***)
Triglycerides	0.33+0.08	0.28+0.04	21.49+1.56(***)
Free fatty acids	0.07+0.01	0.15+0.04	3.20+0.49(***)
Diglycerides	0.73+0.14	0.54+0.08	3.81+0.36(***)
Cholesterol	1.09+0.24	1.34+0.23	2.33+0.35(*)
Monoglycerides	0.17+0.02	0.19+0.03	0.60+0.36
	^{14}C -stearic acid		
Cholesterol esters	2.55+0.82	1.87+0.38	2.97+0.84
Triglycerides	5.07+1.05	5.75+0.92	28.31+4.19(***)
Free fatty acids	0.05+0.02	0.06+0.03	0.37+0.15
Diglycerides	3.97+0.43	3.65+0.57	5.36+0.98
Cholesterol	2.91+0.47	3.04+0.47	2.36+0.51
Monoglycerides	0.50+0.21	0.25+0.08	1.42+0.53

Results are expressed as the percentage dpm in phospholipid ± S.D., in the total lipid. Table reproduced with permission from Biochem. J. (1983) 214, 665-670.

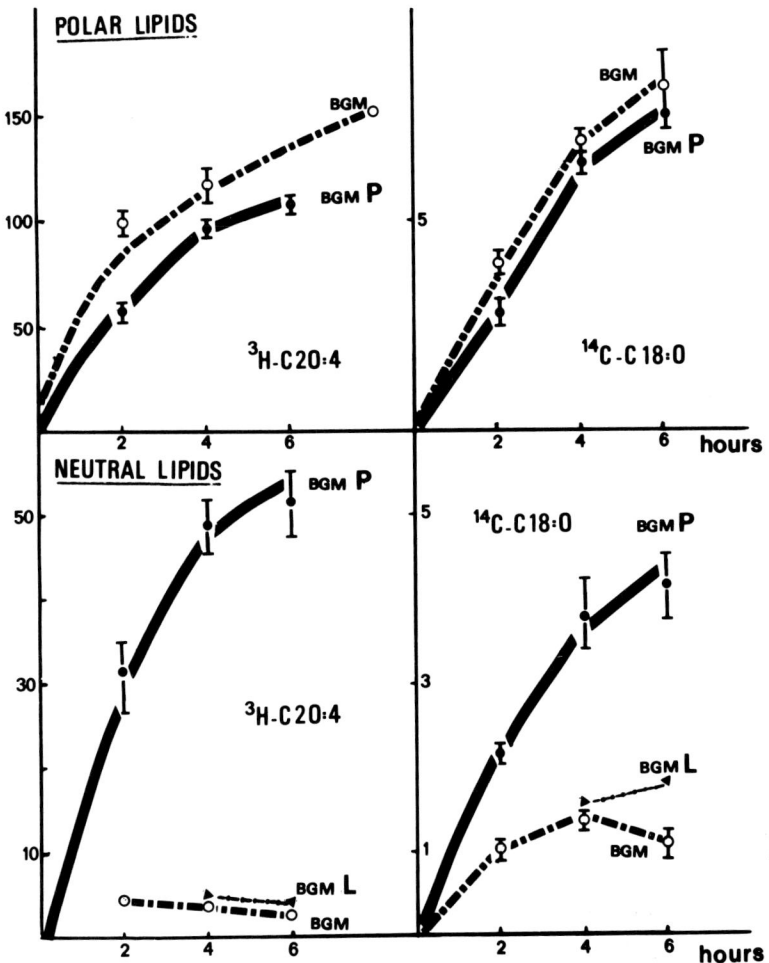

FIGURE 2. Incorporation of ^{3}H-arachidonic acid and ^{14}C-stearic acid into uninfected, lytically and persistently infected BGM cells. Reproduced with permission from Biochem. J. (1983) 214, 665-670.

In BGM-P, there was a 63-67 fold increase in the incorporation of $C_{20:4}$ into the triglycerides and a 37-61 fold increase into the free fatty acids. $C_{18:0}$ was also increased, but not by the same order of magnitude.

Analysis of the distribution of the isotopes into the polar lipid fraction, TABLE 2, showed that compared to BGM, the BGM-P cells had a reduced incorporation of ^3H-$C_{20:4}$ into PC, PE and phosphatidylinosine and ^{14}C-$C_{18:0}$ was also reduced by up to 50 % in PC. To verify that the radioactive precursors were incorporated into the correct position in the diacyl molecule, the radiolabelled phospholipids were hydrolysed with phospholipase A2 and the released fatty acid analysed. 94 % of the ^3H was found at the C-2 position and 87 % of the ^{14}C at the position C-1 position.

TABLE 2
DISTRIBUTION OF ^3H-$C_{20:4}$ AND ^{14}C-$C_{18:0}$ IN PHOSPHOLIPIDS

	BGM	BGM-L	BGM-P
20:4			
Lysophosphocholine	0.1+0.1	0.2+0.1	0.1+0.1
Sphingomyelin	0.1+0.1	0.2+0.1	0.1+0.1
Phosphatidylcholine	31.2+0.9	27.9+0.7	24.6+0.6(**)
Phosphatidylinosine	33.4+0.4	30.5+0.7(*)	21.8+0.4(***)
Phosphatidylserine	3.2+0.2	4.8+0.4	3.8+0.1
Phosphatidylethanolamine	28.9+0.8	33.2+1.6	14.2+1.0(***)
Diphosphoglyceride	0.4+0.1	0.3+0.1	1.1+0.2
18:0			
Lysophosphocholine	0.5+0.1	0.4+0.1	0.4+0.1
Sphingomyelin	1.4+0.1	2.6+0.4	0.4+0.1
Phosphatidylcholine	45.2+1.1	45.6+1.4	19.7+1.1(***)
Phosphatidylinosine	19.4+0.7	19.5+0.7	15.0+1.2(*)
Phosphatidylserine	9.6+0.6	10.0+0.8	6.5+0.8(*)
Phosphatidylethanolamine	18.3+0.4	16.8+0.8	14.6+0.9(**)
Diphosphoglyceride	3.0+1.5	2.1+0.5	7.0+2.1

Results are expressed as the percentage dpm in phospholipid + standard deviation, in the total lipid.

(***) Signifies that there is a significant difference from the corresponding value for uninfected BGM cells.

Reproduced with permission from Biochem. J. (1983) 214, 665-670.

DISCUSSION

Calculation of the relative specific activity of the two fatty acids in the phospholipids, which is a measure of the relative turnover, showed there was a parallel decrease in incorporation into PC of BGM-P for both fatty acids. This presumably results from a slower turnover of the entire diacylglycerol molecule. In contrast in PE of BGM-P, there is a large reduction in the turnover of $C_{20:4}$ compared to $C_{18:0}$. This would explain why we found an accumulation of $C_{20:4}$ in PE of BGM-P cells by chemical analysis. In the BGM-L cells, there was an increase in $C_{20:4}$ turnover.

It has been reported that the transmethylation of PE to PC is rendered insensitive during measles virus persistent infections (5). Although the N-methylation pathway of PE is relatively limited in this type of cell (6) and its inhibition could not account entirely for the accumulation of $C_{20:4}$ in neutral lipids, its biological implications are extremely important. The decrease in secretable $C_{20:4}$ could at the level of the animal lead to a metabolic disease. It may therefore be important to re-investigate the ethiology of certain of the metabolic disorders.

REFERENCES

1. Wild, T. F., and Dugré, R. (1978). J. gen. Virol. 39, 113.
2. Wild, T. F., Bernard, A., and Greenland, T. (1981). Arch. Virol. 67, 297.
3. Giraudon, P., and Wild, T.F. (1981). J. gen. Virol. 54, 325.
4. Anderton, P., Wild, T. F., and Zwingelstein, G. (1981). Biochem. Biophys. Res. Commun. 103, 285.
5. Munzel, P., and Koschel, K. (1982). Proc. Natl Acad. Sci. USA 79, 3692.
6. Hirata, F., Viveros, O H., Dilberto, E. J., and Axelrod, J. (1978). Proc. Natl Acad. Sci. USA 75, 2348.

ASSEMBLY OF MEASLES VIRUS NUCLEOCAPSIDS
DURING LYTIC AND PERSISTENT INFECTIONS[1]

Linda E. Fisher[2] and Elliott Bedows[3]

Department of Natural Sciences
University of Michigan-Dearborn
Dearborn, Michigan 48128[2]
and
Departments of Anesthesiology and Epidemiology
University of Michigan
Ann Arbor, Michigan 48109[3]

Persistence in cell culture may arise from subacute infections involving either temperature sensitive viruses or defective interfering particles. There is evidence, though, that the maintenance of persistence may result from a disruption in normal morphogenesis of virus particles in the cells (1). In this study we describe a model system for studying the morphogenesis of measles virus in both lytic and persistent infections in culture.

The general inhalation anesthetic, halothane (2-bromo-2-chloro-1,1,1-trifluoroethane), reversibly inhibits measles virus replication (2). Halothane vapors do not inactivate measles virions nor do they prevent entry of the virus into cells. During lytic infection in the presence of halothane, however, neither infectious virus nor measles virus nucleocapsids are produced and virus-induced cell fusion is inhibited. It appears that the assembly of measles virus particles is adversely affected by halothane exposure (2). By characterizing virus products synthesized in infected cells in the presence and absence of halothane, we are gaining insight into the mechanism by which measles virus particles are assembled.

LYTIC INFECTION

Vero cells lytically infected with measles virus at low multiplicity of infection were treated with actinomycin D,

---

[1]This work was supported by National Institutes of Health grant number 5 RO1 GM28911 and University of Michigan Rackham and Michigan Memorial Phoenix Project grants.

labeled with [^{3}H] uridine, and treated with 1.8-2.5% halothane in an atmosphere of 95% oxygen and 5% carbon dioxide. Halothane concentrations were assessed by gas chromatography. Concentrations of halothane of 1.8% or above are sufficient to inhibit completely the synthesis of infectious measles virus in Vero cells.

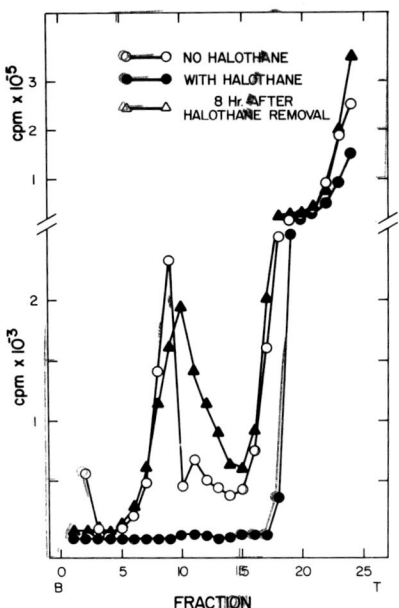

FIGURE 1. Nucleocapsid synthesis in the presence and absence of halothane.

Nucleocapsid Assembly. Nucleocapsids were isolated by the procedure of Kiley et al. (3) by rate zonal centrifugation in 15-40% sucrose gradients. Cells exposed to halothane for 24 hours were harvested without removing the halothane. These were compared with cells treated for 24 hours but harvested 8 hours after removal of halothane, as well as with untreated infected cells (Figure 1). Halothane treatment completely inhibited the incorporation of ^{3}H labeled RNA into nucleocapsids, but nucleocapsids were synthesized soon after halothane removal. There was a significant shift in the size distribution of nucleocapsids to slower sedimenting particles following halothane removal from the acutely infected cells. RNA isolated from the peak fractions of the samples by treatment of the nucleocapsids with 1% SDS at 65C for 30 sec was subjected to centrifugation in 15-30% sucrose gradients

(Figure 2). RNA isolated from the halothane treated infected cells was predominantly of a smaller size (18-28S as compared with 50S) than that of the untreated infected cells.

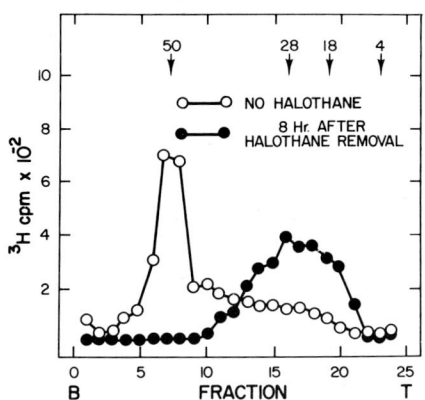

FIGURE 2. Size distribution of measles virus RNA synthesized after halothane treatment.

Preliminary experiments have indicated that measles virus RNA may be synthesized in the presence of halothane but not packaged into nucleocapsids (Figure 3). When [^{3}H] uridine was added to infected cells only after halothane removal, no labeled RNA was incorporated into nucleocapsids. However, if the labeled uridine was present during anesthetic treatment but not after anesthetic removal, the level of labeled nucleocapsid material recovered was comparable to the levels found when [^{3}H] uridine was present throughout the experiment.

Virus Protein Synthesis. In order to get a clearer picture of the effect that halothane has on virus replication and morphogenesis, infected cells were labeled with [^{35}S] methionine. Cell lysates were immunoprecipitated and analyzed by sodium dodecyl sulfate polyacrylamide gel electrophoresis (SDS PAGE) on 9% gels (Figure 4). Such experiments indicated that the amount of the measles virus P protein was greatly reduced in the presence of halothane. In addition, there was a slight decrease in the amount of NP, the nucleocapsid protein, in the presence of halothane. All other virus polypeptides were present in nearly normal amounts. Even though no syncytia formation occurs in the presence of halothane, a polypeptide of the molecular weight of F was detected in normal amounts.

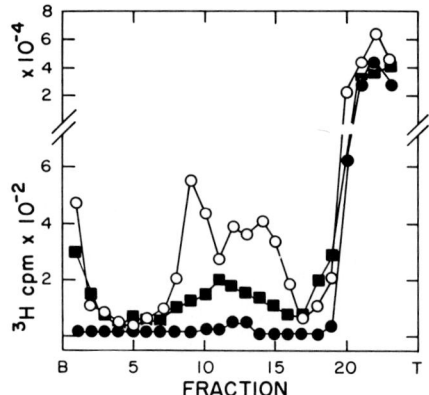

FIGURE 3. RNA synthesis in the presence of halothane. (o) No halothane. (■) Label during halothane treatment. (●) Label following halothane removal.

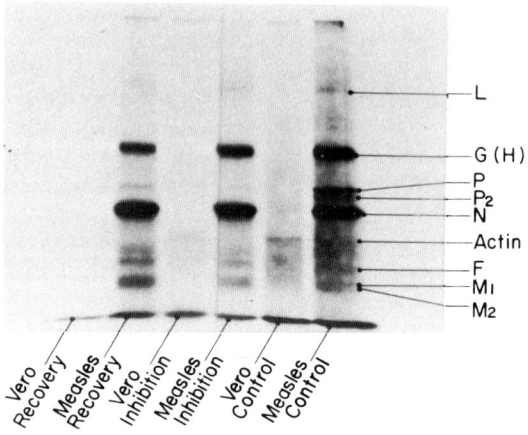

FIGURE 4. Protein synthesis in the presence and absence of halothane.

PERSISTENT INFECTION

Vero cells persistently infected with the Schwarz strain of measles virus (4) were compared with the lytically infected cells to determine what differences might exist in the mechanism of virus morphogenesis and maturation in chronically infected cells. The persistently infected cells were labeled with [^{3}H] uridine in the presence of actinomycin D either with or without exposure to halothane. Nucleocapsids and RNA were isolated as previously described for lytically infected cells. Though the size distribution of nucleocapsids in the persistently infected cells is similar to that found in a lytic infection, there is a shift toward more slowly sedimenting nucleocapsids (Figure 5).

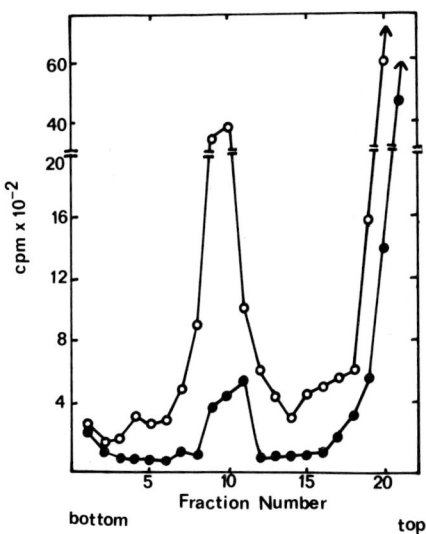

FIGURE 5. Nucleocapsids isolated from cells undergoing persistent (●) and lytic (o) infections with measles virus.

Nucleocapsid assembly. Analysis of [^{3}H] uridine labeled nucleocapsids produced in the persistently infected cells in the presence of actinomycin D and halothane indicated that nucleocapsids were assembled in the presence of the anesthetic for at least 24 hours. If halothane treatment continued for approximately 45 hours, however, there was a significant inhibition in nucleocapsid assembly. (Compare Figure 6A and Figure 7.) Those nucleocapsids assembled in the presence of halothane contain a significant population of slow sedimenting particles (Figure 6A). When halothane was removed after 24 hours, there was an increase in the proportion of slow sedimenting nucleocapsids evident as early as 4.5 hours after removal of the halothane and continuing through 21 hours after anesthetic removal (Figure 6B, C). The size of RNA isolated from the peak fractions of these nucleocapsids appeared to be consistent with the size of the nucleocapsid particles from which the RNA was isolated.

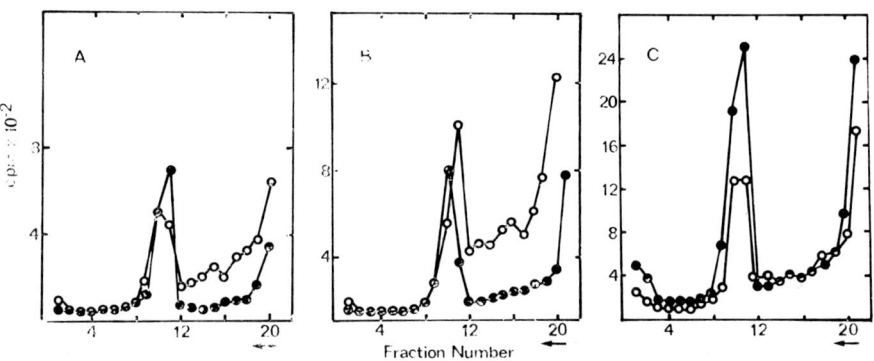

FIGURE 6. Nucleocapsid synthesis in persistently infected cells in the presence of halothane and after its removal. (●) No halothane, (o) with halothane. A. Cells were treated with 2.16% halothane for 24 hours and harvested for nucleocapsid isolation without removal of the anesthetic. B. Treated with halothane and harvested 4.5 hours after halothane removal. C. Treated with halothane and harvested 21 hours after halothane removal. The arrows indicate the direction of sedimentation in the gradients.

ASSEMBLY AND MATURATION OF VIRUS PARTICLES

In the presence of halothane, RNA synthesis continues while the synthesis of the viral protein NP is inhibited to a limited degree. Even though both major components of the ribonucleoprotein are available, they do not assemble into subviral cores. SDS PAGE analysis indicates that the synthesis of the viral P protein is substantially and reversibly inhibited in the presence of halothane. Another possible explanation is that, for some reason, the P protein is more labile than the other viral proteins in the presence of halothane; therefore, P may be preferentially degraded. It is the only viral protein affected to such a large degree. The P protein may be essential in regulating the assembly of viral RNA and NP into the ribonucleoprotein complex.

In addition to nucleocapsid assembly and production of infectious particles, measles virus syncytium formation is inhibited in halothane treated cells. The site of action of the anesthetic in inhibiting cell fusion does not appear to be at the level of polypeptide synthesis since the F protein is found in cell lysates from halothane treated cells. Halothane, though, is known to affect the membrane of cells (5,6). A reorganization of the cell plasma membrane

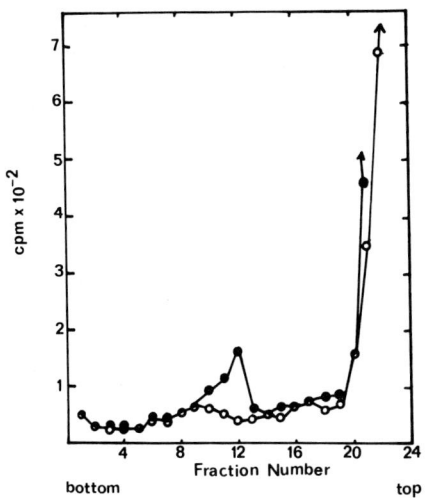

FIGURE 7. Nucleocapsid synthesis in persistently infected cells after prolonged halothane treatment. Cells were treated with halothane for 24 hours. The treatment was repeated at 24 hours and cells were finally harvested 21 hours after the second halothane exposure (45 hours total exposure) for nucleocapsid isolation. [^{3}H] uridine was present only during the last 21 hours of treatment. (●) No halothane, (o) with halothane.

structure in such a way that the F protein is inactive could occur. It is possible that these membrane alterations are influenced by chemical changes in the cellular cytoskeleton. Halothane has been shown to disrupt both the microtubules and microfilaments of certain cell lines (7,8).

In persistently infected cells a different situation occurs. Since preexisting viral proteins are present in the cells before halothane treatment, all proteins needed for the assembly of RNA and protein into nucleocapsids are available. Only after prolonged anesthetic treatment when the proteins that were synthesized before halothane treatment have been utilized in the assembly of ribonucleoprotein cores is further assembly inhibited.

In both lytically and persistently infected cells, when halothane is removed and nucleocapsid assembly occurs, there is an increase in slowly sedimenting RNAs. One possible explanation for the presence of these 18-28S RNAs is that they are encapsidated positive strand RNAs.

Studies aimed at determining the polarity of the slow sedimenting RNA along with studies to clarify the effect of halothane on the synthesis of viral proteins in the persistently infected cells are currently underway. By exploiting the ability of halothane to inhibit measles virus replication reversibly, we have been able to examine various aspects of virus assembly. We anticipate that subsequent studies will provide valuable information on the relationship between mechanisms for virus assembly and maintenance of persistent infections.

ACKNOWLEDGMENTS

We thank Bruce Davidson for his technical assistance and Irene Sayer and Arestea Kakaris for assistance in the preparation of this manuscript.

REFERENCES

1. Fisher, L. E., and Rapp, F. (1979). Virology 94, 55.
2. Knight, P. R., Nahrwold, M. L., and Bedows, E. (1980). Antimicrob. Agents Chemother. 17, 890.
3. Kiley, M. P., Gray, R. H. and Payne, F. E. (1974). J. Virol. 13, 721.
4. Fisher, L. E. (1983). Arch. Virol. in press.
5. Franks, N. P., and Lieb, W. R. (1978). Nature 274, 339.
6. Seeman, P. (1972). Pharmacol. Rev. 24, 583.
7. Hinkley, R. E., and Telser, A. G. (1975). In "Progress in Anesthesiology" (B. R. Fink, ed.), p. 103. Raven Press, New York.
8. Nunn, J. F., and Allison, A. C. (1972). In "Cellular Biology and Toxicity of Anesthetics" (B. R. Fink, ed.) p. 138. Williams and Wilkins, Baltimore.

SYNTHESIS OF MATRIX PROTEIN IN A SUBACUTE [1]
SCLEROSING PANENCEPHALITIS CELL LINE

Michael J. Carter,[2] Margaret M. Willcocks and
Volker ter Meulen

Institute of Virology, University of Würzburg
D-8700 Würzburg, West Germany

INTRODUCTION

Subacute sclerosing panencephalitis (SSPE) is a slowly progressing, fatal disease of the human central nervous system (CNS), associated with measles virus persistence. The disease is characterised by an elevation of anti-measles antibody titres in both serum and CSF (1). Virus nucleocapsids are present as inclusion bodies in the cells of the CNS (2) but budding virus has not been observed and infectious particles have never been detected. The patient shows a low immune response toward matrix (M) protein, and direct examination of infected brain has failed to detect this molecule (3,4). All this evidence points to a defect in virus maturation which is associated with a lesion in the production of M protein. These features of SSPE have been reviewed (5,6). The defect in virus maturation may be to some extent host-controlled, since co-cultivation, or fusion of cells from infected tissue, with cell lines susceptible to measles infection, has sometimes resulted in the rescue of an infectious budding virus (7). More often virus rescue is only partially successful and a cell-associated cytopathic agent is obtained. These cell lines are known as SSPE cell lines, and many reports have confirmed that they also possess a lesion in the synthesis of M protein (8, 9). We have recently described two such SSPE cell lines established by co-cultivation of SSPE patient brain with Vero cells, in which matrix protein was not produced (10). In one (N-1), mRNA specific for M protein was present but unable to give rise to the normal protein by translation (11). In a second cell line (MF), we were unable to

---

[1] Supported by the DFG and Volkswagenstiftung.
[2] Present address, Department of Virology, Royal Victoria Infirmary, Newcastle upon Tyne, England.

detect mRNA sized molecules with matrix protein-related sequences in the polyadenylated RNA fraction. We therefore concluded that failure to produce M protein in the MF cells was accomplished by a defect in mRNA formation or stability (10).

The MF cell line is unusual in that morphological features normally associated with virus maturation have been reported (12). Both the formation of an "inner leaflet" associated with M protein combination with the plasma membrane (13), and budding particles have been observed. These were found to be non-infectious (12). Our previous study (10) utilized high-passage MF cells (above 120) and we were therefore interested in examining this apparent maturational phenomenom in relation to the expression of M protein in early passage cells.

RESULTS

MF cells were revived from liquid nitrogen storage at various passage levels. We examined A(+) RNA, prepared from these cells, by the technique of Southern blotting using nick-translated plasmids containing M protein mRNA sequences (14). A messenger RNA molecule carrying matrix protein sequences was found to be present in one of these cultures (passage 60: figure 1A, track 1) and it was of similar size to that observed in a productive Edmonston virus infection (track 2). We have described another SSPE cell line (N-1) in which M protein mRNA was readily detectable but not utilized correctly in translation reactions, thus it was possible that the mRNA detected in MF cells might not function in protein synthesis. We therefore used the A(+) RNA fraction analysed in Figure 1A, in an in vitro translation reaction. The virus-specific products of this reaction were identified by immune precipitation and separated on an SDS polyacrylamide gel (figure 1B). Polyadenylated RNA extracted from a productive measles virus infection was used as a positive control , and from uninfected Vero cells as a negative control. M protein was detected among the translation products of both MF and Edmonston virus-infected cell mRNAs. Thus we conclude that this mRNA was functional in vitro.

The lesion in M protein production within SSPE patient brain is known to be to some extent host-controlled since cell-fusion or co-cultivation experiments have resulted in the expression of this protein (7). A host- controlled defect in translation has been postulated in a laboratory-produced SSPE virus persistent infection, since matrix protein produced by in vitro translation could not be detected within the persistently infected cells (5). We therefore prepared 35-S methionine labelled extracts from these early passage cells and from uninfected Vero cells and examined them for the presence of

SYNTHESIS OF MATRIX PROTEIN

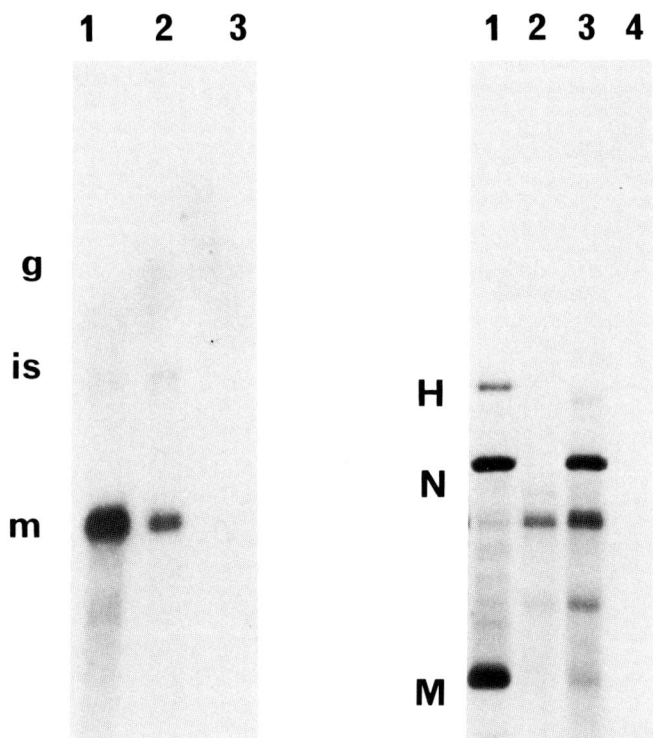

Figure 1.
Panel A. Polyadenylated RNA was extracted from:track 1, Edmonston virus-infected Vero cells; track 2, MF cells; track 3, uninfected Vero cells, and separated by electrophoresis on a Mops buffered 1.5 % agarose gel. RNA was transferred to a nitrocellulose filter and hybridised to 32-P labelled nick-translated plasmid containing sequences derived from the matrix protein mRNA. (g), genome; (is), intermediate-sized RNA; (m), mRNA.
Panel B. Immunoprecipitated protein products formed by translation of the mRNAs analysed in panel A. track 1, Edmonston virus-infected Vero cell mRNA; track 2, uninfected Vero cell mRNA; track 3, MF cell mRNA; track 4, No mRNA addition. Haemagglutinin (H), Nucleocapsid protein (N) and matrix protein (M) were identified by reference to molecular weight markers.

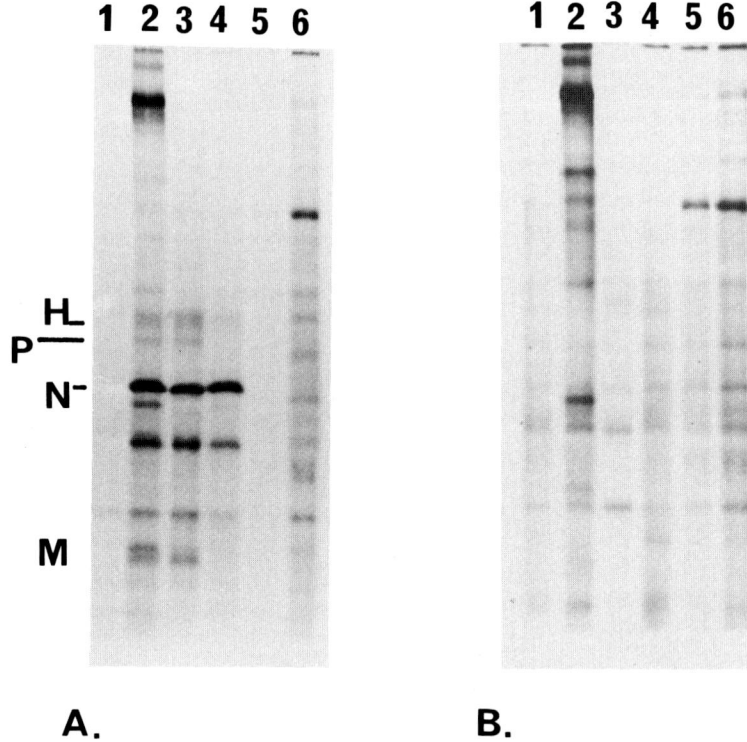

Figure 2.
Immunoprecipitation of proteins from cell lysates. MF cells (panel A) and uninfected Vero cells (panel B) were labelled with 35-S methionine and processed for immunoprecipitation. In each panel proteins were immunoprecipitated by: 1, preimmune rabbit serum; track 2, hyperimmune rabbit anti-measles virus serum; track 3, serum from patient MF; track 4, CSF from patient MF; track 5, control monoclonal antibody; track 6, monoclonal antibody directed against measles virus M protein.

matrix protein by immune precipitation (figure 2). Serum and CSF are still available from the patient from whom the MF cell line was originally derived, and these were used in conjunction with rabbit hyperimmune anti-measles virus serum, and a monoclonal antibody specific for measles virus matrix protein. Rabbit preimmune serum and a monoclonal antibody directed against corona virus antigen provided negative controls.

Matrix protein was produced by the MF cell line and was immunoprecipitated by both the hyperimmune rabbit and human sera. It was not however detected using the monoclonal antibody, or the human CSF. The failure of CSF to immunoprecipitate this peptide was expected since the MF patient suffered from clinically normal SSPE, and antibodies to this protein are lacking in the CSF. We therefore concluded that in this culture of SSPE MF the production of mature functional mRNA for matrix protein led to the successful expression of M protein. In other cultures this protein was not detectable presumably due to the lesion in the production of the corresponding mRNA previously reported (10).

DISCUSSION

The results presented here raise several interesting questions. Firstly the production of M protein during the early passage MF cells that can be inferred from the previous morphological description, is supported by the immunoprecipitation of a matrix protein sized polypeptide formed in infected cells and in <u>in vitro</u> translation reactions. However this protein failed to react with monoclonal antibody directed towards measles virus Edmonston matrix protein. The reasons for this are unclear but this situation could arise from strain differences between the virus which originally established SSPE in the patient, and the Edmonston virus used for monoclonal antibody production. Such differences have been detected between measles virus haemagglutinin proteins by a variety of biological techniques (16). It seems likely that more than one strain of measles virus is capable of giving rise to SSPE, and we have observed nucleoproteins synthesised <u>in vitro</u>, using RNA extracted from two SSPE patients brains, to be of different sizes, a finding we also interpret in terms of virus strain differences .Alternatively this failure to react with monoclonal antibody could have arisen from mutations accumulating during persistence (17), and which have already been demonstrated in the measles virus matrix protein (18).

Budding particles have also been demonstrated in the early passage MF cells (12) but these were found to be non-infectious. It is generally assumed that restoration of matrix protein synthesis is all that is required during attempts to rescue virus from SSPE patient brain, but the observations reported in the early characterization of the MF cell line suggest this may not be the case. Another SSPE cell line is known in which a lack of infectious virus was associated with a defect in H protein expression on the cell surface (9). Defective expression of H protein activity was also observed in the early passage MF cells (12). We have also observed a marked variation in the levels of nucleocapsid protein and its mRNA at various passages. Presumably adequate synthesis of all virus proteins is required for successful release of infectious particles. The MF patient lacked antibodies in the CSF directed against M protein, therefore this protein was probably not produced within the brain. Perhaps this block was overcome when tissue was co-cultivated with Vero cells and the maturation process previously described (12) could then occur. However, cells are not available from the passages used in that report. We observed only one MF culture in which matrix protein was produced. Furthermore other samples of the same passage cells failed to produce M protein when revived. However the data reported here indicate that MF cells possess the capacity to produce a mRNA which is active in translation. The mechanism which normally prevents the expression of this capacity is unknown, but it is possible that physiological trauma involved in the attempts to rescue virus from infected tissue, or cell revival from liquid nitrogen storage, is important in the successful circumvention of this block.

ACKNOWLEDGEMENT

We thank Dr. S. Rozenblatt for making his cloned copies of measles virus mRNAs available to us, and Drs. S. Siddell and H. Wege for provision of additional experimental material. In addition we thank R. Sabando for typing the manuscript.

REFERENCES

1. Server, J. L. and Zeman, W. ((1968). Neurology. 18, 95.
2. ter Meulen, V., Enders-Ruckle, G., Müller, D. and Joppich, G. (1969). Acta. Neuropathologica (Berlin). 12, 244.
3. Hall, W. W., Lamb, R. A. and Choppin, P. W. (1979). Proc. Natl. Acad. Sci. U.S.A. 76, 2047.
4. Hall, W. W. and Choppin, P. W. (1981). New Engl. J. Med. 304, 1152.
5. ter Meulen, V. and Carter, M. J. (1982). In "Virus persistence S. G. M. symposium 33"(B. W. J. Mahy, A. C. Minson and G. K. Darby, eds.), p. 97, Cambridge University Press, England.

6. ter Meulen, V., Stephenson, J. R. and Kreth, H. W. (1983). Comprehensive Virol. 18, 105.
7. Katz, M. and Koprowski, H. (1973). Arch. ges. Virusforsch. 41, 390.
8. Lin, F. H. and Thormar, H. (1980). Nature (Lond.). 285, 490.
9. Machamer, C. E., Hayes, E. C. and Zweerink, H. J. (1981)Virology 108, 515.
10. Carter, M. J., Barrett, P. N., Willcocks, M. M., Koschel, K. and ter Meulen, V. In " Molecular Virology" (Y. Becker, ed.), in press Martinus Nijhoff, New York.
11. Carter, M. J., Willcocks, M. M. and ter Meulen, V. (1983). Nature (Lond.). in press.
12. Kratsch, V., Hall, W. W., Nagashima, K. and ter Meulen, V. (1977). J. Med. Virol. 1, 139.
13. Compans, R. W., Dimmock, N. J. and Meier-Ewart, H. (1970). In " Biology of Large RNA Viruses" (R. D. Barry and B. W. J. Mahy, eds.), p. 87, Academic Press, London.
14. Rozenblatt, S., Gesang, C., Lavie, V. and Neumann, F. S. (1982). J. Virol. 42, 790.
15. Stephenson, J. R., Siddell, S. G. and ter Meulen, V. (1981). J. Gen. Virol. 57, 191.
16. ter Meulen, V., Loeffler, S., Carter, M. J. and Stephenson, J. R. (1981). J. Gen. Virol. 57, 357.
17. Holland, J. J., Grabau, E. A., Jones, C. L. and Semler, B. L. (1979). Cell 16,: 495.
18. Carter, M. J., Willcocks, M. M., Loeffler, S. and ter Meulen, V. (1983). J. Gen. Virol. 64, 1801.
19. Breschkin, A. M., Morgan, E. M., McKimm, J. and Rapp, F. (1979). J. Med. Virol. 4, 67.

MATRIX (M) PROTEIN ALTERATIONS INDUCED BY TLCK IN CELLS ACUTELY AND PERSISTENTLY INFECTED WITH MEASLES VIRUS[1]

Steven L. Wechsler

Department of Molecular Virology
Christ Hospital Institute of Medical Research
Cincinnati, Ohio 45219

TLCK (n-alpha-p-tosyl-L-lysine chloromethyl ketone-HCl) blocks proteolytic cleavages by trypsin and is often used to detect precursor-product relationships of proteins. When measles virus is grown in the presence of TLCK, the viral M (matrix) protein has an apparent molecular weight 1,000-2,000 daltons larger than normal as determined by SDS-PAGE (1). This suggests that M protein might normally be cleaved, by a trypsin like enzyme, from a slightly larger precursor polypeptide. This is particularly interesting, because in some cell lines persistently infected with measles virus, and in some viruses isolated from patients with the slowly progressive neurological disease SSPE (caused by a persistent infection with measles virus) the measles virus M protein has a similarly altered apparent molecular weight (2,3,4). If the normal M protein is cleaved from a larger M protein precursor molecule, then mutations interfering with its presumptive processing might account for the larger M protein often associated with measles virus persistence.

No other evidence for a possible M protein precursor has been reported. Therefore, we studied the effects of TLCK in more detail to see if a precursor polypeptide to the M protein could be conclusively identified. We confirm here that TLCK alters the electrophoretic mobility of the measles virus M protein; however, the mechanism is unrelated to protection of a precursor M protein from proteolytic cleavage. Thus, there is no evidence for a measles virus M protein precursor. More importantly, we found that TLCK affects the mobility of some viral and cellular proteins, even in the cold and in the presence of detergents. Thus, TLCK and related compounds should not be used as protease inhibitors in buffers.

[1] This work was supported by NIH Grant AI-18647

Synthesis of measles virus proteins in the presence of TLCK resulted in changes in the electrophoretic mobilities of the viral M and P proteins and cellular actin (A) (Figure 1). M and actin synthesized in the presence of TLCK (M' and A') migrated more slowly than standard M and A, while the TLCK-induced P (P') migrated more rapidly than standard P. TLCK also caused a decrease in overall protein synthesis.

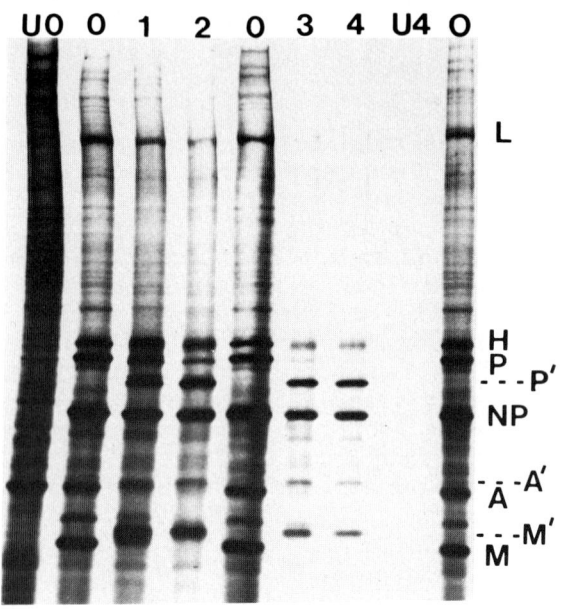

Figure 1. Measles virus proteins synthesized in the presence of TLCK. CV-1 cell monolayers were infected with wild type measles virus and labeled with [^{35}S]methionine (5,6). TLCK and [^{35}S]methionine were added simultaneously. After 15 min. the monolayers were suspended in gel sample buffer, run on SDS-PAGE (8), and processed for fluorography. Viral proteins: H, hemagglutination protein [80,000 daltons (80K)]; P, nucleocapsid associated phosphoprotein (70K); NP, nucleocapsid protein (60K); M, matrix protein (37K); A, cellular actin (43K). P', M' and A' are TLCK-induced forms of P, M, and A. L, involved in transcription (200K) and F_1 (41K) and F_2 (20K), the 2 components of the fusion protein are not consistently detected and were not analyzed. Gel lanes: 1,2,3,4, infected and labeled in the presence of 1,2,3 or 4mM TLCK respectively; 0, infected, no TLCK; U0, uninfected, no TLCK; U4, uninfected, 4mM TLCK.

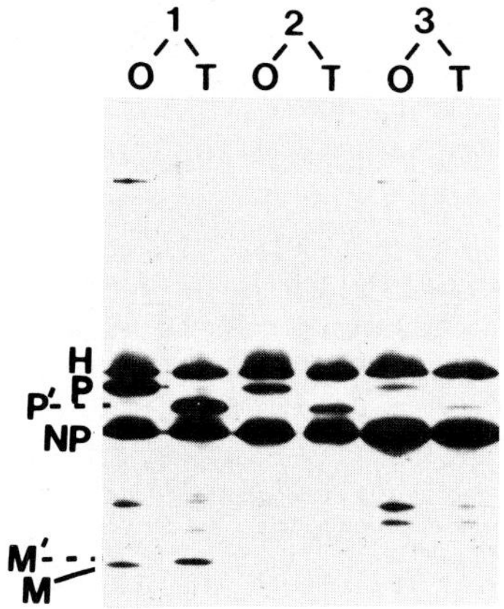

Figure 2. TLCK-induced M' and P' are not immune precipitated by sera with reduced affinity for M and P. Infected CV-1 cell monolayers were labeled, treated with TLCK, suspended in lysing buffer, divided into aliquots, immune precipitated with sera having different specificities for M and P proteins (9), run on SDS-PAGE, and processed for fluorography. Lanes: O, no TLCK; T, 4mM TLCK; 1, hyperimmune rabbit sera with affinity for all measles proteins; 2, a human SSPE sera with reduced affinity for M; 3, a human SSPE sera with reduced affinity for P and M.

Immune precipitations with human sera having selective specificities for different viral proteins demonstrated that the TLCK-induced M' and P' bands were related to M and P (Figure 2). Serum with low affinity to M did not precipitate either M or M' (lanes 2-O, 2-T). Likewise, serum with reduced affinity to P and M precipitated reduced amounts of P' and M' (lanes 3-O, 3-T). The relatedness of M', P' and A' to M, P, and A respectively, has also been demonstrated by peptide mapping with Staphlococus aureus protease V8 (5).

Figure 1, lane 1 shows that 1mM TLCK produces an M protein with a migration rate intermediate between standard M protein and the 4mM TLCK-induced M protein. This was unexpected. If protection from proteolytic cleavage was the sole mechanism of TLCK-induced changes in M, then lower concentrations of TLCK (1mM compared to 4mM) would be expected to produce smaller amounts of the larger (i.e., protected) form of M, rather than an M with an intermediate mobility. The finding of an intermediate sized M suggests a multistep mechanism, thus arguing against protection from a single cleavage event. Further experiments described below confirmed that the TLCK-induced changes were not due to the ability of TLCK to inhibit proteolytic cleavages.

Pulse-chase experiments using a pulse labeling period as short as 20 minutes have not demonstrated any changes in the mobilities of the M or P proteins of measles virus (6,7). Thus, any proteolytic cleavages that alter the electrophoretic mobility of M or P must occur within 20 minutes following synthesis of the presumptive precursor protein. The ability of TLCK to protect against this hypothetical cleavage must also be limited to this time span. Figure 3 shows that TLCK altered the mobilities of M and P proteins synthesized as early as 4 hours prior to the addition of the TLCK. By the above reasoning, the mechanism involved in these alterations cannot be protection of a precursor protein from proteolytic processing.

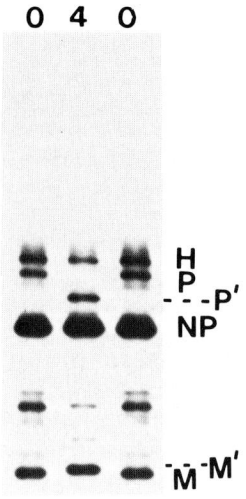

Figure 3. Effect of TLCK on pre-existing measles virus proteins. Infected cells were labeled for 3 hr. and then chased for 90 min. (5). TLCK was added during the final 30 min. of the chase period. The monolayers were harvested, immune precipitated and run on SDS-PAGE. Lanes: 0, no TLCK; 4, 4mM TLCK.

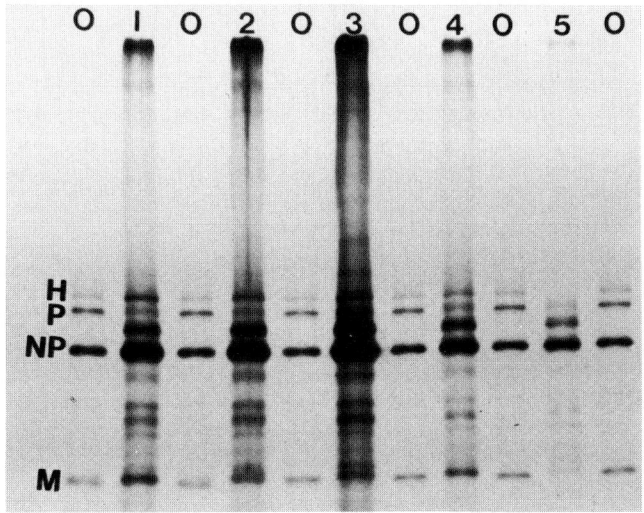

Figure 4. Alterations of viral proteins by TLCK at $0°C$ and in the presence of NP40, SDS, or azide. Measles infected CV-1 cells were labeled with [^{35}S]methionine for 3 hr. The cells were suspended in Tris buffer and aliquots placed in test tubes. The aliquots were all placed in an ice bath. Some aliquots received NP40 at a final concentration of 2%, sodium azide at a final concentration of 0.1M, or SDS at a final concentration of 0.2%. After 10 min. incubation with these additions, TLCK at a final concentration of 4mM was added to all the aliquots except the one shown in lanes O. After 30 min. incubation at either $37°C$ or $0°C$, 1 volume of 2X immune precipitation lysing buffer was added to each tube (9) and the samples were immune precipitated and subjected to SDS-PAGE and fluorography. Gel lanes: 0, no TLCK, $37°C$; 1, TLCK, $37°C$; 2, TLCK, $0°C$; 3, NP40 and TLCK, $37°C$; 4, sodium azide and TLCK, $37°C$; 5, SDS and TLCK, $37°C$.

To further investigate the conditions under which TLCK could induce changes in the mobility of M and P, we incubated previously labeled measles virus proteins with TLCK at $0°C$, in the presence of the detergents NP40 and SDS and in the presence of sodium azide. TLCK-induced changes of P and M were detected in all cases (Figure 4, lanes 1-5).

We have previously reported further evidence that the TLCK-induced changes in M and P are due to factors other than inhibition of a trypsin-like cleavage (5). The general protease inhibitor PMSF, which like TLCK inhibits trypsin, does not affect the electrophoretic mobility of M or P. On the other hand, TPCK and ZPCK which are structurally similar to TLCK, but inhibit alpha-chymotrypsin and not trypsin, do induce TLCK-like changes in P and M (5). This demonstrates that inhibition of a specific protease is not involved in the TLCK-induced changes of P and M and thus that these TLCK-induced changes are not due to protection of a precursor protein from proteolytic cleavage.

The persistently infected HeLa cell line K11 has an M protein with an electrophoretic mobility similar to that of the TLCK-induced M' protein of wild type (wt) virus. We had originally wondered if the alteration of the M protein in these persistently infected cells might be analogous to the TLCK-induced alteration of the wt M protein. Figure 5 shows that TLCK altered the K11 M protein in much the same way as it had the wt M protein, that is, it increased the

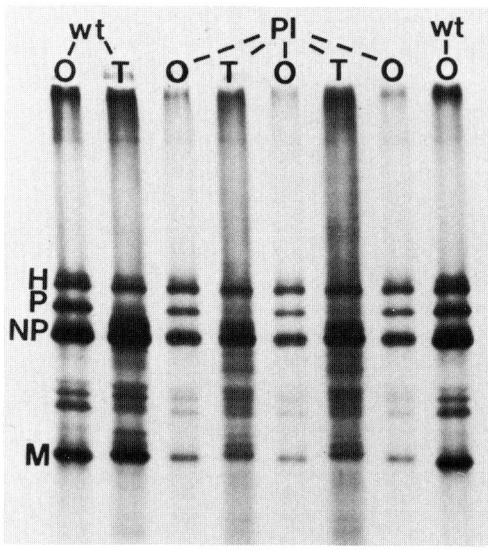

Figure 5. Effect of TLCK on the M protein of persistently infected HeLa cells. Persistently infected HeLa cells (K11) (4) and HeLa cells acutely infected with wt virus were labeled for 3 hr. with [^{35}S]methionine and then incubated with 4mM TLCK for 30 min. The cells were harvested, immune precipitated and run on SDS-PAGE. Gel lanes: wt, wt infected cells; PI, persistently infected cells; O, no TLCK; T, 4mM TLCK.

apparent molecular weight by 1,000-2,000 daltons. Thus, the TLCK-induced alteration of M in wt virus seems to be unrelated to the alteration in mobility sometimes seen in persistently infected cells.

Although TLCK affected the mobility of M protein by an as yet undetermined mechanism, it occurred to us that the TLCK-induced alteration of the M protein might be used as a marker for gel bands related to M. ^{32}P labeling experiments reported several years ago concluded that only the P and NP measles virus proteins were phosphorylated (6). More recent studies have demonstrated a very faint ^{32}P labeled band migrating just above the main [^{35}S]methionine labeled M protein band (7). By analogy with other paramyxoviruses, but without any further evidence, this band was assumed to be a phosphorylated form of M. Figure 6 shows that this faintly phosphorylated band was affected by TLCK in the same way as the main [^{35}S]methionine labeled M protein band, i.e., in the presence of TLCK, the phosphorylated band

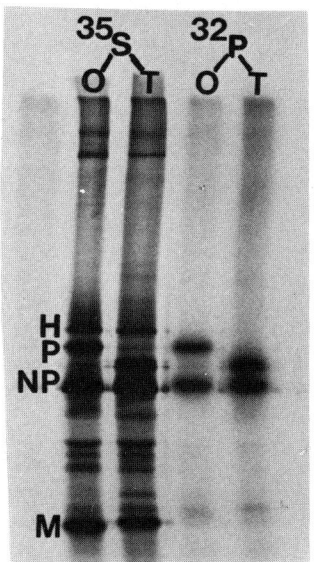

Figure 6. Effect of TLCK on ^{32}P labeled measles virus proteins. Infected cells were labeled for 3 hr. with either [^{35}S]methionine or ^{32}P. The monolayers were then chased for 30 min. either with or without 4mM TLCK added to the medium. The monolayers were harvested, immune precipitated, and run on SDS-PAGE. Lanes: ^{35}S, labeled with [^{35}S]methionine; ^{32}P, labeled with ^{32}P; O, no TLCK; T, 4mM TLCK.

migrates slightly more slowly than the corresponding band not treated with TLCK. To our knowledge, this is the first suggestion, other than co-migration, that this faint ^{32}P labeled band might be related to M. However, this is by no means conclusive evidence, and should be interpreted cautiously.

The results presented here show that TLCK can alter the electrophoretic mobility of several proteins, including the measles virus M protein, by a mechanism that is unrelated to protection of protein precursors from processing by proteolytic cleavage. It is possible that TLCK induces conformational changes of M and P that are reflected in changes in electrophoretic mobility. Binding of TLCK to M and P might be involved. This possibility has not yet been tested due to the unavailablility of isotopically labeled TLCK. In any event, our results indicate that as yet there is no evidence for an M protein precursor.

More importantly our findings should be kept in mind when interpreting past and future reports of precursor-product relationships in which TLCK or related compounds were assumed to act solely by blocking proteolytic cleavages. In addition, since TLCK altered the electrophoretic mobilities of some proteins even in the cold and in the presence of detergents, TLCK or related compounds should not be used as protease inhibitors in buffers.

REFERENCES

1. Vainionpaa, R., (1979). Arch. Virol. 60, 239.
2. Wechsler, S. L., and Meissner, H. C. (1982). Prog. Med. Virol. 28, 65.
3. Wechsler, S. L., and Fields, B. N. (1978). Nature. 272, 458.
4. Wechsler, S. L., Rustigian, R., Stallcup, K. C., Byers, K. B., Winston, S. H., and Fields, B. N. (1979). J. Virol. 31, 677.
5. Wechsler, S. L., Weyand, S., Goosmann, A., and Burge, B. W. (1982). Virology. 121, 204.
6. Wechsler, S. L., and Fields, B. N. (1978). J. Virol. 25, 285.
7. Graves, M. C. (1981). J. Virol. 38, 224.
8. Laemmli, U. K. (1970). Nature. 227, 680.
9. Wechsler, S. L., Weiner, H. L., and Fields, B. N. (1979). J. Immunol. 123, 884.

MECHANISMS OF RSV DI PARTICLE INTERFERENCE[1]

Mary W. Treuhaft

Marshfield Medical Foundation
Marshfield, WI 54449

We have previously shown that defective interfering (DI) particles of respiratory syncytial virus (RSV) not only interfere with replication of standard virus in HEp-2 cells but can also protect cells from virus induced cytopathology (1). And we have used this cell protecting ability to develop a quantitative biological assay for these DI particles (2). During the studies described here defining the role of DI particles in modulating virus infection on undiluted passage, we observed a third biological activity with the potential to attenuate or alter RSV infections - interferon (IFN) induction.

UNDILUTED PASSAGE

Dynamics of Infectious Virus, DI Particle and IFN Production. To determine the role of DI particles in modulating an in vitro model of RSV infection, plaque purified RSV was used to initiate four series of undiluted passages in HEp-2 cells. Culture fluid was examined for infectious virus, DI particles, temperature sensitive (ts) virus and IFN.

The results shown in Fig. 1 are representative of all four series in that infectious virus titer dropped by passage 5 and that this drop coincided with the appearance of DI particles and a marked increase in IFN. A trough of infectious and DI virus was seen at passage 6 followed by a rebound at passage 7. For three of the four series a similar trough in IFN production occurred at passage 6 followed by a rebound at 7. Thereafter oscillating cycles of infectious and DI virus and IFN were produced throughout the remainder of the passages. Both alternate and cocycling of the three

---

[1] This work was supported in part by a grant from the American Heart Association/Wisconsin Affiliate and the Marshfield Medical Foundation.

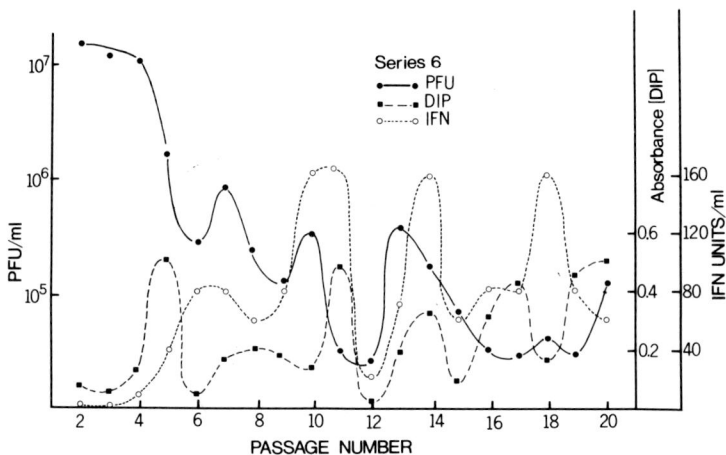

FIGURE 1. Undiluted passage of RSV in HEp-2 cells. Four undiluted passage series were initiated with four different clones of RSV which had each been plaque purified 3X. Flasks containing 2×10^6 HEp-2 cells were infected with 1 ml of culture fluid from the preceding passage for 2 hr at 32°C with rocking. Inoculum was removed. Flasks were fed with 10 ml VGM (MEM + 5% fetal bovine serum). After 72 hr at 37°C, culture fluid was centrifuged to remove cells and the supernatant frozen until it was analyzed for infectious virus by plaque assay at 37°C, and for ts virus by plaque assay at 32, 37 and 39°C. DI particles were analyzed by colorimetric assay (std. virus moi = 1) following 60 sec exposure to UV to inactivate infectious virus which might otherwise mask DI activity (2). Cell protecting activity in colorimetric assay was due to DI particles and not to IFN because in each sample it was inactivated with RSV antiserum but not pre-immune serum. IFN was analyzed by filtering samples through a 0.2 µ filter (Amicon Sterilet) which retains infectious and DI virus but allows IFN to pass through. Two-fold dilutions of the filtrate were evaluated for ability to interfere with VSV cytopathology in human foreskin fibroblasts (3). A 50% end-point determination and internal IFN standards were used to estimate the quantity of IFN present. This filterable interfering activity active against VSV was characterized as βIFN by neutralization with anti-human βIFN serum but not by anti-human αIFN serum.

activities occurred but no consistent pattern was seen beyond passage 7. No ts virus able to grow at 32 °C but not at 37°C or 39°C was detected.

When means of infectious and DI virus and IFN levels at each passage were determined for all four series together, a striking uniformity over the first seven passages is apparent (Fig. 2). Because these series were initiated with four different virus clones obtained by plaque purification (3X), they represent four independent generations of DI particles. Thus in HEp-2 cells, DI can be rapidly and reproducibly generated for RSV and once they are generated their effects are predictable.

This presentation also emphasizes the pattern of appearance of DI particles by passage 5 coinciding with an increase in IFN and drop in infectious virus followed by a further drop in all three activities and then a rebound. Beyond passage 7, DI particles and IFN are produced in relatively stable amounts which are significantly greater than those in initial passages. Infectious virus is also produced at stable levels but in significantly reduced amounts than in initial passages. These results conform to the model proposed by Huang and Baltimore (4,5) for interactions of DI and infectious virus on continued passage in that once DI are generated, further passage results in continued balanced production of both DI and infectious virions.

Our inability to detect ts virus supports DI as the major determinant of infectious virus modulation in these undiluted passages. Further support comes from the observation that both DI particle and IFN levels exhibited a significant negative correlation with infectious virus produced in the subsequent passage (Table 1).

Establishment of Persistent Infections (PI). With continued feeding, two of the four cultures at passage 5 and all of the cell cultures (with two exceptions) beyond passage 5 survived and developed into PI stable in culture for a minimum of eight weeks. These PI were characterized by scattered foci of viral cytopathology, continued shedding of infectious virus and periodic crises. Thus on continuous passage, once DI appear, PI can be readily established. Determination of the role of DI particles and of IFN in establishing and maintaining these PI is in progress.

Role of IFN. Both IFN in the virus inoculum and that produced during the passages might be contributing to the low infectious virus yield and low cytopathology of these undiluted passages. However, it is unlikely that IFN in the inoculum exerts an inhibitory effect because only a 2 hr adsorption period was allowed. Greater than 2 hr exposure to

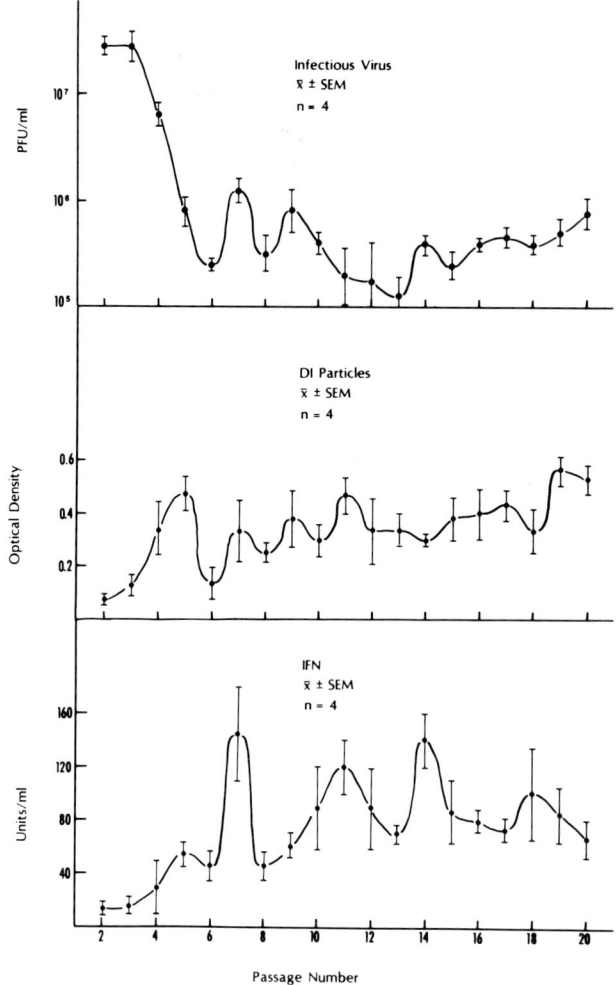

FIGURE 2. Summary of infectious virus, DI particles and IFN production on undiluted passage of RSV in HEp-2 cells. Means of infectious virus, DI particles and IFN at each passage for the four passage series described in Fig. 1 were determined and are shown bracketed by standard error of the mean.

cells was found to be required for the IFN (0.2 μ filtrate) in DI preparations to provide cell protection (Table 2).
To determine whether the IFN produced during the passages was exerting antiviral activity, antiserum to

TABLE 1
CORRELATIONS BETWEEN INFECTIOUS VIRUS, DI PARTICLES AND IFN PRODUCED ON UNDILUTED PASSAGE[a]

	Series	r
Log DI particles vs log subsequent pfu (DF = 34)	1	-.92*
	2	-.67*
	4	-.50*
	6	-.10
Log IFN vs log subsequent pfu (DF = 34)	1	-.71*
	2	-.74*
	4	-.74*
	6	-.78*
Log DIP vs log IFN (DF = 36)	1	.65*
	2	.45*
	4	.41*
	6	.42

[a] Levels of DI and infectious virus and IFN in the same sample and in the subsequent passage for all of the passages of the series described in Fig. 1 were evaluated for association by determining correlation coefficients (r).
* $P < 0.01$

βIFN was added to four of the passages and infectious virus yields and cytopathology evaluated relative to the same passages without antiserum and to cultures infected with similar mois of standard virus alone (Table 3). Neutralization of extracellular IFN did result in an increase in infectious virus yield but not to the levels produced by similar mois of standard virus alone. Cell protection was decreased only minimally. Therefore, we conclude that the IFN induced during the passages does contribute to interference with standard virus replication and cytopathology, but it is not the only mechanism by which this is accomplished.

ENHANCED IFN INDUCTION BY DI RSV

The increase in IFN production coinciding with the appearance and amplification of DI particles in these undiluted passages suggests that the newly generated DI

TABLE 2
EFFECT OF FILTRATION ON RSV DI PREPARATION INTERFERENCE

Challenge virus	DI prep adsorption time	Filtration (0.2 μ)	Percent of maximum cell protection in colorimetric assay[a]
VSV	2 hr	−	3
		+	5
	24 hr	−	100
		+	91
RSV	2 hr	−	100
		+	11
	24 hr	−	100
		+	42

[a] Duplicate wells containing 1.5×10^5 HEp-2 cells were exposed to 0.1 ml RSV DI preparation either prior to or subsequent to filtration through a 0.2 μ (Amicon Sterilet) filter for either 2 or 24 hr. Wells were then challenged with RSV (moi = 1) or VSV (moi = 0.1). Virus inoculum was removed and wells were fed with 1 ml VGM and incubated at 72 hr at 37°C until virus control wells containing standard RSV or VSV alone exhibited maximum cytopathology. Cell protection was assessed by incubating wells with neutral red dye (33 mg/L) for 2 hr at 37°C. Dye was removed, plates were washed and dye taken up by cells remaining on monolayer was extracted and determined spectrophotometrically at 540 nm. Percent of maximum cell protecting activity by the DI preparation for each virus was determined.

particles are responsible for the enhanced IFN induction. The ability of standard RSV alone to induce IFN is uncertain as yet. Initial studies suggest that when standard virus does induce IFN, it does so only at relatively high moi (5 vs 0.05-0.5 pfu/ml), relatively late in replication (>36 hr) and that DI particles can be detected. Standard virus pools contain relatively low amounts of IFN ($\leq$10 U/ml) compared to DI preparations (100 U/ml). Finally, the positive correlation observed for DI particles and IFN levels in the same passage supports an interdependent relationship between IFN and DI particle synthesis (Table 1). These observations suggest that DI particles are better IFN inducers than standard virus alone. Whether these DI particles alone can

TABLE 3
EFFECT OF ANTI-βIFN ANTISERUM ON VIRUS YIELD AND
CELL PROTECTION OF UNDILUTED PASSAGES[a]

Inoculum	Infectious virus input (moi)	Log pfu		Cell protection % of cell control	
		No Ab	+ Ab	No Ab	+ Ab
Std. virus	.5	6.73		1	
	.1	6.49		3	
	.05	6.59		5	
	.01	6.56		38	
Series 2-P14	.3[b]	5.34	5.72	46	40
Series 2-P15	.3	5.58	6.23	51	45
Series 1-P7	.05	5.59	6.08	56	51
Series 2-P12	.03	5.11	5.77	95	95

[a] Wells containing 1.5×10^5 cells were infected with either standard virus or the undiluted passage indicated in 0.1 ml for 2 hr at 32°C with rocking. Inoculum was removed, wells were fed with 1 ml VGM and anti-βIFN antiserum (obtained from J. Vilček, New York University School of Medicine, NY) sufficient to neutralize 500 U of βIFN was added to wells indicated. Wells were incubated at 37°C for 72 hr and then culture fluid analyzed for infectious virus by plaque assay. Cell protection was determined as described in Table 2 and is expressed here as percent of dye taken up by uninfected cell controls.
[b] Determined by plaque assay of inoculum.

induce IFN or whether replication is required remains to be determined.

MULTIPLE INTERFERING ACTIVITIES

In summary, DI particles of RSV exhibit multiple biological activities: reduction of infectious virus yield, reduction of cytopathology and IFN induction. Each of these activities has the potential to alter infection in vivo either by attenuating the infection or establishing a PI just as they do in the in vitro cell culture system described here. Whether these DI particle mediated activities share a common mechanism or pathway is not yet known. All could be due to competition of the DI genome with the standard virus genome for replicase. This could result not only in reduced infectious virus yield but also in reduced synthesis of m-RNAs coding for the viral proteins responsible for cytopathology and shutoff of host macromolecular synthesis.

ACKNOWLEDGMENTS

I thank Dr. Ernest Borden for IFN standards and Dr. Jan Vilček for antisera to human α and β IFNs. I also appreciate the excellent technical assistance of Ms. Joleen Soukup and Ms. Susan Willging.

REFERENCES

1. Treuhaft, M. W., and Beem, M. O. (1982). Infec. Immun. 37, 439.
2. Treuhaft, M. W. (1983). J. Gen. Virol. 64, 1301.
3. Rubinstein, S., Familletti, P. C., and Pestka, S. (1981). J. Virol. 37, 755.
4. Huang, A. S., and Baltimore, D. (1970). Nature 226, 325.
5. Huang, A. S. (1973). Ann. Rev. Microbiol. 27, 101.

THE EFFECT OF VIRUS PERSISTENCE
ON PLASMA MEMBRANE BOUND FUNCTIONS
IN CNS DERIVED CELL LINES.

P. N. Barrett, P. Münzel, C. Winkelkötter,
and K. Koschel

Institut für Virologie und Immunbiologie
der Universität Würzburg
Versbacher Straße 7
D-8700 Würzburg, West-Germany

In many neurological diseases with a virus aetiology it is uncertain whether the virus has a direct cytopathogenic effect on neural cells or whether virus induced cytotoxic immune responses are responsible for the cellular dysfunction characteristic of such diseases. It is known that many negative stranded RNA viruses such as the paramyxoviruses, rhabdoviruses, arenaviruses integrate virus glycoproteins into the host cell membrane before release through budding. It has been postulated that these virus proteins could induce specific or nonspecific disturbances of the normal microstructure of the cell membrane thus creating the molecular basis for an impairment of specialized membrane bound cellular functions (1, 2). We have looked at suitable CNS derived model cells which were persistently infected with different viruses to investigate this possibility. The C6 rat glioma cell line (ATCC CCL107) was chosen because they possess large numbers of ß-adrenergic receptors for catecholamines. They were persistently infected with measles (SSPE), canine distemper (CDV) and lymphocytic choriomeningitis viruses (LCM). The neuron model cell mouse neuroblastoma x rat glioma hybrid (108CC15) created by Amano and Hamprecht (3) were chosen because they possess membrane receptors for prostaglandins, adenosine, catecholamines, acetylcholine, opiates and other signal transmitting substances. These cells were persistently infected with rabies virus. The stimulation of these different receptors in both cell lines with their appropriate agonists is followed by typical shifts of the intracellular cyclic AMP concentrations which can be easily measured. We have concentrated on the study of ß-receptor function in C6 cells and the opiate receptor and prostaglandin E 1 systems in 108CC15 cells. In C6 cells the ß-adrenergic receptor is linked to the adenylate cyclase enzyme by a stimulating regulatory N_s protein. Various effects on the different components of the receptor adenylate cyclase system have been observed following establishment of persistence with different viruses in these cells (4, 5). In 108CC15 cells the cyclic AMP synthesizing enzyme is under two controls: there are stimulating receptors for cAMP synthesis (like PGE_1 receptors) which stimulate after occupation by the specific hormone via a stimulatory coupling protein N_s. Other types of receptors (like opiate receptors) inhibits this stimulation if they are occupied by their specific agonists via an inhibitory coupling protein N_i also connected to the adenylate cyclase. Effects have also been observed on different components of these two systems in persistently rabies virus infected cells (6, 7). The differing effects of SSPE, CDV and LCM virus persistence in C6 cells is of particular interest. These persistently infected cultures had virus antigen inserted in 100 % of cells but the effect of this insertion of foreign antigen on membrane function differed considerably. There was a large reduction in the accumulation of intracellular cyclic AMP following stimulation of ß-adrenergic receptors in paramyxo virus infected cells but no effect in LCM infected cells. The paramyxo viruses also differ in their effects. CDV infection results in a 50 % decrease in the number of ß-adrenergic receptors without any alteration in the binding constant for the ligand to the receptor (5). However, persistence of SSPE virus did not result in a decrease in the number of receptors in the membrane or an alteration in their binding properties. Neither insertion of LCM virus proteins in the membrane did not appear to play any role in cellular dysfunction in respect to the ß-receptor adenylate cyclase system studied. To clarify the role of virus glycoproteins in the C6/SSPE system virus antigen was removed from the cell membrane by the use of antiserum directed against virus proteins as previously described (8).

MODULATION OF VIRUS PROTEINS IN C6/SSPE CELLS

C6 cells persistently infected with the measles SSPE virus (Lec) were grown in medium containing heat inactivated serum from SSPE patients. This serum contains antibodies which recognize all the measles virus proteins with a reduced reaction against M protein. These cells were passaged every 3 to 4 days and the amount of viral antigen present in the membrane and the percentage of cells displaying membrane and intracellular antigen was measured at each passage. The amount of viral antigen present in the membrane was measured by freezing a trypsinized cell suspension with saturating amount of human SSPE antiserum and (125)I-labelled anti-human IgG. Thereafter cell associated radioactivity was measured. The percentage of cells containing intracellular and membrane viral antigen was measured by staining fixed and unfixed cell preparations, respectively by the indirect immunofluorescent method. Our studies demonstrated a gradual loss of viral antigen from the cell membrane over a period of 6 passages (Fig. 1A).

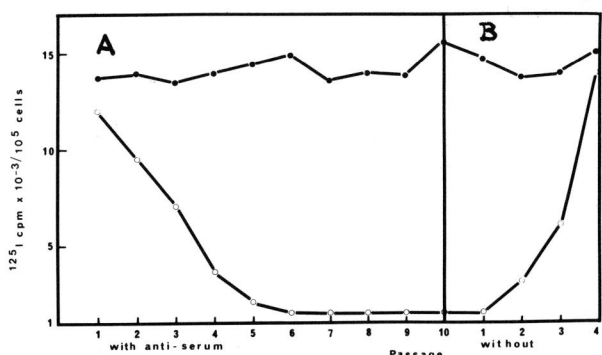

Figure 1
Virus antigen inserted in the cell membrane was measured by counts bound of (125)I-labeled anti-human IgG to cells incubated with human anti SSPE serum for the whole passage time. (A). Amount of virus antigen present after antiserum treatment: (●) C6/SSPE cells grown in antiserum for 1 hr only; (o) C6/SSPE cells maintained in antiserum. (B) Amount of virus antigen present after removal of antiserum. (●) C6/SSPE cells grown in antiserum for 1 hr. only; (o) C6/SSPE cells released from antiserum treatment. Cells were passaged every 3-4 days.

At passage 6 the amount of radioactivity bound to C6 SSPE cells was equal to that binding to uninfected C6 cells grown in medium containing antiserum. At this stage, however, approximately 30 % of cells still contained intracellular viral antigen even though no cells displayed positive staining of the membrane antigens (Fig. 2).

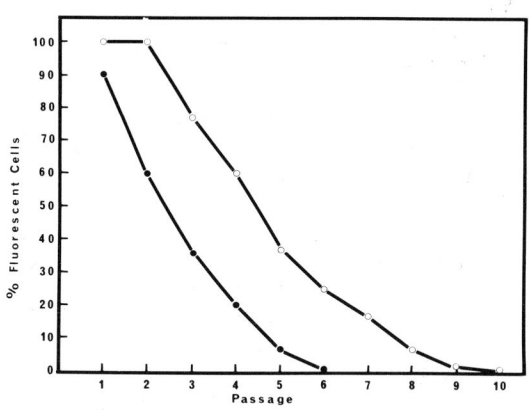

Figure 2
Percentage of cells displaying virus specific immuno fluorescence in (●) unfixed (membrane antigens stained) and (o) acetone-fixed C6/SSPE cells (intracellular viral antigens stained) maintained in antiserum. Cells were passaged every 3-4 days.

When there was no detectable virus antigen of any type present in the antiserum treated cells they were washed in PBS and further incubated in medium without antiserum. This was followed by a rapid recovery in the expression of virus antigen on the cell membrane and intracellularly (Fig. 1B).

EFFECT OF MODULATION ON HORMONE STIMULATED cAMP RESPONSE

To determine if the previously reported impairment of the cyclic AMP response in C6/SSPE cells (5) could be attributed to intracellular or membrane virus antigens we examined cyclic AMP production at various stages during virus antigen modulation. Persistently infected cultures grown with or without

SSPE antiserum were stimulated via ß-adrenergic receptors at every passage, with DL-isoproterenol under the optimal conditions for cAMP synthesis. Intracellular cAMP was expressed as a percentage of the amount of cAMP produced in uninfected C6 cells grown in medium containing an identical concentration of measles antiserum (Fig. 3).

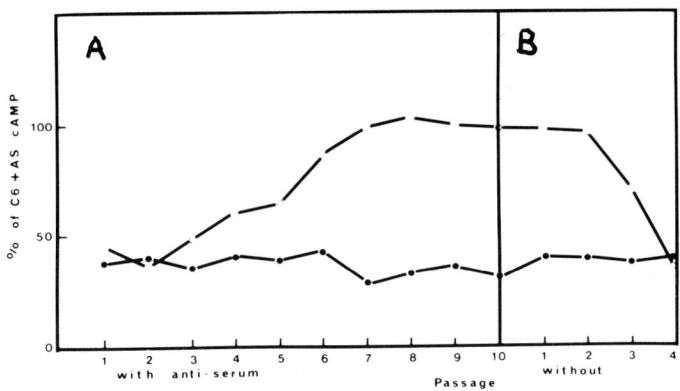

Figure 3
Measurement of isoproterenol stimulated cAMP synthesis expressed as a percentage of control levels in uninfected C6 cells grown in SSPE antiserum. (A) After treatment with antiserum: (●) untreated C6/SSPE cells; (o) C6/SSPE cells maintained in SSPE antiserum. (B) After removal of antiserum: (●) previously untreated; (○) previously treated with antiserum. Cells were passaged every 3-4 days.

Over the initial 3 passages there was little difference between cAMP synthesis in antiserum treated and untreated C6/SSPE cells. This was always less than 50 % of that of uninfected C6 cells. There was an initial recovery in cAMP synthesis at passage 4, where it was for the first time over 50 % of control levels and it reached 100 % at passage 7. At this stage there was no viral antigen present in the cell membrane while approximately 25 % of cells still showed distinct fluorescence in the cytoplasm (Fig. 2). This clearly demonstrates that it is insertion of viral antigen into the cell membrane and not intracellular accumulation which is reponsible for the impairment of the cAMP response in C6/SSPE cells, and this is accomplished without alteration in the total number of receptors or their affinity for DL-isoproterenol. When antiserum treated cells were grown in medium without antiserum as previously described there was a rapid drop in cAMP levels commensurate with the reappearance of virus antigen in the cell membrane (Fig. 1B, 3B).

MEASUREMENT OF ADENYLATE CYCLASE ACTIVITY IN ANTISERUM TREATED CELLS

The mechanism by which virus antigen produced this impairment in membrane function was unclear as there was no alteration in receptor number or affinity. We therefore measured the effect of virus antigen on the cAMP synthesizing enzyme adenylate cyclase. The fluoride ion stimulated activity of this enzyme was determined in particular cell fractions from uninfected C6 cells, C6/SSPE cells grown in antiserum containing medium, and C6/SSPE cells grown in antiserum containing medium until virus antigen negative and then further passaged in normal medium so that virus antigen was again produced. Figure 4 shows that this enzyme activity was drastically reduced in SSPE virus infected cells. However, adenylate cyclase activity had recovered to control levels in antiserum treated cells. Removal of antiserum then led to a renewed depression in this enzyme activity.

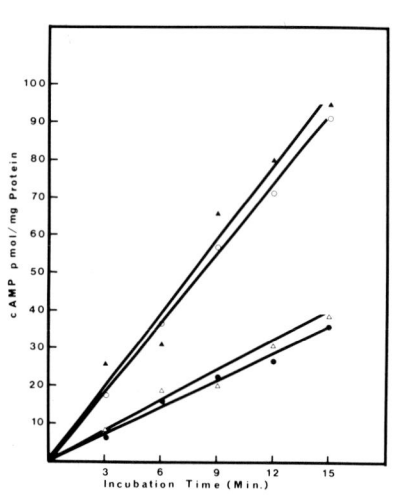

Figure 4
Measurement of the activity of adenylate cyclase enzyme by estimation of F^--ion stimulated cAMP synthesis in membranes isolated from (▲) uninfected C6 cells, (△) C6/SSPE cells, (o) SSPE antiserum-treated C6/SSPE cells, and (●) SSPE antiserum treated C6/SSPE cells removed from antiserum.

CONCLUSIONS

These data indicate that measles virus glycoproteins impair cAMP synthesis through an effect on the adenylate cyclase enzyme on its regulatory component. It is interesting that in C6 cells persistently infected with LCM virus there is no effect on cAMP synthesis although virus glycoproteins are inserted in the cell membrane. This implies that simple insertion of foreign proteins may not be sufficient to cause membrane dysfunctions in virus infected cells. It is likely that a specific interaction must occur between viral membrane proteins and the integral membrane protein components of the signal transmitting receptor adenylate cyclase system. The antibody directed loss of viral antigen from the cell surface of a CNS derived cell described here may be a mechanism for allowing establishment and maintenance of viral persistence in the brain as postulated by Fujinami and Oldstone (9). During such a persistence there would be no insertion of glycoprotein in the membrane causing an impairment of membrane function or stimulating a cytotoxic immune response. However, any drop in titers of modulating antiviral antibody which would allow expression of viral antigen in the membrane would lead to a rapid impairment of membrane function and the possible development of clinical symptoms. Studies on the mechanism of maintenance of virus persistence at the RNA level are described in another communication (Carter et al. this volume).

ACKNOWLEDGEMENTS

This work was supported by the Deutsche Forschungsgemeinschaft, SFB 105, Projekt C 3.

REFERENCES

1. Halbach, M., and Koschel, K. (1979). In "Humoral Immunity in Neurological Diseases. (D. Karcher, A. Loewenthal, and A. D. Strosberg, Eds.) Plenum, New York.
2. Oldstone, M. B. A., Holmstoen, J. and Welsh, R. M (1977). J. Cell. Physiol. $\underline{91}$, 459-472.
3. Hamprecht, B. (1977). Int. Rev. Cytol. $\underline{49}$, 99-170.

4. Halbach, M., and Koschel, J. (1979). J. Gen. Virol. 42, 615-619.
5. Koschel, K. and Münzel, P. (1980). J. Gen. Virol. 47, 513-517.
6. Koschel, K. and Halbach, M. (1979). J. Gen. Virol. 42, 627-632.
7. Münzel, P., and Koschel, K. (1981). Biochem. Biophys. Res. Commun. 101, 1241-1252.
8. Joseph, B. S., and Oldstone, M. B. A. (1975). J. Expt. Med. 142, 864-876.
9. Fujinami, R. S. and Olstone, M. B. A. (1980). J. Immunol. 125, 78-85.

Index

A

AD 2 replication system, 217–220
AMP and ts G16 mutant, 124, 125, 128, 129
ATP, 95–101
AV–WT M protein, 315–320
Acetylcholine receptor, 379
Adenovirus
   and SV40 DNA synthesis, 223–231
   type 2 replication and VSV effects, 215–221
Adenovirus-associated gene, see VA gene
Adenovirus-2 late promoter gene, see LP gene
Amino acid composition
   of L protein, 38–40
   of SVCV and VCV M protein, 46
Amino acid sequences of glycopeptides, 287, 290, 291
Amino acid substitution in glycoprotein of mutants of CVS strain, 295–299
Anti-H monoclonal antibodies and measles virus, 353–355
Anti-HN monoclonal antibodies, epitope assignments of, 309, 310
Anti-N antibodies, microinjection of, 394–396
Anti-NP monoclonal antibodies, 339, 341, 342, 345, 347–350
Antigenic determinants of measles virus, 351–357
Autoimmunity by infection by virus, 427–433

B

BGM–L cells and measles virus, 505–511
BGM–P cells and measles virus, 505–511
BHK cells, 110–112
   and RNA synthesis, 139, 141, 142
BHK-21 cells, 388, 497–504
Bovine rinderpest virus, 167

C

C protein and mumps, 336
cAMP synthesis, 548–551
CCCP, 203
cDNA clones, 17–19, 21–26
   construction and characterization of, 28–30
   of measles virus, 359–364
   and NDV, 161
   with RNA, 22
   and 10 different families, 21–26
cDNA library for RS virus, 28–30
CDV
   and MV, 55–60
   molecular studies, 167–174
CL, Sendai virus-induced in mouse spleen cells, 451–457
CNS
   and measles virus, 233
   and mumps, 443–449
   and SSPE, 521
CNS derived cell lines and virus persistence on plasma membranes, 545–551
CNS tissue and MV, 247–251
CSF and SSPE, 521, 524–526
CTP, 95, 96, 100
   and ts G16 mutant, 124, 125
CVS strain of rabies virus
   and BHK-21 cells, 497–504
   and mutants, 295–299
   virus G, 279–283
CVS variant viruses with antigenically altered glycoproteins, 285–292
Canine distemper virus, see CDV
Cell surface receptors and rabies virus, 387–392
Cellular LA protein and leader RNAs, 103–107
Cellular RNA synthesis, 131–133

Central nervous system, see CNS
Chemiluminescence, see CL
Chimeric gene G–M, 193–199
Cloning
  of genomic RNA, 233–237
  of L gene, 35, 36
  of P Protein of measles virus, 359–364
Cyclic AMP and MV-specific protein synthesis, 247–251
Cyclic AMP response, see cAMP synthesis
Cyclic nucleotides and mouse virus, 247–251

D

DI particle, 147–152, 215
  biological activities, 461–467
  and IFN production, 537–544
  MS–T, 153–160
  and RNA synthesis, 469–474
  of RSV, 537–544
DI RNA
  and morbilliviruses, 167
  varieties of, 483–487
  of VSV, 469–474
DI–LT
  and recombination events, 469–474
  and VSV DI particles, 475–481
  and VSV_{NJ} and VSV_{IND}, 461–467
DI-011, 480
DI-121, 462–467
DI–T, 147–152
DNA synthesis and SV40, 223–231
Defective interfering particles, see DI particle
Deletion mutants of VSV, 475–481
Dose response
  to inhibition of transcription of LP and VA genes, 136, 137
  to inhibition of transcription of RNA leader, 134
Double-label fluorescent antibody techniques, 421–426
Drosophila, host range mutants in, 413–418

E

ERA strain of rabies virus, 279–283, 387–392
  and mutants, 295
ERA G, 280, 281
84 K proteins, 365
Eukaryotic macromolecular synthesis, 215–217

F

F protein
  and comparison with 24 K protein, 367, 368
  and HPIV3, 323, 325–327
  and LEC measles virus, 352, 354, 355
  of measles virus, 421–426
  mitogenic activity of, 406, 408–411
  and NDV, 161–166
  and Sendai virus, 345–350
  and TM treatment, 372–375
F1 protein and mumps, 334
F0 proteins, 333, 334, 337
Fatty acid metabolism and measles virus infections, 505–511
Fusion protein, see F protein

G

G protein, 68, 69
  and comparison with 24 K protein, 367, 368
  and G–M chimeric gene, 193–199
  mitogenic activity of, 406–411
  and RNA synthesis, 93
  of rabies virus, 279–283, 497, 498, 502, 504
  and TM treatment, 372–375
  and ts G16 mutant, 126, 129
  and VSV, 79, 201, 204, 205
  and VSV RNP, 271
G* RNA, 475–481
G–M chimeric gene, 193–199
GTP, 95–97, 99, 100
  and ts G16 mutant, 124, 125, 128, 129
Genomic RNA, cloning of, 233–237
Glycopeptides, amino acid sequences of, 287, 290, 291
Glycoprotein
  and CVS variant virus, 285–292
  HN and NDV, 309–314
  of mutants of CVS strain, 295–299
  and RS virus, 369–375
Glycosylation sites, 289–292
Golgi membranes, 201–205

H

H protein
  and K11 cells, 241, 242, 244, 245
  and K11A cells, 242, 243
  and measles virus, 355
HA
  protein of measles virus, 421–426
  units of virus, 312, 313
HeLa cells, 228, 229
  and K11, 239–242, 244, 245
  and K11A, 240, 242–245
  and measles viral proteins, 239–245

INDEX

HN protein
  and ELISA, 345
  and HPIV3, 323–327
  mitogenic activity of, 408–411
  and monoclonal antibodies, 345–350
  and mumps, 333, 334, 337
  and NDV, 161–166, 201, 203–205, 309–314
Halothane and measles virus, 513–520
Hamsters
  LSH strain, 489–495
  LVG strain, 489–495
  and long-term persistence by VSV, 489–495
Hemagglutinin–neuraminidase protein, see HN protein
Heterologous reconstitution system, 64–67
Human parainfluenza virus, see Parainfluenza virus
Human parainfluenza virus type 3, 321–327
Human paramyxovirus mumps, 333–338
Human respiratory syncytial virus, see RSV
Hydropathic pattern of SVCV and VSV M protein, 46, 47

I

IFN
  and measles RNA synthesis, 183–189
  and mouse L cells, 207–213
  production and DI particles, 537–544
  protein kinase, 211–213
Interferon, see IFN

L

L messenger RNA, 37, 38
L polymerase protein, 93
L protein
  amino acid composition of, 38–40
  comparison to 24 K protein, 367, 368
  and HPIV3, 323, 327
  and LT RNA, 483, 486, 487
  and MS–T, 155, 158
  and measles virus, 351, 354
  and mumps virus, 333
  and NDV, 161, 163–166
  and rabies virus, 502
  and ts G16 mutant, 123, 126, 129
  and VSV, 63–69, 79, 80, 83, 84
  and VSV RNP, 272, 275
La protein, 103–107
LEC measles virus and F protein, 352, 354, 355
LP gene, 133–136

Leader RNA
  and effect of double-stranded regions, 134, 135
  and UV-inactivated virus, 133
Lymphocytes
  murine, 405–411
  subpopulations and measles virus, 435–441
Lytic infection and measles virus, 513–520

M

M protein, 4–8, 41, 42
  amino acid composition of, 46
  and binding of VSV RNP to sonicated phospholipid vesicles, 271–277
  and CDV, 167
  comparison with 24 K protein, 366, 367
  and G-M chimeric gene, 193–199
  and HPIV3, 323, 324, 326, 327
  and hydropathic pattern of, 46, 47
  and inhibition of transcription, 68
  interaction with heterologous RNP, 67–69
  and K11 cells, 241, 244, 245
  and K11A cells, 240, 243, 245
  measles and IFN, 189
  and measles virus, 355, 357, 421–426
  and mumps, 333
  and NDV, 161–166, 315–320
  and plasma membranes, 402–404
  and RNA replication, 148, 150–152
  and SSPE, 167, 233, 245, 247, 248, 521–526
  and TLCK, 529–536
  and ts G16 mutant, 123, 126, 129
  and ts mutant DI, 317, 319, 320
  and VSV, 46, 47, 63–69, 79
M protein antibody, 69
M_1 protein and rabies virus, 497, 498, 501, 502
M_2 protein and rabies virus, 504
Met-Ser-X-Lys sequence, 48
MF cell line, 521–526
mRNAs and PI-3 infected cells, 331–332
MS–T nucleocapsids
  and L protein, 155
  and viral proteins replicating in vitro, 154, 155
Measles infected
  lymphocytes, 183–185, 189
  Vero cells, 185–188
Measles virus, 167
  of cDNA clone, 359–364
  and CDV, 55–60
  and CNS, 233

and H protein, 355
and HeLa cells, 239–245
and L protein, 351, 354
and M protein, 355, 357
and M protein alterations by TLCK, 529–536
M protein and IFN, 18
and molecular cloning of, 359–364
monoclonal antibody, 505–507
and multiple sclerosis, 233, 435, 441
and NP protein, 354, 355
and P protein, 354, 355
persistent infections, 505–511
and SSPE, 233–237
and specific protein synthesis, 247–251
and variations in antigenic determinants of, 351–357
Measles virus glycoproteins and cAMP, 548–551
Measles virus infections
and fatty acid metabolism, 505–511
of human peripheral blood lymphocytes, 435–441
Measles virus nucleocapsids during lytic and persistent infections, 513–520
Measles virus particles and halothane exposure, 513–520
Measles virus phosphoprotein, cross-reaction of, 427–433
Measles virus proteins and double-label fluorescent antibodies, 421–426
Measles virus RNA
in brain tissue, 233–237
characterization of, 49–53
molecular weight, 49–53
synthesis and IFN, 183–189
Microinjection
of anti-N antibody, 394–396
of monoclonal antibodies to VSV N protein, 393–398
Mitogenic activity
and F glycoprotein, 406, 408–411
and G protein, 406–411
and HN glycoprotein, 408–411
Molecular cloning
of measles virus, 359–364
and Sendai virus genome, 3–9, 17
techniques, 21
Molecular mimicry during virus infection, 427–433
Molecular studies and CDV, 167–174
Monoclonal antibodies, 155–158
and measles virus, 505–507

microinjection of, 393–398
and mumps virus, 443–449
and N protein, 393–398
and phosphoprotein of measles virus, 427–433
to P protein, 359–364
and Sendai virus glycoproteins, 345–350
and Sendai virus protein NP, 339–344
and therapeutic potential in fatal viral infections of CNS, 443–449
Morbilliviruses, 167–174
cDNA synthesis and cloning of, 58
messenger RNA and the purification of, 55–57
specific clones, characterization of, 58–60
specific RNA clones, 55–60
Multiple sclerosis and measles virus, 435, 441
Mumps virus
and monoclonal antibodies, 443–449
and RNA, 333–338
Murine lymphocytes, 405–411

N

N protein, 104, 106, 107
compared with 24 K protein, 366–368
and MS–T, 155, 157, 160
and microinjection of monoclonal antibodies, 393–398
and monoclonal antibodies, 393–398
and mumps, 333, 334, 337, 338
and RNA replication, 148, 150–152
and rabies virus, 497, 498, 500, 502
and SSPE, 523
and ts G16 mutant, 123, 126, 129
template, 71–76
and VSV, 79
and VSV RNA, 93, 271, 272, 275
NA activity of HN molecule, 311, 312
N-alpha-p-tosyl-L-lysine chloromethyl ketone-HCl, *see* TLCK and measles virus
NDV, 324
and characterization of polycistronic transcripts, 161–166
and four functional domains on HN, 309–314
and HN, 201, 203–205
and mutant and wild-type M proteins, 315–320
and P proteins, 301–307
and VSV, 201–205
N:NS protein complex, 158, 160
NP clone, 17–19

INDEX

NP protein
  and HPIV3, 323, 324, 327
  and K11 cells, 241, 242
  and K11A cells, 242, 243
  and LT RNA, 483–487
  and measles virus, 354, 355
  measles virus and double-label fluorescent antibodies, 421–426
  and NDV, 161–166
  and Sendai virus, 339–344
N–RNA complex of VSV, 63–69
NS protein, 79–84, 93
  and CNBr, 265–269
  and cleavage, 265–268
  and DEAE–cellulose chromatography, 79, 81
  and MS–T, 155, 157, 158, 160
  and mumps, 334–338
  and phosphorylation site, 255–263
  and RNA replication, 148–152
  and SDS, 255–259, 263
  and ts G16 mutant, 123, 126, 129
  and VSV, 63–69
  and VSV RNP, 272, 275
  and VSV transcription, 265–269
NS protein tryptic peptides, analysis of, 259–263
NS1 and NS2 protein, binding studies of, 79–84
Neuraminidase, see NA activity of HN molecule
Nucleocapsid protein, see NP protein
Nucleocapsids of measles virus, 513–520
Nucleotide sequence analysis, 36, 37
Newcastle disease and P protein, 336
Newcastle disease virus, see NDV

O

OKT4+ T-cell subset and measles virus, 435–441
Old dog encephalitis, 167

P

P multimers from virions, composition of, 304, 305
P protein
  in cell-free protein synthesizing systems, 303, 304
  comparison to 24 K protein, 366–368
  and DI RNAs, 487
  forms of in infected cells, 301–303
  and HPIV3, 323, 324, 327

  and K11 cells, 241
  and measles virus, 354, 355
  of measles virus and double-label fluorescent antibodies, 421–426
  of measles virus and monoclonal antibodies, 427–433
  of measles virus and molecular cloning, 359–364
  molecular anatomy of, 305–307
  and monoclonal antibodies, 359–364
  and mumps, 333–338
  and NDV, 161–166, 301–307
  and TLCK, 530–536
PBLs and measles virus, 435–441
P1 cells and rabies virus, 497–504
P1-3, structure and function of, 329–332
POL R VSV mutants, 71–76
Parainfluenza virus
  HPIV3, 321–327
  P1-3, 329–332
Peripheral blood lymphocytes, see PBLs and measles virus
Phosphoprotein, see P protein
Piry virus, host range mutants in, 413–418
Plasma membranes
  and CNS derived cell lines, 545–551
  glycoproteins and RER, 201–205
  and M protein, 402–404
  and VSV nucleocapsid coiling, 399–404
Polypeptide coding assignments of RS virus mRNAs, 24, 25
Proflavin and leader RNAs, 136
Protein A–gold labeling, 379–382
Protein and state of phosphorylation, 79–84

R

RER and plasma membrane glycoproteins, 201–205
rgD mutants, 414–418
RNA
  of DI particles, 90–92
  and dyad-symmetry in, 87–93
  and L protein, 272, 275
  leader and role in inhibition of transcription by VSV, 131–138
  and mumps virus, 336, 337
  northern blots, 21–23
  polymerase binding site, 87–93
  small synthesized and VSV transcription, 109–114
  and 3' ends of negative strands, 87–92
  and 3' ends of positive strands, 87, 89, 90

RNA replication
   and effect of M and NS proteins, 148–152
   and protein requirement, 148
   and VSV proteins, 147–152
RNA synthesis, 71–76, 87
   and BHK cells, 139, 141, 142
   and DI–LT, 475–481
   during mixed infection, 139–141
   and halothane, 514–520
   leading to DI particle generation, 469–474
   template modification model, 73–76
   of VSV_{NJ}, 115–122
RNA templates, 11, 12
RNase V1 cleavage sites in leader RNAs, 135
RNP, 110–113, 271–277
   and G protein, 271
   and N protein, 271, 272, 275
   and NS protein, 272, 275
   and phospholipid vesicles, 275
   and RNP/M, 274
   reconstitution of, 273, 275–277
RNP–M, 271–277
RSV, 21–26, 175–181, 329
   and cDNA library, 28–30
   characterization of the glycoproteins of, 369–375
   and DNA sequence analysis, 31–33
   and DTT, 176, 177
   in vitro reaction, 175–178
   and mRNAs, 21–26
   and RNA-dependent RNA polymerase, 175–180
   size determination of, 31
   structural analysis of, 27–33
   and synthesized RNA, 177–180
   and TM treatment, 371–373
   and the 24 K protein, 365–368
   virion polypeptides, 369–371
RSV DI particle interference, 537–544
Rabies virus
   and BHK-21 cells, 388, 497–504
   and CNS, 387
   CVS strain of, 295–299
   and cell surface receptors, 387–392
   and ERA strain, 279–283
   and G protein, 279–283, 497, 498, 502, 504
   and infections of BHK-21 cells, 497–504
   and L protein, 502
   M_1 protein, 497, 498, 501, 502
   and N protein, 497, 498, 500, 502
   at peripheral and central synapses, 379–385
   and protein A–gold labeled antigen, 379–382
Rabies virus reception-rich regions, 379–386
Rabies virus G and comparative nucleotide sequence analysis of, 279–283
Rapid growth in *Drosophila, see* rgD mutants
Recombination events and DI RNAs of VSV, 469–474
Respiratory syncytial virus, *see* RSV
Ribonucleocapsid cores, *see* RNP
Ribonucleocapsid cores with M protein, *see* RNP–M
Ribonucleoprotein, *see* RNP
Rough endoplasmic reticulum, *see* RER and plasma membrane glycoproteins

S

SSPE, 167, 174
   and CNS derived model cells, 545–551
   CNS tissues and MV, 247–251
   and HeLa cell lines, 239, 240, 245
   and M protein, 167, 247, 248
   and measles virus, 233–237
   and synthesis of matrix protein in, 521–526
   and TLCK, 529
SV40 DNA synthesis, 223–231
SVCV
   M protein, 41, 42, 46, 47
   and RNA, 43–45
   and VSV, comparison of, 41
Semliki Forest virus, 146
Sendai virus, 321, 323–327
   and biochemical aspects of CL, 451–457
   and C protein, 3, 4
   and cloning of P protein, 363, 364
   and F protein, 345–350
   glycoproteins, 408–411
   in mouse spleen cells, 451–457
   and NP protein, 339–344
   and P protein and mumps, 336
Sendai virus DI RNA, 483–487
Sendai virus genome
   analysis of by molecular cloning, 3–9
   and complete set of signals, 14, 15
   and non-coding regulatory sequences, 11–15
   and overlapping clone technique, 4, 8
   and sequences of I and E regions, 13, 14
   and sequence signals at gene junctions, 12, 13
   and 3' proximal third, 4–6
Sendai virus glycoproteins and monoclonal antibodies, 345–350

INDEX 559

Sendai virus NP gene and structure of a complete clone, 17–19
Specific kinase inhibitory factor, 207
Spring Viremia of Carp virus, see SVCV
Subacute sclerosing panencephalitis, see SSPE
Synthesized RNA and RSV, 177–180
Synthetic subunit viral vaccines, 405, 406

T

TLCK and measles virus, 529–536
TM treatment
  and F protein, 372–375
  and G protein, 372–375
  and RS virus, 371–373
TNC fractions, 127
ts G11, 139–144
ts G16 mutant, 123–126, 128, 129
ts G41, 140–145
ts Mutants, 215
  of *Drosophila*, 415–418
  and wt-VSV, 139–146
ts Mutant DI and M protein, 317, 319, 320
Temperature sensitive mutants, see ts Mutants
Template N protein, 71–76
Transcribing nucleocapsid, see TNC fractions
Transcriptional map for RS virus, 22, 23, 26
Tryptic glycopeptides, analysis of, 287–289
Tunicamycin, see TM treatment

U

UV-irradiation, 215
  and sensitivity of interference to, 144–146

V

VA gene, 133–136
VSV, 41
  and binding studies of NS proteins, 79–84
  and cloning and sequencing of M mRNA, 41–48
  and G protein, 201, 204, 205
  and G–M chimeric gene, 193–199
  and inhibition of adenovirus, 223–231
  and interaction of L, NS, and M proteins, 63–69
  and long-term persistence by, in hamsters, 489–495
  and mouse L cells, 207–213
  and NDV, 201–205
  polymerase gene of, 35–40
  and rabies virus, 387–392
  and SV40 DNA synthesis, 223–231
  and structure and generation of deletion mutants, 475–481
  transcription of and concentration of ATP, 95–101
VSV DI particles, see DI particles
VSV effects on adenovirus type 2 replication, 215–221
VSV glycoprotein, peptides of, 406–408
VSV infection
  and histopathology of hamsters, 491
VSV inhibition of eukaryotic macromolecular synthesis, 215–217
VSV mutants, pol R, 71–76
VSV mutant with an aberrant polyadenylation activity, 123–129
VSV nucleocapsid coiling and plasma membranes, 399–404
VSV proteins, see also specific proteins
  and DI particle RNA, 153–160
  in RNA replication, 147–152
VSV RNAs, see RNAs
VSV RNP, see RNP
VSV role plus strand leader RNA, 220, 221
VSV sequences
  sources of, 88
  in transcription inhibition, 136–138
VSV transcription
  on NS protein, 265–269
  and synthesized small RNAs, 109–114
VSV–HR, 475
VSV_{IND}
  and DI particles, 461–467
  Glasgow strain, 126
  and leader RNA, 131–138
  Mudd–Summers strain, 63–69, 104–106
  and mutant ts G16, 123–126, 128, 129
  San Juan strain, 124, 126, 476
VSV_{NJ}, 105, 106, 115–122
  and leader RNA, 131–138
  Ogden strain, 63–69
  and ts mutants, 115–122
VSV–PI, 207–213
Vaccinia
  superinfection with, 207–213
  and VSV gene expression, 209
Vero cells and SV40, 223–226
Vesicular stomatitis virus, see VSV
Viral RNA replication and monoclonal antibodies, 155, 156
Virion polypeptides and RS, 369–371
Virus persistence in CNS derived cell lines, 545–551
Virus specific mRNAs, 331

Virus specific RNAs
　and CDV, 167–174
　molecular weights of, 173

W

wt-NDV, 146

wt-VSV
　and L cells, 207–213
　and ts mutants, 139–146
wt-VSV$_{NJ}$ and RNA synthesis, 116–118, 120, 121
wild-type VSV, *see* wt-VSV